Matrix Computations
and Semiseparable Matrices

Matrix Computations and Semiseparable Matrices

Volume II: **Eigenvalue and Singular Value Methods**

Raf Vandebril
Department of Computer Science
Catholic University of Leuven

Marc Van Barel
Department of Computer Science
Catholic University of Leuven

Nicola Mastronardi
M. Picone Institute
for Applied Mathematics, Bari

The Johns Hopkins University Press
Baltimore

© 2008 The Johns Hopkins University Press
All rights reserved. Published 2008
Printed in the United States of America on acid-free paper
9 8 7 6 5 4 3 2 1

The Johns Hopkins University Press
2715 North Charles Street
Baltimore, Maryland 21218-4363
www.press.jhu.edu

Library of Congress Cataloging-in-Publication Data

Vandebril, Raf, 1978–
    Matrix computations and semiseparable matrices / Raf Vandebril,
Marc Van Barel, Nicola Mastronardi.
        v. cm.
    Includes bibliographical references and indexes.
    Contents: v. 1. Linear systems
    ISBN-13: 978-0-8018-8714-7 (hardcover : alk. paper)
    ISBN-10: 0-8018-8714-3 (hardcover : alk. paper)
    1. Semiseparable matrices. 2. Matrices—Data processing.
3. Numerical analysis. I. Barel, Marc van, 1960– II. Mastronardi, Nicola,
1962– III. Title.
    QA188.V36 2007
    512.9'434—dc22   2007030657

ISBN-13: 978-0-8018-9052-9 (v. 2, hardcover : alk. paper)
ISBN-10: 0-8018-9052-7 (v. 2, hardcover : alk. paper)

A catalog record for this book is available from the British Library.

*Special discounts are available for bulk purchases of this book. For more
information, please contact Special Sales at 410-516-6936 or specialsales@
press.jhu.edu.*

The Johns Hopkins University Press uses environmentally friendly book
materials, including recycled text paper that is composed of at least 30
percent post-consumer waste, whenever possible. All of our book papers
are acid-free, and our jackets and covers are printed on paper with recycled content.

Contents

Preface

In this volume of the book, eigenvalue problems related to structured rank matrices are studied. The first volume considered mainly two goals: introducing structured rank matrices and investigating their properties, and secondly discussing several methods for solving systems of equations involving these structured rank matrices. Based on the ideas and methods presented in the first volume one can now fully understand the techniques discussed in this volume. Nevertheless most techniques are briefly repeated so that the book can be read independently from Volume I. In this book, two types of problems are considered: some direct eigenvalue problems and some inverse eigenvalue problems. Since many techniques are known for sparse matrices, such as tridiagonal, band and so forth, quite often a comparison between the sparse and structured rank case will be made.

The book is written from a numerical linear algebra viewpoint, meaning that most of the attention is paid to the structure of the matrices involved, the computational complexity, convergence analysis and numerical stability. Besides the rank-revealing chapter, no real applications are considered. However, the volume can be interesting for engineers as well, because in various fields nowadays one has to deal with structured rank matrices. E.g., in signal processing, video processing, state space models, these matrices arise.

A solid basic background and interesting theoretical results are provided. This, combined with the available MATLAB codes and the references to related publications should make it possible to adapt the presented tools to the reader's needs.

As not all readers are interested in the full details of all the material covered, selective reading parts and an extensive index are included. This should make it possible to focus directly on the part the reader is interested in.

The first two parts of the book focus on the development of implicit QR-algorithms for special classes of structured rank matrices. These two parts provide an alternative to the classical eigenvalue/singular value methods based on tridiagonal, Hessenberg and bidiagonal matrices. All the necessities to obtain such an algorithm based on structured rank matrices are included. The following items are covered: orthogonal reductions to semiseparable matrices and its variants, the convergence properties of these reductions, implicit QR-algorithms for these matrices, chasing techniques, an implicit Q-theorem, a multishift QR-method a new iteration (QH-iteration) for computing the eigenvalues and so forth.

The third part of the book contains some miscellaneous topics. The topics fit however completely into the context of eigenvalue problems related to structured rank matrices. A divide-and-conquer method for computing eigenvalues and eigenvectors is discussed, as well as a Lanczos-like semiseparabilization of a symmetric matrix. Finally the rank-revealing properties of the reductions are investigated in more detail.

The final part of this book relates structured rank matrices to orthogonal functions, such as polynomials, polynomial vectors and rational functions. These problems fit into this book because they address in some sense inverse eigenvalue problems: reconstructing a certain structured rank matrix based on its eigenvalues and some additional information.

Bibliography and author/editor index

After each section in the book, references, including a small summary of each cited paper, related to that section, are presented.

The overall bibliography list contains boldface numbers in braces at the end of each source. These numbers point to the pages in which the corresponding manuscript is cited.

Besides a subject index an author/editor index is also included. All cited authors and editors are included with references to their citations and occurrences in the bibliography. An author's name written in uppercase refers to manuscripts in which the author is the first author. The author's name written in lowercase refers to the manuscripts of which the author is a coauthor.

Additional resources

This book comes together with a webpage. This page contains additional resources for interested readers.

Updates of references can be found here, as well as additional references and notes related to them. New, interesting and exciting developments can be found here with the correct bibliographic links.

Errata will be posted here as well as extra material, such as additional examples or numerical experiments.

Implementations of different algorithms proposed or explained in this book can be downloaded from this resource.

This page can be found at:
http://www.cs.kuleuven.be/~mase/books/

Acknowledgments

Writing a book is not an easy task and is not possible without the help of many people. Therefore we would to take this chance to thank several people who were directly or indirectly involved in the writing process.

First of all we thank Gene Golub for bringing us into contact with Trevor Lipscombe from the Johns Hopkins University Press. We are grateful for the patience and trust he gave us.

We do not forget the referee of the book. It must have been a serious task reading our draft manuscripts and commenting on the directions to take. The referee helped us a lot and pointed out important aspects, which led to the book as it is now.

Furthermore, we also thank Gene Golub for being a wonderful host during Raf's visits. We thank the PhD students who started together with Raf and were involved in our initial research related to structured rank matrices: Gianni Codevico and Ellen Van Camp. Currently we are involved in an exciting research environment, the MASE-team (MAtrices having StructurE), consisting of Yvette, Katrijn, Andrey and Iurie.

We would also like to thank all people from The Johns Hopkins University Press and in particular Trevor Lipscombe, Nancy Wachter, Andre Barnett and Juliana McCarthy.

Raf likes to thank Els, his parents Roger and Brigitte, his brother Paul and sister Pascale, who have always created a supporting and pleasant environment.

Marc expresses his gratitude to his wife Maria and his sons Andreas and Lucas for their continuing support, patience and inspiration.

Nicola thanks his wife Teresa for her patience and love.

We would like to emphasize that this book is also due to several fruitful discussions with many friends and colleagues at meetings, conferences and other occasions. We especially want to thank the following persons for their interactions and contributions to our work: G. Ammar, R. Bevilacqua, D. Bini, A. Bultheel, S. Chandrasekaran, P. Dewilde, G. M. Del Corso, S. Delvaux, K. Diepold, Y. Eidelman, D. Fasino, L. Gemignani, I. Gohberg, B. Gragg, M. Gu, W. Hackbusch, A. M. Kaashoek, T. Kailath, P. Kravanja, F. T. Luk, K. Meerbergen, V. Mehrmann, G. Meurant, V. Olshevsky, V. Pan, B. Plestenjak, L. Reichel, T. Sakurai, M. Schuermans, G. Strang, E. Tyrtyshnikov, P. Van Dooren, S. Van Huffel, P. Van gucht, D. S. Watkins and H. Woerdeman.

We also acknowledge all the organizations that supported us during the writing of the book.

- The Fund for Scientific Research–Flanders (FWO–Vlaanderen) for giving a postdoctoral fellowship (Postdoctoraal Onderzoeker) to the first author and for supporting the projects:

    - G.0078.01 (SMA: Structured Matrices and their Applications);

    - G.0176.02 (ANCILA: Asymptotic aNalysis of the Convergence behavior of Iterative methods in numerical Linear Algebra);

    - G.0184.02 (CORFU: Constructive study of Orthogonal Functions);

    - G.0455.0 (RHPH: Riemann-Hilbert problems, random matrices and Padé-Hermite approximation).

- The Research Council K.U.Leuven for the projects:

    - OT/00/16 (SLAP: Structured Linear Algebra Package);

  – OT/05/40 (Large rank structured matrix computations);

  – CoE EF/05/006 Optimization in Engineering (OPTEC).

- The Belgian Programme on Interuniversity Poles of Attraction, initiated by the Belgian State, Prime Minister's Office for Science, Technology and Culture, project IUAPV-22 (Dynamical Systems and Control: Computation, Identification & Modelling)

- The research of the third author was partially supported by

  – MIUR, grant number 2004015437;

  – The short-term mobility program, Consiglio Nazionale delle Ricerche;

  – VII Programma Esecutivo di Collaborazione Scientifica Italia–Comunità Francese del Belgio, 2005–2006.

Notation

Throughout the book we use the following notation, unless stated otherwise.

$\langle \mathbf{x}, \mathbf{y}, \mathbf{z} \rangle$ .... Denotes the vector space spanned by the vectors $\mathbf{x}, \mathbf{y}$ and $\mathbf{z}$.

$A = (a_{i,j})_{i,j}$ ...... A general matrix A with elements $a_{i,j}$.

$A(i:j, k:l)$ ...... Submatrix of A, consisting of rows i up to and including j, and columns k up to and including l.

$A(\alpha; \beta)$ .... Submatrix of A, consisting of indices out of the set α and β.

$\alpha \times \beta$ .... $\alpha \times \beta$ denotes the product set $\{(i,j) | i \in \alpha, j \in \beta\}$, with α and β sets.

$\mathbb{C}$ .... Denotes the set of complex numbers.

$\mathbb{C}_n[z]$ .... Denotes the set of complex polynomials of degree $\leq n$.

$\text{cond}(A)$ .... The condition number of a matrix A.

$d(\mathcal{S}, \mathcal{T})$ .... Distance between the subspaces $\mathcal{S}$ and $\mathcal{T}$.

D .... A diagonal matrix.

$\deg(p)$ .... The degree of a polynomial p.

$\det(A)$ .... The determinant of a matrix A.

$\text{diag}(\mathbf{d})$ .... Denotes a diagonal matrix, with as diagonal elements, the elements from the vector $\mathbf{d}$.

$\mathbf{e}_i$ .... The i-th basis vector.

G .... A Givens transformation.

$.^H$ .... The hermitian conjugate of the involved matrix or vector.

H .... A (generalized) Hessenberg matrix.

$H_{\infty \times \infty} = [h_{j,k}]$ ... An infinite Hessenberg matrix with elements h_{jk}.

I_k .... The identity matrix of size $k \times k$.

$\kappa(A)$ .... The condition number of the matrix A.

$\mathcal{K}_k(A, \mathbf{v})$ .... The Krylov subspace of the matrix A, with vector $\mathbf{v}$ of dimension k.

$\Lambda_i = \{\lambda_1, \ldots, \lambda_n\}$ The λ_i denote the eigenvalues and Λ the spectrum of a matrix.

μ .... The shift in the QR-method.

$\hat{p}_{j:i}(\lambda)$ .... Equals $p_j(\lambda) p_{j-1}(\lambda) \ldots p_i(\lambda)$.

$\mathbf{p}_{0:\infty} = [p_0, p_1, \ldots]$ An infinite vector with elements p_i.

$\mathbb{P}_n$	Denotes the set of polynomials of degree less than n.
$\mathbb{P}_n^M$	Denotes the monic set of polynomials of degree less than n.
Range(A)	Denotes the vector space spanned by the columns of A.
$\mathbb{R}$	Denotes the set of real numbers.
$\mathbb{R}_n[z]$	Denotes the polynomials in $\mathbb{R}$ of degree $\leq n$.
rank (A)	The rank of a matrix A.
$\Sigma = \{\sigma_1, \ldots, \sigma_n\}$ ...	The singular values of a given matrix.
S	A subspace S.
S	A semiseparable matrix.
$S(\mathbf{u}, \mathbf{v})$	A symmetric generator representable semiseparable matrix with generators $\mathbf{u}$ and $\mathbf{v}$.
$.^T$	The transpose of the involved matrix or vector.
T	A tridiagonal matrix.
$\mathbb{T}$	Denotes the unit circle.
tril(A, p)	Denotes the lower triangular part of the matrix A, below and including subdiagonal p.
triu(A, p)	Denotes the upper triangular part of the matrix A, above and including superdiagonal p.
Q	A unitary(orthogonal) matrix (with regard to the QR-factorization).
R	An upper triangular matrix (with regard to the QR-factorization).
$\mathbf{u}^T = [u_1, u_2, \ldots, u_n]$	A column vector $\mathbf{u}$, with entries u_i. All vectors are column vectors and denoted in bold.
$\mathbf{u}(i : j) = \mathbf{u}_{i:j}$	A subvector of $\mathbf{u}$, consisting of the elements i up to and including j.
Z	A Hessenberg-like matrix.
$\bar{z}$	The complex conjugate of z.

Chapter 1

Introduction to semiseparable matrices

This chapter is included to make the book consistent and for those who didn't read the first volume of this book entitled: 'Matrix Computations and Semiseparable Matrices: Linear Systems'. For the people who did read the first volume, it might be useful to refresh the notions and definitions used throughout this book.

The aim of this chapter is to present a concise but essential summary of the results presented in Part I of the first volume. We will briefly recapitulate what is meant by a semiseparable matrix and how we can represent it. Also some interesting properties of these matrices will be discussed. Only the information essential for a full and thorough understanding of the topics covered here is provided. More detailed information can be found in the first volume [169].

We start the first section by introducing what is meant by semiseparable matrices. Two types of matrices are introduced, the class of semiseparable matrices and the class of generator representable semiseparable matrices. In a second section, we investigate more closely the relations between these classes of matrices. Results on the structure of the inverses of these matrices and on the close relation between the class of semiseparable and generator representable semiseparable matrices will be presented. It will be shown that the class of generator representable semiseparable matrices is a specific subclass of the class of semiseparable matrices. In the second section we briefly discuss two possible representations for these classes of matrices, which will be used throughout the book. We introduce the Givens-vector representation and also the generator representation.

✎ *As in the first volume, we introduce in this paragraph the 'selective reading hints'. This paragraph is typeset in italics and marked with a pencil. It can be found at the end of each introduction to a chapter or a part. The intention of this paragraph is to highlight the main goals and issues of the chapter. A list of the most significant theorems and examples will be presented. This is intended for those readers who do not want to read the entire book but want to grasp the main ideas presented in it.*

Let us provide the essential information for this chapter. The aim of this chapter is to deliver a concise summary of the first volume. Only those ingredients essential for an understanding of the coming chapters are included. These 12 pages contain a lot of essential information, and it is advisable to read it from the beginning to the end. It covers different definitions, representations and some important properties such as the structure under inversion and the relation between generator representable semiseparable and semiseparable matrices. People who have read the first volume can skip this chapter without any problem. They can immediately start with Part I: The reduction of matrices.

1.1   Definition of semiseparable matrices

Let us start by defining semiseparable matrices. Semiseparable matrices are so-called structured rank matrices. This means that specific parts out of the matrix need to satisfy a certain restriction on the ranks of subblocks. Let us provide some concrete examples.

We start by defining the symmetric classes of matrices.

Definition 1.1. *A matrix S is called a symmetric semiseparable matrix if and only if all submatrices taken out of the lower and upper triangular part[1] (due to symmetry) of the matrix are of rank 1 at most and the matrix is symmetric.*

This means that a symmetric matrix $S \in \mathbb{R}^{n \times n}$ is semiseparable if and only if the following relation is satisfied:

$$\operatorname{rank}\left(S(i:n, 1:i)\right) \leq 1 \text{ with } i = 1, \ldots, n.$$

This relation is a translation of the definition above into mathematical formulas. With $S(i:j, k:l)$[2], we denote that part of the matrix, ranging from row i up to and including row j, and column k up to and including column l.

If all subblocks marked by the $\boxtimes$ taken out of the lower triangular part in the following 5×5 example have rank less then or equal to 1, we have a semiseparable matrix. This illustrates only the rank structure for the lower triangular part of the matrix, but by symmetry this also holds for the upper triangular part.

$$
\begin{bmatrix}
\boxtimes & \times & \times & \times & \times \\
\boxtimes & \times & \times & \times & \times \\
\boxtimes & \times & \times & \times & \times \\
\boxtimes & \times & \times & \times & \times \\
\boxtimes & \times & \times & \times & \times
\end{bmatrix},
\begin{bmatrix}
\times & \times & \times & \times & \times \\
\boxtimes & \boxtimes & \times & \times & \times \\
\boxtimes & \boxtimes & \times & \times & \times \\
\boxtimes & \boxtimes & \times & \times & \times \\
\boxtimes & \boxtimes & \times & \times & \times
\end{bmatrix},
\begin{bmatrix}
\times & \times & \times & \times & \times \\
\times & \times & \times & \times & \times \\
\boxtimes & \boxtimes & \boxtimes & \times & \times \\
\boxtimes & \boxtimes & \boxtimes & \times & \times \\
\boxtimes & \boxtimes & \boxtimes & \times & \times
\end{bmatrix}
$$

$$
\begin{bmatrix}
\times & \times & \times & \times & \times \\
\times & \times & \times & \times & \times \\
\times & \times & \times & \times & \times \\
\boxtimes & \boxtimes & \boxtimes & \boxtimes & \times \\
\boxtimes & \boxtimes & \boxtimes & \boxtimes & \times
\end{bmatrix}
\text{ and }
\begin{bmatrix}
\times & \times & \times & \times & \times \\
\times & \times & \times & \times & \times \\
\times & \times & \times & \times & \times \\
\times & \times & \times & \times & \times \\
\boxtimes & \boxtimes & \boxtimes & \boxtimes & \boxtimes
\end{bmatrix}.
$$

---

[1] The lower as well as the upper triangular part of the matrix includes the diagonal.

[2] This is MATLAB style notation. MATLAB is a registered trademark of The MathWorks, Inc.

Example 1.2 The following three matrices are symmetric semiseparable matrices:

$$\begin{bmatrix} 5 & 2 & 4 \\ 2 & 1 & 2 \\ 4 & 2 & 5 \end{bmatrix}, \quad \begin{bmatrix} 5 & 0 & 0 \\ 0 & 6 & 0 \\ 0 & 0 & 7 \end{bmatrix} \quad \text{and} \quad \begin{bmatrix} 3 & 3 & 0 \\ 3 & 3 & 0 \\ 0 & 0 & 4 \end{bmatrix}.$$

∎

The following frequently used definition for semiseparable matrices is the so-called generator definition[3] (see, e.g., [38, 73, 115, 156]). Hence we name these matrices satisfying this definition 'generator representable semiseparable' matrices.

Definition 1.3. *A matrix S is called a symmetric generator representable semiseparable matrix if and only if the lower triangular part of the matrix is coming from a rank 1 matrix and the matrix S is symmetric.*

This means that a matrix S is symmetric generator representable semiseparable if (for $i = 1, \ldots, n$):

$$\text{tril}(S) = \text{tril}(\mathbf{uv}^T),$$
$$\text{triu}(S) = \text{triu}(\mathbf{vu}^T),$$

where $\mathbf{u}$ and $\mathbf{v}$ are two column vectors of length n. With $\text{triu}(\cdot)$ and $\text{tril}(\cdot)$[4], we denote, respectively, the upper triangular and the lower triangular part of the matrix.

A generator representable semiseparable matrix has the following structure with $\mathbf{u}^T = [u_1, u_2, \ldots, u_n]$ and $\mathbf{v}^T = [v_1, v_2, \ldots, v_n]$:

$$S = \begin{bmatrix} u_1v_1 & u_2v_1 & u_3v_1 & \cdots & u_nv_1 \\ u_2v_1 & u_2v_2 & u_3v_2 & \cdots & u_nv_2 \\ u_3v_1 & u_3v_2 & u_3v_3 & & \vdots \\ \vdots & \vdots & & \ddots & \\ u_nv_1 & u_nv_2 & u_nv_3 & \cdots & u_nv_n \end{bmatrix}. \tag{1.1}$$

Reconsidering Example 1.2, one can easily see that the first matrix is symmetric generator representable semiseparable, but the second and third matrices are not. This indicates already that both classes do not cover the same set of matrices. We will prove later on that the class of generator representable semiseparable matrices is a specific subclass of the class of semiseparable matrices.

In the remainder of the text, we will often denote a generator representable matrix, with generators $\mathbf{u}, \mathbf{v}$ as $S(\mathbf{u}, \mathbf{v})$.

In this book we will not only work with the class of semiseparable matrices but also with a closely related class, namely the class of semiseparable plus diagonal matrices. This class of matrices will become very useful for rank-revealing applications as will be shown afterwards[5].

---

[3]In our book, we name these matrices in this way. Often these matrices are simply referred to as semiseparable matrices. (See also the notes and references at the end of this section.)

[4]The commands triu($\cdot$) and tril($\cdot$) are defined similarly as the MATLAB commands.

[5]In Chapter 3 we will show that a reduction to semiseparable plus diagonal form has some specific convergence properties, which can be used for performing rank revealing.

Definition 1.4. *A matrix S is called a symmetric semiseparable plus diagonal matrix if and only if it can be written as the sum of a diagonal and a symmetric semiseparable matrix.*

This type of matrix arises for example in the discretization of particular integral equations (see, e.g., [109, 110]).

Throughout this book, we also do work with nonsymmetric matrices, such as upper triangular semiseparable, and Hessenberg-like matrices. The definition of an upper triangular semiseparable matrix is straightforward. An upper triangular semiseparable matrix is an upper triangular matrix for which the upper triangular part is of semiseparable form, which means that all submatrices taken out of that specific part have rank less than or equal to 1.

Definition 1.5. *A matrix Z is called a Hessenberg-like matrix if all submatrices taken out of the lower triangular part of the matrix, including the diagonal, have rank 1 at most.*

In other words, the lower triangular part of the Hessenberg-like matrix is of semiseparable form. These matrices are also called lower semiseparable (as they only have the lower triangular part satisfying the semiseparable constraints). Even though it might sound confusing, lower Hessenberg-like matrices (coming from the inverse of lower Hessenberg matrices) have the upper triangular part of structured rank form. These matrices are sometimes also referred to as upper semiseparable matrices.

Even though we did not explicitly define them, sparse[6] matrices such as tridiagonal, bidiagonal and Hessenberg matrices can also be considered as structured rank matrices. For a tridiagonal matrix all submatrices taken out of the part below the subdiagonal have rank 0. Our convention is the following: when speaking about sparse matrices we mean the tridiagonal, bidiagonal and Hessenberg case; when discussing the structured rank case we consider the semiseparable, upper triangular semiseparable and Hessenberg-like case. This is to distinguish between the classes of matrices considered in the book.

All the above matrices will be used throughout the book. The classes of semiseparable and semiseparable plus diagonal matrices will be used as intermediate matrices for solving the symmetric eigenvalue problem. The class of Hessenberg-like matrices will replace the class of Hessenberg matrices in the unsymmetric eigenvalue problem. Finally, the class of upper triangular semiseparable matrices will be used for computing the singular value decomposition instead of bidiagonal matrices. Moreover in Section 1.2 we will prove that all the classes of matrices mentioned here are closely related to each other via inversion.

---

[6]In some sense there are two mainstreams of naming matrices 'sparse'. Often matrices are named sparse if they have many more zero than nonzero elements. Another definition considers matrices sparse when the zero structure can be exploited to obtain efficient algorithms. We choose to use the last definition.

Notes and references

More information concerning these classes of matrices plus the class of quasiseparable matrices can be found in the following article and in the first volume of this book.

> ☞ R. Vandebril, M. Van Barel, and N. Mastronardi. A note on the representation and definition of semiseparable matrices. *Numerical Linear Algebra with Applications*, 12(8):839–858, October 2005.

> ☞ R. Vandebril, M. Van Barel, and N. Mastronardi. *Matrix Computations and Semiseparable Matrices, Volume I: Linear Systems.* Johns Hopkins University Press, Baltimore, Maryland, USA, 2008.

The following revised book includes the contents of some articles in which there is, to our knowledge, the first appearance of semiseparable matrices called 'one-pair' or 'single-pair' matrices at that time.

> ☞ F. R. Gantmacher and M. G. Kreĭn. *Oscillation Matrices and Kernels and Small Vibrations of Mechanical Systems.* AMS Chelsea Publishing, Providence, Rhode Island, USA, revised edition, 2002.

As mentioned before, many more references related to either generator representable semiseparable and semiseparable plus diagonal matrices can be found in the first volume.

There are many more structured rank matrices not discussed in this section. For example quasiseparable matrices, higher order semiseparable, decoupled semiseparable, generator representable plus band and so forth. All these different classes are discussed in a more general context in the first volume of this book.

Historically, semiseparable matrices have been defined in different, quite often inconsistent ways. One can divide these definitions into two main streams: the 'Linear Algebra' form and the 'Operator Theory' definition. From an operator theoretical viewpoint, semiseparable matrices can be considered as coming from the discretization of semiseparable kernels. The resulting matrix is a generator representable semiseparable matrix. Hence, quite often in the literature one speaks about a semiseparable matrix but one means a 'generator representable' matrix, as defined in this section. From a linear algebra viewpoint, semiseparable matrices are often considered as the inverses of tridiagonal matrices. In this book we use the 'Linear Algebra' formulation. We will however always clearly state which type of semiseparable matrix we are working with.

1.2   Some properties

In this section some properties of the previously defined classes of matrices will be discussed. We will investigate the structure of the inverse of some previously defined matrices. Moreover we will see that the inverse of some of the structured rank matrices gives rise to a sparse matrix.

1.2.1   Relations under inversion

The basic tool for predicting the structure of the inverse of a structured rank matrix is the nullity theorem. This theorem does not provide explicit methods for inverting matrices, but it provides theoretical predictions on the inner structure of the resulting matrices. This theorem was first formulated by Gustafson [103] for matrices over principal ideal domains. In [79], Fiedler and Markham translated this abstract formulation to matrices over a field. Barrett and Feinsilver formulated theorems close to the nullity theorem in [14, 15]. We will not provide a proof of the nullity theorem nor explore its power. We will briefly state the theorem and some of its interesting corollaries. Using this theorem, one can easily calculate the structure of the inverse of some of the previously defined structured rank matrices.

Definition 1.6. *Suppose a matrix $A \in \mathbb{R}^{m \times n}$ is given. The nullity $n(A)$ is defined as the dimension of the right null space of A.*

Theorem 1.7 (The nullity theorem). *Suppose we have the following invertible matrix $A \in \mathbb{R}^{n \times n}$ partitioned as*

$$A = \left[\begin{array}{cc} A_{11} & A_{12} \\ A_{21} & A_{22} \end{array} \right],$$

with A_{11} of size $p \times q$. The inverse B of A is partitioned as

$$B = \left[\begin{array}{cc} B_{11} & B_{12} \\ B_{21} & B_{22} \end{array} \right],$$

with B_{11} of size $q \times p$. Then the nullities $n(A_{11})$ and $n(B_{22})$ are equal.

The proof can be found in [79]. For the following corollaries, we use the following notation. Let A be an $m \times n$ matrix. Denote with M the set of numbers $\{1, 2, \ldots, m\}$ and with N the set of numbers $\{1, 2, \ldots, n\}$. Let α and β be nonempty subset of M and N, respectively. Then, we denote with the matrix $A(\alpha; \beta)$ the submatrix of A with row indices in α and column indices in β. Let us denote with $|\alpha|$ the cardinality of the corresponding set α.

The following corollary is derived from the nullity theorem. The abstract formulation in terms of nullities of subblocks is translated in terms of the ranks of submatrices.

Corollary 1.8 (Corollary 3 in [79]). *Suppose $A \in \mathbb{R}^{n \times n}$ is a nonsingular matrix and α, β are nonempty subsets of N with $|\alpha| < n$ and $|\beta| < n$. Then*

$$\text{rank}\left(A^{-1}(\alpha; \beta) \right) = \text{rank}\left(A(N\backslash\beta; N\backslash\alpha) \right) + |\alpha| + |\beta| - n.$$

When choosing $\alpha = N\backslash\beta$, we get

Corollary 1.9. *For a nonsingular matrix $A \in \mathbb{R}^{n \times n}$ and $\alpha \subseteq N$, we have:*

$$\text{rank}\left(A^{-1}(\alpha; N\backslash\alpha) \right) = \text{rank}\left(A(\alpha; N\backslash\alpha) \right).$$

Based on the nullity theorem and its corollaries the proof of the following theorem is an easy exercise.

Theorem 1.10. *The inverse of an invertible*

- *tridiagonal matrix is a semiseparable matrix;*

- *irreducible tridiagonal matrix is a generator representable semiseparable matrix;*

- *Hessenberg matrix is a Hessenberg-like matrix;*

- *an upper triangular semiseparable matrix is a bidiagonal matrix;*

- *a lower triangular semiseparable matrix is a lower bidiagonal matrix.*

The deduction of the structure of the inverse of a semiseparable plus diagonal matrix is a little more tricky. We do not go into the details of the inversion but provide a simple theorem for the special case of semiseparable plus diagonal matrices for which the diagonal does not contain any zero values. A more comprehensive study of its inverse can be found in [51] and in Volume I.

Theorem 1.11. *The inverse of an invertible semiseparable plus diagonal matrix $S + D$, for which the matrix D is also invertible, is again a semiseparable plus diagonal matrix with matrix D^{-1} as the diagonal.*

The proof can be found in [51, 169].

1.2.2 Generator representable semiseparable matrices

Example 1.2 in Section 1.1 already illustrated that there seem to be some major differences between the class of generator representable semiseparable and the class of semiseparable matrices. For example a diagonal matrix belongs to the class of semiseparable matrices, but one is unable to represent it with generators. The following theorems clearly illustrate how the class of generator representable semiseparable matrices fits into the class of semiseparable matrices.

First we define the pointwise limit.[7]

Definition 1.12. *The pointwise limit of a collection of matrices $A_\epsilon \in \mathbb{R}^{n \times n}$ (if it exists) for $\epsilon \to \epsilon_0$, with $\epsilon, \epsilon_0 \in \mathbb{R}$ and with the matrices A_ϵ as*

$$A_\epsilon = \begin{bmatrix} (a_{1,1})_\epsilon & \cdots & (a_{1,n})_\epsilon \\ \vdots & \ddots & \vdots \\ (a_{n,1})_\epsilon & \cdots & (a_{n,n})_\epsilon \end{bmatrix},$$

is defined as:

$$\lim_{\epsilon \to \epsilon_0} A_\epsilon = \begin{bmatrix} \lim_{\epsilon \to \epsilon_0} (a_{1,1})_\epsilon & \cdots & \lim_{\epsilon \to \epsilon_0} (a_{1,n})_\epsilon \\ \vdots & \ddots & \vdots \\ \lim_{\epsilon \to \epsilon_0} (a_{n,1})_\epsilon & \cdots & \lim_{\epsilon \to \epsilon_0} (a_{n,n})_\epsilon \end{bmatrix}.$$

---

[7]Even though we use the pointwise limit here, any matrix norm suffices to prove similar theorems, see the first volume.

It is clear that the class of generator representable semiseparable matrices is not closed under the pointwise limit; this is illustrated by the following example.

Example 1.13 For all ϵ, the matrices A_ϵ are generator representable semiseparable. The limit of these matrices A_ϵ for $\epsilon \to 0$, however, will not be generator representable semiseparable but just semiseparable.

$$\lim_{\epsilon \to 0} A_\epsilon = \lim_{\epsilon \to 0} \begin{bmatrix} 1 & 1 & \epsilon \\ 1 & 1 & \epsilon \\ \epsilon & \epsilon & 1 \end{bmatrix}$$
$$= \begin{bmatrix} 1 & 1 & 0 \\ 1 & 1 & 0 \\ 0 & 0 & 1 \end{bmatrix}.$$

∎

Before formulating a theorem, which defines the pointwise closure of the class of generator representable semiseparable matrices, we need to introduce some more notation and some preliminaries. As mentioned before, the submatrix of the matrix A, consisting of the rows $i, i+1, \ldots, j-1, j$ and the columns $k, k+1, \ldots, l-1, l$, is denoted using the MATLAB-style notation $A(i:j, k:l)$; the same notation style is used for the elements $i, \ldots, j$ in a vector $\mathbf{u}$: $\mathbf{u}(i:j)$ and the (i,j)-th element in a matrix A: $A(i,j) = a_{i,j}$ (either notation is used, depending on the context).

The proofs of the following theorems and a more elaborate study of them can be found in Volume I [169].

Theorem 1.14. *Suppose S to be a symmetric semiseparable matrix of size n. Then, S is not a generator representable semiseparable matrix if and only if there exist indices i, j with $1 \leq j \leq i \leq n$ such that $S(i,j) = 0$, $S(i, 1:i) \neq 0$ and $S(j:n, j) \neq 0$.*

If a zero in the semiseparable part is present, then that this zero propagates up to a diagonal element, either in a row or in a column. Reconsider for example a diagonal matrix, with nonzero diagonal elements. In this case, it is not true that zero elements in the lower triangular part propagate up to and include the diagonal, hence this matrix is not of generator representable semiseparable form.

The next proposition shows how the class of symmetric generator representable semiseparable matrices can be embedded in the class of symmetric semiseparable matrices.

Proposition 1.15. *Suppose a symmetric semiseparable matrix S is given, which cannot be represented by two generators. Then this matrix can be written as a block diagonal matrix, for which all the blocks are symmetric semiseparable matrices representable with two generators.*

This theorem is of interest when developing QR-algorithms; it gives a characterization of the blocks (they are generator representable) if deflation is needed in the algorithm. More information on this property will be given in forthcoming chapters. Let us conclude with a theorem, which will become crucial in the remainder of the book.

Theorem 1.16. *The pointwise closure of the class of symmetric generator representable semiseparable matrices is the class of symmetric semiseparable matrices, which is closed under pointwise convergence.*

The above theorem will become important for the construction of eigenvalue algorithms based on semiseparable and related matrices. As we want the semiseparable matrices to converge to diagonal matrices containing the eigenvalues, we cannot work with generator representable semiseparable matrices. Because generator representable matrices are not capable of dealing with diagonal matrices, this can lead to numerical instabilities! As this representation does not satisfy our needs, we need another type of representation. This representation is explored in the next section.

Notes and references

As already mentioned, the first formulation of the nullity theorem was due to Gustafson for matrices over principal ideal domains.

☞ W. H. Gustafson. A note on matrix inversion. *Linear Algebra and its Applications*, 57:71–73, 1984.

The nullity theorem, as presented in this section, was formulated by Fiedler and Markham, who translated the abstract formulation of Gustafson.

☞ M. Fiedler and T. L. Markham. Completing a matrix when certain entries of its inverse are specified. *Linear Algebra and its Applications*, 74:225–237, 1986.

Also Barrett and Feinsilver formulated theorems close to the nullity theorem.

☞ W. W. Barrett. A theorem on inverses of tridiagonal matrices. *Linear Algebra and its Applications*, 27:211–217, 1979.

☞ W. W. Barrett and P. J. Feinsilver. Inverses of banded matrices. *Linear Algebra and its Applications*, 41:111–130, 1981.

Barrett (1979) formulates another type of theorem connected to the inverse of tridiagonal matrices. In most of the preceding articles, one assumed the sub- and superdiagonal elements of the corresponding tridiagonal matrix to be different from zero. In this article of 1979 only one condition is left. It is assumed that the diagonal elements of the symmetric semiseparable matrix are different from zero. Moreover, the proof is also suitable for nonsymmetric matrices. The theorems presented in this article are very close to the final version result, stating that the inverse of a tridiagonal matrix is a semiseparable matrix, satisfying the rank definition. Barrett and Feinsilver (1981) should be considered as one of the most important articles concerning the inverse of band matrices. Barrett and Feinsilver provide a general framework as presented in this section. General theorems and proofs considering the vanishing of minors when looking at the matrices and their inverses are given, thereby characterizing the complete class of band and semiseparable matrices, without excluding cases in which there are zeros. The results are a straightforward consequence of Barrett (1979).

Based on the above presented historical references, the structure of the inverses of these matrices can be deduced rather easily. Recently an article covering an extension of the nullity theorem was presented.

☞ S. Delvaux and M. Van Barel. Structures preserved by matrix inversion. *SIAM Journal on Matrix Analysis and Applications*, 28(1):213–228, 2006.

The article of Meurant discusses several articles and methods for inverting band and related matrices.

☞ G. Meurant. A review of the inverse of symmetric tridiagonal and block tridiagonal matrices. *SIAM Journal on Matrix Analysis and Applications*, 13:707–728, 1992.

The inner relations between the class of generator representable semiseparable matrices and the class of semiseparable matrices was extensively studied in the first volume [169] and in [165]. In these articles, not only the relation between the above mentioned classes of matrices, but also the relations between quasiseparable and semiseparable plus diagonal matrices, the symmetric as well as the unsymmetric version, are discussed Also the possible numerical problems as mentioned at the end of this section are covered in the first volume. Moreover, robust and reliable techniques are provided for computing the generators of generator representable semiseparable matrices.

The following articles by Woerdeman contain interesting results on completion problems. These problems address in some sense whether a matrix is generator representable. These general theorems lead to a generalization of the nullity theorem.

☞ H. J. Woerdeman. *Matrix and Operator Extensions*, volume 68 of *CWI Tract*. Centre for Mathematics and Computer Science, Amsterdam, Netherlands, 1989.

☞ H. J. Woerdeman. A matrix and its inverse: revisiting minimal rank completions. *Operator Theory: Advances and Applications*, 179:329–338, 2008.

1.3 The representations

In the previous section, we declared different types of structured rank matrices, which will be used extensively throughout this book. The definition of a semiseparable matrix however was purely based on the rank structure of the matrix. The definition did not say anything about an effective way of dealing with these matrices. Storing the matrices in their full, dense form results in $\mathcal{O}(n^2)$ memory usage. In exploiting the known rank properties more efficient ways exist. In this section we will deduce the two frequently used types of representations in this book, namely, the generator representation and the Givens-vector representation. These are however not the only possible representations. More information can again be found in [169].

1.3.1 The generator representation

In the preceding chapter we discussed the relationship between generator representable semiseparable matrices and semiseparable matrices. The class of generator representable semiseparable matrices is a subclass of the class of semiseparable matrices.

Reconsidering the example presented above we have a generator representable semiseparable matrix of the following form.

Suppose two vectors $\mathbf{u}$ and $\mathbf{v}$ of the following form $\mathbf{u}^T = [u_1, u_2, \ldots, u_n]$ and $\mathbf{v}^T = [v_1, v_2, \ldots, v_n]$ are given. The matrix $S(\mathbf{u}, \mathbf{v})$ of the form

$$
S(\mathbf{u}, \mathbf{v}) = \begin{bmatrix}
u_1 v_1 & u_2 v_1 & u_3 v_1 & \ldots & u_n v_1 \\
u_2 v_1 & u_2 v_2 & u_3 v_2 & \ldots & u_n v_2 \\
u_3 v_1 & u_3 v_2 & u_3 v_3 & & \vdots \\
\vdots & \vdots & & \ddots & \\
u_n v_1 & u_n v_2 & u_n v_3 & \ldots & u_n v_n
\end{bmatrix}
\tag{1.2}
$$

is a generator representable semiseparable matrix. The vectors $\mathbf{u}$ and $\mathbf{v}$ are called the generators of this matrix.

We do not go further in detail on how to construct these generators in a reliable way or how they fit in a more general representation type, namely the type of quasiseparable matrices.

In this book, we will most of the time work with the Givens-vector representation. This representation is capable of representing all semiseparable matrices, not only the generator representable ones.

1.3.2 The Givens-vector representation

The Givens-vector representation for a matrix $S \in \mathbb{R}^{n \times n}$ consists of a sequence of $n-1$ Givens transformations and a vector. The following figures denote how a semiseparable matrix can be reconstructed using this information. The Givens transformations are denoted as $G = [G_1, \ldots, G_{n-1}]$ and the vector as $\mathbf{v}^T = [v_1, \ldots, v_n]$. The elements denoted by $\boxtimes$ make up the semiseparable part of the matrix. Initially, one starts on the first two rows of the matrix. The element v_1 is placed in the upper left position, then a Givens transformation is applied, and finally to complete the first step, element v_2 is added in position $(2, 1)$. Only the first two columns and rows are shown here.

$$
\begin{bmatrix} v_1 & 0 \\ 0 & 0 \end{bmatrix} \rightarrow G_1 \begin{bmatrix} v_1 & 0 \\ 0 & 0 \end{bmatrix} + \begin{bmatrix} 0 & 0 \\ 0 & v_2 \end{bmatrix} \rightarrow \begin{bmatrix} \boxtimes & 0 \\ \boxtimes & v_2 \end{bmatrix}.
$$

The second step consists of applying the Givens transformation G_2 on the second and the third rows; furthermore, v_3 is added in position $(3, 3)$. Here only the first three columns are shown and the second and third row. This leads to

$$
\begin{bmatrix} \boxtimes & v_2 & 0 \\ 0 & 0 & 0 \end{bmatrix} \rightarrow G_2 \begin{bmatrix} \boxtimes & v_2 & 0 \\ 0 & 0 & 0 \end{bmatrix} + \begin{bmatrix} 0 & 0 & 0 \\ 0 & 0 & v_3 \end{bmatrix} \rightarrow \begin{bmatrix} \boxtimes & \boxtimes & 0 \\ \boxtimes & \boxtimes & v_3 \end{bmatrix}.
$$

This process can be repeated by applying the Givens transformation G_3 on the third and the fourth rows of the matrix and afterwards adding the diagonal element v_4. After applying all the Givens transformations and adding all the diagonal elements, the lower triangular part of a symmetric semiseparable matrix is constructed. Because of the symmetry, the upper triangular part is also known.

When denoting a Givens transformation G_l as

$$G_l = \left[\begin{array}{cc} c_l & -s_l \\ s_l & c_l \end{array} \right],$$

the elements $S(i,j) = s_{ij}$ are calculated in the following way:

$$S(i,j) = s_{ij} = c_i s_{i-1} s_{i-2} \cdots s_j v_j \text{ for } n > i \geq j$$
$$S(n,j) = s_{nj} = s_{n-1} s_{n-2} \cdots s_j v_j \text{ for } i = n.$$

The elements in the upper triangular part can be calculated similarly due the symmetry. The elements of the semiseparable matrix can therefore be calculated in a stable way based on the Givens-vector representation. This means that a semiseparable matrix, represented with the Givens-vector representation, is of the following form

$$S = \left[\begin{array}{cccc} c_1 v_1 & & & \\ c_2 s_1 v_1 & c_2 v_2 & & \\ c_3 s_2 s_1 v_1 & c_3 s_2 v_2 & c_3 v_3 & \\ \vdots & \vdots & & \ddots \end{array} \right]. \tag{1.3}$$

The storage costs are $3n - 2$. We store the cosine and sine separately because of numerical efficiency. Theoretically, only storing the cosine (or sine) would be enough, leading to a storage cost of $2n - 1$. As can be seen from the structure of a matrix represented with this Givens-vector representation, the Givens transformations store the dependencies between the rows, while the vector contains a weight posed on each column. An extension of the Givens-vector representation is the Givens-weight representation (see [53]).

The Givens-vector representation, as it was designed here, starts at the top and fills up the matrix gradually. Hence, this is a representation from top to bottom. In fact, one can start in a similar way in the lower right corner and gradually fill up the matrix from the right to the left. This representation is called a Givens-vector representation from right to left, or sometimes a column based, because the columns are filled up, whereas the first version is sometimes named the row-based Givens-vector representation. Both representations can easily be transformed into each other. The procedure for going from a top-bottom to a right-left representation is called the swapping of the representation. An implementation of this swapping can be found in the first volume.

The Givens-vector representation, both the top-bottom as well as the right-left form, admit an easy graphical representation. The first volume of this book makes extensive use of this representation for designing QR-factorizations. In this volume this representation is only used for showing how one can interpret the different algorithms easily by using the Givens-vector representation. They are depicted for the first time in Chapter 4. There it is shown how to depict graphically the Givens-vector representation, and the basic operations, such as the shift through and the fusion, are also shown. Based on these graphical schemes the reduction algorithm is reinterpreted. In Chapter 8, the graphical representation is used for reconstructing the implicit QR-algorithm. Finally in Chapter 9, extensive use is made of this graphical representation because a mathematical description of, e.g., the multishift algorithm would be hard to read.

Notes and references

In the literature much attention is being paid to the class of generator representable semiseparable matrices. Especially as they can be considered as the inverses of irreducible tridiagonal matrices.

> ☞ S. O. Asplund. Finite boundary value problems solved by Green's matrix. *Mathematica Scandinavica*, 7:49–56, 1959.

In this article, S. O. Asplund, the father of E. Asplund, proves the same as Gantmacher and Krein, by calculating the inverse via techniques for solving finite boundary value problems. Also some theoretical results concerning inverses of nonsymmetric tridiagonal (not necessarily irreducible) matrices are included.

The Givens-vector representation was defined in [165] (see also Section 1.1 of this chapter). Vandebril, Van Barel and Mastronardi investigate in detail the difference between generator representable semiseparable and semiseparable matrices. Moreover the relations between the generator and Givens-vector representation are explored. An extension of the Givens-vector representation was proposed by Delvaux and Van Barel. This representation is the so-called Givens-weight representation, which can be used to represent also higher-order structured rank matrices by more Givens transformations and the role of the vector is replaced by the weights.

> ☞ S. Delvaux and M. Van Barel. A Givens-weight representation for rank structured matrices. *SIAM Journal on Matrix Analysis and Applications*, 29(4):1147–1170, 2007.

The representations discussed in this chapter are not the only ones. The diagonal subdiagonal representation stores in fact only the diagonal and the subdiagonal of the semiseparable matrix. This representation is just like the generator representation, not capable of representing all semiseparable matrices. This diagonal subdiagonal representation is mainly investigated by Fiedler and co-authors. Important theorems related to the diagonal and subdiagonal representation can be found in

> ☞ M. Fiedler. Basic matrices. *Linear Algebra and its Applications*, 373:143–151, 2003.

> ☞ M. Fiedler and Z. Vavřín. Generalized Hessenberg matrices. *Linear Algebra and its Applications*, 380:95–105, 2004.

In the articles of 2003 and 2004, the properties of so-called basic matrices and complete basic matrices are investigated. These are matrices representable with the diagonal and subdiagonal representation. Theorems concerning the representation, LU-decompositions, factorizations of these matrices and inversion methods are presented.

Another frequently used representation is the quasiseparable representation. In fact, this representation is the most general one. It covers the Givens-vector representation, as well as the generator representation as special cases. Unfortunately this representation uses more parameters than strictly necessary. Hence one has to choose these parameters

very carefully to avoid numerical problems. The following publications introduce the
quasiseparable representation.

☞ Y. Eidelman and I. C. Gohberg. On a new class of structured matrices.
   Integral Equations and Operator Theory, 34:293–324, 1999.

☞ E. E. Tyrtyshnikov. Mosaic ranks for weakly semiseparable matrices. In
   M. Griebel, S. Margenov, and P. Y. Yalamov, editors, *Large-Scale Scientific
   Computations of Engineering and Environmental Problems II*, volume 73 of
   Notes on Numerical Fluid Mechanics, pages 36–41. Vieweg, Braunschweig,
   Germany, 2000.

In the first article, the authors Eidelman and Gohberg investigate a generalization of the
class of semiseparable matrices, namely, the class of quasiseparable matrices in its most
general form. They show that the class of quasiseparable matrices is closed under inversion,
and they present a linear complexity inversion method. Tyrtyshnikov names the class of
quasiseparable matrices weakly semiseparable matrices.

1.4   Conclusions

This chapter was meant as a brief introduction into the theory of semiseparable
matrices. The goal was to get the reader acquainted as quickly as possible with the
definition of a semiseparable and a generator representable semiseparable matrix.
Based on these definitions, two possible representations, which are frequently used
in the remainder of the book, were covered. Finally we provided some background
information on some inner relations between semiseparable and generator representable semiseparable matrices and also the structure of their inverses was briefly
discussed.

Part I

The reduction of matrices

The first and second part of this book focus on the translation of traditional eigenvalue algorithms, based on sparse matrices, towards a structured rank approach. Let us specify this. For example, a traditional method for computing the eigenvalues of a symmetric matrix first transforms this symmetric matrix to a similar tridiagonal one, and then the eigenvalues of this similar tridiagonal one are computed via the well-known QR-algorithm. A translation of the reduction to a more easy form (tridiagonal in the traditional setting) is the topic of the first part, and a translation of the QR-algorithm (for tridiagonal matrices in the traditional algorithm) is the topic of the second part of this book. Let us provide some more information on the first part.

This part goes out to the development and study of new reduction algorithms, which reduce matrices not to a sparse form such as Hessenberg, bidiagonal or tridiagonal form, but to their analogues in the structured rank setting, i.e., Hessenberg-like, upper triangular semiseparable and semiseparable. The algorithms are proposed. Their properties with regard to the traditional reduction algorithms are investigated and their implementations, computational complexities and some numerical examples will be discussed.

Chapter 2 starts with introducing the different reduction algorithms. The traditional algorithms for reducing matrices to sparse form are also revisited to clarify the correspondences and differences with the new proposed methods. First we discuss transformations of symmetric matrices to two specific forms, namely semiseparable and semiseparable plus diagonal. It will be shown that the reduction to semiseparable plus diagonal form is the most general one, covering in fact also the reduction to tridiagonal and semiseparable form. Applying the same reduction algorithm to nonsymmetric matrices leads to an orthogonal similarity transformation for transforming arbitrary matrices to Hessenberg-like form. Similarity transformations are not necessary if one wants to compute the singular values; hence, a reduction method to transform arbitrary rectangular matrices into an upper triangular semiseparable matrix is also proposed. To conclude this chapter, two extra sections are added proposing methods to go from sparse to structured rank form and vice versa. For example efficient algorithms for reducing a semiseparable plus diagonal matrix to a tridiagonal matrix are proposed.

It was already mentioned that the reductions to structured rank form are slightly more costly than the reductions to sparse form. In Chapter 3, we discuss in more detail the effect of these extra transformations. We start by deducing a general theory stating which similarity transformations have the Lanczos-Ritz values as eigenvalues in the already reduced part. The provided theoretical results give necessary and sufficient conditions that need to be imposed on similarity transformations to obtain this convergence behavior. After this investigation we investigate the extra convergence behavior due to extra operations performed in the reduction procedure. This results in a kind of nested subspace iteration. Moreover in the case of the reduction to semiseparable plus diagonal form, it is a kind of nested multishift iteration, in which the shift depends on the diagonal initially chosen for the reduction procedure. Of course these two types of convergence behavior interact with each other. It is proven that the subspace iteration starts working on the already reduced part if the Lanczos-Ritz values approximate well enough particular values in the spectrum. A tight upper bound is computed predicting rather well the speed of the convergence of the multishift subspace iteration taking into consideration the interaction with the Lanczos-Ritz values.

The last chapter of this first part gives some mathematical details of implementing these methods and shows some numerical experiments. First some tools for operating with Givens transformations are provided. Secondly the mathematical details of implementing some of the reduction procedures, based on the Givens-vector representation, are shown. The chapter is concluded by showing some numerical experiments illustrating the convergence behavior of the proposed reduction procedures.

✎ *The main realization of this part will be the development of some reduction algorithms and the study of their (convergence) behavior.*

Chapter 2 proposes the different reduction algorithms that will be used throughout this book. The reductions to structured rank forms are the essentials for this chapter. These reductions are discussed in Theorems 2.6, 2.9, 2.13 and 2.17.

Chapter 3 discusses the convergence properties of the designed reduction algorithms. Two types of behavior are investigated. First the Lanczos-Ritz values behavior (Theorems 3.2 and 3.3) and the effect on the reduction algorithms (Subsection 3.1.5) is discussed. Secondly the nested subspace iteration (Subsection 3.2.3) and its convergence speed during the reduction (Subsection 3.3.3) are studied.

The final chapter of this part discusses the implementation of the methods and shows some numerical examples. This chapter contains extra material that is not necessary for a full understanding of the remainder of the book. We recommend however considering the computational complexity in the case of deflation (Subsection 4.2.5) showing that the computational complexity is heavily dependent on the convergence behavior. The Experiments 2, 3 and 4 of Subsection 4.3.1 illustrate the interaction between the convergence behaviors. The examples starting from 5 in Subsection 4.3.2 compare the theoretical bound on the convergence speed with the real situation.

Chapter 2

Algorithms for reducing matrices

It is known from the introductory chapter that semiseparable and tridiagonal matrices are closely related to each other. Invertible semiseparable matrices have as an inverse a tridiagonal matrix and vice versa. One might wonder whether this close relation between these two classes of matrices can be extended, for example, whether one can also reduce any symmetric matrix via orthogonal similarity transformations into a similar semiseparable one. In this chapter we will construct algorithms for reducing matrices to different structured rank forms. We will propose algorithms for reducing matrices to upper triangular semiseparable, semiseparable and semiseparable plus diagonal form. The aim of these reductions is to exploit the rank structure in the development of eigenvalue/singular value algorithms.

In the first introductory section we briefly discuss the nomenclature of transformations. In the second section we discuss three types of orthogonal similarity transformations, applied on symmetric matrices. Firstly we will introduce the well-known tridiagonalization of a symmetric matrix. Secondly we will see that the semiseparabilization of a symmetric matrix is closely related to the previously defined tridiagonalization process. Finally we will further explore the semiseparabilization and adapt this procedure to arrive at an orthogonal similarity transformation for reducing a symmetric matrix to semiseparable plus diagonal form.

The third section focuses on applying this reduction scheme on nonsymmetric matrices. We will show the similarity transformation of matrices into Hessenberg and Hessenberg-like forms. A Hessenberg-like matrix can be considered as the analogue of a Hessenberg matrix for the structured rank case.

The second and the third section focused attention on similarity transformations. Similarity transformations are useful if the aim is to compute the eigenvalues of the original matrix. If one wants to compute the singular values, however, there is no reason to stick to similarity transformations. Therefore, we propose in this section orthogonal transformations to come to an easy form, useful for computing the singular values. In correspondence with the reduction of an arbitrary matrix by orthogonal transformations to a bidiagonal one, we construct an algorithm that reduces a matrix into an upper triangular semiseparable one.

In Sections 2.5 and 2.6 algorithms for transforming matrices from sparse form to the structured rank form and vice versa are presented.[1] Similarity transformations as well as the other type of transformation are discussed for transforming tridiagonal to semiseparable (plus diagonal) form and transforming bidiagonal to upper triangular semiseparable form and vice versa.

✎ *The reader familiar with the standard reductions to sparse matrix format such as to tridiagonal, Hessenberg and bidiagonal (described in Theorems 2.5, 2.12 and 2.15) can omit the discussion of these reduction algorithms and proceed immediately to the new reduction methods to structured rank format. The following reductions are the key algorithms for this chapter: the reduction to semiseparable form in Theorem 2.6; the reduction to semiseparable plus diagonal form in Theorem 2.9; the reduction to Hessenberg-like form in Theorem 2.13 and the reduction to upper triangular semiseparable form in Theorem 2.17. These algorithms are the basics of the forthcoming chapters. For completeness also reductions from sparse to structured rank and from structured rank to sparse form are included (Sections 2.5 and 2.6). These are not however essential for the remainder of the book.*

2.1   Introduction

Before focusing attention on the various reduction algorithms, we will briefly explain some tools and notations used throughout this chapter. We start by introducing the different types of transformations that will be used such as similarity and equivalence transformations. Secondly we will introduce two popular types of orthogonal transformations, namely the Givens and the Householder transformations.

2.1.1   Equivalence transformations

We will briefly repeat here the standard nomenclature related to the transformations discussed in this book.

Definition 2.1. *A matrix[2] $A \in \mathbb{R}^{n \times n}$ is said to be equivalent with a matrix $B \in \mathbb{R}^{n \times n}$ if there exist two nonsingular matrices C and D such that:*

$$C^{-1}AD = B.$$

This transformation is called an equivalence transformation.

Equivalence transformations come in handy when computing the singular values of matrices as orthogonally equivalent matrices (with C and D orthogonal) have

---

[1]Remember the convention from the preceding chapter. With sparse form we mean tridiagonal, bidiagonal and Hessenberg, with structured rank we mean semiseparable, upper triangular semiseparable and Hessenberg-like.

[2]Even though in this book, we will mainly focus on real matrices, most of the results presented here are directly generalizable to complex matrices.

the same singular values. In the case of an orthogonal equivalence transformation, we denote this as

$$U^T A V = B,$$

in which both U and V are orthogonal matrices.[3] See the forthcoming Definition 2.4 of an orthogonal matrix.

A similarity transformation is a special equivalence transformation that preserves the eigenvalues, the so-called spectrum of the matrix.

Definition 2.2. *A matrix $A \in \mathbb{R}^{n \times n}$ is said to be similar to a matrix $B \in \mathbb{R}^{n \times n}$ if there exists a nonsingular matrix C such that:*

$$C^{-1} A C = B.$$

This transformation is called a similarity transformation.

In case of an orthogonal similarity transformation, this is denoted as $U^{-1} A U = U^T A U = B$, with U as an orthogonal matrix.[4] A special type of transformation is the congruence transformation.

Definition 2.3. *A matrix $A \in \mathbb{R}^{n \times n}$ is said to be congruent with a matrix $B \in \mathbb{R}^{n \times n}$ if there exists a nonsingular matrix $C \in \mathbb{R}^{n \times n}$ such that:*

$$C^T A C = B.$$

This transformation is called a congruence transformation.

An orthogonal similarity transformation is hence also a congruence transformation.[5]

In this chapter we will explore all these three transformations as a preprocessing step for computing the eigenvalues and/or singular values. The clue is to perform any of these transformations, such that the obtained matrix B is of easy structure, which means that one can more efficiently compute its eigenvalues/singular values.

2.1.2 Orthogonal transformations

In the previous subsection we proposed different kinds of transformations for obtaining a more easy structure of the resulting matrix B. In this chapter we will focus on orthogonal transformation matrices.

Definition 2.4. *A matrix Q is said to be orthogonal[6] if*

$$Q^{-1} = Q^T.$$

---

[3] In the case of unitary matrices we obtain $U^H A V = B$.

[4] For unitary matrices we obtain $U^H A U = B$.

[5] In the case of complex matrices, one also uses the relation $C^H A C$ for defining a congruence.

[6] In the complex case, these matrices are called unitary and $Q^{-1} = Q^H$. The operation $\cdot^H$ is called the hermitian conjugate.

These transformations are used because of their favorable numerical behavior with regard to other transformations. This means that an orthogonal similarity transformation is in fact also a congruence transformation.

Let us briefly explore the orthogonal transformations extensively used throughout this book.

Givens transformations

A Givens transformation is a 2×2 orthogonal transformation applied on a column or row vector of length 2. A Givens transformation is constructed in such a way that one element in the vector is annihilated.

A 2×2 Givens transformation is of the following form:

$$G = \left[\begin{array}{cc} c & -s \\ s & c \end{array} \right],$$

where $c^2 + s^2 = 1$. In fact c and s are, respectively, the sine and cosine of a certain angle. Suppose a vector $[x, y]^T$ is given, then there exists always a Givens transformation such that

$$G \left[\begin{array}{c} x \\ y \end{array} \right] = \left[\begin{array}{c} r \\ 0 \end{array} \right],$$

with $r = \sqrt{x^2 + y^2}$.

Givens transformations are often used in different applications for creating zeros in matrices and so on. A study on how to compute Givens transformations accurately can for example be found in [20]. In the remainder of the book, the Givens transformations will also be working on vectors, not necessary of length 2. Hence we need to embed our Givens transformation (see, e.g., [94]), in a larger matrix. The embedding is an easy procedure. In fact, an embedded Givens transformation acting on a vector does only change two elements in this vector. One element is zeroed out, and the other element incorporates the energy of the element, which is made zero. The notation we use is the following: $\check{G}$ denotes the 2×2 Givens transformation matrix, consisting of all the sines and cosines. And the matrix G is the Givens transformation acting on the complete vector. In fact the matrix $\check{G}$ is embedded in G in the following way:

$$G = \left[\begin{array}{ccccccccc} 1 & 0 & \cdots & & & & & \cdots & 0 \\ 0 & \ddots & & & & & & & \vdots \\ \vdots & & 1 & 0 & \cdots & & & & \\ & & 0 & c & \cdots & -s & & & \\ & & \vdots & \vdots & & \vdots & & & \\ & & s & \cdots & c & & & \vdots \\ \vdots & & & & & & \ddots & 0 \\ 0 & \cdots & & & & & \cdots & 0 & 1 \end{array} \right].$$

The matrix G is the identity matrix, with the elements of the matrix $\check{G}$ in the positions (k, k), (k, l), (l, k) and (l, l), where the l and k correspond to the positions of the elements v_k and v_l in the vector $\mathbf{v}$, which one wants the Givens transformation to act on. Storing a Givens procedure can be done in two ways, based on the following relation[7]:

$$G = \frac{1}{\sqrt{1 + t^2}} \begin{bmatrix} t & -1 \\ 1 & t \end{bmatrix} = \begin{bmatrix} c & -s \\ s & c \end{bmatrix}.$$

We have

$$\tan \theta = \frac{s}{c} = \frac{\sin \theta}{\cos \theta} = \frac{1}{t}.$$

Either one stores one parameter t from which one can easily compute the cosine and the sine, but this costs some operations. Or one can store both the sine and cosine. This costs more memory allocation. In our implementations we chose to store both the sine and the cosine. But essentially a Givens transformation is only determined by one parameter.

Householder transformations

A Householder transformation is an orthogonal transformation working on a vector. This transformation creates many zeros in the vector by one matrix vector multiplication. This section is based on [94, Subsection 5.1.2]. If $\mathbf{v}$ is a nonzero column vector of length n, then the matrix H, with

$$H = I - \frac{2}{\mathbf{v}^T \mathbf{v}} \mathbf{v}\mathbf{v}^T,$$

defines a Householder reflection. These matrices reflect any vector with respect to the hyperplane orthogonal to $\langle \mathbf{v} \rangle$[8].

Suppose now a vector $\mathbf{x}$ is given. Setting $\mathbf{v} = \mathbf{x} \pm \|\mathbf{x}\|_2 \mathbf{e}_1$, with $\mathbf{e}_1$ as the first basis vector, gives us the following relation:

$$P\mathbf{x} = \left(I - 2\frac{\mathbf{v}\mathbf{v}^T}{\mathbf{v}^T \mathbf{v}} \right) \mathbf{x} = \mp \|\mathbf{x}\|_2 \mathbf{e}_1.$$

The matrix P projects all the energy of the vector $\mathbf{x}$ via an orthogonal transformation into the first element. Efficient manners to store Householder transformations or to compute them are discussed in [94].

Notes and references

Householder transformations have already been known for quite some time. They are named after A. S. Householder. Givens transformations are named after W. Givens.[9] Many more references on Givens and Householder transformations can be found in

---

[7]Remark that this relation also admits the value $t = \infty$!

[8]With $\langle \mathbf{x}, \mathbf{y}, \mathbf{z} \rangle$, we denote the subspace spanned the vectors $\mathbf{x}, \mathbf{y}$ and $\mathbf{z}$.

[9]Givens transformations are also sometimes referred to as Jacobi rotations.

☞ G. H. Golub and C. F. Van Loan. *Matrix Computations*. Johns Hopkins
University Press, Baltimore, Maryland, USA, third edition, 1996.

Often 2×2 unitary transformations are also referred to as Givens transformations. Unless stated otherwise, when speaking about Givens transformations we mean rotations in this book. There is, however, no loss of generality when using arbitrary 2×2 unitary matrices.

Givens transformations are frequently used in this volume. Results on how to compute them accurately can be found in.

☞ D. Bindel, J. W. Demmel, W. Kahan, and O. A. Marques. On computing
Givens rotations reliably and efficiently. *ACM Transactions on Mathematical Software*, 28(2):206–238, June 2002.

2.2   Orthogonal similarity transformations of symmetric matrices

As already mentioned in the introduction of this chapter, this section focuses on orthogonal similarity transformations of symmetric matrices to easier forms. The aim of the proposed reduction methods is to exploit the structure of the reduced matrices in the eigenvalue computations.

First we briefly repeat the traditional reduction of symmetric matrices to tridiagonal form, followed by the reduction to semiseparable form. The final subsection focuses on an orthogonal similarity transformation to semiseparable plus diagonal form. The reduction to semiseparable plus diagonal form has a special, tunable convergence behavior, which will be investigated in the next chapter.

2.2.1   To tridiagonal form

The reduction of a symmetric matrix to a semiseparable one is similar to the reduction to tridiagonal form. Therefore the reduction to tridiagonal form will briefly be given in this section. More detailed information concerning this reduction can be found in [129, 58, 94, 145]. We present this well-known method to get the reader familiar with the notation. Understanding the notation and proof of this section will make the understanding of the reduction to semiseparable form in the next subsection easier.

Theorem 2.5. *Let A be a symmetric matrix. Then there exists an orthogonal matrix U such that*

$$U^T A U = T,$$

where T is a tridiagonal matrix.

Proof. We will construct a tridiagonal matrix by using Givens transformations. These transformations will introduce the zeros in the correct places. The same can of course be done via Householder reflectors. Let[10] $A_0^{(1)} = A$. Often, we denote

---

[10]In fact one can perform an arbitrary initial orthogonal similarity transformation and set $A_0^{(1)} = U_0^T A U_0$, this initial transformation does not change the reduction procedure, but it has a significant influence on the convergence behavior. We will come back to this in Chapter 3.

$A_0^{(i)}$ as $A^{(i)}$. Let $G_i^{(l)}$ be a Givens transformation, such that the product $A_{i-1}^{(l)} G_i^{(l)}$ has the entry $(n - l + 1, i)$ annihilated and the i-th and the $(i + 1)$-th columns of $A_{i-1}^{(l)}$ modified. Zero elements are not denoted in the figures.

- **Step** 1. We will start by making all the elements in the last row, except for the last two, zero. Just multiply $A_0^{(1)}$ to the left by $G_1^{(1)T}$ and to the right by $G_1^{(1)}$ to annihilate the elements in the positions $(1, n)$ and $(n, 1)$ in $A_0^{(1)}$, respectively. The arrows denote the columns and rows, which will be affected by transforming the matrix $A_0^{(1)}$ into $G_1^{(1)T} A_0^{(1)} G_1^{(1)}$:

$$
\begin{array}{c} \downarrow \quad \downarrow \\ \rightarrow \\ \rightarrow \end{array}
\begin{bmatrix}
\times & \times & \cdots & \times & \otimes \\
\times & \ddots & & \times & \times \\
\vdots & & \ddots & \vdots & \vdots \\
\times & \times & \cdots & \times & \times \\
\otimes & \times & \cdots & \times & \times
\end{bmatrix}
\xrightarrow{G_1^{(1)T} A_0^{(1)} G_1^{(1)}}
\begin{bmatrix}
\times & \times & \cdots & \times & 0 \\
\times & \ddots & & \times & \times \\
\vdots & & \ddots & \vdots & \vdots \\
\times & \times & \cdots & \times & \times \\
0 & \times & \cdots & \times & \times
\end{bmatrix}
$$

$$
\Updownarrow
$$

$$
A_0^{(1)} \xrightarrow{G_1^{(1)T} A_0^{(1)} G_1^{(1)}} A_1^{(1)}.
$$

Let us remark that the elements denoted with $\times$ are arbitrary elements of the matrix, while the elements $\otimes$ denote the elements that will be annihilated by the similarity transformation. Summarizing, we obtain that $A_1^{(1)} = G_1^{(1)T} A_0^{(1)} G_1^{(1)}$. Continuing this process of annihilating all the elements in the last row (column), except for the elements in position $(n, n - 1)$ $((n - 1, n))$ and (n, n), we get

$$
A_{n-2}^{(1)} =
\begin{bmatrix}
\times & \times & \cdots & \times & 0 \\
\times & \ddots & & \vdots & \vdots \\
\vdots & & \ddots & \times & 0 \\
\times & \cdots & \times & \times & \times \\
0 & \cdots & 0 & \times & \times
\end{bmatrix} .
$$

We remark that the lower right 2×2 block already satisfies the tridiagonal structure. Let us define now $A_0^{(2)}$ as $A_0^{(2)} = A_{n-2}^{(1)}$.

- **Step** k. Let $k = n - j$, $1 < j < n$. Assume by induction that $A_0^{(k)}$ has the lower right $k \times k$ block already in tridiagonal form.

$$
A_0^{(k)} = \begin{bmatrix}
\times & \cdots & \times & \times & & & & \\
\vdots & \ddots & \vdots & \vdots & & & & \\
\times & \cdots & \times & \times & & & & \\
\times & \cdots & \times & \times & \times & & & \\
& & & \times & \times & \ddots & & \\
& & & & \ddots & \ddots & \times & \\
& & & & & \times & \times &
\end{bmatrix}
\begin{matrix}
\leftarrow 1 \\
\vdots \\
\leftarrow j \\
\leftarrow j+1 \\
\leftarrow j+2 \\
\vdots \\
\leftarrow n
\end{matrix}
$$

We can now construct a Givens transformation $G_1^{(k)}$ such that multiplying $A_0^{(k)}$ to the left by $G_1^{(k)T}$ will annihilate element $(1, j+1)$ without destroying the tridiagonal structure of the lower $k \times k$ block. In a similar way, we can construct the Givens transformations $G_2^{(k)}, \ldots, G_{j-2}^{(k)}$ and $G_{j-1}^{(k)}$ such that multiplying $A_1^{(k)}$ on the left with $G_{j-1}^{(k)T} \ldots G_2^{(k)T}$ will make zero elements in rows 2 up to $j-1$ in the column $j+1$. Again, applying these transformations does not involve the lower right tridiagonal block starting from row and column $j+1$.

$$
A_{j-1}^{(k)} = G_{j-1}^{(k)T} \cdots G_1^{(k)T} A_0^{(k)} G_1^{(k)} \cdots G_{j-1}^{(k)}
$$

$$
= \begin{bmatrix}
\times & \cdots & & \times & 0 & & & \\
\vdots & \ddots & & \vdots & \vdots & & & \\
& & \ddots & \times & 0 & & & \\
\times & \cdots & & \times & \times & \times & & \\
0 & \cdots & 0 & \times & \times & \ddots & & \\
& & & & \ddots & \ddots & \times & \\
& & & & & \times & \times &
\end{bmatrix}
\begin{matrix}
\leftarrow 1 \\
\vdots \\
\vdots \\
\leftarrow j \\
\leftarrow j+1 \\
\vdots \\
\leftarrow n
\end{matrix}
$$

One can clearly see that the latter sequence of Givens transformations expanded the lower right tridiagonal block by adding a row and a column to it. This proves the inductive procedure. If we define our matrix U as follows:

$$
U = G_1^{(1)} \cdots G_{n-2}^{(1)} G_1^{(2)} \cdots \cdots G_2^{(n-3)} G_1^{(n-2)},
$$

we have constructed an orthogonal matrix U as the product of individual Givens transformations such that $U^T A U$ is a tridiagonal matrix.

The same procedure, of course, can be performed by using Householder transformations instead of Givens transformations.   □

One can see that the overall reduction algorithm involves either $n-2$ Householder reflectors or $3(n-1)(n-2)/2$ Givens transformations, leading to a complexity of $4/3n^3 + \mathcal{O}(n^2)$ or $2n^3 + \mathcal{O}(n^2)$, with, respectively, Householder and Givens transformations.

This process of tridiagonalization is well-known. In the following section, a procedure will be constructed for reducing arbitrary symmetric matrices to semiseparable form. Part of this reduction is similar to the reduction to tridiagonal form. Moreover it will be shown that one can decouple the reduction to semiseparable form into two parts. A first part will make the matrix tridiagonal and a second part will transform this tridiagonal matrix to semiseparable form. The difference between the uncoupled algorithm and the decoupled algorithm, in which we first reduce to tridiagonal and then to semiseparable form, has to do with the convergence properties and will be explained in Chapter 3. In the last section of this chapter we will present also another reduction to semiseparable form, which one cannot decouple in a reduction to tridiagonal and afterwards to semiseparable form.

2.2.2   To semiseparable form

An algorithm to transform a symmetric matrix into a semiseparable one by orthogonal similarity transformations is presented in this section. The constructive proof of the next theorem provides the algorithm. The original algorithm was presented in [154].

Theorem 2.6. *Let A be a symmetric matrix. Then there exists an orthogonal matrix U such that*

$$U^T A U = S,$$

where S is a semiseparable matrix.

Proof. The constructive proof is made by induction on the rows of the matrix A. Let $A_0^{(1)} = A$. We remark that an extra initial orthogonal transformation can be applied on the matrix A, similar to the tridiagonal case. We will often briefly denote $A_0^{(i)}$ as $A^{(i)}$. Let $G_i^{(l)}$ be a Givens transformation, such that the product $A_{i-1}^{(l)} G_i^{(l)}$ has the entry $(n - l + 1, i)$ annihilated and the i-th and the $(i + 1)$-th columns of $A_{i-1}^{(l)}$ modified, unless stated otherwise.

- **Step 1.** We start by constructing a similarity transformation that makes the last two rows (columns) linearly dependent in the lower (upper) triangular part. To this end, we multiply $A_0^{(1)}$ to the left by $G_1^{(1)^T}$ and to the right by $G_1^{(1)}$ to annihilate the elements in position $(1, n)$ and $(n, 1)$ in $A_0^{(1)}$, respectively. The arrows denote the columns and rows that will be affected by transforming the matrix $A_0^{(1)}$ into $G_1^{(1)^T} A_0^{(1)} G_1^{(1)}$:

$$\begin{array}{c} \rightarrow \\ \rightarrow \\ \\ \\ \\ \end{array} \begin{bmatrix} \times & \times & \cdots & \times & \otimes \\ \times & \ddots & & \times & \times \\ \vdots & & \ddots & \vdots & \vdots \\ \times & \times & \cdots & \times & \times \\ \otimes & \times & \cdots & \times & \times \end{bmatrix} \xrightarrow{G_1^{(1)^T} A_0^{(1)} G_1^{(1)}} \begin{bmatrix} \times & \times & \cdots & \times & 0 \\ \times & \ddots & & \times & \times \\ \vdots & & \ddots & \vdots & \vdots \\ \times & \times & \cdots & \times & \times \\ 0 & \times & \cdots & \times & \times \end{bmatrix}$$

$$\Updownarrow$$

$$A_0^{(1)} \xrightarrow{G_1^{(1)^T} A_0^{(1)} G_1^{(1)}} A_1^{(1)}.$$

This process of annihilation is exactly the same as the one used for tri-diagonalizing a symmetric matrix. Continuing the annihilation process of the nonzero elements in the last row gives us

$$A_{n-2}^{(1)} = \begin{bmatrix} \times & \times & \cdots & \times & 0 \\ \times & \ddots & & \vdots & \vdots \\ \vdots & & \ddots & \times & 0 \\ \times & \cdots & \times & \times & \times \\ 0 & \cdots & 0 & \times & \times \end{bmatrix}.$$

In the tridiagonalization procedure, the first step is finished, and all the desired zeros are created in the bottom row. Because we want to obtain a semisep-arable matrix in this case, we have to apply another, extra transformation. Multiplying $A_{n-2}^{(1)}$ to the left by $G_{n-1}^{(1)^T}$, leads to the following situation:

$$\begin{array}{c} \\ \\ \\ \rightarrow \\ \rightarrow \end{array} \begin{bmatrix} \times & \times & \cdots & \times & 0 \\ \times & \ddots & & \vdots & \vdots \\ \vdots & & \ddots & \times & 0 \\ \times & \cdots & \times & \times & \otimes \\ 0 & \cdots & 0 & \times & \times \end{bmatrix} \xrightarrow{G_{n-1}^{(1)^T} A_{n-2}^{(1)}} \begin{bmatrix} \times & \times & \cdots & \times & 0 \\ \times & \ddots & & \vdots & \vdots \\ \vdots & & \ddots & \times & 0 \\ \boxtimes & \cdots & \boxtimes & \times & 0 \\ \boxtimes & \cdots & \boxtimes & \times & \times \end{bmatrix}$$

$$\Updownarrow$$

$$A_{n-2}^{(1)} \xrightarrow{G_{n-1}^{(1)^T} A_{n-2}^{(1)}} \tilde{A}_{n-2}^{(1)},$$

i.e., the last two rows are proportional to each other, with exception of the entries in the last two columns (to emphasize the linear dependency among the rows [columns], we denote by $\boxtimes$ the elements of the matrix belonging to these rows [columns]; in fact the elements marked with $\boxtimes$ denote the part of the matrix already satisfying the semiseparable structural constraints). Let

$\tilde{A}_{n-2}^{(1)} = {G_{n-1}^{(1)}}^{T} A_{n-2}^{(1)}$. Multiplying now $\tilde{A}_{n-2}^{(1)}$ to the right by $G_{n-1}^{(1)}$, the last two columns become linearly dependent above and on the main diagonal, i.e., the last two columns form an upper semiseparable part. Because of symmetry, the last two rows become linearly dependent below and on the main diagonal and form a lower semiseparable part:

$$
\begin{bmatrix}
\times & \times & \cdots & \times & 0 \\
\times & \ddots & & \vdots & \vdots \\
\vdots & & \ddots & \times & 0 \\
\boxtimes & \cdots & \boxtimes & \times & 0 \\
\boxtimes & \cdots & \boxtimes & \times & \times
\end{bmatrix}
\xrightarrow{\tilde{A}_{n-2}^{(1)} G_{n-1}^{(1)}}
\begin{bmatrix}
\times & \times & \cdots & \boxtimes & \boxtimes \\
\times & \ddots & & \vdots & \vdots \\
\vdots & & \ddots & \boxtimes & \boxtimes \\
\boxtimes & \cdots & \boxtimes & \boxtimes & \boxtimes \\
\boxtimes & \cdots & \boxtimes & \boxtimes & \boxtimes
\end{bmatrix}
$$

$$
\Updownarrow
$$

$$
\tilde{A}_{n-2}^{(1)} \xrightarrow{\tilde{A}_{n-2}^{(1)} G_{n-1}^{(1)}} A_{n-1}^{(1)}.
$$

To start the next step, $A_0^{(2)}$ is defined as $A_0^{(2)} = A_{n-1}^{(1)}$.

- **Step k.** Let $k = n - j$, $1 < j < n$. Assume by induction that $A_0^{(k)}$ has the lower (upper) triangular part of semiseparable form, from row n up to row $j + 1$ (column $j + 1$ to n). We will now prove that we can make the lower (upper) triangular part semiseparable up to row j (column j). Let us denote the lower (and upper) triangular elements that form a semiseparable part with $\boxtimes$. Matrix $A_0^{(k)}$ looks like:

$$
A_0^{(k)} =
\begin{bmatrix}
\times & \cdots & \times & \boxtimes & \cdots & \boxtimes \\
\vdots & \ddots & \vdots & \vdots & & \vdots \\
\times & \cdots & \times & \boxtimes & \cdots & \boxtimes \\
\boxtimes & \cdots & \boxtimes & \boxtimes & \cdots & \boxtimes \\
\vdots & & \vdots & \vdots & \ddots & \vdots \\
\boxtimes & \cdots & \boxtimes & \boxtimes & \cdots & \boxtimes
\end{bmatrix}
\begin{matrix}
\leftarrow 1 \\
\vdots \\
\leftarrow j \\
\leftarrow j+1 \\
\vdots \\
\leftarrow n
\end{matrix} \; .
$$

We remark that the lower right $k \times k$ block of $A_0^{(k)}$ is already semiseparable. Because the submatrix of $A_0^{(k)}$ consisting of the first j columns and the last $n - j$ rows is already semiseparable, it has rank ≤ 1. Hence, we can construct a Givens transformation $G_1^{(k)}$ such that multiplying $A_0^{(k)}$ to the left by ${G_1^{(k)}}^{T}$ will annihilate all elements $(1, j+1)$, $(1, j+2)$ up to $(1, n)$. In a similar way, we can construct the Givens transformations $G_2^{(k)}, \ldots, G_{j-2}^{(k)}$ and $G_{j-1}^{(k)}$ such that multiplying $A_1^{(k)}$ on the left with ${G_{j-1}^{(k)}}^{T} \ldots {G_2^{(k)}}^{T}$, will make zero elements in rows 2 up to $j - 1$, the columns $j + 1$ up to n. Applying these similarity

$$
\begin{array}{r}
\\ \\ \\ \\ j \rightarrow \\ j+1 \rightarrow \\ \\ \\ \\
\end{array}
\begin{bmatrix}
\times & \cdots & \times & \times & 0 & 0 & \cdots & 0 \\
\vdots & \ddots & \vdots & \vdots & \vdots & \vdots & & \vdots \\
\times & \cdots & \times & \times & 0 & 0 & \cdots & 0 \\
\times & \cdots & \times & \boxtimes & \otimes & \otimes & \cdots & \otimes \\
0 & \cdots & 0 & \boxtimes & \boxtimes & \boxtimes & \cdots & \boxtimes \\
0 & \cdots & 0 & \boxtimes & \boxtimes & \boxtimes & \cdots & \boxtimes \\
\vdots & & \vdots & \vdots & \vdots & \vdots & \ddots & \vdots \\
0 & \cdots & 0 & \boxtimes & \boxtimes & \boxtimes & \cdots & \boxtimes
\end{bmatrix}
\xrightarrow{G_j^{(k)^T} A_{j-1}^{(k)}}
\begin{array}{c}
\begin{matrix} & & & j & j+1 & & & \\ & & & \downarrow & \downarrow & & & \end{matrix} \\
\begin{bmatrix}
\times & \cdots & \times & \times & 0 & 0 & \cdots & 0 \\
\vdots & \ddots & \vdots & \vdots & \vdots & \vdots & & \vdots \\
\times & \cdots & \times & \times & 0 & 0 & \cdots & 0 \\
\boxtimes & \cdots & \boxtimes & \times & 0 & 0 & \cdots & 0 \\
\boxtimes & \cdots & \boxtimes & \boxtimes & \boxtimes & \boxtimes & \cdots & \boxtimes \\
0 & \cdots & 0 & \otimes & \boxtimes & \boxtimes & \cdots & \boxtimes \\
\vdots & & \vdots & \vdots & \vdots & \vdots & \ddots & \vdots \\
0 & \cdots & 0 & \otimes & \boxtimes & \boxtimes & \cdots & \boxtimes
\end{bmatrix}
\end{array}
$$

$$
\xrightarrow{G_j^{(k)^T} A_{j-1}^{(k)} G_j^{(k)}}
\begin{bmatrix}
\times & \cdots & \times & \boxtimes & \boxtimes & 0 & \cdots & 0 \\
\vdots & \ddots & \vdots & \vdots & \vdots & \vdots & & \vdots \\
\times & \cdots & \times & \boxtimes & \boxtimes & 0 & \cdots & 0 \\
\boxtimes & \cdots & \boxtimes & \boxtimes & \boxtimes & 0 & \cdots & 0 \\
\boxtimes & \cdots & \boxtimes & \boxtimes & \boxtimes & \boxtimes & \cdots & \boxtimes \\
0 & \cdots & 0 & 0 & \boxtimes & \boxtimes & \cdots & \boxtimes \\
\vdots & & \vdots & \vdots & \vdots & \vdots & \ddots & \vdots \\
0 & \cdots & 0 & 0 & \boxtimes & \boxtimes & \cdots & \boxtimes
\end{bmatrix}.
$$

Figure 2.1. *The transformation* $A_j^{(k)} = G_j^{(k)^T} A_{j-1}^{(k)} G_j^{(k)}$ *makes the first* j *entries of rows (columns)* j *and* $j+1$ *proportional.*

transformations, we obtain the following matrix:[11]

$$
A_{j-1}^{(k)} = G_{j-1}^{(k)^T} \cdots G_1^{(k)^T} A_0^{(k)} G_1^{(k)} \cdots G_{j-1}^{(k)}
$$

$$
= \begin{bmatrix}
\times & \cdots & & \times & 0 & \cdots & 0 \\
\vdots & \ddots & & \vdots & \vdots & & \vdots \\
 & & \ddots & \times & 0 & \cdots & 0 \\
\times & \cdots & \times & \times & \boxtimes & \cdots & \boxtimes \\
0 & \cdots & 0 & \boxtimes & \boxtimes & \cdots & \boxtimes \\
\vdots & & \vdots & \vdots & \vdots & \ddots & \vdots \\
0 & \cdots & 0 & \boxtimes & \boxtimes & \cdots & \boxtimes
\end{bmatrix}
\begin{array}{l}
\leftarrow 1 \\ \vdots \\ \vdots \\ \leftarrow j \\ \leftarrow j+1\,. \\ \vdots \\ \leftarrow n
\end{array}
$$

Since the rows j and $j+1$ are proportional for the indexes of columns greater than j, the Givens transformation $G_j^{(k)^T}$ also can be constructed, annihilating all the entries in row j with column index greater than j. Furthermore, the product of $G_j^{(k)^T} A_{j-1}^{(k)}$ to the right by $G_j^{(k)}$ makes the columns j and $j+1$ proportional in the first j entries. The latter similarity transformation is depicted in Figure 2.1.

To retrieve the semiseparable structure, below the $(j+1)$-th row and to the right of the $(j+1)$-th column, $n-j-1$ more similarity Givens transformations G_i, $i = j+1, j+2, \ldots, n-1$ are considered. Each of these is chosen in such a way that when the Givens transformation G_i^T is applied to the left, the elements $i+1, i+2, \ldots, n$ are annihilated in row i, using the corresponding

[11] Later on, we will prove that these transformations are essentially the same as the transformations used for tridiagonalizing this matrix.

elements of row $i + 1$. To obtain a similar matrix, we apply G_i also on the right, thereby adding one more row (column) to the semiseparable part. The whole process for adding one more row/column to the semiseparable structure is summarized in Figure 2.2.

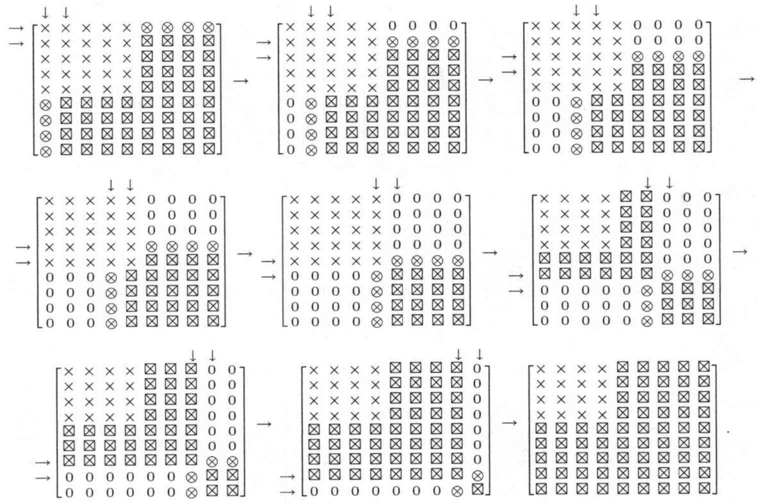

Figure 2.2. *Description of the similarity transformations used to retrieve the semiseparable structure.*

Combining, similarly as in the tridiagonal case, all Givens transformations together in an orthogonal matrix U, we proved the theorem. □

Note 2.7. *In the remainder of the text we will often refer to this reduction as starting at the bottom (bottom-right) of the matrix, since during the reduction the part already of semiseparable form is located in the bottom right position of the matrix. Similarly one can easily adapt the procedure by creating zeros in the upper-left position (see Theorem 2.8) and hence the growing semiseparable matrix can be found in the upper-left corner. This second reduction is often referred to as the reduction starting at the top (upper-left). For example the reduction to bidiagonal or upper triangular semiseparable form traditionally starts on the upper-left corner. Hence we will refer to this reduction as starting at the top or at the upper-left corner of the matrix.*

In Chapter 4, the details are given on how to implement this algorithm using the Givens-vector representation for the semiseparable part of the matrices involved. This will result in an $\mathcal{O}(n^3)$ algorithm to compute the Givens-vector representation of a semiseparable matrix similar to the original symmetric matrix. The algorithm can in fact be divided into two main parts. First we have the zero creating step, similar to the tridiagonalization process, and secondly we have to perform a chasing step to bring the matrix back to the semiseparable form. Taking a closer look at the algorithm reveals that the Givens transformations applied in the first part of each step in the reduction algorithm can be replaced by a corresponding Householder transformation. In step $k = n - j$ of the reduction a similarity transformation of the following form is performed ${G_{j-1}^{(k)}}^T \ldots {G_1^{(k)}}^T A_0^{(k)} G_1^{(k)} \ldots G_{j-1}^{(k)}$, which can be replaced by a single Householder transformation $H^{(k)}$, such that ${H^{(k)}}^T A_0^{(k)} H^{(k)}$ has the same entries annihilated. Moreover we obtain also a semiseparable matrix, using the same number of operations, when transforming the symmetric matrix into a tridiagonal one, which costs $4n^3/3 + \mathcal{O}(n^2)$ with Householders, $2n^3 + \mathcal{O}(n^2)$ with Givens transformations and then transforming the tridiagonal matrix to a semiseparable matrix by using Givens transformations, which costs an extra $9n^2 + \mathcal{O}(n)$, when storing the semiseparable part of the matrix efficiently. This means that we have, in comparison with the tridiagonalization, an extra cost of $9n^2 + \mathcal{O}(n)$. When first reducing matrices to tridiagonal form, intermediate matrices look like:

$$
\begin{bmatrix}
\times & \times & 0 & 0 & 0 & 0 & 0 & 0 & 0 \\
\times & \times & \times & 0 & 0 & 0 & 0 & 0 & 0 \\
0 & \times & \times & \times & 0 & 0 & 0 & 0 & 0 \\
0 & 0 & \times & \times & \boxtimes & \boxtimes & \boxtimes & 0 & 0 \\
0 & 0 & 0 & \boxtimes & \boxtimes & \boxtimes & \boxtimes & 0 & 0 \\
\hline
0 & 0 & 0 & \boxtimes & \boxtimes & \boxtimes & \boxtimes & 0 & 0 \\
0 & 0 & 0 & \boxtimes & \boxtimes & \boxtimes & \boxtimes & \otimes & \boxtimes \\
0 & 0 & 0 & 0 & 0 & 0 & \otimes & \boxtimes & \boxtimes \\
0 & 0 & 0 & 0 & 0 & 0 & \boxtimes & \boxtimes & \boxtimes
\end{bmatrix} .
$$

Because the algorithm from Theorem 2.6 and the algorithm that first tridiagonalizes a symmetric matrix and afterwards makes a semiseparable matrix from it are essentially the same, one might doubt the usefulness of the reduction to semiseparable form. However, the algorithm of Theorem 2.6 has some advantages, namely different convergence properties, which will become apparent in Chapter 3.

Once the semiseparable matrix has been computed, several algorithms can be considered to compute its eigenvalues (see, for instance, [38, 40, 77, 117] and the implicit QR-algorithm, as considered in this book; see Part II, Chapter 7).

Before deducing another type of reduction algorithm, we briefly propose a theorem that will come in handy for investigating the convergence properties of these reduction algorithms. The reduction algorithm as presented here constructs a semiseparable block in the lower-right part of the matrix. During the reduction algorithm, this block increases in size. In fact one can deduce a similar method that constructs a block of increasing size starting from the upper-left part of the matrix.

Theorem 2.8. *Let $A \in \mathbb{R}^{n \times n}$. Then there exists an orthogonal similarity reduction to semiseparable form, such that the intermediate matrices $A^{(m)}$ have the upper left $m \times m$ block of semiseparable form.*

Proof. Apply the reduction of Theorem 2.6 to the matrix JAJ, where J is the counteridentity matrix. The counteridentity matrix has zeros everywhere, except ones on the antidiagonal. The intermediate matrices $A^{(m)}$ arising from the reduction in Theorem 2.6 have the lower right block semiseparable. Denote with $U^{(m)}$ an intermediate orthogonal matrix such that the following equation is satisfied:

$$A^{(m)} = U^{(m)^T} JAJU^{(m)}.$$

This means that

$$
\begin{aligned}
\tilde{A}^{(m)} &= JA^{(m)}J \\
&= JU^{(m)^T} JAJU^{(m)} J \\
&= \left(\tilde{U}^{(m)}\right)^T A \tilde{U}^{(m)}
\end{aligned}
$$

and $\tilde{A}^{(m)}$ has the upper left block of semiseparable form. $\square$

We remark that this algorithm is easy to implement. Instead of starting the reduction on the last row and annihilating every element except the last one, one has to start on the first row and annihilate all the elements except for the first one.

2.2.3   To semiseparable plus diagonal form

In the following chapter it is proved that the chasing step inside of the reduction procedure corresponds to performing a QR-step *without* shift on the already semiseparable part of the matrix. A natural question arises here. Is it possible to change the algorithm in such a manner that QR-steps with shift are performed inside the algorithm and what will the outcome be in this case. Intuitively, one expects the outcome of such an algorithm to be a semiseparable plus diagonal matrix, as this shift will create the diagonal matrix. The algorithm having these properties will be constructed in this section. It is similar to the reduction to semiseparable form but changes in every step the diagonal elements slightly to obtain finally not a semiseparable but a semiseparable plus diagonal matrix. The theorem will be proved similarly as the other ones in a constructive way and was firstly presented in [172].

Theorem 2.9. *Let A be a symmetric matrix and $d_1, \ldots, d_n$ are n arbitrary elements. Then there exists an orthogonal matrix U, such that*

$$U^T A U = S + D,$$

where S is a semiseparable matrix and D is a diagonal matrix containing the elements $d_1, \ldots, d_n$ as diagonal elements.

Proof. The proof is again by finite induction and illustrated on a 5×5 example. The more general case is completely similar. Notations are the same as before. We remark that there is the possibility to perform an initial extra similarity transformation.

- **Step** 1. Exactly the same Givens transformations are performed to annihilate all the elements in the last row except for the last two ones (this can also be done by performing specific Householder transformations). Let $A_0^{(1)} = A$.

$$
\begin{bmatrix}
\times & \times & \times & \times & \otimes \\
\times & \times & \times & \times & \otimes \\
\times & \times & \times & \times & \otimes \\
\times & \times & \times & \times & \times \\
\otimes & \otimes & \otimes & \times & \times
\end{bmatrix}
\longrightarrow
\begin{bmatrix}
\times & \times & \times & \times & 0 \\
\times & \times & \times & \times & 0 \\
\times & \times & \times & \times & 0 \\
\times & \times & \times & \times & \times \\
0 & 0 & 0 & \times & \times
\end{bmatrix}
$$

$$\Updownarrow$$

$$
A_0^{(1)} \xrightarrow{\ {G_3^{(1)}}^T {G_2^{(1)}}^T {G_1^{(1)}}^T A_0^{(1)} G_1^{(1)} G_2^{(1)} G_3^{(1)}\ } A_3^{(1)}.
$$

In the reduction to semiseparable form one would immediately perform another Givens transformation, thereby making the last two rows and columns dependent. Here however we do not want a semiseparable but a semiseparable plus diagonal matrix, and therefore we first have to change the diagonal elements in the matrix $A_3^{(1)}$. Before doing so, some new notation is introduced. As in the previous proofs the elements surrounded by $\bigcirc$ denote the elements that will be annihilated by the upcoming transformation and the elements surrounded by $\square$ denote the elements already satisfying the semiseparable structural constraints. The elements themselves were denoted with $\times$. As such an element changes now, e.g., by subtraction or by addition we will denote this new element with a $+$ sign.

Rewrite the matrix $A_3^{(1)}$ as the sum of a matrix $\hat{A}_3^{(1)}$, having the same elements equal to zero and a diagonal matrix $D_1^{(1)}$:

$$
A_3^{(1)} =
\begin{bmatrix}
\times & \times & \times & \times & 0 \\
\times & \times & \times & \times & 0 \\
\times & \times & \times & \times & 0 \\
\times & \times & \times & + & \times \\
0 & 0 & 0 & \times & +
\end{bmatrix}
+
\begin{bmatrix}
0 & & & & \\
& 0 & & & \\
& & 0 & & \\
& & & d_1 & \\
& & & & d_1
\end{bmatrix}
$$
$$
= \hat{A}_3^{(1)} + D_1^{(1)}.
$$

Determining now the Givens transformation $G_4^{(1)}$, to annihilate the element $(5,4)$ in the matrix $\hat{A}_3^{(1)}$ gives us the following transformation:

$$
\begin{bmatrix}
\times & \times & \times & \times & 0 \\
\times & \times & \times & \times & 0 \\
\times & \times & \times & \times & 0 \\
\times & \times & \times & + & \times \\
0 & 0 & 0 & \times & +
\end{bmatrix}
\xrightarrow{\; G_4^{(1)^T} \hat{A}_3^{(1)} G_4^{(1)} \;}
\begin{bmatrix}
\times & \times & \times & \boxtimes & \boxtimes \\
\times & \times & \times & \boxtimes & \boxtimes \\
\times & \times & \times & \boxtimes & \boxtimes \\
\boxtimes & \boxtimes & \boxtimes & \boxtimes & \boxtimes \\
\boxtimes & \boxtimes & \boxtimes & \boxtimes & \boxtimes
\end{bmatrix}
$$

$$
\hat{A}_3^{(1)} \xrightarrow{\; G_4^{(1)^T} \hat{A}_3^{(1)} G_4^{(1)} \;} \hat{A}_4^{(1)}.
$$

For ease of notation, after the transformation we changed the elements $\boxplus$ again to $\boxtimes$. Applying now this transformation $G_4^{(1)}$ to the matrix $A_3^{(1)}$ instead of $\hat{A}_3^{(1)}$ we get the following matrix $G_4^{(1)^T} A_3^{(1)} G_4^{(1)} = A_4^{(1)} = A_0^{(2)}$, which can be written as

$$
A_0^{(2)} =
\begin{bmatrix}
\times & \times & \times & \boxtimes & \boxtimes \\
\times & \times & \times & \boxtimes & \boxtimes \\
\times & \times & \times & \boxtimes & \boxtimes \\
\boxtimes & \boxtimes & \boxtimes & \boxtimes & \boxtimes \\
\boxtimes & \boxtimes & \boxtimes & \boxtimes & \boxtimes
\end{bmatrix}
+
\begin{bmatrix}
0 & & & & \\
& 0 & & & \\
& & 0 & & \\
& & & d_1 & \\
& & & & d_1
\end{bmatrix}.
$$

We rewrite the matrix $A_0^{(2)}$ now in the following form:

$$
A_0^{(2)} =
\begin{bmatrix}
\times & \times & \times & \boxtimes & \boxtimes \\
\times & \times & \times & \boxtimes & \boxtimes \\
\times & \times & \times & \boxtimes & \boxtimes \\
\boxtimes & \boxtimes & \boxtimes & \boxtimes & \boxtimes \\
\boxtimes & \boxtimes & \boxtimes & \boxtimes & \boxplus
\end{bmatrix}
+
\begin{bmatrix}
0 & & & & \\
& 0 & & & \\
& & 0 & & \\
& & & d_1 & \\
& & & & d_2
\end{bmatrix}
$$

$$
= \hat{A}_0^{(2)} + D_0^{(2)}.
$$

This completes the first step in the proof. We have now two rows and two diagonal elements satisfying the desired semiseparable plus diagonal structure. The remainder of the proof proceeds by induction. An extra row and diagonal element will be added to the existing semiseparable plus diagonal structure.

- **Step k.** Assume that the last k rows of the matrix $A^{(k)}$ are already in semiseparable form and the corresponding diagonal elements are $d_1, \ldots, d_k$. We will add now one row (column) to this structure such that $A^{(k+1)}$ has $k+1$ rows satisfying the semiseparable plus diagonal structure with corresponding diagonal elements $d_1, \ldots, d_{k+1}$. For simplicity we assume here $k = 3$ in our example. Our matrix is of the following form:

$$
A_0^{(3)} =
\begin{bmatrix}
\times & \times & \boxtimes & \boxtimes & \boxtimes \\
\times & \times & \boxtimes & \boxtimes & \boxtimes \\
\boxtimes & \boxtimes & \boxtimes & \boxtimes & \boxtimes \\
\boxtimes & \boxtimes & \boxtimes & \boxtimes & \boxtimes \\
\boxtimes & \boxtimes & \boxtimes & \boxtimes & \boxtimes
\end{bmatrix}
+
\begin{bmatrix}
0 & & & & \\
& 0 & & & \\
& & d_1 & & \\
& & & d_2 & \\
& & & & d_3
\end{bmatrix}
$$

$$
= \hat{A}_0^{(3)} + D_0^{(3)}.
$$

Similarly as in the reduction to semiseparable form we introduce some zeros in the structure by applying a similarity Givens transformation $G_1^{(3)}$ to the first two rows and columns. We remark that applying this transformation does not affect the diagonal matrix as the first two rows and columns of the diagonal matrix $D_0^{(3)}$ equal zero; therefore we only demonstrate this Givens transformation on the matrix $\hat{A}_0^{(3)}$:

$$
\begin{bmatrix}
\times & \times & \otimes & \otimes & \otimes \\
\times & \times & \boxtimes & \boxtimes & \boxtimes \\
\otimes & \boxtimes & \boxtimes & \boxtimes & \boxtimes \\
\otimes & \boxtimes & \boxtimes & \boxtimes & \boxtimes \\
\otimes & \boxtimes & \boxtimes & \boxtimes & \boxtimes
\end{bmatrix}
\xrightarrow{G_1^{(3)^T} \hat{A}_0^{(3)} G_1^{(3)}}
\begin{bmatrix}
\times & \times & 0 & 0 & 0 \\
\times & \times & \boxtimes & \boxtimes & \boxtimes \\
0 & \boxtimes & \boxtimes & \boxtimes & \boxtimes \\
0 & \boxtimes & \boxtimes & \boxtimes & \boxtimes \\
0 & \boxtimes & \boxtimes & \boxtimes & \boxtimes
\end{bmatrix}.
$$

Applying this transformation onto the matrix $A_0^{(3)}$ gives us $G_1^{(3)^T} A_0^{(3)} G_1^{(3)} = A_1^{(3)}$, which can be rewritten as the sum of a new partially semiseparable matrix $\hat{A}_1^{(3)}$ and a diagonal $D_1^{(3)}$. This matrix $A_1^{(3)}$ is then again rewritten by changing diagonal element $(2,2)$.

$$
\begin{aligned}
A_1^{(3)} &=
\begin{bmatrix}
\times & \times & 0 & 0 & 0 \\
\times & \times & \boxtimes & \boxtimes & \boxtimes \\
0 & \boxtimes & \boxtimes & \boxtimes & \boxtimes \\
0 & \boxtimes & \boxtimes & \boxtimes & \boxtimes \\
0 & \boxtimes & \boxtimes & \boxtimes & \boxtimes
\end{bmatrix}
+
\begin{bmatrix}
0 & & & & \\
& 0 & & & \\
& & d_1 & & \\
& & & d_2 & \\
& & & & d_3
\end{bmatrix} \\
&=
\begin{bmatrix}
\times & \times & 0 & 0 & 0 \\
\times & + & \boxtimes & \boxtimes & \boxtimes \\
0 & \boxtimes & \boxtimes & \boxtimes & \boxtimes \\
0 & \boxtimes & \boxtimes & \boxtimes & \boxtimes \\
0 & \boxtimes & \boxtimes & \boxtimes & \boxtimes
\end{bmatrix}
+
\begin{bmatrix}
0 & & & & \\
& d_1 & & & \\
& & d_1 & & \\
& & & d_2 & \\
& & & & d_3
\end{bmatrix} \\
&= \hat{A}_1^{(3)} + D_1^{(3)}.
\end{aligned}
$$

Consecutively we use the matrix $\hat{A}_1^{(3)}$ to determine the next Givens transformation $G_2^{(3)}$, which will transform the matrix $\hat{A}_1^{(3)}$ in the following way:

$$
\begin{bmatrix}
\times & \times & 0 & 0 & 0 \\
\times & + & \boxtimes & \boxtimes & \boxtimes \\
0 & \boxtimes & \boxtimes & \boxtimes & \boxtimes \\
0 & \boxtimes & \boxtimes & \boxtimes & \boxtimes \\
0 & \boxtimes & \boxtimes & \boxtimes & \boxtimes
\end{bmatrix}
\xrightarrow{G_2^{(3)^T} \hat{A}_1^{(3)} G_2^{(3)}}
\begin{bmatrix}
\times & \boxtimes & \boxtimes & 0 & 0 \\
\boxtimes & \boxtimes & \boxtimes & 0 & 0 \\
\boxtimes & \boxtimes & \boxtimes & \boxtimes & \boxtimes \\
0 & 0 & \boxtimes & \boxtimes & \boxtimes \\
0 & 0 & \boxtimes & \boxtimes & \boxtimes
\end{bmatrix}.
$$

Applying this transformation on the matrix $A_1^{(3)}$ and rewriting the matrix by changing diagonal element $(3,3)$ gives us

$$
A_2^{(3)} = \begin{bmatrix} \times & \boxtimes & \boxtimes & 0 & 0 \\ \boxtimes & \boxtimes & \boxtimes & 0 & 0 \\ \boxtimes & \boxtimes & \boxtimes & \boxtimes & \boxtimes \\ 0 & 0 & \boxtimes & \boxtimes & \boxtimes \\ 0 & 0 & \boxtimes & \boxtimes & \boxtimes \end{bmatrix} + \begin{bmatrix} 0 & & & & \\ & d_1 & & & \\ & & d_1 & & \\ & & & d_2 & \\ & & & & d_3 \end{bmatrix}
$$

$$
= \begin{bmatrix} \times & \boxtimes & \boxtimes & 0 & 0 \\ \boxtimes & \boxtimes & \boxtimes & 0 & 0 \\ \boxtimes & \boxtimes & \boxplus & \boxtimes & \boxtimes \\ 0 & 0 & \boxtimes & \boxtimes & \boxtimes \\ 0 & 0 & \boxtimes & \boxtimes & \boxtimes \end{bmatrix} + \begin{bmatrix} 0 & & & & \\ & d_1 & & & \\ & & d_2 & & \\ & & & d_2 & \\ & & & & d_3 \end{bmatrix}
$$

$$
= \hat{A}_2^{(3)} + D_2^{(3)}.
$$

Transformation $G_3^{(3)}$ is determined in a similar way such that we have the following equations for the matrix $A_3^{(3)} = G_3^{(3)^T} \hat{A}_2^{(3)} G_3^{(3)}$:

$$
A_3^{(3)} = \begin{bmatrix} \times & \boxtimes & \boxtimes & \boxtimes & 0 \\ \boxtimes & \boxtimes & \boxtimes & \boxtimes & 0 \\ \boxtimes & \boxtimes & \boxtimes & \boxtimes & 0 \\ \boxtimes & \boxtimes & \boxtimes & \boxtimes & \boxtimes \\ 0 & 0 & 0 & \boxtimes & \boxtimes \end{bmatrix} + \begin{bmatrix} 0 & & & & \\ & d_1 & & & \\ & & d_2 & & \\ & & & d_2 & \\ & & & & d_3 \end{bmatrix}
$$

$$
= \begin{bmatrix} \times & \boxtimes & \boxtimes & \boxtimes & 0 \\ \boxtimes & \boxtimes & \boxtimes & \boxtimes & 0 \\ \boxtimes & \boxtimes & \boxtimes & \boxtimes & 0 \\ \boxtimes & \boxtimes & \boxtimes & \boxplus & \boxtimes \\ 0 & 0 & 0 & \boxtimes & \boxtimes \end{bmatrix} + \begin{bmatrix} 0 & & & & \\ & d_1 & & & \\ & & d_2 & & \\ & & & d_3 & \\ & & & & d_3 \end{bmatrix}
$$

$$
= \hat{A}_3^{(3)} + D_3^{(3)}.
$$

Applying the last transformation $G_4^{(3)}$ gives us the desired result $A_4^{(3)}$, which completes step k in the iterative procedure:

$$
A_4^{(3)} = \begin{bmatrix} \times & \boxtimes & \boxtimes & \boxtimes & \boxtimes \\ \boxtimes & \boxtimes & \boxtimes & \boxtimes & \boxtimes \\ \boxtimes & \boxtimes & \boxtimes & \boxtimes & \boxtimes \\ \boxtimes & \boxtimes & \boxtimes & \boxtimes & \boxtimes \\ \boxtimes & \boxtimes & \boxtimes & \boxtimes & \boxtimes \end{bmatrix} + \begin{bmatrix} 0 & & & & \\ & d_1 & & & \\ & & d_2 & & \\ & & & d_2 & \\ & & & & d_3 \end{bmatrix}
$$

$$
= \begin{bmatrix} \times & \boxtimes & \boxtimes & \boxtimes & \boxtimes \\ \boxtimes & \boxtimes & \boxtimes & \boxtimes & \boxtimes \\ \boxtimes & \boxtimes & \boxtimes & \boxtimes & \boxtimes \\ \boxtimes & \boxtimes & \boxtimes & \boxtimes & \boxtimes \\ \boxtimes & \boxtimes & \boxtimes & \boxtimes & \boxplus \end{bmatrix} + \begin{bmatrix} 0 & & & & \\ & d_1 & & & \\ & & d_2 & & \\ & & & d_3 & \\ & & & & d_4 \end{bmatrix}
$$

$$
= \hat{A}_4^{(3)} + D_4^{(3)}. \tag{2.1}
$$

To obtain the complete semiseparable matrix one has to perform one extra chasing step. $\square$

Note 2.10. *We remark that the last step in the inductive procedure, namely step n can be done in two different ways, both for the reduction to semiseparable as well as for the reduction to semiseparable plus diagonal form.*

A first way is to perform an extra chasing procedure, as indicated in the proofs of both methods.

Or one can stop after step $n-1$. This can be seen by looking at Equation (2.1): in fact the matrix $\hat{A}_4^{(3)}$ is already in semiseparable form, i.e., the matrix $A_4^{(3)}$ is already a semiseparable plus diagonal matrix. The diagonal is however not of the desired form. This problem can be solved rather easily by initially starting with diagonal element d_2 instead of the element d_1. In this way Equation (2.1) becomes:

$$A_4^{(3)} = \begin{bmatrix} \boxtimes & \boxtimes & \boxtimes & \boxtimes & \boxtimes \\ \boxtimes & \boxtimes & \boxtimes & \boxtimes & \boxtimes \\ \boxtimes & \boxtimes & \boxtimes & \boxtimes & \boxtimes \\ \boxtimes & \boxtimes & \boxtimes & \boxtimes & \boxtimes \\ \boxtimes & \boxtimes & \boxtimes & \boxtimes & \boxtimes \end{bmatrix} + \begin{bmatrix} 0 & & & & \\ & d_2 & & & \\ & & d_3 & & \\ & & & d_4 & \\ & & & & d_5 \end{bmatrix}$$
$$= \hat{A}_4^{(3)} + D_4^{(3)}.$$

This can easily be rewritten as

$$A_4^{(3)} = \begin{bmatrix} \boxplus & \boxtimes & \boxtimes & \boxtimes & \boxtimes \\ \boxtimes & \boxtimes & \boxtimes & \boxtimes & \boxtimes \\ \boxtimes & \boxtimes & \boxtimes & \boxtimes & \boxtimes \\ \boxtimes & \boxtimes & \boxtimes & \boxtimes & \boxtimes \\ \boxtimes & \boxtimes & \boxtimes & \boxtimes & \boxtimes \end{bmatrix} + \begin{bmatrix} d_1 & & & & \\ & d_2 & & & \\ & & d_3 & & \\ & & & d_4 & \\ & & & & d_5 \end{bmatrix},$$

which gives us the desired semiseparable plus diagonal matrix. This trick can of course also be applied when reducing a matrix to semiseparable form. In this case one does not even need to worry about starting with the correct diagonal element.

It is clear that the reduction to semiseparable plus diagonal is the most general one. If we choose the diagonal elements equal to zero we obtain the reduction to semiseparable form, and if we omit the extra chasing procedure, we obtain the reduction to tridiagonal form. In fact we have even more. One can see the reduction to tridiagonal form as the reduction to semiseparable form with all the diagonal elements equal to $-\infty$.

Theorem 2.11. *The orthogonal similarity reduction to tridiagonal form can be seen as the orthogonal similarity reduction to semiseparable plus diagonal form, where we choose the diagonal equal to $-\infty$.*

Proof. If we prove that the performed Givens transformations in the chasing step equal the identity matrices, we know that the resulting matrix will be of tridiagonal form.

Let us define the Givens transformation as follows:

$$\begin{bmatrix} c & s \\ -s & c \end{bmatrix} \begin{bmatrix} x \\ y \end{bmatrix} = \begin{bmatrix} 0 \\ r \end{bmatrix}.$$

Where c, s and r are defined as

$$r = \sqrt{x^2 + y^2},$$
$$c = y/r,$$
$$s = -x/r.$$

The c and s are, respectively, the cosine and sine of a specific angle.

Assume we would like to obtain a semiseparable plus diagonal matrix with diagonal $-\epsilon$. This means that we have to perform the following Givens transformations in the chasing technique on the right to annihilate the element x, where $y + \epsilon$ is the diagonal element:

$$\begin{bmatrix} c & s \\ -s & c \end{bmatrix} \begin{bmatrix} x \\ y + \epsilon \end{bmatrix} = \begin{bmatrix} 0 \\ r \end{bmatrix},$$

with the parameters defined as

$$r = \sqrt{x^2 + (y + \epsilon)^2},$$
$$c = (y + \epsilon)/r,$$
$$s = -x/r.$$

Taking the limit now for $\epsilon \to \infty$ leads to the Givens transformation equal to the identity:

$$\lim_{\epsilon \to \infty} \left(\frac{1}{\sqrt{x^2 + (y + \epsilon)^2}} \begin{bmatrix} (y + \epsilon) & -x \\ x & (y + \epsilon) \end{bmatrix} \right) = \begin{bmatrix} 1 & 0 \\ 0 & 1 \end{bmatrix}.$$

□

We remark that this previous theorem implies that we can construct for every tridiagonal matrix a sequence of semiseparable plus diagonal matrices converging to this tridiagonal matrix. This implies that the class of semiseparable plus diagonal matrices is not pointwise closed, as a tridiagonal matrix cannot always be written as the sum of a semiseparable and a diagonal matrix. We briefly illustrate this with a symmetric 3×3 matrix:

$$\lim_{\epsilon \to \infty} \left(\begin{bmatrix} a & b & \frac{1}{\epsilon} \\ b & \epsilon bd & d \\ \frac{1}{\epsilon} & d & e \end{bmatrix} + \begin{bmatrix} 0 & 0 & 0 \\ 0 & c - \epsilon bd & 0 \\ 0 & 0 & 0 \end{bmatrix} \right) = \begin{bmatrix} a & b & \\ b & c & d \\ & d & e \end{bmatrix}$$

The sequence of matrices on the left are all semiseparable plus diagonal matrices, and their limit is tridiagonal. There is a larger class of structured rank matrices named the class of quasiseparable matrices, which covers both the class of semiseparable plus diagonal and tridiagonal matrices. This class is also closed under pointwise convergence. Also reductions to quasiseparable form and QR-algorithms for this type of matrices exist. We will not however cover this class of matrices in detail. Information on them can be found in a forthcoming chapter in which related references are discussed. The ideas and theorems provided here present enough information for understanding the ideas presented in these articles. More information concerning quasiseparable matrices can also be found in Volume I.

Notes and references

The reduction to tridiagonal form can be found in different textbooks, such as the book of Golub and Van Loan [94] and

☞ B. N. Parlett. *The Symmetric Eigenvalue Problem*, volume 20 of *Classics in Applied Mathematics*. SIAM, Philadelphia, Pennsylvania, USA, 1998.

☞ G. W. Stewart. *Matrix Algorithms, Volume II: Eigensystems*. SIAM, Philadelphia, Pennsylvania, USA, 2001.

☞ L. N. Trefethen and D. Bau. *Numerical Linear Algebra*. SIAM, Philadelphia, Pennsylvania, USA, 1997.

The original articles in which the reduction to semiseparable and semiseparable plus diagonal form were described are the following ones (see also [155]):

☞ M. Van Barel, R. Vandebril, and N. Mastronardi. An orthogonal similarity reduction of a matrix into semiseparable form. *SIAM Journal on Matrix Analysis and Applications*, 27(1):176–197, 2005.

☞ R. Vandebril, E. Van Camp, M. Van Barel, and N. Mastronardi. Orthogonal similarity transformation of a symmetric matrix into a diagonal-plus-semiseparable one with free choice of the diagonal. *Numerische Mathematik*, 102:709–726, 2006.

In the following article, the authors Van Barel, Van Camp and Mastronardi present an adaptation of the algorithm for reducing matrices to semiseparable form. They present a method for reducing a matrix to a higher order semiseparable one. In fact an orthogonal similarity reduction to a block semiseparable matrix of rank k is proposed.

☞ M. Van Barel, E. Van Camp, and N. Mastronardi. Orthogonal similarity transformation into block-semiseparable matrices of semiseparability rank k. *Numerical Linear Algebra with Applications*, 12:981–1000, 2005.

The authors Bevilacqua and Del Corso investigate the properties of the reduction to semiseparable form if the resulting matrix is of generator representable semiseparable form. Existence and uniqueness results are proved.

☞ R. Bevilacqua and G. M. Del Corso. Structural properties of matrix unitary reduction to semiseparable form. *Calcolo*, 41(4):177–202, 2004.

2.3   Orthogonal similarity transformation of (unsymmetric) matrices

In this section three different reduction algorithms will be presented. Arbitrary matrices, not necessary symmetric, will be reduced via orthogonal similarity transformations to a more simple form. These reduced matrices can afterwards be used, e.g., for calculating the eigenvalues of the original matrices. First, we will present the well-known reduction to Hessenberg form, and secondly we deduce the transformation to Hessenberg-like form. Hessenberg-like matrices are the inverses of Hessenberg matrices and have the lower triangular part of semiseparable form. The results in this section are a straightforward extension of the results presented in the previous section on symmetric matrices; hence they are not covered in detail.

2.3.1   To Hessenberg form

Even though this is a well-known algorithm, we will briefly repeat the reduction procedure.

Theorem 2.12. *Let A be an $n \times n$ matrix. There exists an orthogonal matrix U such that the following equation is satisfied:*

$$U^T A U = H,$$

for which H is a Hessenberg matrix.

Proof. In fact we have to perform exactly the same transformations as for the symmetric case. We will briefly illustrate what happens. Instead of using Givens transformations as we used in the tridiagonal reduction, we will illustrate the procedure here with Householder reflectors. Of course one can also use Givens transformations. Let $A_0^{(1)} = A$. Often, we denote $A_0^{(i)}$ as $A^{(i)}$. Let $U_i^{(l)}$ be a Householder reflector, such that the product $A_0^{(l)} H_1^{(l)}$ has the entries 1 up to $n-l-1$ annihilated in the $n-l+1$-th row, thereby changing the columns 1 up to $n-l$ of the matrix.

- **Step 1.** We will start by making all the elements in the last row, except for the last two, zero. Just multiply $A_0^{(1)}$ to the right by $H_1^{(1)}$ to create the wanted zeros. To complete the similarity transformation, we also perform the transformation $H_1^{(1)^T}$ to the left of $A_0^{(1)} H_1^{(1)}$. This last transformation on the left does not destroy the zeros created by the transformation on the right, as this transformation only interacts with rows 1 to $n-1$. Graphically we have the following figure: on the left the elements that will be annihilated are marked as $\otimes$:

$$\begin{bmatrix} \times & \times & \cdots & \times & \times \\ \times & \ddots & & \vdots & \vdots \\ \vdots & & \ddots & \times & \times \\ \times & \cdots & \times & \times & \times \\ \otimes & \cdots & \otimes & \times & \times \end{bmatrix} \xrightarrow{H_1^{(1)^T} A_0^{(1)} H_1^{(1)}} \begin{bmatrix} \times & \times & \cdots & \times & \times \\ \times & \ddots & & \vdots & \vdots \\ \vdots & & \ddots & \times & \times \\ \times & \cdots & \times & \times & \times \\ 0 & \cdots & 0 & \times & \times \end{bmatrix}$$

$$\Updownarrow$$

$$A_0^{(1)} \xrightarrow{H_1^{(1)^T} A_0^{(1)} H_1^{(1)}} A_1^{(1)}.$$

The initialization step of the inductive procedure is completed.

- **Step k.** Let $k = n - j$, $1 < j < n$. Assume by induction that $A_0^{(k)}$ has the lower right $k \times k$ block already in Hessenberg form:

$$A_0^{(k)} = \begin{bmatrix} \times & \cdots & & \times & \times & \times & \cdots & \times \\ \vdots & \ddots & & \vdots & \vdots & \vdots & & \vdots \\ & & \ddots & \times & \times & \times & \vdots & \times \\ \times & \cdots & \times & \times & \times & \times & \cdots & \times \\ \otimes & \cdots & \otimes & \times & \times & \times & \cdots & \times \\ & & & & \times & \times & & \\ & & & & & \ddots & \ddots & \vdots \\ & & & & & & \times & \times \end{bmatrix}.$$

Performing now the Householder reflector $H_1^{(k)}$ on the matrix $A_0^{(k)}$, it will annihilate the elements $\otimes$ and we will get the following matrix:

$$A_0^{(k)} = \begin{bmatrix} \times & \cdots & & \times & \times & \times & \cdots & \times \\ \vdots & \ddots & & \vdots & \vdots & \vdots & & \vdots \\ & & \ddots & \times & \times & \times & \vdots & \times \\ \times & \cdots & \times & \times & \times & \times & \cdots & \times \\ 0 & \cdots & 0 & \times & \times & \times & \cdots & \times \\ & & & & \times & \times & & \\ & & & & & \ddots & \ddots & \vdots \\ & & & & & & \times & \times \end{bmatrix}.$$

The latter matrix has now a $(k+1) \times (k+1)$ block in Hessenberg form. This proves the induction step.

Combining all the orthogonal transformations, i.e., the Householder transformations into one single orthogonal matrix U gives us the desired orthogonal transformation matrix.   □

Similarly as in the symmetric case, the reduction to Hessenberg-like form has similarities with this reduction. First we have the zero-making transformations, and secondly we have to perform a kind of chasing to obtain the Hessenberg-like form.

2.3.2 To Hessenberg-like form

It is well-known how any square matrix can be transformed into a Hessenberg matrix by orthogonal similarity transformations. As the inverse of a Hessenberg matrix is a Hessenberg-like matrix, one might wonder whether it is possible to transform a matrix into a similar Hessenberg-like one. By using an algorithm similar to that described in the previous section, it is possible to do so, for every matrix.

Theorem 2.13. *Let A be an $n \times n$ matrix. There exists an orthogonal matrix U such that the following equation is satisfied:*

$$U^T A U = Z,$$

for which Z is a Hessenberg-like matrix.

Proof. In Theorem 2.6, it was unimportant whether one first annihilated the elements in the last row or in the last column because the matrix was symmetric. Here, however, our matrix is not necessarily symmetric anymore. Therefore the transformations determined for annihilating elements in the upper or in the lower triangular part are not necessarily the same anymore.

To obtain a Hessenberg-like matrix, the transformation matrix U should be composed of Householder and/or Givens transformations, constructed to annihilate elements in the lower triangular part of A, when performing this transformation on the right of A. Applying then the transformation $U^T A U$ results in a matrix whose lower triangular part is semiseparable. Hence the matrix is a Hessenberg-like matrix.[12]   □

The algorithm to reduce an arbitrary matrix by means of orthogonal transformations into a similar Hessenberg-like matrix also requires $\mathcal{O}(n^3)$ operations. The previous theorem can also be formulated for a lower Hessenberg-like matrix.[13]

For the sake of completeness, the following theorem is also included:

Theorem 2.14. *Let A be an $n \times n$ matrix. Then there exist a symmetric semiseparable matrix S, a strictly upper triangular matrix R and an orthogonal matrix U such that the following equation is satisfied:*

$$A = U(S + R)U^T,$$

---

[12]Constructing the matrix U^T such that application on the left annihilates elements in the upper triangular part results in a lower Hessenberg-like matrix.

[13]Remark that a lower Hessenberg-like matrix has the upper triangular part of structured rank form.

i.e., A is similar to the sum of a symmetric semiseparable and a strictly upper triangular matrix R by means of orthogonal transformations.

Proof. The same as from Theorem 2.12. By adding a strictly upper triangular matrix to the Hessenberg-like matrix, one can make this matrix symmetric; hence one can write the Hessenberg-like matrix as the sum of a symmetric semiseparable plus a strictly upper triangular matrix.   □

This last theorem will come in handy for certain proofs further in the text.

2.3.3   To Hessenberg-like plus diagonal form

The transformation of a matrix via similarity transformations into a similar Hessenberg-like plus diagonal matrix is left to the reader as an exercise. Using the knowledge of the reduction to Hessenberg-like form and the similarity reduction to semiseparable plus diagonal form, this should be an easy exercise.

Notes and references

The transformation to Hessenberg form is widespread and can be found in many general textbooks. The reduction to Hessenberg-like form, as well as the development of an implicit QR-algorithm for Hessenberg-like matrices, is discussed in the following PhD dissertation.

> ☞ R. Vandebril. *Semiseparable Matrices and the Symmetric Eigenvalue Problem.* PhD thesis, Department of Computer Science, Katholieke Universiteit Leuven, Celestijnenlaan 200A, 3000 Leuven (Heverlee), Belgium, May 2004.

2.4   Orthogonal transformations of matrices

The preceding two sections focused on orthogonal similarity transformations. Orthogonal similarity transformations are useful for transforming matrices to a similar easier form, which can be exploited then for effectively computing the eigenvalues. If one however wants to compute the singular values instead of the eigenvalues, one does not necessarily need to use similarity transformations. Hence we propose in this section some methods for transforming matrices to an easier form, without changing the singular values of the transformed matrices. In the final subsection, we discuss the relation of the orthogonal transformations proposed in this section, with the orthogonal similarity transformations for reducing symmetric matrices to semiseparable form. We will show that both reduction procedures are closely related to each other.

2.4.1   To upper (lower) bidiagonal form

In this section we will apply the Householder bidiagonalization process [93] to a matrix A. The algorithm can also be written in terms of Givens transformations.

Theorem 2.15. *Let $A \in \mathbb{R}^{m \times n}$, $m \geq n$. There exist two orthogonal matrices $U \in \mathbb{R}^{m \times m}$ and $V \in \mathbb{R}^{n \times n}$ such that*

$$U^T A V = \begin{bmatrix} B \\ 0 \end{bmatrix},$$

where B is an upper bidiagonal matrix.

The case $n \geq m$ is considered afterwards.

Proof. We will present a proof by finite induction, constructing thereby the upper bidiagonal matrix.

The proof is illustrated for a matrix $A \in \mathbb{R}^{m \times n}$, with $m = 6$ and $n = 5$, as this illustrates the general case. The side on which the operations are performed plays an important role. Starting from $A^{(1)} = A_{0,0}^{(1)} = A$, we apply the following rules, giving information about the performed transformations:

$$U_{k+1}^{(l)} A_{k,j}^{(l)} = A_{k+1,j}^{(l)}, \quad A_{k,j}^{(l)} V_{j+1}^{(l)} = A_{k,j+1}^{(l)}.$$

Furthermore, we define $A_{0,0}^{(l)} = A^{(l)}$ and $A^{(l+1)} = A_{1,1}^{(l)}$. In this reduction procedure, we only consider Householder reflectors.

- **Step 1.** The initialization step consists of making the first row and the first column in the desired upper bidiagonal form. The first transformation ${U_1^{(1)}}^T$ is performed to the left of the matrix $A_{0,0}^{(1)}$, thereby annihilating all the elements in the first column except for the top left diagonal element:

$$\begin{bmatrix} \times & \times & \times & \times & \times \\ \otimes & \times & \times & \times & \times \\ \otimes & \times & \times & \times & \times \\ \otimes & \times & \times & \times & \times \\ \otimes & \times & \times & \times & \times \\ \otimes & \times & \times & \times & \times \end{bmatrix} \xrightarrow{{U_1^{(1)}}^T A_{0,0}^{(1)}} \begin{bmatrix} \times & \times & \times & \times & \times \\ 0 & \times & \times & \times & \times \\ 0 & \times & \times & \times & \times \\ 0 & \times & \times & \times & \times \\ 0 & \times & \times & \times & \times \\ 0 & \times & \times & \times & \times \end{bmatrix}$$

$$\Updownarrow$$

$$A_{0,0}^{(1)} = A^{(1)} \xrightarrow{{U_1^{(1)}}^T A_{0,0}^{(1)}} A_{1,0}^{(1)}.$$

As we want the first column and first row to be in upper bidiagonal form, another transformation has to be performed on the right. This Householder transformation, will create zeros in the first row, except for the first two elements[14]:

---

[14]The elements not shown in the graphical representation are assumed to be zero.

$$
\begin{bmatrix}
\times & \times & \otimes & \otimes & \otimes \\
 & \times & \times & \times & \times \\
 & \times & \times & \times & \times \\
 & \times & \times & \times & \times \\
 & \times & \times & \times & \times \\
 & \times & \times & \times & \times
\end{bmatrix}
\xrightarrow{A_{1,0}^{(1)}V_1^{(1)}}
\begin{bmatrix}
\times & \times & 0 & 0 & 0 \\
 & \times & \times & \times & \times \\
 & \times & \times & \times & \times \\
 & \times & \times & \times & \times \\
 & \times & \times & \times & \times \\
 & \times & \times & \times & \times
\end{bmatrix}
$$

$$\Updownarrow$$

$$A_{1,0}^{(1)} \xrightarrow{A_{1,0}^{(1)}V_1^{(1)}} A_{1,1}^{(1)}.$$

Then we put $A^{(2)} = A_{1,1}^{(1)}$, as the first step is completed.

- **Step k.** By induction, for $k > 1$. The first $k - 1$ rows and columns of $A^{(k)}$ already satisfy the bidiagonal structure. Let us assume $k = 3$. This means that $A^{(3)}$ has the following structure:

$$
A^{(3)} =
\begin{bmatrix}
\times & \times & & \\
 & \times & \times & \\
 & & \times & \times & \times \\
 & & \times & \times & \times \\
 & & \times & \times & \times \\
 & & \times & \times & \times
\end{bmatrix}.
$$

In this step we will expand this bidiagonal structure with one row and one column. Therefore we apply a Householder transformation $U_1^{(3)^T}$ to the left of $A^{(3)}$, annihilating all elements below the diagonal in the third column:

$$
\begin{bmatrix}
\times & \times & & \\
 & \times & \times & \\
 & & \times & \times & \times \\
 & & \otimes & \times & \times \\
 & & \otimes & \times & \times \\
 & & \otimes & \times & \times
\end{bmatrix}
\xrightarrow{U_1^{(3)^T}A_{0,0}^{(3)}}
\begin{bmatrix}
\times & \times & & \\
 & \times & \times & \\
 & & \times & \times & \times \\
 & & & \times & \times \\
 & & & \times & \times \\
 & & & \times & \times
\end{bmatrix}
$$

$$\Updownarrow$$

$$A_{0,0}^{(3)} \xrightarrow{U_1^{(3)^T}A_{0,0}^{(3)}} A_{1,0}^{(3)}.$$

One more element has to be annihilated to get all three first columns and rows in the correct form. This is done by a Householder transformation performed on the right of the matrix:

$$
\begin{bmatrix}
\times & \times & & & \\
& \times & \times & & \\
& & \times & \times & \otimes \\
& & \times & \times & \\
& & \times & \times & \\
& & \times & \times &
\end{bmatrix}
\xrightarrow{A_{1,0}^{(3)} V_1^{(3)}}
\begin{bmatrix}
\times & \times & & & \\
& \times & \times & & \\
& & \times & \times & \\
& & & \times & \times \\
& & & \times & \times \\
& & & \times & \times
\end{bmatrix}
$$

$$\Updownarrow$$

$$A_{1,0}^{(3)} \xrightarrow{A_{1,1}^{(3)}} A_{1,1}^{(3)}.$$

This completes the induction step.

- **Step** $n + 1$. When the number of rows exceeds the number of columns one extra transformation needs to be performed to annihilate the remaining elements in the last column. This is done by a single Householder transformation on the left of the matrix $A^{(5)}$:

$$
\begin{bmatrix}
\times & \times & & & \\
& \times & \times & & \\
& & \times & \times & \\
& & & \times & \times \\
& & & & \times \\
& & & & \otimes
\end{bmatrix}
\xrightarrow{U_1^{(5)^T} A_{0,0}^{(5)}}
\begin{bmatrix}
\times & \times & & & \\
& \times & \times & & \\
& & \times & \times & \\
& & & \times & \times \\
& & & & \times \\
& & & & 0
\end{bmatrix}
$$

$$\Updownarrow$$

$$A^{(5)} \xrightarrow{U_1^{(5)^T} A_{0,0}^{(5)}} A_{1,0}^{(5)}.$$

This last transformation completes the proof as we have an upper bidiagonal matrix now. $\qquad\Box$

The case considered in the theorem is just a special case. The proof of any of the following cases proceeds similarly.

Theorem 2.16. *Let $A \in \mathbb{R}^{m \times n}$. There exist two orthogonal matrices $U \in \mathbb{R}^{m \times m}$ and $V \in \mathbb{R}^{n \times n}$ such that*

- *for $m \geq n$:*

$$U^T A V = \begin{bmatrix} B_l \\ 0 \end{bmatrix},$$

where B_l is a lower bidiagonal matrix,

- *for $m \leq n$:*

$$U^T A V = \begin{bmatrix} B_u & 0 \end{bmatrix},$$

where B_u is an upper bidiagonal matrix,

- *for $m \leq n$:*

$$U^T A V = \begin{bmatrix} B_l & 0 \end{bmatrix},$$

  where B_l is a lower bidiagonal matrix.

Proof. Using the transpose operation and permuting rows or columns, these results can easily be obtained from Theorem 2.17.   □

In the following section the reduction to upper triangular semiseparable or lower triangular semiseparable form is described.

2.4.2   To upper (lower) triangular semiseparable form

Similar to the Householder bidiagonalization process as described above, the algorithm given here makes use of orthogonal transformations, i.e., Givens and Householder transformations to reduce the matrix A into an upper triangular semiseparable matrix. The algorithm can be retrieved from the constructive proof of the following theorem.

Theorem 2.17.   *Let $A \in \mathbb{R}^{m \times n}$, $m \geq n$. There exist two orthogonal matrices $U \in \mathbb{R}^{m \times m}$ and $V \in \mathbb{R}^{n \times n}$ such that*

$$U^T A V = \begin{bmatrix} S_u \\ 0 \end{bmatrix},$$

where S_u is an upper triangular semiseparable matrix.

The case $n \geq m$ is formulated in Theorem 2.18.

Proof. The proof given here is constructive. We prove the existence of such a transformation by reducing the matrix A to the appropriate form using Givens and Householder transformations. Of course one can replace a single Householder transformation by a number of Givens transformation.
The proof is given by finite induction. The proof is outlined for a matrix $A \in \mathbb{R}^{m \times n}$, with $m = 6$ and $n = 5$. We use the same notation and conventions as in the previous proof:

$$U^{(l)}_{k+1} A^{(l)}_{k,j} = A^{(l)}_{k+1,j}, \quad A^{(l)}_{k,j} V^{(l)}_{j+1} = A^{(l)}_{k,j+1}.$$

Let $A^{(l)}_{0,0} = A^{(l)}$ and define $A^{(l+1)} = A^{(l)}_{l+1,l}$. The applied transformations to the left and to the right of $A^{(l)}_{k,j}$ are Givens and/or Householder transformations.
Step 1 illustrates the basic idea of the reduction procedure. One starts by creating zeros as in the reduction to (lower) bidiagonal form. Having obtained this (lower) bidiagonal structure in the first column, one performs a Givens transformation for creating the structured rank part. The next steps of the reduction proceed similarly. First create zeros, followed by a procedure enlarging the rank structure.

- **Step 1.** In this first step of the algorithm three orthogonal matrices $U_1^{(1)^T}$, $U_2^{(1)^T}$ and $V_1^{(1)}$ are to be found, such that the matrix

$$A^{(2)} = U_2^{(1)^T} U_1^{(1)^T} A^{(1)} V_1^{(1)},$$

with $A^{(1)} = A$, has the following properties: the first two rows of the matrix $A^{(2)}$ already satisfy the semiseparable structure and the first column of $A^{(2)}$ is zero below the first element. A Householder transformation $V_1^{(1)}$ is applied to the right of $A^{(1)} = A_{0,0}^{(1)}$ to annihilate all the elements in the first row except for the first one. The elements denoted with $\otimes$ will be annihilated, and the ones denoted with $\boxtimes$ mark the part of the matrix having already a semiseparable structure:

$$
\begin{bmatrix}
\times & \otimes & \otimes & \otimes & \otimes \\
\times & \times & \times & \times & \times \\
\times & \times & \times & \times & \times \\
\times & \times & \times & \times & \times \\
\times & \times & \times & \times & \times \\
\times & \times & \times & \times & \times
\end{bmatrix}
\xrightarrow{A_{0,0}^{(1)} V_1^{(1)}}
\begin{bmatrix}
\times & 0 & 0 & 0 & 0 \\
\times & \times & \times & \times & \times \\
\otimes & \times & \times & \times & \times \\
\otimes & \times & \times & \times & \times \\
\otimes & \times & \times & \times & \times \\
\otimes & \times & \times & \times & \times
\end{bmatrix}
$$

$$\Updownarrow$$

$$A_{0,0}^{(1)} = A^{(1)} \xrightarrow{A_{0,0}^{(1)} V_1^{(1)}} A_{0,1}^{(1)}.$$

A Householder transformation $U_1^{(1)^T}$ is now applied to the left of $A_{0,1}^{(1)}$ to annihilate all the elements in the first column except the first two ones as depicted in the previous equation[15], $A_{1,1}^{(1)} = U_1^{(1)^T} A_{0,1}^{(1)}$. This is followed by a Givens transformation $U_2^{(1)^T}$ applied to the left of $A_{1,1}^{(1)}$ to annihilate the second element in the first column $A_{2,1}^{(1)} = U_2^{(1)^T} A_{1,1}^{(1)}$. As a consequence, the first two rows of $A_{2,1}^{(1)}$ already have a semiseparable structure:

$$
\begin{bmatrix}
\times & 0 & 0 & 0 & 0 \\
\otimes & \times & \times & \times & \times \\
0 & \times & \times & \times & \times \\
0 & \times & \times & \times & \times \\
0 & \times & \times & \times & \times \\
0 & \times & \times & \times & \times
\end{bmatrix}
\xrightarrow{U_2^{(1)^T} A_{1,1}^{(1)}}
\begin{bmatrix}
\boxtimes & \boxtimes & \boxtimes & \boxtimes & \boxtimes \\
0 & \boxtimes & \boxtimes & \boxtimes & \boxtimes \\
0 & \times & \times & \times & \times \\
0 & \times & \times & \times & \times \\
0 & \times & \times & \times & \times \\
0 & \times & \times & \times & \times
\end{bmatrix}
$$

$$\Updownarrow$$

$$A_{1,1}^{(1)} \xrightarrow{U_2^{(1)^T} A_{1,1}^{(1)}} A_{2,1}^{(1)}.$$

Then we put $A^{(2)} = A_{2,1}^{(1)}$.

---

[15] This reduction procedure can be seen as first reducing the matrix to a lower bidiagonal form after which it is transformed to an upper triangular semiseparable form.

There is little difference with the standard bidiagonalization (to lower bidi-agonal form). The only difference is the performance of one extra Givens transformations. The induction steps proceed similarly. First zeros are created as in the bidiagonalization procedure; next some Givens transformations are performed for enlarging the rank structure.

- **Step k.** By induction, for $k > 1$. The first k rows of $A^{(k)}$ have a semiseparable structure and the first $k - 1$ columns are already in an upper triangular form. In fact, the upper left k by k block is already an upper triangular semiseparable matrix. Without loss of generality, let us assume $k = 3$. This means that $A^{(3)}$ has the following structure:

$$A^{(3)} = \begin{bmatrix} \boxtimes & \boxtimes & \boxtimes & \boxtimes & \boxtimes \\ 0 & \boxtimes & \boxtimes & \boxtimes & \boxtimes \\ 0 & 0 & \boxtimes & \boxtimes & \boxtimes \\ 0 & 0 & \times & \times & \times \\ 0 & 0 & \times & \times & \times \\ 0 & 0 & \times & \times & \times \end{bmatrix}.$$

The aim of this step is to make the upper triangular semiseparable structure in the first four rows and the first three columns of the matrix. To this end, a Householder transformation $V_1^{(3)}$ is applied to the right of $A^{(3)}$, chosen in order to annihilate the last two elements of the first row of $A^{(3)}$, $A_{0,1}^{(3)} = A_{0,0}^{(3)} V_1^{(3)}$. Note that because of the dependency between the first three rows, $V_1^{(3)}$ annihilates the last two entries of the second and third row, too. Furthermore, a Householder transformation is performed to the left of the matrix $A_{0,1}^{(3)}$ to annihilate the last two elements in column 3:

$$\begin{bmatrix} \boxtimes & \boxtimes & \boxtimes & 0 & 0 \\ 0 & \boxtimes & \boxtimes & 0 & 0 \\ 0 & 0 & \boxtimes & 0 & 0 \\ 0 & 0 & \times & \times & \times \\ 0 & 0 & \otimes & \times & \times \\ 0 & 0 & \otimes & \times & \times \end{bmatrix} \xrightarrow{U_1^{(3)^T} A_{0,1}^{(3)}} \begin{bmatrix} \boxtimes & \boxtimes & \boxtimes & 0 & 0 \\ 0 & \boxtimes & \boxtimes & 0 & 0 \\ 0 & 0 & \boxtimes & 0 & 0 \\ 0 & 0 & \times & \times & \times \\ 0 & 0 & 0 & \times & \times \\ 0 & 0 & 0 & \times & \times \end{bmatrix}$$

$$\Updownarrow$$

$$A_{0,1}^{(3)} = A_{0,0}^{(3)} V_1^{(3)} \xrightarrow{U_1^{(3)^T} A_{0,1}^{(3)}} A_{1,1}^{(3)}.$$

The Givens transformation $U_2^{(3)^T}$ is now applied to the left of the matrix $A_{1,1}^{(3)}$, annihilating the element marked with a circle:

$$
\begin{bmatrix}
\times & \boxtimes & \boxtimes & 0 & 0 \\
0 & \boxtimes & \boxtimes & 0 & 0 \\
0 & 0 & \boxtimes & 0 & 0 \\
0 & 0 & \otimes & \times & \times \\
0 & 0 & 0 & \times & \times \\
0 & 0 & 0 & \times & \times
\end{bmatrix}
\xrightarrow{U_2^{(3)^T} A_{1,1}^{(3)}}
\begin{bmatrix}
\times & \boxtimes & \boxtimes & 0 & 0 \\
0 & \boxtimes & \boxtimes & 0 & 0 \\
0 & 0 & \boxtimes & \boxtimes & \boxtimes \\
0 & 0 & 0 & \boxtimes & \boxtimes \\
0 & 0 & 0 & \times & \times \\
0 & 0 & 0 & \times & \times
\end{bmatrix}
$$

$$\Updownarrow$$

$$
A_{1,1}^{(3)} \xrightarrow{U_2^{(3)^T} A_{1,1}^{(3)}} A_{2,1}^{(3)}.
$$

Dependency is now created between the fourth and the third rows. Neverthe-less, as it can be seen in the figure above, the upper part does not satisfy the semiseparable structure, yet. A chasing technique is used in order to chase the nonsemiseparable structure upwards and away, by means of Givens trans-formations. Applying $V_2^{(3)}$ to the right to annihilate the entries $(2,3)$ and $(1,3)$ of $A_{2,1}^{(3)}$, a nonzero element is introduced in the third row on the second column (i.e., in position $(3,2)$). Because of the semiseparable structure, this operation introduces two zeros in the third column, namely in the first and second row. Annihilating the element just created in the third row with a Givens transformation to the left, the semiseparable structure holds between the second and the third row:

$$
\begin{bmatrix}
\times & \boxtimes & 0 & 0 & 0 \\
0 & \boxtimes & 0 & 0 & 0 \\
0 & \otimes & \times & \boxtimes & \boxtimes \\
0 & 0 & 0 & \boxtimes & \boxtimes \\
0 & 0 & 0 & \times & \times \\
0 & 0 & 0 & \times & \times
\end{bmatrix}
\xrightarrow{U_3^{(3)^T} A_{2,2}^{(3)}}
\begin{bmatrix}
\times & \boxtimes & 0 & 0 & 0 \\
0 & \boxtimes & \boxtimes & \boxtimes & \boxtimes \\
0 & 0 & \boxtimes & \boxtimes & \boxtimes \\
0 & 0 & 0 & \boxtimes & \boxtimes \\
0 & 0 & 0 & \times & \times \\
0 & 0 & 0 & \times & \times
\end{bmatrix}
$$

$$\Updownarrow$$

$$
A_{2,2}^{(3)} = A_{2,1}^{(3)} V_2^{(3)} \xrightarrow{U_3^{(3)^T} A_{2,2}^{(3)}} A_{3,2}^{(3)}.
$$

This up-chasing of the semiseparable structure can be repeated to create a complete upper semiseparable part starting from row 4 to row 1. First we annihilate the element in position $(1,2)$: $A_{3,3}^{(3)} = A_{3,2}^{(3)} V_3^{(3)}$.

$$
\begin{bmatrix}
\times & 0 & 0 & 0 & 0 \\
\otimes & \times & \boxtimes & \boxtimes & \boxtimes \\
0 & 0 & \boxtimes & \boxtimes & \boxtimes \\
0 & 0 & 0 & \boxtimes & \boxtimes \\
0 & 0 & 0 & \times & \times \\
0 & 0 & 0 & \times & \times
\end{bmatrix}
\xrightarrow{U_4^{(3)T} A_{3,3}^{(3)}}
\begin{bmatrix}
\boxtimes & \boxtimes & \boxtimes & \boxtimes & \boxtimes \\
0 & \boxtimes & \boxtimes & \boxtimes & \boxtimes \\
0 & 0 & \boxtimes & \boxtimes & \boxtimes \\
0 & 0 & 0 & \boxtimes & \boxtimes \\
0 & 0 & 0 & \times & \times \\
0 & 0 & 0 & \times & \times
\end{bmatrix}
$$

$$\Updownarrow$$

$$
A_{3,3}^{(3)} = A_{3,2}^{(3)} V_3^{(3)} \xrightarrow{U_4^{(3)T} A_{3,3}^{(3)}} A_{4,3}^{(3)}.
$$

Then we put $A^{(4)} = A_{4,3}^{(3)}$. This proves the induction step.

- **Step $n + 1$.** This step is only performed if $m > n$. A Householder transformation has to be performed to the left to have a complete upper triangular semiseparable structure. Suppose, the matrix has already the semiseparable structure in the first n rows, then one single Householder transformation is needed to annihilate all the elements in the n-th column below the n-th row:

$$
\begin{bmatrix}
\boxtimes & \boxtimes & \boxtimes & \boxtimes & \boxtimes \\
0 & \boxtimes & \boxtimes & \boxtimes & \boxtimes \\
0 & 0 & \boxtimes & \boxtimes & \boxtimes \\
0 & 0 & 0 & \boxtimes & \boxtimes \\
0 & 0 & 0 & 0 & \boxtimes \\
0 & 0 & 0 & 0 & \times
\end{bmatrix}
\xrightarrow{U_1^{(5)T} A_{0,0}^{(5)}}
\begin{bmatrix}
\boxtimes & \boxtimes & \boxtimes & \boxtimes & \boxtimes \\
0 & \boxtimes & \boxtimes & \boxtimes & \boxtimes \\
0 & 0 & \boxtimes & \boxtimes & \boxtimes \\
0 & 0 & 0 & \boxtimes & \boxtimes \\
0 & 0 & 0 & 0 & \boxtimes \\
0 & 0 & 0 & 0 & 0
\end{bmatrix}
$$

$$\Updownarrow$$

$$
A^{(5)} \xrightarrow{U_1^{(5)T} A_{0,0}^{(5)}} A_{1,0}^{(5)}.
$$

After the latter Householder transformation, the desired upper triangular semiseparable structure is created and the theorem is proved.   □

The reduction just described is obtained by applying Givens and Householder transformations to the matrix A. Note that the computational complexity of applying the Householder transformations is the same as of the standard procedure that reduces matrices into bidiagonal form using Householder transformations. This complexity is $4mn^2 - 4/3n^3$ (see [94]). At step k of the reduction procedure, described above, k Givens transformations $U_j^{(k)}, j = 2, 3, \ldots, k + 1$ are applied to the left and $k - 1$ Givens transformations $V_j^{(k)}$, $j = 2, 3, \ldots, k$ are applied to the right of the almost upper triangular semiseparable part of $A_{1,1}^{(k)}$. The purpose of these Givens transformations is to chase the bulge on the subdiagonal upwards while maintaining the upper triangular semiseparable structure in the upper part of the matrix. Applying the $k - 1$ Givens transformations $V_j^{(k)}$ on the first $(k+1)$ columns of $A_{1,1}^{(k)}$ requires only $\mathcal{O}(k)$ flops using the Givens-vector representation of the upper left upper triangular semiseparable matrix. Applying the k Givens transformations $U_j^{(k)}$ on the upper left part of the matrix $A_{1,1}^{(k)}$ requires also only $\mathcal{O}(k)$ flops using the Givens-vector representation of the upper triangular semiseparable matrix. Because the upper right part of the matrix $A_{1,1}^{(k)}$ can be written as a $(k+1) \times (n-k-1)$ matrix of rank 1, applying the Givens transformations $U_j^{(k)}$ on this part of $A_{1,1}^{(k)}$ also requires only $\mathcal{O}(k)$ flops. Hence applying the Givens transformations during the whole reduction requires $\mathcal{O}(n^2)$ flops.

A generalization of the previous result towards $m \leq n$ and to lower triangular semiseparable matrices is formulated in the following theorem.

Theorem 2.18. *Let $A \in \mathbb{R}^{m \times n}$. There exist two orthogonal matrices $U \in \mathbb{R}^{m \times m}$ and $V \in \mathbb{R}^{n \times n}$ such that*

- *for $m \geq n$:*

$$U^T A V = \left[\begin{array}{c} S_l \\ 0 \end{array} \right],$$

  where S_l is a lower triangular semiseparable matrix,

- *for $m \leq n$:*

$$U^T A V = \left[\begin{array}{cc} S_u & 0 \end{array} \right],$$

  where S_u is an upper triangular semiseparable matrix,

- *for $m \leq n$:*

$$U^T A V = \left[\begin{array}{cc} S_l & 0 \end{array} \right],$$

  where S_l is a lower triangular semiseparable matrix.

Proof. Using the transpose operation and permuting rows or columns, these results can easily be obtained from Theorem 2.17. □

2.4.3 Relation with the reduction to semiseparable form

As the reduction to upper triangular semiseparable form, and the orthogonal similarity transformation to semiseparable form are closely related, we present here some theorems, relating both reduction procedures.

The first theorem connects the full reduction procedure of both methods.

Theorem 2.19. *Suppose A is an $m \times n$ with $m \geq n$ matrix and U and V are orthogonal matrices such that*

$$U^T A V = \left[\begin{array}{c} S_u \\ 0 \end{array} \right],$$

with S_u an upper triangular semiseparable matrix. Denote with $\tilde{A}$ the symmetric matrix $\tilde{A} = AA^T$ and $\tilde{U}$ the orthogonal matrix such that

$$\tilde{U}^T \tilde{A} \tilde{U} = S,$$

with S a symmetric semiseparable matrix, with the matrix $\tilde{U}$ coming from Theorem 2.8.

Then we have that $\tilde{U} = U$ and $S = S_u S_u^T$, in case no initial transformation is performed!

Proof. The statement $\tilde{U} = U$ can be checked rather easily by looking at the structure of the orthogonal transformations. The second statement is justified by the following equation:

$$S_u S_u^T = U^T A V V^T A^T U = U^T A A^T U = \tilde{U}^T \tilde{A} U = S,$$

which proves the theorem.[16]   □

The following theorem is a generalization of the previous one connected with the intermediate steps in the reductions.

Theorem 2.20. *Suppose we have the matrix A and the matrix $\tilde{A}$ from the previous theorem. At step k of the reduction we have the following equations:*

$$U^{(k)^T} A V^{(k)} = A^{(k)} \tag{2.2}$$

$$\left(\tilde{U}^{(k)} \right)^T \tilde{A} \tilde{U}^{(k)}. \tag{2.3}$$

In Equation (2.2) we have that the upper k rows are already of semiseparable structure; denote them with $S^{(k)}$. In Equation (2.3) we have that the upper left $k \times k$ block is of semiseparable form; denote this block with $\tilde{S}^{(k)}$. Then we have that:

$$\tilde{S}^{(k)} = S^{(k)} S^{(k)^T}$$

Proof. Similar to the proof of Theorem 2.19.   □

Notes and references

The original algorithm for computing the singular value decomposition, based on an adaptation of the QR-algorithm for computing the eigenvalue decomposition, was proposed by Golub and Kahan.

---

[16]In some sense we neglected some details. The orthogonal transformations are not uniquely determined; some sign changes might occur. The resulting matrices are what is called essentially unique, meaning that the matrices are identical up to the signs. For a formal definition of essentially unique we refer to Definition 6.20.

☞ G. H. Golub and W. Kahan. Calculating the singular values and pseudo-inverse of a matrix. *SIAM Journal on Numerical Analysis*, 2:205–224, 1965.

The reduction procedure to upper triangular semiseparable form, was described in.

☞ R. Vandebril, M. Van Barel, and N. Mastronardi. A QR-method for computing the singular values via semiseparable matrices. *Numerische Mathematik*, 99:163–195, November 2004.

In this article a complete algorithm based on semiseparable matrices instead of sparse matrices is described to compute the singular value decomposition. It includes the reduction to upper triangular semiseparable form as well as an implicit QR-algorithm for these matrices.

We remark that in the reduction to upper triangular semiseparable form, one can still reduce the complexity of the proposed method. The algorithm was kept consistent with the reduction to upper bidiagonal form, which is why we first reduced the intermediate results to upper bidiagonal form. If one takes a closer look at the reduction one can see that in fact one first reduces the upper bidiagonal form to lower bidiagonal form, and then the semiseparable structure is created. To reduce the complexity one can therefore reduce the matrix instead to upper bidiagonal form immediately to lower bidiagonal form and then create the upper triangular semiseparable matrix from this structure. This procedure is cheaper than the procedure proposed in the text; moreover the convergence properties, as they will be discussed in the remainder of the text, are still valid.

2.5 Transformations from sparse to structured rank form

In the previous sections we considered various reduction algorithms. Algorithms were considered for reducing matrices either to tridiagonal/bidiagonal form (named sparse matrices in this book) and algorithms were considered for transforming matrices to structured rank form, such as semiseparable or upper triangular semiseparable. In fact we discussed transformations from unstructured matrices to either structured rank or sparse form. These reductions to structured rank or the sparse form involved always $\mathcal{O}(n^3)$ operations.

In this section and in the forthcoming section, we will investigate possibilities for transforming the structured rank matrices to the sparse form and vice versa. Moreover we will show that these reduction algorithms are cheaper to perform.

In this section we start with the reductions from sparse to structured rank form. The transformations involve $\mathcal{O}(n^2)$ operations going from a tridiagonal to a semiseparable (plus diagonal) and also $\mathcal{O}(n^2)$ when going from a bidiagonal to an upper triangular semiseparable. In this section we do not discuss the transition from Hessenberg to Hessenberg-like form as it is similar to the reduction to semiseparable form. Transition costs are due to the unsymmetric structure, $\mathcal{O}(n^3)$.

2.5.1 From tridiagonal to semiseparable (plus diagonal)

The previously proposed reduction procedure from symmetric to semiseparable (plus diagonal) form involved two parts in each iterative step of the reduction procedure. One part consisted of creating zeros in the matrix, and a second part consisted of chasing the semiseparable structure downwards. Taking a closer look at this reduction procedure, one can clearly see that one can decouple these two parts, as already mentioned before in Subsection 2.2.2. This means one can first create all the zeros to obtain a tridiagonal matrix, and then one can start the chasing procedure to create a full semiseparable matrix.

Hence, the computational cost of reducing a tridiagonal matrix to semiseparable form involves only the cost of performing the chasing step, which is $9n^2 + \mathcal{O}(n)$ flops and for reducing the matrix to semiseparable plus diagonal form, which takes $10n^2 + \mathcal{O}(n)$ flops.

2.5.2   From bidiagonal to upper triangular semiseparable

Similarly as the orthogonal similarity reduction of a symmetric matrix becomes a similar semiseparable one, we can decompose the reduction of a matrix to upper triangular semiseparable form into two main parts. A first part consists of creating zeros. A second part consists of performing the chasing technique for making the matrix upper triangular semiseparable.

We do not present the complete algorithm but show how to decompose the two parts when transforming a matrix to upper triangular semiseparable form. Hence applying only the last step to a bidiagonal matrix solves the problem.

- The first part consists of transforming the matrix via orthogonal similarity transformations to a "lower" bidiagonal matrix.

- The second part consists of performing the chasing techniques as described for making the matrix of upper triangular semiseparable form.

As an upper and a lower bidiagonal matrix can easily be transformed into each other by performing a sequence of $n - 1$ Givens transformations we can perform any kind of transformation from upper/lower bidiagonal to upper/lower triangular semiseparable form. These transformations involve $\mathcal{O}(n^2)$ operations because each time a chasing has to be performed.

2.6   From structured rank to sparse form

In the preceding section we discussed the other way around: from sparse to structured rank form. This was rather easy, as we could decompose the reductions from symmetric to structured rank into two parts. The first part reduced the symmetric matrix to sparse form and then performed a chasing technique on the sparse matrix.

The reductions discussed in the previous section involved $\mathcal{O}(n^2)$ operations. One might wonder if the reductions from the dense structured rank matrices to the sparse form can also be performed in $\mathcal{O}(n^2)$ operations. In this section we will address this question and design the appropriate reduction methods. Again we do not discuss the nonsymmetric problem as this is similar to the proposed techniques.

2.6.1 From semiseparable (plus diagonal) to tridiagonal

As the semiseparable plus diagonal case is completely similar to the semiseparable case, we will only discuss the semiseparable plus diagonal case. The results presented in this section are based on [115]. We will first present the algorithm, which involves $\mathcal{O}(n^2)$ operations, for transforming the semiseparable plus diagonal matrix to tridiagonal form.

Theorem 2.21. *Let $S + D$ be a semiseparable plus diagonal matrix. Then there exists an orthogonal matrix U such that*

$$U^T(S + D)U = T,$$

where T is a tridiagonal matrix. Moreover T can be computed in $\mathcal{O}(n^2)$ operations.

Proof. We will illustrate the reduction procedure on a semiseparable plus diagonal matrix of dimension 6×6. The proof is by finite induction. Let $A = S + D$ and $A_0^{(1)} = A$. Often, we briefly denote $A^{(i)}$ as $A_0^{(i)}$. Elements that are zero are not shown in the figures.

- **Step** 1. Suppose a semiseparable plus diagonal matrix is given, and one wants to make it tridiagonal. The first orthogonal similarity transformation consists of making the complete last row and column zero, except for the subdiagonal, the diagonal and the superdiagonal element. Because of the specific rank structure of the matrix this can be achieved by performing a similarity Givens transformation involving the last two rows and the last two columns of the matrix $A_0^{(1)}$. As we have a semiseparable plus diagonal matrix, the diagonal is not includable in the semiseparable structure; hence, the elements $\boxtimes$ do not include the diagonal. This is what happens (remark that the elements $\otimes$ in this figure are also includable in the rank structure):

$$
\begin{bmatrix}
\times & \boxtimes & \boxtimes & \boxtimes & \boxtimes & \otimes \\
\boxtimes & \times & \boxtimes & \boxtimes & \boxtimes & \otimes \\
\boxtimes & \boxtimes & \times & \boxtimes & \boxtimes & \otimes \\
\boxtimes & \boxtimes & \boxtimes & \times & \boxtimes & \otimes \\
\boxtimes & \boxtimes & \boxtimes & \boxtimes & \times & \boxtimes \\
\otimes & \otimes & \otimes & \otimes & \boxtimes & \times
\end{bmatrix}
\xrightarrow{\ {G_1^{(1)}}^T A_0^{(1)} G_1^{(1)}\ }
\begin{bmatrix}
\times & \boxtimes & \boxtimes & \boxtimes & \boxtimes & 0 \\
\boxtimes & \times & \boxtimes & \boxtimes & \boxtimes & 0 \\
\boxtimes & \boxtimes & \times & \boxtimes & \boxtimes & 0 \\
\boxtimes & \boxtimes & \boxtimes & \times & \boxtimes & 0 \\
\boxtimes & \boxtimes & \boxtimes & \boxtimes & \times & \times \\
0 & 0 & 0 & 0 & \times & \times
\end{bmatrix}
$$

$$\Updownarrow$$

$$A_0^{(1)} \xrightarrow{\ {G_1^{(1)}}^T A_0^{(1)} G_1^{(1)}\ } A_1^{(1)}.$$

The Givens transformation on the left exploits the specific rank structure in the last two rows to make all the elements except the subdiagonal and the diagonal element zero. In fact we have already completed one step of the induction process now. The last row and column satisfy already the tridiagonal structure. Pose now $A_0^{(2)} = A_1^{(1)}$.

- **Step** k. For simplicity reasons we assume we are in step $k = 3$ of the algorithm, as this illustrates the general case. This means that our intermediate matrix $A_0^{(3)}$ has already the last two rows and columns satisfying the tridiagonal structure. The matrix $A_0^{(3)}$ looks like:

$$
A_0^{(3)} =
\begin{bmatrix}
\times & \boxtimes & \boxtimes & \boxtimes & & \\
\boxtimes & \times & \boxtimes & \boxtimes & & \\
\boxtimes & \boxtimes & \times & \boxtimes & & \\
\boxtimes & \boxtimes & \boxtimes & \times & \times & \\
& & & \times & \times & \times \\
& & & & \times & \times
\end{bmatrix} .
$$

We perform now a similarity Givens transformation on row 3 and 4 (column 3 and 4) to annihilate the first two elements in row 4. This can be done by one single Givens transformation, exploiting the semiseparable structure.

$$
\begin{bmatrix}
\times & \boxtimes & \boxtimes & \otimes & & \\
\boxtimes & \times & \boxtimes & \otimes & & \\
\boxtimes & \boxtimes & \times & \boxtimes & & \\
\otimes & \otimes & \boxtimes & \times & \times & \\
& & & \times & \times & \times \\
& & & & \times & \times
\end{bmatrix}
\xrightarrow{\ G_1^{(3)^T} A_0^{(3)} G_1^{(3)}\ }
\begin{bmatrix}
\times & \boxtimes & \boxtimes & & & \\
\boxtimes & \times & \boxtimes & & & \\
\boxtimes & \boxtimes & \times & \times & \times & \\
& & \times & \times & \times & \\
& & \times & \times & \times & \times \\
& & & & \times & \times
\end{bmatrix}
$$

$$
\Updownarrow
$$

$$
A_0^{(3)} \quad \xrightarrow{\ G_1^{(3)^T} A_0^{(3)} G_1^{(3)}\ } \quad A_1^{(3)} .
$$

We remark that performing this transformation creates fill-in in the positions $(3,5)$ and $(5,3)$ of our matrix. This can be seen as a bulge in the tridiagonal structure. We can remove this bulge by standard chasing techniques (see, e.g., [181, 183]). To remove this bulge we apply a similarity Givens rotation to the rows 4 and 5 (and columns 4 and 5). This is illustrated in the following figure:

$$
\begin{bmatrix}
\times & \boxtimes & \boxtimes & & & \\
\boxtimes & \times & \boxtimes & & & \\
\boxtimes & \boxtimes & \times & \times & \otimes & \\
& & \times & \times & \times & \\
& & \otimes & \times & \times & \times \\
& & & & \times & \times
\end{bmatrix}
\xrightarrow{\ G_2^{(3)^T} A_1^{(3)} G_2^{(3)}\ }
\begin{bmatrix}
\times & \boxtimes & \boxtimes & & & \\
\boxtimes & \times & \boxtimes & & & \\
\boxtimes & \boxtimes & \times & \times & & \\
& & \times & \times & \times & \times \\
& & & \times & \times & \times \\
& & & \times & \times & \times
\end{bmatrix}
$$

$$
\Updownarrow
$$

$$
A_1^{(3)} \quad \xrightarrow{\ G_2^{(3)^T} A_1^{(3)} G_2^{(3)}\ } \quad A_2^{(3)} .
$$

This transformation created however another bulge, and therefore the process needs to be repeated. We perform another Givens rotation to the rows 5 and

6 (and columns 5 and 6). This gives us

$$
\begin{bmatrix}
\times & \boxtimes & \boxtimes & & & \\
\boxtimes & \times & \boxtimes & & & \\
\boxtimes & \boxtimes & \times & \times & & \\
 & & \times & \times & \times & \otimes \\
 & & & \times & \times & \times \\
 & & & \otimes & \times & \times
\end{bmatrix}
\xrightarrow{G_3^{(3)T} A_2^{(3)} G_3^{(3)}}
\begin{bmatrix}
\times & \boxtimes & \boxtimes & & & \\
\boxtimes & \times & \boxtimes & & & \\
\boxtimes & \boxtimes & \times & \times & & \\
 & & & \times & \times & \times \\
 & & & \times & \times & \times \\
 & & & & \times & \times
\end{bmatrix}
$$

$$
\Updownarrow
$$

$$
A_2^{(3)} \xrightarrow{G_3^{(3)T} A_2^{(3)} G_3^{(3)}} A_3^{(3)}.
$$

This completes the induction step.

One can clearly see that the final step in the algorithm only consists of chasing the bulge away.   □

The proof of the theorem gives an algorithm for performing the reduction involving $\mathcal{O}(n^2)$ operations, however this is not the only possible way for reducing a semiseparable plus diagonal matrix to tridiagonal form. One can for example perform also a two-way chasing algorithm. This reduction can in fact be performed parallel on two processors. The idea is simple: in the algorithm presented above, we started creating zeros in the bottom row by performing a transformation on the left. In fact one can at the same time perform a transformation on the left, creating zeros in the top row starting from the second superdiagonal element. This means that as in our method we start creating zeros which go from the bottom to the top, but at the same time we also create zeros which go from the top to bottom. More details on this algorithm and on the implementation can be found in [115]. See also the notes and references for other variants.

2.6.2   From semiseparable to bidiagonal

In this section we will derive an orthogonal transformation, which will transform an upper triangular semiseparable matrix into an upper bidiagonal one. A similar procedure can be constructed for reducing a lower triangular semiseparable to a lower bidiagonal matrix.[17]

Theorem 2.22. *Suppose $S \in \mathbb{R}^{n \times n}$ to be an upper triangular semiseparable matrix, then there exist two orthogonal matrices $U \in \mathbb{R}^{n \times n}$ and $V \in \mathbb{R}^{n \times n}$ such that*

$$
U^T S V = B,
$$

where B is an upper bidiagonal matrix.

---

[17]The transitions from a lower triangular semiseparable to an upper bidiagonal (and an upper triangular semiseparable to a lower bidiagonal) are possible.

Proof. The proof given here is constructive, i.e., we will again explicitly show which Givens transformations are performed in order to get the desired upper bidiagonal matrix.

The proof is outlined for an upper triangular semiseparable matrix of size 6×6. We use again the following rules for $A = S$:

$$U_{k+1}^{(l)} A_{k,j}^{(l)} = A_{k+1,j}^{(l)}, \quad A_{k,j}^{(l)} V_{j+1}^{(l)} = A_{k,j+1}^{(l)}.$$

We define $A_{0,0}^{(l)} = A^{(l)}$ and $A^{(l+1)} = A_{l,l}^{(l)}$.

The applied transformations to the left and to the right of $A_{k,j}^{(l)}$ are all Givens transformations in this case.

- **Step 1.** In this first step of the algorithm, we will bring the first row of our upper triangular semiseparable matrix A to upper bidiagonal form. We know that the upper part of the matrix $A_{0,0}^{(1)}$ is of semiseparable form, i.e., by performing one Givens transformation on the left, working on rows 1 and 2, of the matrix we can introduce all zeros in the upper row, except for the diagonal element. The elements $\boxtimes$ indicate the part of semiseparable form. The elements $\otimes$ will be annihilated but are also includable in the semiseparable structure.

$$
\begin{bmatrix}
\boxtimes & \otimes & \otimes & \otimes & \otimes & \otimes \\
 & \boxtimes & \boxtimes & \boxtimes & \boxtimes & \boxtimes \\
 & & \boxtimes & \boxtimes & \boxtimes & \boxtimes \\
 & & & \boxtimes & \boxtimes & \boxtimes \\
 & & & & \boxtimes & \boxtimes \\
 & & & & & \boxtimes
\end{bmatrix}
\xrightarrow{U_1^{(1)} A_{0,0}^{(1)}}
\begin{bmatrix}
\times & 0 & 0 & 0 & 0 & 0 \\
\times & \boxtimes & \boxtimes & \boxtimes & \boxtimes & \boxtimes \\
 & & \boxtimes & \boxtimes & \boxtimes & \boxtimes \\
 & & & \boxtimes & \boxtimes & \boxtimes \\
 & & & & \boxtimes & \boxtimes \\
 & & & & & \boxtimes
\end{bmatrix}
$$

$$\Updownarrow$$

$$A_{0,0}^{(1)} = A^{(1)} \xrightarrow{U_1^{(1)^T} A_{0,0}^{(1)}} A_{1,0}^{(1)}.$$

As one can see we have introduced a bulge now in position $(2, 1)$; this bulge can be removed by performing one Givens transformation on the first and the second column of the matrix $A_{1,0}^{(1)}$. In this way we get:

$$
\begin{bmatrix}
\times & 0 & 0 & 0 & 0 & 0 \\
\otimes & \boxtimes & \boxtimes & \boxtimes & \boxtimes & \boxtimes \\
 & & \boxtimes & \boxtimes & \boxtimes & \boxtimes \\
 & & & \boxtimes & \boxtimes & \boxtimes \\
 & & & & \boxtimes & \boxtimes \\
 & & & & & \boxtimes
\end{bmatrix}
\xrightarrow{A_{1,0}^{(1)} V_1^{(1)}}
\begin{bmatrix}
\times & \times & 0 & 0 & 0 & 0 \\
 & \boxtimes & \boxtimes & \boxtimes & \boxtimes & \boxtimes \\
 & & \boxtimes & \boxtimes & \boxtimes & \boxtimes \\
 & & & \boxtimes & \boxtimes & \boxtimes \\
 & & & & \boxtimes & \boxtimes \\
 & & & & & \boxtimes
\end{bmatrix}
$$

$$\Updownarrow$$

$$A_{1,0}^{(1)} \xrightarrow{A_{1,0}^{(1)} V_1^{(1)}} A_{1,1}^{(1)}.$$

Then we put $A^{(2)} = A^{(1)}_{1,1}$ and we have completed the first step in the reduction process, as the first row of the matrix is in bidiagonal form.

- **Step k.** By induction, for $k > 1$. Let us assume that the first $k - 1$ rows of $A^{(k)}$ are in bidiagonal form and then we will add now one more row to this upper bidiagonal structure. Without loss of generality, let us assume $k = 3$. This means that $A^{(3)}$ has the following structure:

$$A^{(3)} = \begin{bmatrix} \times & \times & & & & \\ & \times & \times & & & \\ & & \boxtimes & \boxtimes & \boxtimes & \boxtimes \\ & & & \boxtimes & \boxtimes & \boxtimes \\ & & & & \boxtimes & \boxtimes \\ & & & & & \boxtimes \end{bmatrix}.$$

We start in a similar manner as in the first step. A Givens transformation is applied on the left acting on rows 3 and 4 to annihilate the last three elements in the third row. The annihilation of these three elements by one operation is possible due to the semiseparable structure:

$$\begin{bmatrix} \times & \times & & & \\ & \times & \times & & \\ & & \boxtimes & \otimes & \otimes & \otimes \\ & & \boxtimes & \boxtimes & \boxtimes & \\ & & & \boxtimes & \boxtimes & \\ & & & & \boxtimes \end{bmatrix} \xrightarrow{U_1^{(3)^T} A_{0,0}^{(3)}} \begin{bmatrix} \times & \times & & & \\ & \times & \times & & \\ & & \times & 0 & 0 & 0 \\ & & \times & \boxtimes & \boxtimes & \boxtimes \\ & & & \boxtimes & \boxtimes & \\ & & & & \boxtimes \end{bmatrix}$$

$$\Updownarrow$$

$$A_{0,0}^{(3)} \xrightarrow{U_1^{(3)^T} A_{0,0}^{(3)}} A_{1,0}^{(3)}.$$

Unfortunately we created again a bulge. This bulge will be chased away by a sequence of Givens transformations performed alternating on the right and left side of the matrix. The first transformation will be performed on the right and it will work on columns 3 and 4.

$$\begin{bmatrix} \times & \times & & & \\ & \times & \times & & \\ & & \times & & & \\ & & \otimes & \boxtimes & \boxtimes & \boxtimes \\ & & & \boxtimes & \boxtimes & \\ & & & & \boxtimes \end{bmatrix} \xrightarrow{A_{1,0}^{(3)} V_1^{(3)}} \begin{bmatrix} \times & \times & & & \\ & \times & \times & \times & \\ & & \times & \times & \\ & & & \boxtimes & \boxtimes & \boxtimes \\ & & & \boxtimes & \boxtimes & \\ & & & & \boxtimes \end{bmatrix}$$

$$\Updownarrow$$

$$A_{1,0}^{(3)} \xrightarrow{A_{1,0}^{(3)} V_1^{(3)}} A_{1,1}^{(3)}.$$

A new bulge is created that is removed by a Givens transformation on the left involving the second and the third row.

$$
\begin{bmatrix}
\times & \times & & & & \\
& \times & \times & \otimes & & \\
& & \times & \times & & \\
& & & \boxtimes & \boxtimes & \boxtimes \\
& & & & \boxtimes & \boxtimes \\
& & & & & \boxtimes
\end{bmatrix}
\xrightarrow{U_2^{(3)} A_{1,1}^{(3)}}
\begin{bmatrix}
\times & \times & & & & \\
& \times & \times & & & \\
& \times & \times & \times & & \\
& & & \boxtimes & \boxtimes & \boxtimes \\
& & & & \boxtimes & \boxtimes \\
& & & & & \boxtimes
\end{bmatrix}
$$

$$
\Updownarrow
$$

$$
A_{1,1}^{(3)} \xrightarrow{U_2^{(3)} A_{1,1}^{(3)}} A_{2,1}^{(3)}.
$$

The latter bulge is removed by a transformation on the right involving columns 2 and 3; next the following bulge is removed by a Givens transformation on the left involving the first two rows.

$$
\begin{bmatrix}
\times & \times & & & & \\
& \times & \times & & & \\
\otimes & \times & \times & & & \\
& & & \boxtimes & \boxtimes & \boxtimes \\
& & & & \boxtimes & \boxtimes \\
& & & & & \boxtimes
\end{bmatrix}
\xrightarrow{A_{2,1}^{(3)} V_2^{(3)}}
\begin{bmatrix}
\times & \times & \times & & & \\
& \times & \times & & & \\
& & \times & \times & & \\
& & & \boxtimes & \boxtimes & \boxtimes \\
& & & & \boxtimes & \boxtimes \\
& & & & & \boxtimes
\end{bmatrix}
$$

$$
\Updownarrow
$$

$$
A_{2,1}^{(3)} \xrightarrow{A_{2,1}^{(3)} V_2^{(3)}} A_{2,2}^{(3)}.
$$

$$
\begin{bmatrix}
\times & \times & \otimes & & & \\
& \times & \times & & & \\
& & \times & \times & & \\
& & & \boxtimes & \boxtimes & \boxtimes \\
& & & & \boxtimes & \boxtimes \\
& & & & & \boxtimes
\end{bmatrix}
\xrightarrow{U_3^{(3)} A_{2,2}^{(3)}}
\begin{bmatrix}
\times & \times & & & & \\
& \times & \times & \times & & \\
& & \times & \times & & \\
& & & \boxtimes & \boxtimes & \boxtimes \\
& & & & \boxtimes & \boxtimes \\
& & & & & \boxtimes
\end{bmatrix}
$$

$$
\Updownarrow
$$

$$
A_{2,2}^{(3)} \xrightarrow{U_3^{(3)} A_{2,2}^{(3)}} A_{3,2}^{(3)}.
$$

A final Givens transformation needs to be performed on the first and second columns to remove the element in position $(2,1)$.

$$
\begin{bmatrix}
\times & \times & & & & \\
\otimes & \times & \times & & & \\
& \times & \times & & & \\
& & \boxtimes & \boxtimes & \boxtimes & \\
& & & \boxtimes & \boxtimes & \\
& & & & \boxtimes &
\end{bmatrix}
\xrightarrow{A_{3,2}^{(3)} V_3^{(3)}}
\begin{bmatrix}
\times & \times & & & & \\
& \times & \times & & & \\
& & \times & \times & & \\
& & & \boxtimes & \boxtimes & \boxtimes \\
& & & & \boxtimes & \boxtimes \\
& & & & & \boxtimes
\end{bmatrix}
$$

$$\Updownarrow$$

$$A_{3,2}^{(3)} \xrightarrow{A_{3,2}^{(3)} V_3^{(3)}} A_{3,3}^{(3)}.$$

This completes the induction step. In a similar fashion one can proceed and make also the last part of the matrix upper bidiagonal.

$\square$

Similarly as in the reduction from semiseparable to tridiagonal, we can consider another, faster chasing technique acting on two sides of the matrix at the same time. More information can be found in the notes and references.

Notes and references

Originally these reductions to sparse form were used for solving systems of equations involving semiseparable (plus diagonal) matrices or for computing their spectra. At that time these methods were the fastest available for solving systems of equations involving semiseparable and semiseparable plus diagonal matrices. Of course these methods were of complexity $\mathcal{O}(n^2)$. Nowadays, the complexity of solving these systems of equations has significantly decreased involving only $\mathcal{O}(n)$ operations.

Let us present in this section some of the references on which the results in this section are based. We have to make an important remark concerning the titles used by the different authors in the following articles. The nomenclature concerning structured rank matrices is not so unified. This means that sometimes matrices named semiseparable plus diagonal are named quasiseparable by other authors. The reader has to keep this always in mind and carefully check which type of matrices the authors consider. We will always carefully point out the type of structured rank matrices the authors use, with regard to the conventions used in our book.

The reduction technique as presented in this section for bringing diagonal plus semiseparable matrices via orthogonal similarity transformations to tridiagonal form was originally proposed in the following article.

☞ S. Chandrasekaran and M. Gu. Fast and stable eigendecomposition of symmetric banded plus semi-separable matrices. *Linear Algebra and its Applications*, 313:107–114, 2000.

Chandrasekaran and Gu present a way to compute the eigenvalues of what they name 'semiseparable plus band' matrices. The matrices considered in this article are of the following form: they are symmetric and can be written as the sum of a band matrix plus two matrices, which fill up the remaining zeros of the band matrix with a rank r matrix.

Hence the parts above and below the band are coming from a matrix of the form VU^T and UV^T, both matrices U, V of dimension $n \times r$.

The idea to compute the eigendecomposition of these matrices is to reduce them in an effective way to tridiagonal form. Then one can use in fact any suitable method for computing the eigenvalues of the resulting tridiagonal matrix.

The reduction to tridiagonal form proceeds as follows: a part of the low rank structure is removed, after which one has to chase the created bulge away. The complexity of the proposed method is approximately $3n^2(b + 2r)$ in which b denotes the bandwidth and r the rank of the parts above and below the band.

☞ N. Mastronardi, S. Chandrasekaran, and S. Van Huffel. Fast and stable reduction of diagonal plus semi-separable matrices to tridiagonal and bidiagonal form. *BIT*, 41(1):149–157, 2003.

The authors Mastronardi, Chandrasekaran and Van Huffel, provide an algorithm to transform a symmetric generator representable quasiseparable matrix into a similar tridiagonal one, by using Givens transformations. A generator representable quasiseparable matrix, has the parts below and above the diagonal coming from a rank 1 matrix. The presented method is demonstrated for matrices having semiseparability rank 1 but is also applicable for the matrices considered in the article above. The Givens transformations remove gently the rank structure. After each transformation also a step of chasing is performed to remove the created bulge, similarly as in the previous article by Gu and Chandrasekaran. The main adaptation is that the algorithm can easily be run on two sides, a so-called two-way algorithm. Instead of only removing the rank structure starting from the bottom, one also starts by removing the rank structure at the top. This leads to a complexity that is halve of the one presented above. Moreover the presented method can be run in parallel on two processors. Also an algorithm to reduce a generator representable quasiseparable matrix to a bidiagonal one by means of orthogonal transformations is included. Also this bidiagonalization procedure admits a two-way version.

The following articles (and the references therein) are related with the reduction of arrowhead matrices to tridiagonal form. Arrowhead matrices can also be considered as special types of structured rank matrices.

☞ A. Abdallah and Y. Hu. Parallel *VLSI* computing array implementation for signal subspace updating algorithm. *IEEE Transactions on Acoustics, Speech and Signal Processing*, 37:742–748, 1989.

☞ S. Van Huffel and H. Park. Efficient reduction algorithms for bordered band matrices. *Journal of Numerical Linear Algebra and Applications*, 2(2):95–113, 1995.

☞ H. Zha. A two-way chasing scheme for reducing a symmetric arrowhead matrix to tridiagonal form. *Numerical Linear Algebra with Applications*, 1(1):49–57, 1992.

In the following article, the authors Fasino, Mastronardi and Van Barel propose two new algorithms for transforming a generator representable quasiseparable matrix to tridiagonal or bidiagonal form.

☞ D. Fasino, N. Mastronardi, and M. Van Barel. Fast and stable algorithms for reducing diagonal plus semiseparable matrices to tridiagonal and bidiagonal form. In *Fast Algorithms for Structured Matrices: Theory and Applications*, volume 323 of *Contemporary Mathematics*, pages 105–118. American Mathematical Society, Providence, Rhode Island, USA, 2003.

Instead of performing one Givens transformation as in [115, 38] and then chasing the bulge away, $n/2$ Givens transformations are performed at once. The algorithm is demonstrated for the easy class of matrices having semiseparability rank 1. After having performed $n/2$ Givens transformations Q (assume n to be even) on the matrix A we obtain

$$Q^T A Q = \left[\begin{array}{c|c} A_1 & 0 \\ \hline 0 & A_2 \end{array} \right],$$

which is a block diagonal matrix, with both blocks A_1 and A_2 of generator representable quasiseparable form. Hence we can work on both blocks separately. This procedure can also be rewritten in terms of a two-way form and a variant is also presented for bidiagonalizing a semiseparable plus diagonal matrix.

Recently also more hybrid structured rank matrices are investigated.

☞ Y. Eidelman, L. Gemignani, and I. C. Gohberg. On the fast reduction of a quasiseparable matrix to Hessenberg and tridiagonal forms. *Linear Algebra and its Applications*, 420(1):86–101, January 2007.

In this article, the authors present a method for effectively reducing a quasiseparable matrix (not symmetric) to Hessenberg form. The authors exploit also the rank structure of the upper triangular part to come to a fast reduction. This reduction is important as in practice the reduction to Hessenberg and the corresponding QR-method applied on the Hessenberg matrix can be much faster (above a certain rank) than immediately computing the eigenvalues of the quasiseparable matrix by the QR-method for quasiseparable matrices [70].

☞ S. Delvaux and M. Van Barel. A Hessenberg reduction algorithm for rank structured matrices. *SIAM Journal on Matrix Analysis and Applications*, 29(3):895–926, 2007.

In this article the authors Delvaux and Van Barel present a method for reducing arbitrary structured rank matrices, based on the 'structured rank blocks' characterization, to Hessenberg form. This is a very general method as all types of structured rank matrices nicely fit into this class. The matrices are represented with a unitary weight or a Givens-weight representation.

☞ V. Y. Pan. A reduction of the matrix eigenproblem to polynomial rootfinding via similarity transforms into arrow-head matrices. Technical Report TR-2004009, Department of Computer Science, City University of New York, New York, USA, July 2004.

In this article Pan makes use of the reduction to Hessenberg-like form, for computing the eigensystem of a matrix. After having computed the similar Hessenberg-like matrix, this matrix is transformed via nonunitary similarity transforms into an arrowhead matrix. The eigenvalues of the arrowhead matrix are computed via polynomial rootfinding algorithms.

2.7 Conclusions

In this chapter different transformation techniques were discussed. The main aim of the proposed transformation techniques was to obtain sparse 'representable' structures, for which the structure can be efficiently exploited for computing the eigenvalues or singular values.

For completeness we also discussed the traditional reductions to tridiagonal and bidiagonal form. Moreover, also the transformations from sparse to structured rank form and vice versa were investigated.

In the following chapters we will investigate in more detail the properties of the reductions to structured rank form. These reductions cost more operations than the corresponding reductions to sparse form, but the extra performed operations induce an extra convergence behavior. This extra convergence behavior might come in handy in certain applications when only parts of the spectrum are needed, or it can even become faster than the traditional methods if there are gaps in the spectrum.

Chapter 3

Convergence properties of the reduction algorithms

In the previous chapter, several types of reduction algorithms to either sparse or structured rank matrices were presented. It was stated there that the orthogonal similarity reduction to a similar semiseparable matrix could also be achieved by first transforming the matrix to a similar tridiagonal one and then transforming the tridiagonal matrix into a similar semiseparable one. It is a known result that the already reduced part in the tridiagonal reduction contains the Lanczos-Ritz values. In this chapter we will see to which extent these classical results for tridiagonal matrices are valid for the semiseparable case. Secondly there was also a remark, which stated that the reduction, although it costs an extra $9n^2 + \mathcal{O}(n)$ with respect to the tridiagonalization, inherited an extra kind of convergence behavior. These extra operations create a subspace iteration convergence behavior, which also will be addressed here.

This chapter contains three sections. The layout of all three sections is similar. First we describe the property we want to investigate for the reduction to semiseparable form. Then we extend this property to a general framework, and finally we apply this general framework to all the reduction algorithms as proposed in the previous chapter, i.e., the reduction to semiseparable, semiseparable plus diagonal, Hessenberg-like and so forth.

In the first section we deduce a general theorem, providing necessary and sufficient conditions, for an orthogonal similarity transformation to obtain the Arnoldi-Ritz values in the already reduced part of the matrix. The theorem provided is very simple, only two conditions have to be placed on the reduction algorithm, but it is completely general. This means, that using this theorem the convergence properties of the transformations to tridiagonal, semiseparable, Hessenberg and Hessenberg-like can be proved in one line.

In the second section we will investigate in detail the extra convergence behavior of the reductions to structured rank form, with regard to the reductions to sparse form. We will prove that the extra performed Givens transformations cause a convergence behavior, which can be interpreted as a type of nested subspace iteration.

In the last section the interaction between the Lanczos convergence behavior and the subspace iteration will be investigated. We will prove that the subspace iteration will start converging as soon as the Lanczos-Ritz values approximate well enough the eigenvalues the subspace iteration tends to converge to. The interaction behavior as proposed in this section is affirmed by the derivation of the convergence speed. Moreover a detailed complexity analysis, if deflation is possible, will be given. The theoretical results in this chapter will be confirmed by numerical experiments in the following chapter.

✎ *In this chapter, the convergence properties of the reduction algorithms proposed in the previous chapter are examined. Two types of behavior are covered. The Lanczos-Ritz values convergence behavior and the subspace iteration convergence behavior.*

An introduction to Lanczos-Ritz values and an explanation of the notation used can be found in Subsection 3.1.1. The Lanczos-type of convergence behavior is summarized into two theorems, namely Theorem 3.2 and Theorem 3.3, providing the necessary and sufficient conditions, respectively. The case of invariant subspaces is left to the interested reader and is rather similar to the standard case (Subsection 3.1.5). Important for our analysis is the application of these theorems (Theorems 3.2 and 3.3) towards the proposed reduction algorithms. This is done in Subsection 3.1.7.

The second convergence behavior is analyzed in its most general form in Subsection 3.2.2 and Subsection 3.2.3. Especially Theorem 3.6 covers the global picture. The different formulations of the theorem leading to different interpretations covered in Corollaries 3.7, 3.8 and 3.9 help to understand the convergence behavior.

The interaction of both convergence behaviors is neatly summarized in Subsection 3.3.2. Estimates on the convergence speed towards a block diagonal matrix are given in Subsection 3.3.3. The paragraph titled "the convergence speed of the nested multishift iteration" contains the theoretical upper bound on the convergence speed.

3.1   The Arnoldi(Lanczos)-Ritz values

It is well-known that while reducing a symmetric matrix into a similar tridiagonal one the intermediate tridiagonal matrices contain the Lanczos-Ritz values as eigenvalues. Or for a Hessenberg matrix they contain the so-called Arnoldi-Ritz values. More information can be found in the following books [43, 58, 94, 129, 136, 145] and the references therein.

In this section we will investigate to which extent this is true for the reductions discussed in the previous chapter, namely the reduction to semiseparable, semiseparable plus diagonal, Hessenberg-like and so on. After defining what is meant with Ritz values, we will provide, in a first subsection, an easy proof stating that the reduced parts in the reduction to semiseparable form have the same eigenvalues as the reduction to tridiagonal form. Hence we know that both reduced parts will have the same eigenvalues, namely the Lanczos-Ritz values. Instead of proving for every type of reduction which eigenvalues are located in the already reduced part, we provide a general framework in this section. In the following subsection, we will briefly recapitulate what is meant by the Ritz values in this book, and we provide a general scheme for orthogonal similarity transformations, which we will investigate. The Subsections 3.1.3 to 3.1.6 deal with these orthogonal similarity transformations and provide necessary and sufficient conditions such that the already reduced parts in the matrix have the Ritz values. The final subsection discusses all the different reduction algorithms from the previous chapter and proves which type of eigenvalues are contained in the already reduced block.

3.1.1 Ritz values and Arnoldi(Lanczos)-Ritz values

We will briefly introduce here the notion of 'Ritz values', related to the orthogonal similarity transformation. The orthogonal similarity transformations we consider are based on finite induction. In each induction step a row and a column are added to the desired structure. In this way all the columns and rows are transformed, such that the resulting matrix satisfies the desired structure. Suppose we have a matrix $A^{(0)} = A$, which is transformed via an initial orthogonal similarity transformation into the matrix $A^{(1)} = Q^{(0)^T} A^{(0)} Q^{(0)}$. The initial transformation $Q^{(0)}$ is not essential in the following scheme, as it does not affect the reduction algorithms. It does, however, have an effect on the convergence behavior of the reduction, as will be shown. We remark that in real applications often a matrix $Q^{(0)}$ is chosen in such a way to obtain a specific convergence behavior. The other orthogonal transformations $Q^{(k)}$, $1 \leq k \leq n - 1$, are constructed by the reduction algorithms. Let us denote the orthogonal transformation to go from $A^{(k)}$ to $A^{(k+1)}$ as $Q^{(k)}$, and we denote with $Q_{0:k}$ the orthogonal matrix equal to the product $Q^{(0)} Q^{(1)} \ldots Q^{(k)}$. This means that

$$
\begin{aligned}
A^{(k+1)} &= Q^{(k)^T} A^{(k)} Q^{(k)} \\
&= Q^{(k)^T} Q^{(k-1)^T} \ldots Q^{(1)^T} Q^{(0)^T} A Q^{(0)} Q^{(1)} \ldots Q^{(k-1)} Q^{(k)} \\
&= Q_{0:k}^T A Q_{0:k}.
\end{aligned}
$$

The matrix $A^{(k+1)}$ is of the following form:

$$
\left[\begin{array}{c|c} R_{k+1} & \times \\ \hline \times & A_{k+1} \end{array} \right]
$$

where R_{k+1} (Reduced) stands for that part of the matrix of dimension $(k+1) \times (k+1)$, which is already transformed to the appropriate form, e.g., tridiagonal, semiseparable and Hessenberg.[18] The matrix A_{k+1} is of dimension $(n-k-1) \times (n-$

---

[18]In this section the already reduced part of the matrix can be found in the top-left corner. The reduction algorithms as proposed in the previous chapter have the reduced part in the bottom-right corner. Essentially this does not make any difference. We chose not to stick to either formulation, as in practice both descriptions occur. In several books, the reduction to tridiagonal form starts at the bottom, whereas often the reduction to bidiagonal is illustrated starting at the top. See Note 2.7.

$k-1$). The $\times$ denote arbitrary matrices. They are unimportant in the remaining part of the exposition. The matrices $A^{(k)}$ are not necessarily symmetric, as the elements $\times$ may falsely indicate.

Let us partition the matrix $Q_{0:k}$ as follows:

$$Q_{0:k} = \left[\overleftarrow{Q}_{0:k} \,|\, \overrightarrow{Q}_{0:k}\right] \text{ with } \begin{cases} \overleftarrow{Q}_{0:k} \in \mathbb{R}^{n\times(k+1)} \\ \overrightarrow{Q}_{0:k} \in \mathbb{R}^{n\times(n-k-1)}. \end{cases}$$

This means

$$A\left[\overleftarrow{Q}_{0:k} \,|\, \overrightarrow{Q}_{0:k}\right] = \left[\overleftarrow{Q}_{0:k} \,|\, \overrightarrow{Q}_{0:k}\right] \left[\begin{array}{c|c} R_{k+1} & \times \\ \hline \times & A_{k+1} \end{array}\right].$$

The eigenvalues of R_{k+1} are called the Ritz values of A with respect to the subspace spanned by the columns of $\overleftarrow{Q}_{0:k}$ (see, e.g., [58]).

Suppose we have now the Krylov subspace of order k with initial vector $\mathbf{v}$:

$$\mathcal{K}_k(A,\mathbf{v}) = \langle \mathbf{v}, A\mathbf{v}, \ldots, A^{k-1}\mathbf{v}\rangle,$$

where $\langle \mathbf{x},\mathbf{y},\mathbf{z}\rangle$ denotes the vector space spanned by the vectors $\mathbf{x},\mathbf{y}$ and $\mathbf{z}$. For simplicity we assume in Subsections 3.1.3 and 3.1.4 that the Krylov subspaces we are working with are not invariant, i.e., that for every k: $\mathcal{K}_k(A,\mathbf{v}) \neq \mathcal{K}_{k+1}(A,\mathbf{v})$, where $k = 1,2,\ldots,n-1$. The special case of invariant subspaces is dealt with in Subsection 3.1.5.

If the columns of the matrix $\overleftarrow{Q}_{0:k}$ form an orthonormal basis of the Krylov subspace $\mathcal{K}_{k+1}(A,\mathbf{v})$, then we say that the eigenvalues of R_{k+1} are called the Arnoldi-Ritz values of A with respect to the initial vector $\mathbf{v}$. If the matrix A is symmetric, one often calls the Ritz values the Lanczos-Ritz values.

3.1.2　The reduction to semiseparable form

We will briefly formulate a theorem valid for the reductions to either tridiagonal, semiseparable and semiseparable plus diagonal form.

Theorem 3.1. *Let A be a symmetric matrix and U is the orthogonal matrix (from Theorems 2.9,2.6 or 2.5) such that*

$$U^T A U = B,$$

where B is either a tridiagonal, semiseparable or semiseparable plus diagonal matrix. If we consider the reduction algorithms like in the proofs of Theorems 2.9,2.6 or 2.5, the intermediate matrices at step m of the reduction have as eigenvalues of the lower right $m\times m$ block the Lanczos-Ritz values with regard to $U_0\mathbf{e}_n$, where U_0 is an initial transformation applied on A, before starting the reduction procedure, as indicated in Theorem 2.6.

Proof. We will not prove this statement for the tridiagonal matrices, as this is a well-known, and classical results (see, e.g., [94]). Taking a closer look at the algorithms from the proofs of Theorems 2.9 and 2.6, we see that these latter two orthogonal similarity transformations perform always a chasing step after performing the Householder transformation. This chasing step is applied on the lower right $k \times k$ block. In fact an orthogonal similarity transformation is performed on the lower right block, and hence it does not change the eigenvalues of this block. The eigenvalues of this block are in fact essentially the same eigenvalues as the ones coming from the reduction to tridiagonal form, as all three reduction methods use exactly the same (up to the sign) Householder transformations.   □

Even though the proof above was very easily based on the knowledge that the reduction to tridiagonal form had the Lanczos-Ritz values, we are not yet completely satisfied. We based ourselves on the properties of the reduction to tridiagonal form. In the following sections we will not use this prior knowledge anymore and derive a general formulation for all kinds of orthogonal similarity transformations.

3.1.3   Necessary conditions to obtain the Ritz values

In this section, we investigate the properties of orthogonal similarity transformations, where the eigenvalues in the already reduced block of the matrix are the Arnoldi-Ritz values, with respect to the starting vector $\mathbf{v}$, where $\mathbf{v}/\|\mathbf{v}\| = \pm Q^{(0)}\mathbf{e}_1$. This makes clear that the initial transformation can change the convergence behavior, as it changes the Krylov subspace and hence also the Ritz values. We remark once more that this initial transformation does not change the reduction algorithm as the actual algorithm reduces the matrix $A^{(1)} = Q^{(0)^T}AQ^{(0)}$ to the desired form. However, in practice a good choice of the vector $\mathbf{v}$, can have important consequences for the convergence behavior in applications.

Suppose that our orthogonal similarity reduction of the matrix into another matrix has the following form after step $k-1$ (with $k = 1, 2, \ldots, n-1$):

$$\begin{bmatrix} R_k & \times \\ \times & \times \end{bmatrix} = Q_{0:k-1}^T A Q_{0:k-1}.$$

This means that we start with this matrix at step k of the reduction: with R_k a square matrix of dimension k, which has as eigenvalues the Arnoldi-Ritz values. Hence, we have the following properties for the orthogonal matrix $Q_{0:k-1}$:

1. The columns of $\overleftarrow{Q}_{0:k-1}$ form an orthogonal basis for $\mathcal{K}_k(A, \mathbf{v})$.

2. The columns of $\overrightarrow{Q}_{0:k-1}$ form an orthogonal basis for the orthogonal complement of $\mathcal{K}_k(A, \mathbf{v})$.

As already mentioned before, for simplicity reasons we assume here that we work with one noninvariant Krylov subspace $\mathcal{K}_k(A, \mathbf{v})$. The more general case is dealt with in Subsection 3.1.5.

After the next step in the transformation we have that the block R_{k+1} has the Ritz values as eigenvalues with respect to $\mathcal{K}_{k+1}(A, \mathbf{v})$. This results in two easy conditions, similar to the ones described above. After step k, in the beginning of step $k+1$ we have:

1. The columns of $\overleftarrow{Q}_{0:k}$ form an orthogonal basis for $\mathcal{K}_{k+1}(A, \mathbf{v}) = \mathcal{K}_k(A, \mathbf{v}) + \langle A^k \mathbf{v} \rangle$.

2. The columns of $\overrightarrow{Q}_{0:k}$ form an orthogonal basis for the orthogonal complement of $\mathcal{K}_{k+1}(A, \mathbf{v})$.

We have the following equalities:

$$A = Q_{0:k-1} A^{(k)} Q_{0:k-1}^T$$
$$= Q_{0:k} A^{(k+1)} Q_{0:k}^T.$$

This means that the transformation to go from matrix $A^{(k)}$ to matrix $A^{(k+1)}$ can also be written in the following form:

$$Q_{0:k}^T Q_{0:k-1} A^{(k)} Q_{0:k-1}^T Q_{0:k} = A^{(k+1)}.$$

Using the fact that $Q^{(k)}$ denotes the orthogonal matrix to go from matrix $A^{(k)}$ to matrix $A^{(k+1)}$, we get:

$$Q^{(k)^T} = Q_{0:k}^T Q_{0:k-1}$$
$$= \begin{bmatrix} \overleftarrow{Q}_{0:k}^T \\ \overrightarrow{Q}_{0:k}^T \end{bmatrix} \left[\overleftarrow{Q}_{0:k-1} \middle| \overrightarrow{Q}_{0:k-1} \right]$$
$$= \begin{bmatrix} \left(Q^{(k)}\right)_{11}^T & \left(Q^{(k)}\right)_{12}^T \\ \left(Q^{(k)}\right)_{21}^T & \left(Q^{(k)}\right)_{22}^T \end{bmatrix},$$

where the $\left(Q^{(k)}\right)_{11}^T, \left(Q^{(k)}\right)_{12}^T, \left(Q^{(k)}\right)_{21}^T$ and $\left(Q^{(k)}\right)_{22}^T$ denote a partitioning of the matrix $Q^{(k)^T}$. These blocks have the following dimensions: $\left(Q^{(k)}\right)_{11}^T \in \mathbb{R}^{(k+1) \times k}$, $\left(Q^{(k)}\right)_{12}^T \in \mathbb{R}^{(k+1) \times (n-k)}$, $\left(Q^{(k)}\right)_{21}^T \in \mathbb{R}^{(n-k-1) \times k}$ and $\left(Q^{(k)}\right)_{22}^T \in \mathbb{R}^{(n-k-1) \times (n-k)}$. It can be seen rather easily, by combining the properties of the matrices $Q_{0:k-1}$ and $Q_{0:k}$ from above, that the block $\left(Q^{(k)}\right)_{21}^T$ has to be zero. *This zero block in the matrix $Q^{(k)}$ is the first necessary condition.*

To obtain a second condition, we will investigate the structure of an intermediate matrix $\tilde{A}^{(k)}$ satisfying

$$\tilde{A}^{(k)} = Q^{(k)^T} A^{(k)}$$
$$= Q^{(k)^T} Q_{0:k-1}^T A \, Q_{0:k-1}$$
$$= Q_{0:k}^T A \, Q_{0:k-1},$$

which can be rewritten as

$$Q_{0:k} \tilde{A}^{(k)} = A Q_{0:k-1}. \tag{3.1}$$

Rewriting Equation (3.1) gives us:

$$A \left[\overleftarrow{Q}_{0:k-1} | \overrightarrow{Q}_{0:k-1} \right] = \left[\overleftarrow{Q}_{0:k} | \overrightarrow{Q}_{0:k} \right] \tilde{A}^{(k)}.$$

Because the columns of $A \overleftarrow{Q}_{0:k-1}$ belong to the Krylov subspace: $\mathcal{K}_{k+1}(A, \mathbf{v})$, which is spanned by the columns of $\overleftarrow{Q}_{0:k}$, $\tilde{A}^{(k)}$ has a zero block of dimension $(n-k-1) \times k$ in the lower left corner. *This zero block provides us a second necessary condition.*

The two conditions presented here, namely the condition on $\tilde{A}^{(k)}$ and the condition on $Q^{(k)}$, are necessary to have the desired convergence properties in the reduction. In the next section we will prove that they are also sufficient. We will formulate this as a theorem:

Theorem 3.2. *Suppose we apply an orthogonal similarity transformation on the matrix A (as described in Subsection 3.1.1), such that the already reduced part R_k in the matrix has the Arnoldi-Ritz values in each step of the reduction algorithm. Then we have the following two properties for every $1 \leq k \leq n-1$:*

- *The matrix $Q^{(k)^T}$, which is the orthogonal matrix to transform $A^{(k)}$ into the matrix $A^{(k+1)} = Q^{(k)^T} A^{(k)} Q^{(k)}$, has a zero block of dimension $(n-k-1) \times k$ in the lower left corner.*

- *The matrix $\tilde{A}^{(k)} = Q^{(k)^T} A^{(k)}$ has a zero block of dimension $(n-k-1) \times k$ in the lower left corner.*

3.1.4 Sufficient conditions to obtain the Ritz values

We prove that the properties from Theorem 3.2 connected to the matrices $Q^{(k)}$ and $\tilde{A}^{(k)}$ are sufficient to have the Arnoldi-Ritz values as eigenvalues in the blocks R_k.

Theorem 3.3. *Suppose, we apply an orthogonal similarity transformation on the matrix A (as described in Subsection 3.1.1), such that we have for $A^{(0)} = A$:*

$$Q^{(0)} \mathbf{e}_1 = \pm \frac{\mathbf{v}}{\|\mathbf{v}\|} \quad and \quad Q^{(0)^T} A^{(0)} Q^{(0)} = A^{(1)}.$$

Assume that the corresponding Krylov subspace $\mathcal{K}_k(A, \mathbf{v})$ will not become invariant for $k \leq n-1$. Suppose that for every step $1 \leq k \leq n-1$ of the reduction algorithm we have the following two properties:

- *The matrix $Q^{(k)^T}$, which is the orthogonal matrix to transform $A^{(k)}$ into the matrix $A^{(k+1)} = Q^{(k)^T} A^{(k)} Q^{(k)}$, has a zero block of dimension $(n-k-1) \times k$ in the lower left corner;*

- *The matrix $\tilde{A}^{(k)} = Q^{(k)^T} A^{(k)}$ has a zero block of dimension $(n-k-1) \times k$ in the lower left corner.*

Then we have that for the matrix $A^{(k+1)}$ partitioned as

$$A^{(k+1)} = \left[\begin{array}{c|c} R_{k+1} & \times \\ \hline \times & A_{k+1} \end{array} \right],$$

the matrix R_{k+1} of dimension $(k+1) \times (k+1)$ has the Ritz values with respect to the Krylov space $\mathcal{K}_{k+1}(A, \mathbf{v})$.

Proof. We will prove the theorem by induction on k.

- **Step 1.** The theorem is true for $k = 1$ because $Q^{(0)^T} A Q^{(0)}$ contains clearly the Arnoldi-Ritz value in the upper left 1×1 block.

- **Step k.** Suppose the theorem is true for $A^{(1)}, A^{(2)}, \ldots, A^{(k)}$, with $k \leq n - 1$. This means that the columns of $\overleftarrow{Q}_{0:k-1}$ span the Krylov subspace $\mathcal{K}_k(A, \mathbf{v})$. Then we will prove now that the conditions are sufficient to have that the columns of $\overleftarrow{Q}_{0:k}$ span the Krylov subspace of $\mathcal{K}_{k+1}(A, \mathbf{v})$. We have the following equalities

$$\begin{aligned} \tilde{A}^{(k)} &= Q^{(k)^T} A^{(k)} \\ &= Q^{(k)^T} Q^T_{0:k-1} A \, Q_{0:k-1} \\ &= Q^T_{0:k} A \, Q_{0:k-1}. \end{aligned}$$

Therefore,

$$A \, Q_{0:k-1} = Q_{0:k} \, \tilde{A}^{(k)}$$
$$A \left[\overleftarrow{Q}_{0:k-1} | \overrightarrow{Q}_{0:k-1} \right] = \left[\overleftarrow{Q}_{0:k} | \overrightarrow{Q}_{0:k} \right] \tilde{A}^{(k)}.$$

Hence, we have already that the columns of $A\overleftarrow{Q}_{0:k-1}$ are part of the space spanned by the columns of $\overleftarrow{Q}_{0:k}$. Note that the columns of $A\overleftarrow{Q}_{0:k-1}$ span the same space as $A\mathcal{K}_k(A, \mathbf{v})$. We have the following relation:

$$A\mathcal{K}_k(A, \mathbf{v}) \subseteq \text{Range}(\overleftarrow{Q}_{0:k}). \tag{3.2}$$

With $\text{Range}(A)$ we denote the vector space spanned by the columns of the matrix A. Since $Q^{(k)^T}$ has a zero block in the lower left position, we have that:

$$Q_{0:k} = Q_{0:k-1} Q^{(k)}$$
$$Q_{0:k} Q^{(k)^T} = Q_{0:k-1}.$$

Hence,

$$\left[\ \overleftarrow{Q}_{0:k} | \overrightarrow{Q}_{0:k} \ \right] Q^{(k)^T} = \left[\ \overleftarrow{Q}_{0:k-1} | \overrightarrow{Q}_{0:k-1} \ \right].$$

Using the zero structure of the matrix $Q^{(k)^T}$ we have

$$\text{Range}(\overleftarrow{Q}_{0:k-1}) = \mathcal{K}_k(A, \mathbf{v}) \subseteq \text{Range}(\overleftarrow{Q}_{0:k}).$$

When we combine this with Equation (3.2) and the fact that our subspace $\mathcal{K}_k(A, \mathbf{v})$ is not invariant, i.e., $\mathcal{K}_{k+1}(A, \mathbf{v}) \neq \mathcal{K}_k(A, \mathbf{v})$, we get

$$\mathrm{Range}(\overleftarrow{Q}_{0:k}) = \mathcal{K}_{k+1}(A, \mathbf{v}).$$

This proves the theorem for $A^{(k+1)}$.

Hence the theorem is proved.   □

3.1.5   The case of invariant subspaces

The theorems and proofs of the previous sections were based on the fact that the Krylov subspace $\mathcal{K}_k(A, \mathbf{v})$ was never invariant. In the case of an invariant subspace, we can apply deflation and continue working on the deflated part, or we can derive similar theorems as in the previous sections but for combined Krylov subspaces. We will investigate these two possibilities in deeper detail in this section.

Some notation

We will introduce some extra notation to be able to work in an efficient way with these combined Krylov subspaces. Let us denote with $K_k(A)$ the following subspace:

$$K_k(A) = \mathcal{K}_{l_1}(A, \mathbf{v}_1) \oplus \mathcal{K}_{l_2}(A, \mathbf{v}_2) \oplus \ldots \oplus \mathcal{K}_{l_{t-1}}(A, \mathbf{v}_{t-1}) \oplus \mathcal{K}_{k_t}(A, \mathbf{v}_t).$$

The $\oplus$ denotes the direct sum of vector spaces. This means that $K_k(A)$ is the union of t different Krylov subspaces. We make the following assumptions.

- The dimension of $K_k(A)$ equals k, and $k = l_1 + l_2 + l_3 + \ldots + l_{t-1} + k_t$.

- With l_i we denote the maximum dimension the Krylov subspace with matrix A and initial vector $\mathbf{v}_i$ can reach before becoming invariant. This means that the Krylov subspaces $\mathcal{K}_p(A, \mathbf{v}_i)$ with $p \geq l_i$ are all equal to $\mathcal{K}_{l_i}(A, \mathbf{v}_i) \neq \mathcal{K}_{l_i-1}(A, \mathbf{v}_i)$. The k_i is an index $1 \leq k_i \leq l_i$ for the Krylov subspace with matrix A and initial vector $\mathbf{v}_i$:

$$\mathcal{K}_{k_i}(A, \mathbf{v}_i) = \langle \mathbf{v}_i, A\mathbf{v}_i, \ldots, A^{k_i-1}\mathbf{v}_i \rangle$$

- The starting vectors $\mathbf{v}_i$ of the different Krylov subspaces are chosen in such a way that they are not part of the previous Krylov subspaces. This means

$$\mathbf{v}_j \notin \bigoplus_{i=1}^{j-1} \mathcal{K}_{l_i}(A, \mathbf{v}_i).$$

In the following two subsections we will note what changes in the theoretical derivations for obtaining the necessary and sufficient conditions. In a final subsection we will investigate in more detail the case of deflation.

The necessary conditions

For the derivation of the necessary conditions we assumed, for simplicity reasons that we were working with only one Krylov subspace. In the case of invariant subspaces, however, we have to work with a union of these Krylov subspaces. This means that our orthogonal transformations $Q_{0:k-1}$ satisfy the following conditions:

1. The columns of $\overleftarrow{Q}_{0:k-1}$ form an orthogonal basis for $K_k(A) = \mathcal{K}_{l_1}(A, \mathbf{v}_1) \oplus \ldots \oplus \mathcal{K}_{k_t}(A, \mathbf{v}_t)$.

2. The columns of $\overrightarrow{Q}_{0:k-1}$ form an orthogonal basis for the orthogonal complement of $K_k(A)$.

In fact this is the only thing that changes in these derivations, all the remaining statements stay valid. The conditions put on the orthogonal matrices $Q^{(k)}$ and on the matrices $\tilde{A}^{(k)}$ remain valid.

However in the case of invariant subspaces we can derive also the following property. The occurrence of invariant subspaces creates zero blocks below the diagonal in the matrices $A^{(k)}$. Suppose that for a certain k the space $K_k(A, \mathbf{v})$ becomes invariant. Due to the invariance we have $AK_k(A) \subset K_k(A)$. As the matrix $\overleftarrow{Q}_{0:k-1}$ forms an orthogonal basis for the space $K_k(A)$, we get the following equations:

$$Q_{0:k-1}^T A Q_{0:k-1} = A^{(k)},$$

which can be rewritten as

$$AQ_{0:k-1} = Q_{0:k-1}A^{(k)}$$
$$= Q_{0:k-1}\left[\begin{array}{c|c} R_k & \times \\ \hline 0 & A_k \end{array}\right].$$

In case of an invariant subspace the matrix $A^{(k)}$ has a zero block of dimension $(n-k) \times k$ in the lower left position. In fact one can apply deflation now and continue working with the lower right block A_k, e.g., for finding eigenvalues. If one applies deflation after every invariant subspace, one does in fact not work with the whole space $K_k(A)$, but on separate (invariant) subspaces. Indeed, after the first invariant subspace $K_{l_1}(A) = \mathcal{K}_{l_1}(A, \mathbf{v}_1)$ one applies deflation and one starts iterating this procedure on a new matrix A_{l_1}.

Sufficient conditions

In this subsection, we will take a closer look at the proof of Theorem 3.3 and investigate the changes in case of an invariant Krylov subspace.

Suppose at step k of the reduction algorithm we encounter an invariant subspace $K_k(A)$, i.e., $K_k(A) = \mathcal{K}_{l_1}(A, \mathbf{v}_1) \oplus \ldots \oplus \mathcal{K}_{l_t}(A, \mathbf{v}_t)$. Only the last lines of the proof of Theorem 3.3 do not hold anymore. The following equation however remains valid:

$$K_k(A) \subset \text{Range}(\overleftarrow{Q}_{0:k}).$$

This leads to the following equation:

$$\left[\overleftarrow{Q}_{0:k} | \overrightarrow{Q}_{0:k}\right] = \left[\overleftarrow{Q}_{0:k-1} | \mathbf{v}_{t+1} | \overrightarrow{Q}_{0:k}\right] \left[\begin{array}{c|c} W & 0 \\ \hline 0 & I \end{array}\right].$$

With W a matrix of dimension $(k+1) \times (k+1)$. Therefore

$$\text{Range}(\overleftarrow{Q}_{0:k}) = K_k(A) \oplus \langle \mathbf{v}_{t+1} \rangle,$$

with

$$\mathbf{v}_{t+1} \notin \text{Range}(\overleftarrow{Q}_{0:k-1}) = \bigoplus_{i=1}^{t} \mathcal{K}_{l_i}(A, \mathbf{v}_i).$$

So, defining $K_{k+1}(A) = \mathcal{K}_{l_1}(A, \mathbf{v}_1) \oplus \ldots \oplus \mathcal{K}_{l_t}(A, \mathbf{v}_t) \oplus \mathcal{K}_1(A, \mathbf{v}_{t+1})$ proves the theorem.

One might wonder whether it is possible to choose any vector $\mathbf{v}$ not in $K_k(A)$ for defining the new space $K_{k+1}(A)$ because our matrix $Q^{(k)^T}$ still has to satisfy some conditions. Let us define the vector $\mathbf{w}$ as the orthogonal projection of the vector $\mathbf{v}$ onto $\overrightarrow{Q}_{0:k-1}$. Then we have that

$$K_k(A) \oplus \langle \mathbf{v} \rangle = K_k(A) \oplus \langle \mathbf{w} \rangle, \tag{3.3}$$

with $\mathbf{w}$ orthogonal to $\overleftarrow{Q}_{0:k-1}$. We have the relation:

$$\left[\overleftarrow{Q}_{0:k} | \overrightarrow{Q}_{0:k}\right] Q^{(k)^T} = \left[\overleftarrow{Q}_{0:k-1} | \overrightarrow{Q}_{0:k-1}\right],$$

so if we choose now $\overleftarrow{Q}_{0:k} = [\overleftarrow{Q}_{0:k-1} | \mathbf{w}]$, we see that the conditions put on $Q^{(k)}$ and $\tilde{A}^{(k)}$ still are satisfied and moreover we have chosen $\mathbf{v}$ to be an arbitrary vector not in $K_k(A)$ because the matrix $\overleftarrow{Q}_{0:k}$ spans the space $K_{k+1}(A) = K_k(A) \oplus \langle \mathbf{v} \rangle$, due to Equation (3.3).

The case of deflation

Finally we will take a closer look at the case of deflation. Suppose now that for a certain k we get an invariant space $K_k(A) = \mathcal{K}_{l_1}(A, \mathbf{v}_1) \oplus \ldots \oplus \mathcal{K}_{l_t}(A, \mathbf{v}_t)$. Suppose we apply deflation, and we can continue iterating the procedure on the matrix A_k. Assume we start on this matrix with vector $\mathbf{w}$ by applying an initial transformation W_2. This means that $W_2 \mathbf{e}_1 = \mathbf{w}$. In fact this corresponds to applying the similarity transformation (with W_1 an arbitrary orthogonal matrix):

$$Q^{(k)} = \left[\begin{array}{c|c} W_1 & 0 \\ \hline 0 & W_2 \end{array}\right],$$

which satisfies the desired condition (the lower left $(n-k-1) \times k$ block is zero), on the matrix $A^{(k)}$. Looking closer at the matrix $Q_{0:k}$ we get the following relations:

$$\begin{aligned} Q_{0:k} &= Q_{0:k-1} Q^{(k)} \\ &= \left[\overleftarrow{Q}_{0:k-1} | \overrightarrow{Q}_{0:k-1}\right] Q^{(k)}. \end{aligned}$$

This means that

$$\overleftarrow{Q}_{0:k} = \left[\overleftarrow{Q}_{0:k-1} W_1 | \mathbf{v}_{t+1} \right].$$

Clearly the vector $\mathbf{v}_{t+1} = \overrightarrow{Q}_{0:k-1}\mathbf{w}$ does not belong to the space $K_k(A)$; moreover the vector is perpendicular to $K_k(A)$. This last deduction clearly shows the relation between the different Krylov spaces when applying deflation and the continuation of the reduction process on the complete matrix.

3.1.6   Some general remarks

When we take a closer look at the matrix equation:

$$\begin{aligned} Q^{(k)T} &= Q_{0:k}^T Q_{0:k-1} \\ &= \left[\begin{array}{c} \overleftarrow{Q}_{0:k}^T \\ \overrightarrow{Q}_{0:k}^T \end{array} \right] \left[\overleftarrow{Q}_{0:k-1} | \overrightarrow{Q}_{0:k-1} \right], \end{aligned}$$

we can see that the matrix $Q^{(k)T}$ has the upper right $(k+1) \times (n-k)$ block of rank less than or equal to 1. The upper right $(k+1) \times (n-k)$ block corresponds to the product $\overleftarrow{Q}_{0:k}^T \overrightarrow{Q}_{0:k-1}$. The columns of the matrix $\overleftarrow{Q}_{0:k}$ span the subspace $K_{k+1}(A) = K_k(A) + \langle A^{k_t}\mathbf{v}_t \rangle$ (assuming that $K_k(A)$ is not invariant) and the columns of $\overrightarrow{Q}_{0:k-1}$ span the space orthogonal to $K_k(A)$, which leads directly to the fact that the product $\overleftarrow{Q}_{0:k}^T \overrightarrow{Q}_{0:k-1}$, has rank less than or equal to 1. The invariant case can be dealt with in a similar way.

The reader can easily verify that the similarity reductions of a symmetric matrix into a similar tridiagonal or a semiseparable one, and the similarity reduction of a matrix into a similar Hessenberg or a matrix having the lower triangular part of semiseparable form [154], perfectly fit in this scheme. Moreover one can derive that the vector $\mathbf{v}$ equals $\mathbf{e}_1$, if of course the initial transformation $Q^{(0)}$ equals the identity matrix as we silently assumed in the previous chapter, when presenting the different reduction algorithms.

3.1.7   The different reduction algorithms revisited

In this subsection we will briefly discuss the different reduction algorithms presented in the previous chapter. We will first discuss the orthogonal similarity transformations, and secondly we will take a closer look at the transformations to upper triangular semiseparable and bidiagonal form.

The orthogonal similarity transformations

We remark once more that the different reduction algorithms as presented in the previous chapter, start reducing the matrices from the bottom-right up to the upper-left corner. The orthogonal transformations as discussed in this chapter deal with reduction algorithms starting at the top-left corner. There is no loss in generality, because all the transformation algorithms as presented in the first chapter can easily be translated to start working on the upper-left part (see Theorem 2.8).[19]

This gives us

Theorem 3.4. *The orthogonal similarity transformations, as proposed in the previous chapter to the*

- *tridiagonal*

- *semiseparable*

- *semiseparable plus diagonal*

- *Hessenberg*

- *Hessenberg-like*

form, have the Ritz values as eigenvalues in the already reduced part of the matrix.

Proof. The proof is straightforward by looking at the structure of the orthogonal similarity transformations involved. If one wants to rewrite Theorem 3.3, for matrices starting to reduce at the bottom-right position, one has to consider that the initial starting vector for the Krylov subspace changes from $Q^{(0)}\mathbf{e}_1$ into $Q^{(0)}\mathbf{e}_n$. $\qquad\square$

The orthogonal transformations

In this section we will discuss the orthogonal transformations to either bidiagonal or upper triangular semiseparable form.

Let us formulate the following theorem:

Theorem 3.5. *The orthogonal transformations of a matrix A, as proposed in the previous chapter, to either*

- *upper(lower) bidiagonal;*

- *upper(lower) triangular semiseparable;*

form, have as singular values in the already reduced part of the matrix the square roots of the Ritz values originating from the reduction of the matrix AA^T $(A^T A)$ to respectively

- *tridiagonal;*

- *semiseparable;*

---

[19]If one translates the previous deductions to matrices starting the reduction at the bottom right, the Krylov subspace changes and has starting vector $Q^{(0)}\mathbf{e}_n$ instead of $Q^{(0)}\mathbf{e}_1$.

form.

Proof. We only formulate the proof for the reduction to upper triangular semiseparable form.

We know from the reduction algorithm as presented in Theorem 2.17 that the intermediate matrices $A^{(k)}$ have the upper $k \times n$ matrix $S^{(k)}$ of upper triangular semiseparable form. Using the relations provided in Theorem 2.20, we know that the singular values of this matrix $S^{(k)}$ are the square roots of the eigenvalues of the matrix $\tilde{S}^{(k)} = S^{(k)} S^{(k)^T}$, and the eigenvalues of the matrix $\tilde{S}^{(k)}$ as can be seen when combining Theorem 2.20 and the results from Subsection 3.1.7 are the Lanczos-Ritz values. $\qquad\square$

This concludes the results of this section. We know now which eigenvalues are contained in the already reduced block using any of the orthogonal transformations discussed in the previous chapter.

Notes and references

The main results in this section concerning the necessary and sufficient conditions for obtaining the Lanczos-Ritz values, were published in the following article.

> ☞ R. Vandebril and M. Van Barel. Necessary and sufficient conditions for orthogonal similarity transformations to obtain the Arnoldi(Lanczos)-Ritz values. *Linear Algebra and its Applications*, 414:435–444, 2006.

The following books and the references therein contain a lot of material related to Lanczos,Arnoldi,...algorithms.

> ☞ J. K. Cullum and R. A. Willoughby. *Lanczos algorithms for large symmetric eigenvalue computations.* Birkhäuser, Boston, Massachusetts, USA, 1985.

> ☞ J. W. Demmel. *Applied Numerical Linear Algebra.* SIAM, Philadelphia, Pennsylvania, USA, 1997.

> ☞ Y. Saad. *Numerical Methods for Large Eigenvalue Problems.* Manchester University Press, Manchester, United Kingdom, 1992.

The main goal of this section was to prove that these submatrices have very close connections with the Lanczos/Arnoldi-Ritz values. More information concerning the convergence behavior of the Ritz values themselves can be found in the following publications by the authors Kuijlaars, Helsen and Van Barel.

> ☞ S. Helsen, A. B. J. Kuijlaars, and M. Van Barel. Convergence of the isometric Arnoldi process. *SIAM Journal on Matrix Analysis and Applications*, 26(3):782–809, 2005.

> ☞ A. B. J. Kuijlaars. Which eigenvalues are found by the Lanczos method? *SIAM Journal on Matrix Analysis and Applications*, 22(1):306–321, 2000.

3.2 Subspace iteration inside the reduction algorithms

In this section we will investigate in detail the reduction to semiseparable (plus diagonal form), with regard to the extra chasing step that is not present in the reduction to the tridiagonal form. We know already that the reduction to semiseparable and the reduction to tridiagonal form can be seen as special cases of the reduction to semiseparable plus diagonal form. In fact we showed that one can interpret the reduction to tridiagonal form as a reduction to a semiseparable plus diagonal matrix with the diagonal equal to ∞. We also know already that all three reduction methods have as eigenvalues in the already reduced part of the matrix the Lanczos-Ritz values. In this section we will prove that the reduction to semiseparable plus diagonal form has an extra convergence behavior, which we can interpret as a nested multishift iteration on the original (untransformed) matrix. Having some information on the spectrum of the matrix, the diagonal can be chosen in order to enforce the convergence to possible clusters present in the spectrum.

This section is organized as follows. First we will deduce the subspace iteration convergence theory for the reduction to semiseparable form, as this is the most easy case. Secondly we will adapt this theoretical consideration towards the semiseparable plus diagonal case, which leads to a nested subspace with multishift interpretation.

3.2.1 The reduction to semiseparable form

In Subsection 3.1.7 we showed that the intermediate semiseparable matrices have as eigenvalues the Lanczos-Ritz values. This behavior is completely the same as the one observed in an orthogonal similarity transformation to reduce a symmetric matrix into a similar tridiagonal one. One might therefore doubt the usefulness of this reduction because it costs $9n^2 + \mathcal{O}(n)$ more and has no advantage over the tridiagonal approach. In this section we will prove that the extra cost of $9n^2 + \mathcal{O}(n)$ operations provides an extra interesting property for this reduction.

At each step of the algorithm introduced in Theorem 2.6, one more row (column) is added by means of orthogonal transformations to the set of the rows (columns) of the matrix already proportional to each other. In this section, using arguments similar to those considered in [177, 178, 179, 182, 191], we show that this algorithm can be interpreted as a kind of nested subspace iteration method [94], where the size of the vector subspace is increased by one and a change of coordinate system is made at each step of the algorithm. As a consequence, depending on the gaps between the eigenvalues, the semiseparable part of the matrix will converge to a block diagonal matrix, and the eigenvalues of these blocks converge to the largest eigenvalues in absolute value of the original symmetric matrix.

Given a matrix A and an initial subspace $\mathsf{S}^{(0)}$, a subspace iteration method [94] can be described as follows:

$$\mathsf{S}^{(k)} = A\mathsf{S}^{(k-1)}, \quad k = 1, 2, 3, \ldots$$

Under weak assumptions on A and $\mathsf{S}^{(0)}$, the $\mathsf{S}^{(k)}$ converge to an invariant subspace.

(More details on these assumptions will be investigated in the next subsection, because the subspace iteration will interact with the Lanczos convergence behavior.) We will see that the reduction algorithm from a symmetric to a semiseparable matrix can be interpreted as a kind of subspace iteration, where the dimension of the subspace grows by one at each step of the algorithm. Let $A^{(1)} = Q^{(0)^T} A Q^{(0)}$. For the reduction algorithm as presented in Subsection 2.2.2 in Chapter 2 we have $Q^{(0)} = I$. Suppose we have only performed the first orthogonal similarity transformations such that the rows (columns) n and $n-1$ are already proportional. (Let us capture all the necessary Givens and Householder transformations to go from matrix $A^{(k)}$ to matrix $A^{(k+1)}$ in one orthogonal matrix $Q^{(k)}$.):

$$A^{(2)} = Q^{(1)^T} A^{(1)} Q^{(1)}, \tag{3.4}$$

where $A^{(2)}$ has the semiseparable structure in the rows (columns) n and $n-1$ and $Q^{(1)} = \left[\mathbf{q}_1^{(1)}, \ldots, \mathbf{q}_n^{(1)} \right]$. We know that $Q^{(1)^T}$ applied to the left of the matrix $A^{(1)}$ annihilates all elements in the last column of the matrix, except for the last one (the element on the diagonal). Combining this knowledge with (3.4) leading to $A^{(2)} Q^{(1)^T} = Q^{(1)^T} A^{(1)}$, we can write

$$A^{(1)} = Q^{(1)} \left(A^{(2)} Q^{(1)^T} \right) \tag{3.5}$$

$$= Q^{(1)} \begin{bmatrix} \times & \cdots & \times & 0 \\ \vdots & \ddots & \vdots & \vdots \\ \times & \cdots & \times & 0 \\ \times & \cdots & \times & \times \end{bmatrix}$$

$$= Q^{(1)} L^{(1)}.$$

Let $\mathbf{e}_1, \ldots, \mathbf{e}_n$ be the standard basis vectors of $\mathbb{R}^n$. From (3.5), because of the structure of $L^{(1)}$, it can clearly be seen that:

$$A^{(1)} \langle \mathbf{e}_n \rangle = \langle \mathbf{q}_n^{(1)} \rangle.$$

This means that the last column of $A^{(1)}$ and $\mathbf{q}_n^{(1)}$ span the same one-dimensional space. In fact one subspace iteration step is performed on the vector $\mathbf{e}_n$. The first step of the algorithm is completed when the following orthogonal transformation is performed:

$$A^{(2)} = Q^{(1)^T} A^{(1)} Q^{(1)}.$$

The latter transformation can be interpreted as a change of coordinate system: $A^{(1)}$ and $A^{(2)}$ represent the same linear transformation with respect to different coordinate systems. Let $\mathbf{y} \in \mathbb{R}^n$. Then $\mathbf{y}$ is represented in the new system by $Q^{(1)^T} \mathbf{y}$. This means that for the vector $\mathbf{q}_n^{(1)}$ we get $Q^{(1)^T} \mathbf{q}_n^{(1)} = \mathbf{e}_n$. Summarizing, this means that one step of subspace iteration on the subspace $\langle \mathbf{e}_n \rangle$ is performed, resulting in a new subspace $\langle \mathbf{q}_n^{(1)} \rangle$, and then, by means of a coordinate transformation, it is transformed back into the subspace $\langle \mathbf{e}_n \rangle$. Therefore, denoting by $\mathbf{z}^{(k)}$ the eigenvector corresponding to the largest eigenvalue in absolute value λ of $A^{(k)}$, $k = 1, 2, \ldots$, we can say that, if $\mathbf{z}^{(k)}$ has a nonzero last component and if the largest eigenvalue is unique, the sequence $\{\mathbf{z}^{(k)}\}$ converges to $\mathbf{e}_n$, and, consequently, the lower right element of $A^{(k)}$ converges to λ. Note that the last assumption of the nonzero component will play an important role in Section 3.3, where the interaction between the Lanczos behavior and the subspace iteration is investigated.

The second step can be interpreted in a completely analogous way. Suppose we have already the semiseparable structure in the last two rows (columns). Then we perform the following similarity transformation on $A^{(2)}$,

$$A^{(3)} = Q^{(2)^T} A^{(2)} Q^{(2)}, \qquad (3.6)$$

in order to make the rows (columns) n up to $n-2$ dependent. Using (3.6), $A^{(2)}$ can be written as follows:

$$A^{(2)} = Q^{(2)} \begin{bmatrix} \times & \cdots & \times & 0 & 0 \\ \vdots & \ddots & \vdots & \vdots & \vdots \\ \times & \cdots & \times & 0 & 0 \\ \times & \cdots & \times & \times & 0 \\ \times & \cdots & \times & \times & \times \end{bmatrix}$$

$$= Q^{(2)} L^{(2)}.$$

Considering the subspace $\langle \mathbf{e}_{n-1}, \mathbf{e}_n \rangle$ and using the same notation as above, we have

$$A^{(2)} \langle \mathbf{e}_{n-1}, \mathbf{e}_n \rangle = \langle \mathbf{q}_{n-1}^{(2)}, \mathbf{q}_n^{(2)} \rangle.$$

This means that the second step of the algorithm is a step of subspace iteration performed on a larger subspace. For every new dependency that is created in the symmetric matrix $A^{(k)}$, the dimension of the subspace is increased by one.

One can also see the algorithm as performing in each iteration step a QL-step without shift on the semiseparable bottom-right submatrix. Let us partition the matrix $A_0^{(k)}$ (as in the proof of Theorem 2.6) as follows:

$$A_0^{(k)} = A^{(k)} = \left[\begin{array}{c|c} A_k & \mathbf{u}_k \mathbf{v}_k^T \\ \hline \mathbf{v}_k \mathbf{u}_k^T & S_k \end{array} \right],$$

with $A_k \in \mathbb{R}^{(n-k)\times(n-k)}$, $\mathbf{u}_k \in \mathbb{R}^{(n-k)\times 1}$, $\mathbf{v}_k \in \mathbb{R}^{k\times 1}$ and $S_k \in \mathbb{R}^{k\times k}$. From the semiseparable structure, we know that $[\mathbf{v}_k, S_k \mathbf{e}_1]$ has rank one. In the first part of the k-th iteration step, an $(n-k) \times (n-k)$ orthogonal matrix $Q_1^{(k)}$ is constructed such that

$$Q_1^{(k)^T} \mathbf{u}_k = [0, 0, \ldots, 0, \eta]^T,$$

with $\eta = \pm \|\mathbf{u}_k\|$. Hence, our matrix $A^{(k)}$ is transformed into the following orthogonally similar matrix

$$
\left[\begin{array}{c|c} Q_1^{(k)^T} & 0 \\ \hline 0 & I_k \end{array}\right] \left[\begin{array}{c|c} A_k & \mathbf{u}_k\mathbf{v}_k^T \\ \hline \mathbf{v}_k\mathbf{u}_k^T & S_k \end{array}\right] \left[\begin{array}{c|c} Q_1^{(k)} & 0 \\ \hline 0 & I_k \end{array}\right] \quad (3.7)
$$
$$
= \left[\begin{array}{c|c} \tilde{A}_k & \eta\mathbf{e}_{n-k}\mathbf{v}_k^T \\ \hline \eta\mathbf{v}_k\mathbf{e}_{n-k}^T & S_k \end{array}\right].
$$

Let us partition the matrix in a slightly different way, adding one more column and row to the semiseparable part. Let

$$
\tilde{A}_k = \left[\begin{array}{cc} A_{k+1} & \mathbf{a} \\ \mathbf{a}^T & \alpha \end{array}\right],
$$

then we can define the semiseparable matrix $\tilde{S}_k$ as follows

$$
\tilde{S}_{k+1} = \left[\begin{array}{cc} \alpha & \eta\mathbf{v}_k^T \\ \eta\mathbf{v}_k & S_k \end{array}\right].
$$

In the second part of the k-th iteration step, we perform a QL-step on the semi-separable matrix $\tilde{S}_{k+1}$ of order $k+1$, i.e., we construct an orthogonal matrix $Q_2^{(k)}$ such that

$$
\tilde{S}_{k+1} = Q_2^{(k)}L^{(k)},
$$

with $L^{(k)}$ lower triangular. Hence, the matrix of (3.7) is transformed into the orthogonal similar matrix

$$
\left[\begin{array}{c|c} I_{n-k-1} & 0 \\ \hline 0 & Q_2^{(k)^T} \end{array}\right] \left[\begin{array}{c|c} A_{k+1} & \mathbf{a}\mathbf{e}_1^T \\ \hline \mathbf{e}_1\mathbf{a}^T & \tilde{S}_{k+1} \end{array}\right] \left[\begin{array}{c|c} I_{n-k-1} & 0 \\ \hline 0 & Q_2^{(k)} \end{array}\right]
$$
$$
= \left[\begin{array}{c|c} A_{k+1} & \mathbf{u}_{k+1}\mathbf{v}_{k+1}^T \\ \hline \mathbf{v}_{k+1}\mathbf{u}_{k+1}^T & S_{k+1} \end{array}\right].
$$

The execution of the two steps is illustrated in Figure 3.1. At the beginning of the iteration the last four rows and columns already have a semiseparable structure. In (1), a similarity transformation (either a Householder matrix or a product of Givens transformations) is considered in order to annihilate the entries indicated by $\otimes$. Due to the semiseparable structure, all the entries in the first four rows (columns) with column (row) indexes greater than 5 are annihilated, too. In steps (2), (3), (4) and (5), the Givens rotations are only applied to the left, transforming the (5×5) principal submatrix in the right-bottom corner to lower triangular form. To retrieve the semiseparable structure in the last five rows and columns, the transpose of the latter Givens rotation must be applied to the right.

This means that from step i, $i = 1, \ldots, n$, all the consecutive steps perform subspace iterations on the subspace of dimension i. From [182], we know that these consecutive iterations on subspaces tend to create lower block triangular matrices.

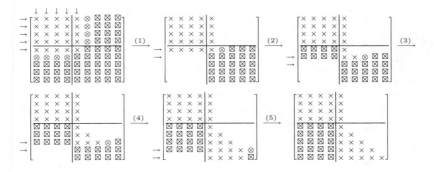

Figure 3.1. *Description of QL-step without shift applied in each iteration of the semiseparable reduction.*

Hence, for a symmetric matrix these are block diagonal. Furthermore, the process works for all the nested subspaces at the same time, and so the semiseparable part of the matrices $A^{(k)}$ generated by the proposed algorithm becomes more and more block diagonal, and the blocks contain eigenvalues of the original matrix. This explains why in the numerical examples (see Chapter 4) the lower right block already gives a good estimate of the largest eigenvalues, since they are connected to a subspace on which the subspace iteration is performed most. In this subsection we discussed the subspace iteration interpretation and the QL-interpretation, in fact they are the same, just looking from a different viewpoint towards the convergence properties. Also in the coming subsection we will give different interpretations to the convergence behavior.

This insight also opens several new perspectives. In many problems, only the few largest eigenvalues (in absolute value) need to be computed [29, 108, 173, 174, 189, 188]. In such cases, the proposed algorithm gives the required information after only a few steps without running the algorithm to completion. Moreover, because the sequence of similar matrices generated at each step of the algorithm converges to a block diagonal matrix, the original problem can be divided into smaller independent subproblems.

We know that at each step of the algorithm a step of a QL-iteration is performed on the semiseparable part of the matrix. This insight opens the perspective to a new research topic. Is it possible to incorporate a shift in this reduction, thereby performing a QL-step with shift on the already semiseparable part? In fact this viewpoint gave rise to the development of the reduction to semiseparable plus diagonal form as it was proposed before.

As the reduction to semiseparable plus diagonal form is the most general one from the reductions to either tridiagonal or semiseparable form, we will reconsider the previously discussed theory and come to subspace iteration interpretation for the reduction to semiseparable plus diagonal.

3.2.2   The reduction to semiseparable plus diagonal form

The reduction to semiseparable and to semiseparable plus diagonal form performs more operations than the corresponding reduction to tridiagonal form. More precisely, at every step m of the reduction algorithm m extra Givens transformations are performed. These extra Givens transformations create the "nested subspace iteration" behavior. In this section we will investigate more in detail what is meant by this behavior. Firstly we will derive an analogue of the convergence theory as described in the former subsection.

The nested subspace iteration, connected to the orthogonal similarity transformation of a matrix to semiseparable form, was investigated in the previous subsection. As proved before, the reduction to semiseparable form is a special case of the reduction to semiseparable plus diagonal form, therefore its convergence behavior is also a special case of the more general convergence behavior. Hence, we will derive the nested subspace iteration, related to the transformation into semiseparable plus diagonal form and afterwards translate this to the reduction to semiseparable and the reduction to tridiagonal form.

Denote by $D^{(m)}$ the diagonal matrix of dimension n, with the lower right m diagonal elements equal to $[d_1, d_2, \ldots, d_m]$.

When looking at the first step of the reduction algorithm, we can state that

$$
\begin{aligned}
Q^{(0)^T} A Q^{(0)} = A^{(1)} &= Q^{(1)}(Q^{(1)^T} A^{(1)}) \\
&= Q^{(1)}(Q^{(1)^T}(D^{(1)} + S^{(1)})) \\
&= Q^{(1)}\left(Q^{(1)^T} \left[\begin{array}{ccc|c} 0 & & & \\ & \ddots & & \\ & & 0 & \\ \hline & & & d_1 \end{array} \right] + \left[\begin{array}{ccc|c} \times & \cdots & \times & 0 \\ \vdots & & \vdots & \vdots \\ \times & \cdots & \times & 0 \\ \hline \times & \cdots & \times & \times \end{array} \right] \right).
\end{aligned}
$$

Multiplying both sides of the former equality to the right by $\langle \mathbf{e}_n \rangle$ leads to

$$
\begin{aligned}
A^{(1)}\langle \mathbf{e}_n \rangle &= d_1 I_n \langle \mathbf{e}_n \rangle + Q^{(1)}\langle \mathbf{e}_n \rangle \\
\Rightarrow (A^{(1)} - d_1 I_n)\langle \mathbf{e}_n \rangle &= Q^{(1)}\langle \mathbf{e}_n \rangle = \langle \mathbf{q}_n^{(1)} \rangle,
\end{aligned}
\tag{3.8}
$$

with $\mathbf{q}_n^{(1)}$ in the last column of $Q^{(1)}$. We assume that the lower-right element in the matrix $Q^{(1)^T} S^{(1)}$ is different from zero, otherwise d_1 would be an eigenvalue and $\mathbf{e}_n$ would be an eigenvector. This brings us to the case of invariant subspaces, which is in fact good news. We will however not go into these details and assume, throughout the remainder of the text, that the subspaces we work with are not invariant. The invariant case naturally splits up in blocks, and the blocks can be dealt with completely similar as in the remainder of the book.

To complete the first step of the algorithm, the following transformation is performed:

$$
A^{(2)} = Q^{(1)^T} A^{(1)} Q^{(1)}.
$$

This can also be interpreted as a transformation of the basis used. The vector $\mathbf{q}_n^{(1)}$ becomes $Q^{(1)^T} \mathbf{q}_n^{(1)} = \mathbf{e}_n$, and hence, because of Equation (3.8), $(A^{(1)} - d_1 I_n)\langle \mathbf{e}_n \rangle$

becomes $\langle \mathbf{e}_n \rangle$. This means that the subspace $\langle \mathbf{e}_n \rangle$ we are working on stays the same, and only the matrix we use changes.

Looking now at the m-th step, $m = 2, \ldots, n-1$, of the reduction algorithm, we get

$$A^{(m)} = Q^{(m)} (Q^{(m)^T} A^{(m)})$$
$$= Q^{(m)} (Q^{(m)^T} (D^{(m)} + S^{(m)}))$$

$$= Q^{(m)} \left(Q^{(m)^T} \left(\left[\begin{array}{ccc|ccc} 0 & & & & & \\ & \ddots & & & 0 & \\ & & 0 & & & \\ \hline & & & d_1 & & \\ & & & & \ddots & \\ & & & & & d_m \end{array} \right] + \left[\begin{array}{ccc|ccc} \times & \cdots & \times & 0 & \cdots & 0 \\ \vdots & & \vdots & \vdots & & \vdots \\ \times & \cdots & \times & 0 & \cdots & 0 \\ \times & \cdots & \times & \times & \cdots & 0 \\ \vdots & & \vdots & \vdots & \ddots & \vdots \\ \times & \cdots & \times & \times & \cdots & \times \end{array} \right] \right) \right).$$

Multiplying both sides of the latter equality to the right by the subspace $\langle \mathbf{e}_{n-(m-1)}, \mathbf{e}_{n-(m-2)}, \ldots, \mathbf{e}_{n-1}, \mathbf{e}_n \rangle$ leads to

$$A^{(m)} \langle \mathbf{e}_{n-(m-1)}, \mathbf{e}_{n-(m-2)}, \ldots, \mathbf{e}_n \rangle =$$
$$D^{(m)} \langle \mathbf{e}_{n-(m-1)}, \mathbf{e}_{n-(m-2)}, \ldots, \mathbf{e}_n \rangle + Q^{(m)} \langle \mathbf{e}_{n-(m-1)}, \mathbf{e}_{n-(m-2)}, \ldots, \mathbf{e}_n \rangle.$$

This implies that

$$(A^{(m)} - D^{(m)}) \langle \mathbf{e}_{n-(m-1)}, \ldots, \mathbf{e}_n \rangle = \langle \mathbf{q}^{(m)}_{n-(m-1)}, \ldots, \mathbf{q}^{(m)}_n \rangle, \qquad (3.9)$$

with $\mathbf{q}^{(m)}_i$ the i-th column of $Q^{(m)}$. The left-hand side can be rewritten as

$$(A^{(m)} - D^{(m)}) \langle \mathbf{e}_{n-(m-1)}, \mathbf{e}_{n-(m-2)}, \ldots, \mathbf{e}_n \rangle$$
$$= \langle (A^{(m)} - D^{(m)}) \mathbf{e}_{n-(m-1)}, \ldots, (A^{(m)} - D^{(m)}) \mathbf{e}_n \rangle$$
$$= \langle (A^{(m)} - d_1 I_n) \mathbf{e}_{n-(m-1)}, \ldots, (A^{(m)} - d_m I_n) \mathbf{e}_n \rangle.$$

Hence, the completion of each m-th step ($m = 2, \ldots, n-1$):

$$A^{(m+1)} = Q^{(m)^T} A^{(m)} Q^{(m)}$$

can also be considered as a change of coordinate system: transform any vector $\mathbf{y}$ of the old system into $Q^{(m)^T} \mathbf{y}$ for the new system. Then $A^{(m)}$ will be transformed into $A^{(m+1)}$ and the subspace of Equation (3.9): $\langle \mathbf{q}^{(m)}_{n-(m-1)}, \mathbf{q}^{(m)}_{n-(m-2)}, \ldots, \mathbf{q}^{(m)}_n \rangle$ will become

$$Q^{(m)^T} \langle \mathbf{q}^{(m)}_{n-(m-1)}, \mathbf{q}^{(m)}_{n-(m-2)}, \ldots, \mathbf{q}^{(m)}_n \rangle = \langle \mathbf{e}_{n-(m-1)}, \mathbf{e}_{n-(m-2)}, \ldots, \mathbf{e}_n \rangle.$$

Therefore, at each step the basis remains the same, but the matrix used changes. It is called a nested subspace iteration because the subspace involved increases in each step of the algorithm.

The subspace iteration involved for the semiseparable reduction is just a regular subspace iteration without the shift (see, e.g., [154]), as the chasing technique is not involved in the tridiagonal case, no subspace iteration is involved for the tridiagonal matrices. In the next subsection, we will investigate in more detail the resulting convergence behavior, subject to this subspace iteration, as the interpretation of these results is not straightforward because we do not have a shift matrix anymore but really a diagonal which is subtracted from our matrix.

3.2.3   Nested multishift iteration

In this subsection we will give some theoretical results that might help us to choose the diagonal in the reduction to semiseparable plus diagonal form in such a way that we can tune the convergence behavior. In the following chapter numerical experiments are given to illustrate these results. The results we will give here are based on the convergence properties of a generic GR-algorithm[20] as derived in [182, 179].

In this section we will rewrite the subspace iteration as presented in the previous section, such that it can be interpreted as a nested multishift iteration.

Related to the diagonal elements d_i, used in the reduction algorithm we define the following monic polynomials $p_i(\lambda) = \lambda - d_i$. The monic polynomial $\hat{p}_i(\lambda)$ of degree i represents a multiplication of all the polynomials $p_i, \ldots, p_1$, i.e.,

$$\hat{p}_i(\lambda) = p_i(\lambda)p_{i-1}(\lambda) \ldots p_1(\lambda)$$
$$= (\lambda - d_i)(\lambda - d_{i-1}) \ldots (\lambda - d_1).$$

Moreover we also need partial combinations of these polynomials. Define the polynomials $\hat{p}_{j:i}(\lambda)$ with $(j \geq i)$ in the following way:

$$\hat{p}_{j:i}(\lambda) = p_j(\lambda)p_{j-1}(\lambda) \ldots p_i(\lambda).$$

Note that $\hat{p}_i(\lambda) = \hat{p}_{i:1}(\lambda)$ and assume $\hat{p}_0 = \hat{p}_{0:0} = 1$.

Let us prove now the following theorem, which rewrites the subspace iteration behavior in terms of the original matrix. This theorem is an extension of the results in the previous subsection (see also Lemma 3.1 in [155]).

Theorem 3.6. *Let us use the notation as defined before. At step $m = 1, 2, \ldots, n-1$ of the algorithm we have for every $n \geq k \geq n - m$, that (denote $\eta = k - n + m$)*

$$Q_{0:m}\langle \mathbf{e}_k, \ldots, \mathbf{e}_n \rangle = \hat{p}_\eta(A)\langle \mathbf{f}_k^{(m)}, \hat{p}_{\eta+1:\eta+1}(A)\mathbf{f}_{k+1}^{(m)}, \ldots$$
$$\ldots, \hat{p}_{j-n+m:\eta+1}(A)\mathbf{f}_j^{(m)}, \ldots, \hat{p}_{m:\eta+1}(A)\mathbf{f}_n^{(m)}\rangle,$$

where the vectors $\mathbf{f}_k^{(m)}$ are defined as follows. For every m, $\mathbf{f}_n^{(m)} = \mathbf{e}_n$. For $n \geq k > n - m$, with $\eta = k - n + m$:

$$\mathbf{f}_k^{(m)} \in \langle \mathbf{f}_k^{(m-1)}, \hat{p}_{\eta:\eta}(A)\mathbf{f}_{k+1}^{(m-1)}, , \ldots, \hat{p}_{m-1:\eta}(A)\mathbf{f}_n^{(m-1)} \rangle,$$

---

[20]A GR-algorithm is based on a factorization $p(A^{(i)}) = GR$, in which $p(\cdot)$ is a polynomial, G is assumed invertible and R upper triangular. A GR-algorithm performs iterates of the form $A^{(i+1)} = G^{-1}A^{(i)}G$ on the sequence of matrices $A^{(i)}$.

and the vector $\mathbf{f}_{n-m}^{(m)}$ equals $Q_{0:m}\mathbf{e}_{n-m}$ and is hence orthogonal to the subspace

$$\hat{p}_1(A)\langle \mathbf{f}_{n-m+1}^{(m)}, \hat{p}_{\eta+1:2}(A)\mathbf{f}_{k+1}^{(m)}, \ldots, \hat{p}_{j-n+m:2}(A)\mathbf{f}_j^{(m)}, \ldots, \hat{p}_{m:2}(A)\mathbf{f}_n^{(m)}\rangle.$$

Proof. We will prove the theorem by induction on m.

- **Step $m = 0$.** We will prove that for every $n \geq k \geq n - 0$ (i.e., $k = n$) the following holds:
$$Q_{0:0}\langle \mathbf{e}_n \rangle = \langle \mathbf{f}_n^{(0)} \rangle.$$

  This is straightforward by choosing $\mathbf{f}_n^{(0)}$ equal to the last column of $Q_{0:0}$. We remark that $Q_{0:0}$ is an initial transformation, which in fact does not explicitly needs to be applied on the matrix A, to reduce it to semiseparable plus diagonal form. In the reduction method we proposed $Q_{0:0} = I$ and hence $\mathbf{f}_n^{(0)} = \mathbf{e}_n$.

- **Step 1.** Before starting the induction procedure on m, we will demonstrate the case $m = 1$. We have to prove two things: for $k = n$,
$$\begin{aligned}Q_{0:1}\langle \mathbf{e}_n \rangle &= (A - d_1 I)\langle \mathbf{f}_n^{(1)} \rangle \\ &= \hat{p}_1(A)\langle \mathbf{f}_n^{(1)} \rangle\end{aligned}$$

  and for $k = n - 1$,
$$\begin{aligned}Q_{0:1}\langle \mathbf{e}_{n-1}, \mathbf{e}_n \rangle &= \langle \mathbf{f}_{n-1}^{(1)}, (A - d_1 I)\mathbf{f}_n^{(1)} \rangle \\ &= \langle \mathbf{f}_{n-1}^{(1)}, \hat{p}_1(A)\mathbf{f}_n^{(1)} \rangle.\end{aligned}$$

We will first prove that the equation holds for $k = n$.

When the transformation $Q^{(1)}$ used at the first step is only applied on the rows, the matrix $A^{(1)} = Q_{0:0}^T A Q_{0:0}$ is transformed into

$$Q^{(1)^T} A^{(1)} = Q^{(1)^T}(D^{(1)} + S^{(1)}) = Q^{(1)^T} D^{(1)} + L^{(1)}$$

$$= Q^{(1)^T} \begin{bmatrix} 0 & & & 0 \\ & \ddots & & \vdots \\ & & 0 & 0 \\ \hline 0 & \cdots & 0 & d_1 \end{bmatrix} + \begin{bmatrix} \times & \cdots & \times & 0 \\ \vdots & & \vdots & \vdots \\ \times & \cdots & \times & 0 \\ \hline \times & \cdots & \times & \times \end{bmatrix}, \quad (3.10)$$

with all the elements of the strictly upper-triangular part of the last column of $L^{(1)}$ zero.

Hence, combining Equation (3.10) with

$$Q^{(1)^T} A^{(1)} = Q^{(1)^T}(Q_{0:0}^T A Q_{0:0}) = Q_{0:1}^T A Q_{0:0},$$

implies that

$$AQ_{0:0} - Q_{0:0}D^{(1)} = Q_{0:1}L^{(1)}. \quad (3.11)$$

Multiplying both sides of Equation (3.11) to the right by $\langle \mathbf{e}_n \rangle$ and using the knowledge for $m = 0$, leads to

$$(AQ_{0:0} - Q_{0:0}D^{(1)})\langle \mathbf{e}_n \rangle = (Q_{0:1}L^{(1)})\langle \mathbf{e}_n \rangle = Q_{0:1}\langle \mathbf{e}_n \rangle$$
$$= (AQ_{0:0} - Q_{0:0}d_1 I)\langle \mathbf{e}_n \rangle$$
$$= (A - d_1 I)Q_{0:0}\langle \mathbf{e}_n \rangle$$
$$= (A - d_1 I)\langle \mathbf{f}_n^{(0)} \rangle$$
$$= \hat{p}_1(A)\langle \mathbf{f}_n^{(0)} \rangle = \hat{p}_1(A)\langle \mathbf{f}_n^{(1)} \rangle,$$

with $\mathbf{f}_n^{(0)} = \mathbf{f}_n^{(1)}$. This completes the case $k = n$. Using this equation, the case $k = n - 1$ is straightforward. Taking $\mathbf{f}_{n-1}^{(1)} = Q_{0:1}\mathbf{e}_{n-1}$ immediately leads to:

$$Q_{0:1}\langle \mathbf{e}_{n-1}, \mathbf{e}_n \rangle = \langle \mathbf{f}_{n-1}^{(1)}, \hat{p}_1(A)\mathbf{f}_n^{(1)} \rangle.$$

Moreover, we also have that $\mathbf{f}_{n-1}^{(1)}$ is orthogonal to $\hat{p}_1(A)\mathbf{f}_n^{(1)}$.

- **Step m.** We will prove now the general formulation, assuming that the case $m - 1$ holds for every $n \geq k \geq n - m - 1$. So we know that for every $n \geq k \geq n - m - 1$, the following equation is true (denote $\eta = k - n + m$)[21]:

$$Q_{0:m-1}\langle \mathbf{e}_k, \ldots, \mathbf{e}_n \rangle$$
$$= \hat{p}_{\eta-1}(A)\langle \mathbf{f}_k^{(m-1)}, \hat{p}_{\eta:\eta}(A)\mathbf{f}_{k+1}^{(m-1)}, \ldots, \hat{p}_{m-1:\eta}(A)\mathbf{f}_n^{(m-1)} \rangle.$$

To prove the case m, we will distinguish two cases, namely $n \geq k > n - m$ and $k = n - m$.

We start with the case $n \geq k > n - m$, in a similar way as for $m = 1$. When the transformation $Q^{(m)}$ used at the m-th step, is only applied on the rows, the matrix $A^{(m)}$ is transformed into:

$$Q^{(m)^T} A^{(m)}$$
$$= Q^{(m)^T}(D^{(m)} + S^{(m)}) = Q^{(m)^T}D^{(m)} + L^{(m)} \tag{3.12}$$

$$= Q^{(m)^T}
\left[\begin{array}{ccc|ccc}
0 & & & & & \\
& \ddots & & & & \\
& & 0 & & & \\
\hline
& & & d_1 & & \\
& & & & \ddots & \\
& & & & & d_m
\end{array}\right]
+
\left[\begin{array}{ccc|ccc}
\times & \cdots & \times & 0 & \cdots & 0 \\
\vdots & & \vdots & \vdots & & \vdots \\
\times & \cdots & \times & 0 & \cdots & 0 \\
\times & \cdots & \times & \times & \cdots & 0 \\
\vdots & & \vdots & \vdots & \ddots & \vdots \\
\times & \cdots & \times & \times & \cdots & \times
\end{array}\right],$$

with all the elements of the strictly upper-triangular part of the last m columns of $L^{(m)}$ zero.

---

[21] The definition of η is slightly different than the one from the theorem. This is done to obtain the final formulation in the correct form.

Hence, combining Equation (3.12) with

$$Q^{(m)^T}A^{(m)} = Q^{(m)^T}(Q_{0:m-1}^T AQ_{0:m-1}) = Q_{0:m}^T AQ_{0:m-1},$$

implies that

$$AQ_{0:m-1} - Q_{0:m-1}D^{(m)} = Q_{0:m}L^{(m)}. \tag{3.13}$$

Multiplying now Equation (3.13) on the right by $\langle \mathbf{e}_k, \ldots, \mathbf{e}_n \rangle$ leads to:

$$
\begin{aligned}
&Q_{0:m}\langle \mathbf{e}_k, \ldots, \mathbf{e}_n \rangle \\
&= (Q_{0:m}L^{(m)})\langle \mathbf{e}_k, \ldots, \mathbf{e}_n \rangle \\
&= (AQ_{0:m-1} - Q_{0:m-1}D^{(m)})\langle \mathbf{e}_k, \ldots, \mathbf{e}_n \rangle \\
&= \langle (AQ_{0:m-1} - Q_{0:m-1}D^{(m)})\mathbf{e}_k, \ldots, (AQ_{0:m-1} - Q_{0:m-1}D^{(m)})\mathbf{e}_n \rangle \\
&= \langle (A - d_\eta I)Q_{0:m-1}\mathbf{e}_k, \ldots, (A - d_m I)Q_{0:m-1}\mathbf{e}_n \rangle. \tag{3.14}
\end{aligned}
$$

We know by induction, that for every k, with $k \leq j \leq n$ the following equation is true (denote $\eta_j = j - n + m$):

$$
\begin{aligned}
&Q_{0:m-1}\langle \mathbf{e}_j, \ldots, \mathbf{e}_n \rangle \\
&= \hat{p}_{\eta_j-1}(A)\langle \mathbf{f}_j^{(m-1)}, \hat{p}_{\eta_j:\eta_j}(A)\mathbf{f}_{j+1}^{(m-1)}, \ldots, \hat{p}_{m-1:\eta_j}(A)\mathbf{f}_n^{(m-1)} \rangle.
\end{aligned}
$$

So we can write:

$$Q_{0:m-1}\langle \mathbf{e}_j \rangle = \hat{p}_{\eta_j-1}(A)\langle \mathbf{f}_j^{(m)} \rangle$$

where $\mathbf{f}_j^{(m)}$ is a suitably chosen vector such that

$$\mathbf{f}_j^{(m)} \in \langle \mathbf{f}_j^{(m-1)}, \hat{p}_{\eta_j:\eta_j}(A)\mathbf{f}_{j+1}^{(m-1)}, \ldots, \hat{p}_{m-1:\eta_j}(A)\mathbf{f}_n^{(m-1)} \rangle.$$

Using now this relation for every vector $Q_{0:m-1}\mathbf{e}_j$ in Equation (3.14), we get the following relations:

$$
\begin{aligned}
&\langle (A - d_\eta I)Q_{0:m-1}\mathbf{e}_k, \ldots, (A - d_m I)Q_{0:m-1}\mathbf{e}_n \rangle \\
&= \langle (A - d_\eta I)\hat{p}_{\eta_k-1}(A)\mathbf{f}_k^{(m)}, \ldots, (A - d_m I)\hat{p}_{\eta_n-1}(A)\mathbf{f}_n^{(m)} \rangle, \\
&= \langle \hat{p}_{\eta_k}(A)\mathbf{f}_k^{(m)}, \ldots, \hat{p}_{\eta_n}(A)\mathbf{f}_n^{(m)} \rangle, \\
&= \hat{p}_\eta(A)\langle \mathbf{f}_k^{(m)}, \hat{p}_{\eta+1:\eta+1}(A)\mathbf{f}_{k+1}^{(m)}, \ldots, \hat{p}_{m:\eta+1}(A)\mathbf{f}_n^{(m)} \rangle.
\end{aligned}
$$

Proving thereby the theorem for $k > n - m$. The case $k = n - m$ is again straightforward by defining $\mathbf{f}_{n-m}^{(m)}$ as $Q_{0:m}\mathbf{e}_{n-m}$. □

This means that at step m of the reduction algorithm we perform for every $n \geq k \geq n-m$ a multishift iteration on the subspace $\langle \mathbf{f}_k^{(m)}, \ldots, \mathbf{f}_n^{(m)} \rangle$. This is called

a nested type of multishift iteration. Under mild assumptions we will therefore
get a similar convergence behavior as in the multishift case. Before giving some
reformulations of Theorem 3.6, we will give a first intuitive interpretation to this
convergence behavior. Let us write down for some $k = n, n-1, n-2$, the different
formulas:

$$Q_{0:m}\langle \mathbf{e}_n \rangle = \langle \hat{p}_m(A)\mathbf{f}_n^{(m)} \rangle, \tag{3.15}$$

$$Q_{0:m}\langle \mathbf{e}_{n-1}, \mathbf{e}_n \rangle = \langle \hat{p}_{m-1}(A)\mathbf{f}_{n-1}^{(m)}, \hat{p}_m(A)\mathbf{f}_n^{(m)} \rangle, \tag{3.16}$$

$$Q_{0:m}\langle \mathbf{e}_{n-2}, \mathbf{e}_{n-1}, \mathbf{e}_n \rangle = \langle \hat{p}_{m-2}(A)\mathbf{f}_{n-2}^{(m)}, \hat{p}_{m-1}(A)\mathbf{f}_{n-1}^{(m)}, \hat{p}_m(A)\mathbf{f}_n^{(m)} \rangle.$$

We will assume, for simplicity reasons, that the vectors $\mathbf{f}_k^{(m)}$ do not have a significant
influence on the convergence behavior[22]. This means that they do not have a small
or zero component in the direction we want them to converge too. Further on in the
text we will investigate in more detail the influence of these vectors $\mathbf{f}_k^{(m)}$, with regard
to the convergence speed. Under this assumption, according to Equation (3.15), the
last vector of $Q_{0:m}$ will converge (for increasing m) towards the eigenvector of the
matrix $\hat{p}_m(A)$ corresponding to the dominant eigenvalue of $\hat{p}_m(A)$. Combining
this with Equation (3.16) shows us that $Q_{0:m}\mathbf{e}_{n-1}$ will converge to the eigenvector
perpendicular to the vector $Q_{0:m}\mathbf{e}_n$ and corresponding to the dominant eigenvalue
for $\hat{p}_{m-1}(A)$. Similarly a combination of all above equations reveals that $Q_{0:m}\mathbf{e}_{n-2}$
converges to an eigenvector perpendicular to the above two and corresponding to
the dominant eigenvalue for $\hat{p}_{m-2}(A)$. More details on the convergence behavior
will be given in Subsection 3.3.3.

Below, we have rewritten Theorem 3.6 in different forms, to illustrate more
clearly what happens, and to make different interpretations of the method possible:

- **Formulation 1: The shift through form.** We drag the polynomial in
  front of the subspace completely through the subspace.

  Corollary 3.7. *Let us use the notation as defined before. At step m of the
  algorithm we have for every $n \geq k \geq n - m$: (denote $\eta = k - n + m$):*

  $$Q_{0:m}\langle \mathbf{e}_k, \ldots, \mathbf{e}_n \rangle = \langle \hat{p}_\eta(A)\mathbf{f}_k^{(m)}, \hat{p}_{\eta+1}(A)\mathbf{f}_{k+1}^{(m)}, \ldots, \hat{p}_m(A)\mathbf{f}_n^{(m)} \rangle.$$

  *This means that on every vector, a form of the shifted power method is applied
  and the vectors are re-orthogonalized with regard to each other.*

- **Formulation 2: The nested shift formulation.** We can also reformulate
  the theorem such that we apply on each nested subspace an iteration with
  shift.

  Corollary 3.8. *Let us use the notation as defined before. At step m of the*

---

[22] Further in the text we will show that these vectors have an influence that is not neglectable for
the convergence behavior. This is due to the interaction with the Lanczos convergence behavior.

algorithm we have for every $n \geq k \geq n - m$ (denote $\eta = k - n + m$)

$$Q_{0:m}\langle \mathbf{e}_k, \ldots, \mathbf{e}_n \rangle$$
$$= \hat{p}_\eta(A)\langle \mathbf{f}_k^{(m)}, p_{\eta+1}(A)\langle \mathbf{f}_{k+1}^{(m)}, p_{\eta+2}(A)\langle \mathbf{f}_{k+2}^{(m)}, \ldots, p_m(A)\mathbf{f}_n^{(m)} \rangle \ldots \rangle\rangle,$$

which can be rewritten as

$$Q_{0:m}\langle \mathbf{e}_k, \ldots, \mathbf{e}_n \rangle$$
$$= \hat{p}_\eta(A)\langle \mathbf{f}_k^{(m)}, (A - d_{\eta+1}I)\langle \mathbf{f}_{k+1}^{(m)}, (A - d_{\eta+2}I)\langle \mathbf{f}_{k+2}^{(m)}, \ldots$$
$$\ldots, (A - d_m I)\mathbf{f}_n^{(m)} \rangle \ldots \rangle\rangle.$$

So we can see that on each nested subspace an iteration with shift is performed.

- **Formulation 3: The nested QL-iteration with shift.** Theorem 3.6 as presented above incorporates all the transformations in one orthogonal matrix $Q_{0:m}$. However, if we perform the basis transformation after each step in the reduction algorithm, this corresponds to performing a similarity transformation, leading to a different form of the theorem. This formulation corresponds to a shifted QL-iteration on the matrix A.

We already know from the results in Subsection 3.2.2 that we can interpret the reduction method as a nested subspace iteration, as follows:

$$(A^{(m)} - D^{(m)})\langle \mathbf{e}_{n-(m-1)}, \ldots, \mathbf{e}_n \rangle = \langle \mathbf{q}_{n-(m-1)}^{(m)}, \ldots, \mathbf{q}_n^{(m)} \rangle.$$

The interpretation of this type of iteration is not straightforward as we are not subtracting a shift matrix from the matrix $A^{(m)}$ but a diagonal matrix. This interpretation says that at every step m a subspace iteration of the matrix $(A^{(m)} - D^{(m)})$ is performed on the space $\langle \mathbf{e}_{n-(m-1)}, \ldots, \mathbf{e}_n \rangle$, which gives us part of the orthogonal matrix $Q^{(m)}$. If a column of this matrix $Q^{(m)}$ is close enough to an eigenvector of the matrix $A^{(m)}$, this will be visible after performing the basis transformation $Q^{(m)T} A^{(m)} Q^{(m)}$. (For more information see [182].)

Using Theorem 3.6, we can reformulate this nested iteration towards an iteration with a shift and another subspace on which one iterates.

We know that $Q_{0:m-1}^T A Q_{0:m-1} = A^{(m)}$. One can also easily prove the following relations:

$$Q_{0:m-1}^T p_i(A) Q_{0:m-1} = p_i(A^{(m)}),$$
$$Q_{0:m-1}^T \hat{p}_i(A) Q_{0:m-1} = \hat{p}_i(A^{(m)}),$$
$$Q_{0:m-1}^T \hat{p}_{j:i}(A) Q_{0:m-1} = \hat{p}_{j:i}(A^{(m)}).$$

If we perform now the basis transformation corresponding to the orthogonal matrix $Q_{0:m-1}$ on the subspace $Q_{0:m}\langle \mathbf{e}_k, \ldots, \mathbf{e}_n \rangle$, we get the following relations:

$$
\begin{aligned}
Q^{(m)} &\langle \mathbf{e}_k, \dots, \mathbf{e}_n \rangle \\
&= Q_{0:m-1}^T Q_{0:m} \langle \mathbf{e}_k, \dots, \mathbf{e}_n \rangle \\
&= Q_{0:m-1}^T \hat{p}_\eta(A) \langle \mathbf{f}_k^{(m)}, \hat{p}_{\eta+1:\eta+1}(A) \mathbf{f}_{k+1}^{(m)}, \dots, \hat{p}_{m:\eta}(A) \mathbf{f}_n^{(m)} \rangle \\
&= \hat{p}_\eta(A^{(m)}) \langle Q_{0:m-1}^T \mathbf{f}_k^{(m)}, \hat{p}_{\eta+1:\eta+1}(A^{(m)}) Q_{0:m-1}^T \mathbf{f}_{k+1}^{(m)}, \dots \\
&\qquad \dots, \hat{p}_{m:\eta}(A^{(m)}) Q_{0:m}^T \mathbf{f}_n^{(m)} \rangle.
\end{aligned}
$$

We can formulate the following equivalent corollary with $\hat{\mathbf{f}}_j^{(m)} = Q_{0:m-1}^T \mathbf{f}_j^{(m)}$.

Corollary 3.9. *Let us use the notation as defined before. At step m of the algorithm we have for every $n \geq k \geq n - m$ (denote $\eta = k - n + m$)*

$$
\begin{aligned}
Q^{(m)} \langle \mathbf{e}_k, \dots, \mathbf{e}_n \rangle = \hat{p}_\eta(A^{(m)}) \langle \hat{\mathbf{f}}_k^{(m)}, \hat{p}_{\eta+1:\eta+1}(A^{(m)}) \hat{\mathbf{f}}_{k+1}^{(m)}, \dots \\
\dots, \hat{p}_{j-n+m:\eta+1}(A^{(m)}) \hat{\mathbf{f}}_j^{(m)}, \dots, \hat{p}_{m:\eta}(A^{(m)}) \hat{\mathbf{f}}_n^{(m)} \rangle.
\end{aligned}
$$

In this way we know that a partial QL-iteration, by partial we mean using a subspace of dimension less than n, is performed on the subspace defined by the vectors $\hat{\mathbf{f}}_j^{(m)}$ and the last columns of $Q^{(m)}$ span this space.

This means that if a column of $Q^{(m)}$ is close to an eigenvector of $A^{(m)}$ the basis transformation will reveal it. But of course the convergence behavior is heavily dominated by the vectors $\hat{\mathbf{f}}^{(m)}$. This will be investigated in the next subsection. For a traditional convergence behavior, related to QR and QL-iterations, these subspaces are always equal to $\langle \mathbf{e}_1, \dots, \mathbf{e}_n \rangle$, and one can assume (in most cases), that these vectors do not heavily influence the convergence speed. Here however these vectors of the subspace are constructed in a specific way and do have an important impact on the convergence behavior.

Of course we can also reformulate this last theorem with regard to the first two formulations in this list.

Before investigating the convergence speed in more detail and the interaction between the Ritz-value convergence behavior and the subspace iteration, we will translate the theorem towards the semiseparable and tridiagonal case.

1. In the tridiagonal case, there is no chasing step involved, as all the performed Givens transformations are equal to the identity. Hence the lower right already reduced part of the matrix, contains the Lanczos-Ritz values, and no subspace iteration is performed.

2. In a previous subsection (Subsection 3.2.1) it was stated that on the lower right part of the semiseparable matrix always a step of nonshifted subspace iteration was performed. We get exactly this behavior if we take the shift

equal to zero: Theorem 3.6 is therefore an extension of the results in the previous section (see also Lemma 3.1 in [155]).

Theorem 3.10. *Let us use the notation as defined before. At step $m = 1, 2, \ldots, n-1$ of the reduction to semiseparable form we have for every $n \geq k \geq n - m$, that (denote $\eta = k - n + m$)*

$$Q_{0:m}\langle \mathbf{e}_k, \ldots, \mathbf{e}_n \rangle$$
$$= A^{k-1}\langle \mathbf{f}_k^{(m)}, A\mathbf{f}_{k+1}^{(m)}, \ldots, A^{j-n+m}\mathbf{f}_j^{(m)}, \ldots, A^{(m)}\mathbf{f}_n^{(m)} \rangle.$$

Several examples illustrating the convergence behavior will be presented in the next chapter. Using Corollary 3.9 one can easily explain the convergence behavior as observed in [172]. In this article, it was observed, that the reduction of a symmetric matrix into a similar semiseparable plus diagonal one, where the first values of the diagonal were chosen equal to eigenvalues of the symmetric matrix, revealed these eigenvalues. More precisely the upper left $k \times k$ block of the transformed matrix is of diagonal form, where k is the number of top left diagonal elements equal to eigenvalues of the original matrix. This is natural, as after the complete reduction method on the complete matrix a step of the QL-iteration with shift d_1 is performed. If d_1 equals an eigenvalue, we have a perfect shift, and this will be revealed in the upper left position. If we continue now with the trailing $(n-1) \times (n-1)$ block of this matrix, we know that this matrix has the same eigenvalues as the original matrix without the eigenvalue d_1. As on this matrix a QL-iteration with shift d_2 is performed. With d_2 and eigenvalue, the procedure will again reveal this eigenvalue. This process continues as long as the first $d_1, \ldots, d_k$ diagonal elements are equal to the eigenvalues of the original matrix. As soon as one diagonal element does not correspond anymore to an eigenvalue, the procedure stops.

Let us first investigate how the reduction to upper triangular semiseparable form can be interpreted with regard to the subspace iteration. In the next section we will discuss the interaction.

3.2.4 The different reduction algorithms revisited

In this subsection we will briefly investigate the behavior of the other reduction algorithms as proposed in Chapter 2.

The orthogonal similarity transformations

The general reduction theory as applied for the symmetric matrices can easily be translated towards the reduction to Hessenberg-like or Hessenberg-like plus diagonal form. As the complete proofs remain almost exactly the same we do not include details concerning these interpretations.

The orthogonal transformations

It is clear that the reduction to upper/lower bidiagonal form does not involve any kind of subspace iteration. For the reduction to upper triangular semiseparable form, however, we can prove in a similar way as in Subsection 3.2.1 that on the already semiseparable part of the matrix $A^{(k)}$ some kind of nested subspace iteration is performed. The proof is a straightforward extension of the results presented in Subsection 3.2.1. It can be summarized as follows.

At each step of the algorithm introduced in Theorem 2.17, one more row is added to the set of the rows of the matrix already in upper triangular semiseparable form by means of orthogonal transformations. In this section, using arguments similar to those considered in [179, 182], we show that this algorithm can be interpreted as a kind of nested subspace iteration method, where the size of the vector subspace is increased by one and a change of coordinate system is made at each step of the algorithm. As a consequence, the part of the matrix already in semiseparable form reveals information on the gaps in the singular value distribution.

Let $A^{(1)} = U^{(0)^T} A V^{(0)}$; the matrices $U^{(0)}$ and $V^{(0)}$ are initial transformations, which in the reduction procedures in the previous chapter were chosen equal to I. Let $U_1^{(1)}, U_2^{(1)}$ and $V_1^{(1)}$ be the orthogonal matrices, described in step 1 of the proof of Theorem 2.17. Hence the matrix

$$A^{(2)} = U_2^{(1)^T} U_1^{(1)^T} A^{(1)} V_1^{(1)}$$

has the upper triangular semiseparable structure in the first two rows and in the first column.

Define $U^{(1)^T} = U_2^{(1)^T} U_1^{(1)^T}$; in general we define $U^{(i)}$ as the combinations of all matrices $U_j^{(i)}$ performed on the left of the matrix $A^{(i)}$ at that step. We use the same notation for $V^{(i)}$. Let $U^{(1)} = [\mathbf{u}_1^{(1)}, \mathbf{u}_2^{(1)}, \ldots, \mathbf{u}_m^{(1)}]$. Then, because $A^{(1)} V_1^{(1)} = U^{(1)} A^{(2)}$ has the first row, except for the first element equal to zero (see Theorem 2.17) we get:

$$A^{(1)} A^{(1)^T} = U^{(1)} A^{(2)} A^{(2)^T} U^{(1)^T} = U^{(1)} \begin{bmatrix} \times & \times & \cdots & \times \\ 0 & \times & \cdots & \times \\ \vdots & \vdots & \vdots & \vdots \\ 0 & \times & \cdots & \times \end{bmatrix} = U^{(1)} R_1. \quad (3.17)$$

Let $\mathbf{e}_1, \ldots, \mathbf{e}_n$ be the standard basis vectors of $\mathbb{R}^n$. From (3.17), because of the structure of R_1, it turns out that:

$$A^{(1)} A^{(1)^T} \langle \mathbf{e}_1 \rangle = \langle \mathbf{u}_1^{(1)} \rangle.$$

This means that the first column of $A^{(1)} A^{(1)^T}$ and $\mathbf{u}_1^{(1)}$ span the same one-dimensional space. In fact, one subspace iteration step is performed on the vector $\mathbf{e}_1$.

The first step of the algorithm is completed when the following orthogonal transformation is performed:

$$A^{(2)} = U^{(1)^T} A^{(1)} V_1^{(1)} \quad \Leftrightarrow \quad A^{(2)} A^{(2)^T} = U^{(1)^T} A^{(1)} A^{(1)^T} U^{(1)}.$$

The latter transformation is a change of coordinate system. This means that for the vector $\mathbf{u}_1^{(1)}$ we get $U^{(1)^T}\mathbf{u}_1^{(1)} = \mathbf{e}_1$. Summarizing, this means that one step of subspace iteration on the subspace $\langle \mathbf{e}_1 \rangle$ is performed, resulting in a new subspace $\mathbf{u}_1^{(1)}$, and then, by means of a coordinate transformation, it is transformed back into the subspace $\langle \mathbf{e}_1 \rangle$. Therefore, if $\mathbf{z}^{(i)}$ denotes the singular vector corresponding to the largest singular value of $A^{(i)}$, $i = 1, 2, \ldots$, the sequence $\{\mathbf{z}^{(i)}\}$ converges to $\mathbf{e}_1$, and, consequently, the entry $A^{(i)}(1, 1)$ converges to the largest singular value of A. The second step can be interpreted in a completely analogous way. Let

$$A^{(3)} = U^{(2)^T} A^{(2)} V^{(2)}, \tag{3.18}$$

where $U^{(2)}$ and $V^{(2)}$ are orthogonal matrices such that $A^{(3)}$ is an upper triangular semiseparable matrix in the first three rows and the first two columns. Denote $U^{(2)} = [\mathbf{u}_1^{(2)}, \mathbf{u}_2^{(2)}, \ldots, \mathbf{u}_m^{(2)}]$. From (3.18), $A^{(2)} A^{(2)^T}$ can be written as follows:

$$A^{(2)} A^{(2)^T} = U^{(2)} A^{(3)} A^{(3)^T} U^{(2)^T} = U^{(2)} \begin{bmatrix} \times & \times & \cdots & \times & \times \\ 0 & \times & \cdots & \times & \times \\ 0 & 0 & \times & \cdots & \times \\ \vdots & \vdots & \vdots & \vdots & \vdots \\ 0 & 0 & \times & \cdots & \times \end{bmatrix} = U^{(2)} R_2.$$

Considering the subspace $\langle \mathbf{e}_1, \mathbf{e}_2 \rangle$ and using the same notation as above, we have:

$$A^{(2)} A^{(2)^T} \langle \mathbf{e}_1, \mathbf{e}_2 \rangle = \langle \mathbf{u}_1^{(2)}, \mathbf{u}_2^{(2)} \rangle.$$

This means that the second step of the algorithm is a step of subspace iteration performed on a slightly grown subspace. For every new row that is added to the semiseparable structure in the matrix $A^{(i)}$, the dimension of the subspace is increased by one.

This means that from step i, $i = 1, \ldots, n$ (so the semiseparable structure is satisfied for the rows 1 up to i), all the consecutive steps perform subspace iterations on the subspace of dimension i. From [182], we know that these consecutive iterations on subspaces tend to create block-diagonal matrices with upper triangular blocks. Furthermore, the process works for all the nested subspaces at the same time, and so the semiseparable submatrices will tend to become more and more block-diagonal with upper triangular blocks, where the blocks contain the largest singular values. This explains why the upper-left block already gives a good estimate of the largest singular values, as they are connected to a subspace on which the most subspace iterations were performed.

This insight also opens a lot of new perspectives. In many problems, only few largest singular values need to be computed. In such cases, the proposed algorithm gives the required information after only a few steps, without running the algorithm to the completion. Moreover, because the semiseparable submatrices generated at each step of the algorithm converge to a block-diagonal matrix with upper triangular blocks, the original problem can be divided into smaller independent subproblems.

More information on the so-called rank-revealing problems of this method will be given in Section 11.2 in Chapter 11.

In the next section we will investigate the close relation between the Lanczos convergence behavior and the multishift convergence behavior. Moreover also the convergence speed related to this interaction will be investigated.

Notes and references

The results in this section are based on the articles [162, 154], which contain the reduction algorithms to upper triangular semiseparable and semiseparable form. These articles describe separately the subspace iteration in the reduction to semiseparable form and the subspace iteration in the reduction to upper triangular semiseparable form. A general combination of these results can be found in the following articles in which the subspace iteration multishift interpretation is described and its interaction with the Lanczos convergence behavior. Moreover this article also provides interesting convergence bounds for these reduction algorithms.

☞ E. Van Camp. *Diagonal-Plus-Semiseparable Matrices and Their Use in Numerical Linear Algebra*. PhD thesis, Department of Computer Science, Katholieke Universiteit Leuven, Celestijnenlaan 200A, 3000 Leuven (Heverlee), Belgium, May 2005.

☞ R. Vandebril, E. Van Camp, M. Van Barel, and N. Mastronardi. On the convergence properties of the orthogonal similarity transformations to tridiagonal and semiseparable (plus diagonal) form. *Numerische Mathematik*, 104:205–239, 2006.

More information on these convergence bounds and a more detailed analysis on how the interaction with the Lanczos is already included in the subspace iteration are the subject of the next section.

More information concerning subspace iteration methods and the interpretation as the QR-iteration (with shift) as a subspace iteration method can be found in the following articles.

☞ D. S. Watkins. Understanding the QR algorithm. *SIAM Review*, 24(4):427–440, 1982.

☞ D. S. Watkins. Some perspectives on the eigenvalue problem. *SIAM Review*, 35(3):430–471, 1993.

☞ D. S. Watkins. QR-like algorithms—an overview of convergence theory and practice. In J. Renegar, M. Shub, and S. Smale, editors, *The Mathematics of Numerical Analysis*, volume 32 of *Lectures in Applied Mathematics*, pages 879–893. American Mathematical Society, Providence, Rhode Island, USA, 1996.

☞ D. S. Watkins and L. Elsner. Convergence of algorithms of decomposition type for the eigenvalue problem. *Linear Algebra and its Applications*, 143:19–47, 1991.

☞ T. Zhang, G. H. Golub, and K. H. Law. Subspace iterative methods for eigenvalue problems. *Linear Algebra and its Applications*, 294(1–3):239–258, 1999.

The derivations in this section were presented in an analogue manner as the interpretations of the QR-iterations in the articles by Watkins.

3.3 Interaction of the convergence behaviors

In this section we will investigate the close relation between the two previously described convergence behaviors. We will see how the subspace iteration and the appearance of the Lanczos-Ritz values interact with each other.

In a first subsection we will prove that for the reduction to semiseparable form, the subspace iteration will start converging as soon as the Lanczos-Ritz values approximate the dominant eigenvalues well enough. Secondly we will investigate in more detail this relation for the reduction to semiseparable plus diagonal form. As the reduction to semiseparable plus diagonal form inherits a more special type of subspace iteration the interaction will also be more complex. But in the end it comes down to the same: as soon as the Ritz values approximate specific parts of the spectrum well enough, the multishift iteration technique can start working on these parts. Finally we derive some results on the convergence speed of the reduction. This convergence speed also captures the interaction with the Lanczos convergence behavior. Various numerical examples show in the next chapter that the bound is tight.

3.3.1 The reduction to semiseparable form

Taking a look at the numerical examples in the following Chapter 4, sometimes it seems that there is no sign of the convergence of the subspace iteration.

This is due to the interaction with the Lanczos behavior of the algorithm, which will sometimes lead to a slower subspace iteration convergence than expected. This can only happen when all the vectors $\langle \mathbf{e}_{n-k}, \mathbf{e}_{n-k+1}, \ldots, \mathbf{e}_n \rangle$ have a small component in one or more directions of the eigenspace connected to the dominant eigenvalues. Before we show by examples in the following chapter that this happens in practice, we first give a condition under which we are sure that the subspace iteration lets the matrix converge to one having a diagonal block containing the dominant eigenvalues. As soon as some of the Ritz values approximate all of the dominant eigenvalues, this convergence behavior appears. To show this we assume that the initial matrix A has two clusters of eigenvalues, $\Lambda_1 = \{\lambda_{1,j} | j \in J\}$ and $\Lambda_2 = \{\lambda_{2,i} | i \in I\}$, with $\#J = n_1$ and $\#I = n_2$. Suppose also that there is a gap between cluster 2 and cluster 1, $\min_{i \in I} |\lambda_{2,i}| \gg \max_{j \in J} |\lambda_{1,i}|$. Using the following notations $\Delta_1 = \text{diag}(\Lambda_1)$ and $\Delta_2 = \text{diag}(\Lambda_2)$, we can write the matrix $A^{(k)}$ in the following way:

$$\begin{bmatrix} A_k & \times \\ \times & S_k \end{bmatrix} = \begin{bmatrix} V_{11} & V_{12} \\ V_{21} & V_{22} \end{bmatrix} \begin{bmatrix} \Delta_1 & 0 \\ 0 & \Delta_2 \end{bmatrix} \begin{bmatrix} V_{11}^T & V_{21}^T \\ V_{12}^T & V_{22}^T \end{bmatrix}, \tag{3.19}$$

with $S_k \in \mathbb{R}^{k \times k}$, $V_{11} \in \mathbb{R}^{(n-k) \times n_1}$, $V_{12} \in \mathbb{R}^{(n-k) \times n_2}$, $V_{21} \in \mathbb{R}^{k \times n_1}$ and $V_{22} \in \mathbb{R}^{k \times n_2}$. From Equation (3.19), we have the following equation for S_k:

$$S_k = V_{21} \Delta_1 V_{21}^T + V_{22} \Delta_2 V_{22}^T.$$

Let the eigenvalue decomposition of S_k be denoted as

$$S_k = \begin{bmatrix} V_1^{S_k} & V_2^{S_k} \end{bmatrix} \begin{bmatrix} \Delta_1^{S_k} & 0 \\ 0 & \Delta_2^{S_k} \end{bmatrix} \begin{bmatrix} V_1^{S_k^T} \\ V_2^{S_k^T} \end{bmatrix}, \tag{3.20}$$

where we assume that some of the eigenvalues of S_k (in fact the Ritz values) already approximate the dominant eigenvalues Λ_2. Let us denote $\Delta_2^{S_k} = \operatorname{diag}\left(\Lambda_2^{S_k}\right)$, with $\Lambda_2^{S_k}$ as a set of eigenvalues $\Lambda_2^{S_k} = \{\lambda_{2,i}^{S_k} | \forall i \in I\}$; then we assume that

$$\lambda_{2,i}^{S_k} \approx \lambda_{2,i} \quad \forall i \in I.$$

Subspace iteration needs weak conditions before convergence can occur, namely V_{22} has full rank and is far from being not of full rank. This corresponds to the demand that the last vectors $\{\mathbf{e}_{n-k+1}, \ldots, \mathbf{e}_n\}$ projected on the invariant subspace connected to the dominant eigenvalues $\lambda_{2,i}$ have a large component in every direction of this subspace.

Via a Householder transformation we can transform Equation (3.19) into

$$\left[\begin{array}{c|c} \tilde{A} & 0 \\ & \tilde{\mathbf{v}}_k^T \\ \hline 0 \quad \tilde{\mathbf{v}}_k & S_k \end{array}\right] = \begin{bmatrix} \tilde{V}_{11} & \tilde{V}_{12} \\ V_{21} & V_{22} \end{bmatrix} \begin{bmatrix} \Delta_1 & 0 \\ 0 & \Delta_2 \end{bmatrix} \begin{bmatrix} \tilde{V}_{11}^T & V_{21}^T \\ \tilde{V}_{12}^T & V_{22}^T \end{bmatrix}.$$

Then we can substitute S_k by its eigenvalue decomposition (3.20):

$$\left[\begin{array}{c|c} \tilde{A} & 0 \\ & \hat{\mathbf{v}}_k^T \\ \hline 0 \quad \hat{\mathbf{v}}_k & \begin{array}{cc} \Delta_1^{S_k} & 0 \\ 0 & \Delta_2^{S_k} \end{array} \end{array}\right] = \begin{bmatrix} \tilde{V}_{11} & \tilde{V}_{12} \\ \tilde{V}_{21} & \tilde{V}_{22} \end{bmatrix} \begin{bmatrix} \Delta_1 & 0 \\ 0 & \Delta_2 \end{bmatrix} \begin{bmatrix} \tilde{V}_{11}^T & \tilde{V}_{21}^T \\ \tilde{V}_{12}^T & \tilde{V}_{22}^T \end{bmatrix}.$$

Note that V_{22} is of full rank if and only if $\tilde{V}_{22}$ is of full rank. We get

$$\Delta_2^{S_k} = \begin{bmatrix} P\tilde{V}_{21}\Delta_1 & P\tilde{V}_{22}\Delta_2 \end{bmatrix} \begin{bmatrix} \tilde{V}_{12}^T P^T \\ \tilde{V}_{22}^T P^T \end{bmatrix},$$

where P is the projection matrix $[0, I]$ of size $n_2^{S_k} \times n$ where I is the identity matrix of size $n_2^{S_k} \times n_2^{S_k}$, with $n_2^{S_k}$ the dimension of $\Delta_2^{S_k}$. This means that

$$\Delta_2^{S_k} - P\tilde{V}_{12}\Delta_1\tilde{V}_{12}^T P^T = P\tilde{V}_{22}\Delta_2\tilde{V}_{22}^T P^T. \tag{3.21}$$

Note that $P\tilde{V}_{22}$ is a square matrix.

Because there is a gap between the eigenvalues $\lambda_{1,j}, j \in J$ and $\lambda_{2,i}, i \in I$, there will be a comparable gap between $\lambda_{2,j}^{S_k}, j \in J$ and $\lambda_{1,i}, i \in I$. Hence the matrix at the left-hand side of Equation (3.21) is far from singular. Therefore this is also true for the right-hand side, and $\tilde{V}_{22}$ is of full rank. Hence, V_{22} has full rank.

In this paragraph, a brief explanation is given about why in certain situations the subspace iteration does not work immediately from the start. Suppose the size of the block Δ_2 is 2. We can write the matrix $A^{(1)} = Q^{(0)^T} A Q^{(0)}$ in decomposed form:

$$A^{(1)} = \begin{bmatrix} A_1 & \mathbf{v}_n \\ \mathbf{v}_n^T & \alpha \end{bmatrix} = \begin{bmatrix} V_{11} & V_{12} \\ V_{21} & V_{22} \end{bmatrix} \begin{bmatrix} \Delta_1 & 0 \\ 0 & \Delta_2 \end{bmatrix} \begin{bmatrix} V_{11}^T & V_{21}^T \\ V_{12}^T & V_{22}^T \end{bmatrix}. \tag{3.22}$$

Note that α equals S_1. Applying already the Householder transformation on the matrix A gives us the following decomposition:

$$\begin{bmatrix} \tilde{H} & 0 \\ \mathbf{q}^T & 0 \\ 0 & 1 \end{bmatrix} \begin{bmatrix} V_{11} & V_{12} \\ V_{21} & V_{22} \end{bmatrix} \begin{bmatrix} \Delta_1 & 0 \\ 0 & \Delta_2 \end{bmatrix} \begin{bmatrix} V_{11}^T & V_{21}^T \\ V_{12}^T & V_{22}^T \end{bmatrix} \begin{bmatrix} \tilde{H}^T & \mathbf{q} & 0 \\ 0 & 0 & 1 \end{bmatrix}$$

$$= \begin{bmatrix} \tilde{A} & \mathbf{a}^T & 0 \\ \mathbf{a} & \gamma & \beta \\ 0 & \beta & \alpha \end{bmatrix} = \tilde{A}^{(1)}. \tag{3.23}$$

The Ritz values are the eigenvalues of the two by two matrix:

$$\begin{bmatrix} \gamma & \beta \\ \beta & \alpha \end{bmatrix}.$$

In fact, to complete the previous step a Givens transformation should still be performed on the matrix $\tilde{A}^{(1)}$ to make the last two rows and columns dependent, thereby transforming the matrix into $A^{(2)}$. Note that this last Givens transformation does not change the eigenvalues of the bottom right submatrix.

We show now that the weak assumptions to let the subspace iteration converge are not satisfied. We prove that $[\mathbf{e}_{n-1}, \mathbf{e}_n]$ does not have two large linearly independent components in the span of

$$\begin{bmatrix} \tilde{H} & 0 \\ \mathbf{q}^T & 0 \\ 0 & 1 \end{bmatrix} \begin{bmatrix} V_{12} \\ V_{22} \end{bmatrix}.$$

We can write $[\mathbf{e}_{n-1}, \mathbf{e}_n]$ as a linear combination of the eigenvectors:

$$[\mathbf{e}_{n-1}, \mathbf{e}_n] = \begin{bmatrix} \tilde{H} & 0 \\ \mathbf{q}^T & 0 \\ 0 & 1 \end{bmatrix} \begin{bmatrix} V_{11} & V_{12} \\ V_{21} & V_{22} \end{bmatrix} \begin{bmatrix} C_1 \\ C_2 \end{bmatrix}.$$

The coordinates C_2 of $[\mathbf{e}_{n-1}, \mathbf{e}_n]$ with regard to the dominant eigenvectors are the following:

$$C_2 = \begin{bmatrix} V_{12}^T & V_{22}^T \end{bmatrix} \begin{bmatrix} \tilde{H}^T & \mathbf{q} & 0 \\ 0 & 0 & 1 \end{bmatrix} \begin{bmatrix} \mathbf{e}_{n-1} & \mathbf{e}_n \end{bmatrix}$$

$$= \begin{bmatrix} V_{12}^T & V_{22}^T \end{bmatrix} \begin{bmatrix} \mathbf{q} & 0 \\ 0 & 1 \end{bmatrix} = \begin{bmatrix} V_{12}^T \mathbf{q} & V_{22}^T \end{bmatrix}.$$

Using Equations (3.22) and (3.23), we get that

$$
\begin{bmatrix} V_{11}^T & V_{21}^T \\ V_{12}^T & V_{22}^T \end{bmatrix} \begin{bmatrix} \mathbf{v}_n \\ \alpha \end{bmatrix} = \begin{bmatrix} \Delta_1 & 0 \\ 0 & \Delta_2 \end{bmatrix} \begin{bmatrix} V_{21}^T \\ V_{22}^T \end{bmatrix}.
$$

Hence,

$$
V_{12}^T \mathbf{v}_n + V_{22}^T \alpha = \Delta_2 V_{22}^T.
$$

Because of the Householder transformation we know that $\mathbf{q} = \mathbf{v}_n / \|\mathbf{v}_n\|$, hence

$$
V_{12}^T \mathbf{q} = [\Delta_2 - \alpha I] \frac{V_{22}^T}{\|\mathbf{v}_n\|}.
$$

Therefore,

$$
\begin{aligned}
C_2 &= \begin{bmatrix} V_{12}^T \mathbf{q} & V_{22}^T \end{bmatrix} \\
&= \begin{bmatrix} [\Delta_2 - \alpha I] \frac{V_{22}^T}{\|\mathbf{v}_n\|} & V_{22}^T \end{bmatrix}.
\end{aligned}
$$

This means that if $\Delta_2 V_{22}^T$ lies almost in the same direction as V_{22}^T then C_2 is almost singular. If the two largest eigenvalues $\lambda_{2,1} \approx \lambda_{2,2}$ this is the case. Note however that when $\lambda_{2,1} \approx -\lambda_{2,2}$, the subspace iteration will work immediately because they are both extreme and immediately located by the Lanczos procedure.

Briefly spoken we have the following interaction between the Lanczos-Ritz value behavior, and the subspace iteration: the subspace iteration will start converging as soon as the Lanczos-Ritz values have converged close enough to the dominant eigenvalues.

3.3.2   The reduction to semiseparable plus diagonal form

In the previous two sections, we investigated two convergence behaviors of the reduction to semiseparable plus diagonal form. In this section we will prove the following behavior:

> The nested multishift iteration will start converging as soon as the Lanczos-Ritz values approximate the dominant eigenvalues well enough with regard to the multishift iteration.

Let us use the notation as defined in the previous section. At step $m = 1, 2, \ldots, n - 1$ of the algorithm we have for every $n \geq k \geq n - m$, that (denote $\eta = k - n + m$)

$$
Q_{0:m}\langle \mathbf{e}_k, \ldots, \mathbf{e}_n \rangle = \hat{p}_\eta(A)\langle \mathbf{f}_k^{(m)}, \hat{p}_{\eta+1:\eta+1}(A)\mathbf{f}_{k+1}^{(m)}, \ldots, \hat{p}_{m:\eta+1}(A)\mathbf{f}_n^{(m)} \rangle.
$$

Moreover, we also have that, due to the Lanczos-Ritz value convergence

$$
Q_{0:m}\langle \mathbf{e}_{n-m}, \ldots, \mathbf{e}_n \rangle = \mathcal{K}_{m+1}(A, Q^{(0)}\mathbf{e}_n).
$$

Clearly the following relation holds between the two above presented subspaces, for all k:

$$Q_{0:m}\langle \mathbf{e}_k, \ldots, \mathbf{e}_n \rangle \subset Q_{0:m}\langle \mathbf{e}_{n-m}, \ldots, \mathbf{e}_n \rangle.$$

These relations do exactly explain the behavior as presented above. The multishift subspace iteration works on the vectors $\mathbf{f}_j^{(i)}$, but they are constructed in such a way that after the subspace iteration we get a subspace that is part of the Krylov subspace $\mathcal{K}_{m+1}(A, Q^{(0)}\mathbf{e}_n)$. As long as this Krylov subspace is not large enough to contain the eigenvectors corresponding to the dominant eigenvalues of the matrix polynomials $\hat{p}_{j-n+m:1}(A)$, the subspace iteration can simply not converge to these eigenvectors.

As soon as the dominant eigenvectors, with regard to the multishift polynomial, will be present in the Krylov subspace, the Ritz values will approximate the corresponding eigenvalues and this means that the multishift iteration can start converging to these eigenvalues/eigenvectors. This behavior will be illustrated in the numerical experiments in the following chapter.

3.3.3 Convergence speed of the nested multishift iteration

In this subsection we will present some theorems concerning the speed of convergence, using the nested multishift QL-iteration as presented in this chapter. In a first part we present some theorems from [182, 179], which are useful for traditional GR-algorithms. In the second part, we apply these theorems to our nested subspace formulation.

General subspace iteration theory

First we will reconsider some general results concerning the distances between subspaces. A more elaborate study can be found in [182, 94]. Given two subspaces $\mathcal{S}$ and $\mathcal{T}$ in $\mathbb{R}^n$ and denote with $P_{\mathcal{S}}$ and $P_{\mathcal{T}}$ the orthonormal projector onto the subspace $\mathcal{S}$ and $\mathcal{T}$, respectively. The standard metric between subspaces (see [94]) is defined as

$$d(\mathcal{S}, \mathcal{T}) = \|P_{\mathcal{S}} - P_{\mathcal{T}}\|_2 = \sup_{\substack{s \in \mathcal{S} \\ \|s\|_2 = 1}} \quad d(s, \mathcal{T}) = \sup_{\substack{s \in \mathcal{S} \\ \|s\|_2 = 1}} \inf_{t \in \mathcal{T}} \|s - t\|_2,$$

if $\dim(\mathcal{S}) = \dim(\mathcal{T})$ and $d(\mathcal{S}, \mathcal{T}) = 1$ otherwise.

The next theorem states how the distance between subspaces changes when performing subspace iteration with shifted polynomials. The following theorem is slightly changed with regard to the original one to fit into our setting.

Theorem 3.11 (Theorem 5.1 from [182]). *Given a semisimple[23] matrix $A \in \mathbb{R}^{n \times n}$ with eigenvalues $\lambda_1, \lambda_2, \ldots, \lambda_n$ and associated linearly independent eigenvectors $\mathbf{v}_1, \mathbf{v}_2, \ldots, \mathbf{v}_n$. Let $V = [\mathbf{v}_1, \mathbf{v}_2, \ldots, \mathbf{v}_n]$ and κ_V is the condition number of V,*

---

[23]A matrix is called semiimple if it has n linearly independent eigenvectors.

with regard to the spectral[24] norm. Let l be an integer $1 \leq l \leq n - 1$, and define the invariant subspaces $\mathcal{U} = \langle \mathbf{v}_1, \ldots, \mathbf{v}_{l-1} \rangle$ and $\mathcal{T} = \langle \mathbf{v}_l, \ldots, \mathbf{v}_n \rangle$. Denote with (p_i) a sequence of polynomials and let $\hat{p}_i = p_i \ldots p_2 p_1$. Suppose that

$$\hat{p}_i(\lambda_j) \neq 0 \quad j = l, \ldots, n$$

for all i, and let

$$r_i = \frac{\max_{1 \leq j \leq l-1} |\hat{p}_i(\lambda_j)|}{\min_{l \leq j \leq n} |\hat{p}_i(\lambda_j)|}.$$

Let $\mathcal{S}$ be a k-dimensional subspace of $\mathbb{R}^n$ (with $k = n - l + 1$), satisfying

$$\mathcal{S} \cap \mathcal{U} = \{0\}.$$

Let $\mathcal{S}_i = \hat{p}_i(A)\mathcal{S}, i = 1, 2, \ldots$. Then there exists a constant C (depending on $\mathcal{S}$) such that for all i,

$$d(\mathcal{S}_i, \mathcal{T}) \leq C \, \kappa_V \, r_i.$$

In particular $\mathcal{S}_i \to \mathcal{T}$ if $r_i \to 0$. More precisely we have that

$$C = \frac{d(V^{-1}\mathcal{S}, V^{-1}\mathcal{T})}{\sqrt{1 - d(V^{-1}\mathcal{S}, V^{-1}\mathcal{T})}}.$$

We remark that similar theorems exist for defective[25] matrices. Also more information concerning the conditions put on the matrices in Theorem 3.11 can be found in [182]. We will however not go into these details.

The following lemma relates the subspace convergence towards the vanishing of certain subblocks in a matrix.

Lemma 3.12 (Lemma 6.1 from [182]). *Suppose $A \in \mathbb{R}^{n \times n}$ is given, and let $\mathcal{T}$ be a k-dimensional subspace, which is invariant under A. Assume G to be a nonsingular matrix and assume $\mathcal{S}$ to be the subspace spanned by the last k columns of G. (The subspace $\mathcal{S}$ can be seen as an approximation of the subspace $\mathcal{T}$.) Assume $B = G^{-1}AG$, and consider the matrix B, partitioned in the following way:*

$$B = \left[\begin{array}{cc} B_{11} & B_{12} \\ B_{21} & B_{22} \end{array} \right],$$

where $B_{21} \in \mathbb{R}^{k \times (n-k)}$. Then we have:

$$\|B_{21}\|_2 \leq 2 \sqrt{2} \, \kappa_G \, \|A\|_2 \, d(\mathcal{S}, \mathcal{T}),$$

where κ_G denotes the condition number of the matrix G.

We are now ready to use these theorems to derive an upper bound on the norm of the subblocks, while reducing a matrix to semiseparable plus diagonal form.

---

[24]The spectral norm is naturally induced by the $\|.\|_2$ norm on vectors.

[25]A matrix is defective if it does not have a complete basis of eigenvectors.

The convergence speed of the nested multishift iteration

Let us apply the above theorems to our specific case and see how we can derive convergence results for the reduction to semiseparable plus diagonal form.

Let as assume we are working with a symmetric matrix A (which is naturally simple), with eigenvalues $\lambda_1, \lambda_2, \ldots, \lambda_n$. The associated linear independent eigenvectors are denoted by $\mathbf{v}_1, \mathbf{v}_2, \ldots, \mathbf{v}_n$. As we proved before, in Subsection 3.3.2, the subspace iteration will only start working as soon as the Lanczos-Ritz values approximate the dominant eigenvalues well enough, with regard to the multishift polynomial. In this section, however, we do not need to worry about the Lanczos convergence behavior. Our theoretical upper bound for the convergence speed will naturally incorporate this Lanczos influence on the convergence.

Let $\mathcal{T} = \langle \mathbf{v}_l, \mathbf{v}_{l+1}, \ldots, \mathbf{v}_n \rangle$ and $\mathcal{U} = \langle \mathbf{v}_1, \mathbf{v}_2, \ldots, \mathbf{v}_{l-1} \rangle$. In this section we will derive an upper bound for the convergence towards the subspace $\mathcal{T}$.

The outcome of step m in the reduction algorithm is the matrix $A^{(m+1)} = Q_{0:m}^T A^{(0)} Q_{0:m}$, and we are interested in small subblocks of this matrix. (Assume $m \geq n - l + 1$, otherwise there are not yet subspace iteration steps performed on the lower right $(n - l + 1) \times (n - l + 1)$ block.) Using Lemma 3.12, we know that this is related to the orthogonal transformation matrix $Q_{0:m}$. Partition the matrix $A^{(m+1)}$ in the following way:

$$A^{(m+1)} = \begin{bmatrix} A_{11}^{(m+1)} & A_{12}^{(m+1)} \\ A_{21}^{(m+1)} & A_{22}^{(m+1)} \end{bmatrix},$$

where $A_{22}^{(m+1)}$ is of size $(n - l + 1) \times (n - l + 1)$. Denote with $\hat{\mathcal{S}}$ the space spanned by the last $n - l + 1$ components of $Q_{0:m}$. (Hence $\hat{\mathcal{S}} = Q_{0:m} \langle \mathbf{e}_l, \ldots, \mathbf{e}_n \rangle$.) Then we have by Lemma 3.12 that

$$\|A_{21}^{(m+1)}\|_2 \leq 2 \sqrt{2} \, \|A^{(0)}\|_2 \, d(\hat{\mathcal{S}}, \mathcal{T})$$

as $\kappa_2 = 1$ because $Q_{0:m}$ is an orthogonal matrix.

To determine the distance between $\hat{\mathcal{S}}$ and $\mathcal{T}$ one can apply Theorem 3.11. As we are in step m of the reduction algorithm, we can apply Theorem 3.6 for $k = l$; this means that (with $\eta = l - n + m$):

$$Q_{0:m} \langle \mathbf{e}_l, \ldots, \mathbf{e}_n \rangle = \hat{p}_\eta(A) \langle \mathbf{f}_l^{(m)}, \hat{p}_{\eta+1:\eta+1}(A) \mathbf{f}_{l+1}^{(m)}, \ldots, \hat{p}_{m:\eta+1}(A) \mathbf{f}_n^{(m)} \rangle.$$

This gives us

$$\hat{\mathcal{S}} = \hat{p}_\eta(A) \mathcal{S},$$

with

$$\mathcal{S} = \langle \mathbf{f}_l^{(m)}, \hat{p}_{\eta+1:\eta+1}(A) \mathbf{f}_{l+1}^{(m)}, \ldots, \hat{p}_{m:\eta+1}(A) \mathbf{f}_n^{(m)} \rangle.$$

Applying Theorem 3.11 gives us the following upper bound for the distance between $\hat{\mathcal{S}}$ and $\mathcal{T}$. For

$$r = \frac{\max_{1 \leq j \leq l-1} |\hat{p}_\eta(\lambda_j)|}{\min_{l \leq j \leq n} |\hat{p}_\eta(\lambda_j)|},$$

the following upper bound is obtained

$$d(\hat{\mathcal{S}}, \mathcal{T}) \leq C\, r,$$

where

$$C = \frac{d(V^{-1}\mathcal{S}, V^{-1}\mathcal{T})}{\sqrt{1 - d(V^{-1}\mathcal{S}, V^{-1}\mathcal{T})}}.$$

Summarizing this deduction we get that the norm $\|A_{21}^{(m+1)}\|_2$ is bounded as follows:

$$\|A_{21}^{(m+1)}\|_2 \tag{3.24}$$
$$\leq 2\sqrt{2}\,\|A^{(0)}\|_2 \left(\frac{d(V^{-1}\mathcal{S}, V^{-1}\mathcal{T})}{\sqrt{1 - d(V^{-1}\mathcal{S}, V^{-1}\mathcal{T})}} \right) \left(\frac{\max_{1 \leq j \leq l-1} |\hat{p}_\eta(\lambda_j)|}{\min_{l \leq j \leq n} |\hat{p}_\eta(\lambda_j)|} \right).$$

If one is interested in the bound for the next iterate $m + 1$, one has to use in fact another subspace $\tilde{\mathcal{S}}$. But, due to the specific structure of the vectors $\mathbf{f}_j^{(m)}$ (see Theorem 3.6), the subspaces $\mathcal{S}$ and $\tilde{\mathcal{S}}$ span the same space. Hence, the distance remains the same, and only the polynomials in Equation (3.24) determine the change in norm of the subblock. This means that once the subspace iteration starts working on a specific part, one can calculate the constant C, and it will not change anymore.

In practice, the constant C can be very large as long as the dominant eigenvectors, with regard to the polynomial p_η, are not present in the Krylov subspace, and hence the Lanczos-Ritz values are not close enough to the dominant eigenvalues, with regard to the polynomial p_η. This constant can create a delay in the convergence of the subspace iteration behavior. The influence of the Lanczos convergence behavior on the subspace iteration is therefore captured in the constant C.

Let us give a traditional example on the convergence speed. We will only present the results. More information can be found in [182]. The polynomial considered here, namely, $\hat{p}_\eta(\lambda)$, is in fact a multiplication between several polynomials:

$$\hat{p}_\eta(\lambda) = p_\eta(\lambda) p_{\eta-1}(\lambda) \cdots p_2(\lambda) p_1(\lambda).$$

Assume all $p_i(\lambda) = p(\lambda) = \lambda - d$. This means that we always consider the same shift d. If $d = 0$, we get the power method. We order the eigenvalues such that $|p(\lambda_1)| \leq |p(\lambda_2)| \leq \ldots \leq |p(\lambda_n)|$. Assume l to be chosen such that

$$\rho = \frac{\max_{1 \leq j \leq l-1} |p(\lambda_j)|}{\min_{l \leq j \leq n} |p(\lambda_j)|} = \frac{|p(\lambda_{l-1})|}{|p(\lambda_l)|} < 1,$$

then we get that $r = \rho^\eta$, and hence we get linear convergence.

In the next chapter we will show numerical experiments: we calculate some of these bounds in real experiments, and we observe that this is a valuable and useful upper bound in practice. Moreover, we will see that one can use Equation (3.24) to predict possible convergence behavior to eigenvalues. Moreover a computational complexity analysis if deflation is possible is also presented.

3.3.4   The other reduction algorithms

The other reduction algorithms either to Hessenberg-like or to upper triangular semiseparable form can be discussed similarly. Equivalent bounds on the convergence speed can be derived. The details are left to the reader.

Notes and references

These results on the interaction between the convergence behaviors and predictions concerning the convergence speed were previously presented in the article [171], also discussed at the end of the previous section.

3.4   Conclusions

In this chapter we investigated the convergence behavior of the reduction algorithms in detail. Two types of convergence behavior were analyzed: first the Lanczos convergence behavior and second the subspace iteration convergence behavior.

For the Lanczos convergence behavior, a general framework was provided, which enables us to classify similarity transformations. Two easy-to-check conditions should be placed on such a similarity transformation in order to have the Arnoldi-Ritz values in the already reduced part of the matrix. We showed that the tridiagonalization and the reduction to semiseparable form satisfy the desired properties, and therefore also have the predicted convergence behavior. All the other reductions as previously presented were investigated with regard to this relation.

Moreover we indicated in this chapter that the reduction to semiseparable form has an additional convergence behavior with respect to the tridiagonalization procedure. It was proved that during the reduction to semiseparable plus diagonal form some kind of nested subspace iteration is performed. Therefore we presented theoretical results explaining the convergence behavior of the reduction to semiseparable plus diagonal form. As we proved that the reduction to semiseparable and tridiagonal form can be seen as special cases of the reduction to semiseparable plus diagonal form, we know that also the presented theorems are valid for these reduction methods.

A final section of this chapter concentrated on the interaction between both previously mentioned convergence behaviors. Also a theoretical bound for the convergence speed was given.

In the following chapter we will describe some possible implementations for the reduction algorithms, based on the Givens-vector representation. Second also some numerical examples will be presented illustrating our theoretical findings of this chapter.

Chapter 4

Implementation of the algorithms and numerical experiments

In Chapter 2 we proposed several algorithms to reduce matrices to semiseparable form. To calculate the computational complexity of these algorithms we always assumed that the part of the matrix already in semiseparable form could be represented in an efficient way. This efficient representation will be used in this chapter, and the corresponding implementation will be given. Moreover also some tools for working in an efficient way with the Givens-vector representation will be developed. In Chapter 3 we investigated in detail the interaction between the subspace iteration and the Lanczos convergence behavior of the proposed reduction algorithms. In this chapter several numerical examples will illustrate these theoretical investigations.

In the first section we will not yet discuss implementation details nor show numerical experiments. We will provide some tools for figuring out algorithms and implementations related to the Givens-vector represented structured rank matrices. A graphical scheme of how the Givens transformations operate on the vector in the representation is given. Based on this graphical representation and the so-called shift through lemma we will be able to efficiently update the representation.

In Section 4.2 we will discuss in a mathematical manner the implementation of the reduction algorithms deduced in the previous chapter. No real implementations are given, just the mathematical ideas with formulas. As in the previous chapters only insight was created due to operations on figures. At the end of this section a complexity analysis of these algorithms is also presented. Two types of complexities are considered. The case in which the reduction is performed from the beginning to the end, not exploiting the deflation possibilities and a second case in which we exploit the deflation. A theoretical test case is shown to compare this method for computing the eigenvalues with the traditional method based on tridiagonal matrices.

In Section 4.3, several numerical experiments are performed to illustrate the theoretical results of the previous chapters. We start by examining the convergence behavior of the reduction to semiseparable form. Four different experiments are performed to illustrate the interaction between the subspace iteration and the Lanczos convergence behavior. Also examples illustrating the convergence behavior due to the subspace iteration are presented. This might lead to deflation possibilities as mentioned before.

The last subsection of this chapter shows a numerical experiment in which the reduction to semiseparable plus diagonal form is investigated. First the interaction between the Lanczos-Ritz values and the subspace iteration is examined. Secondly we investigate in detail the upper bound as it was derived in the previous chapter. The examples clearly illustrate that the presented bound is rather tight.

✎ *This chapter discusses issues related to the implementation of the reduction algorithms and also examples related to the convergence properties as proved in the previous chapter are shown. If the reader is not interested in these topics, he can readily proceed with the next part. Details on the computational complexity and some examples are however interesting.*

Section 4.1 discusses the graphical patterns resulting from applying sequences of Givens transformations. These patterns were discussed extensively in the first volume of the book. These patterns can be useful when implementing algorithms related to structured rank matrices.

Details on the implementation are given in Section 4.2.

As already mentioned in the introduction to the part, reading of Subsection 4.2.5, considering the computational complexity in case of deflation, is recommended. Also the following experiments are worth reading as they illustrate the theoretical results of the previous chapter. The Experiments 2, 3 and 4 of Subsection 4.3.1 illustrate the interaction between the convergence behaviors. The examples starting from 5 in Subsection 4.3.2 compare the theoretical bound on the convergence speed with the real situation.

4.1   Working with Givens transformations

In this section we will discuss some tools for working in an efficient way with Givens transformations. These tools might come in handy when designing algorithms for structured rank matrices as well as for implementing algorithms based on the Givens-vector representation.

4.1.1   Graphical schemes

All the implementations considered in this book are based on the so-called Givens-vector representation. This is an efficient representation for representing low rank parts in matrices. As a low rank part is not necessarily sparse, transformations involving these low rank parts do change the structure of the involved part. Hence we need to be able to work with this Givens-vector representation, i.e., we need to have the ability to update this representation if an operation is performed on the left or on the right of the matrix.

We will provide here some tools for working with these Givens transformations. We remark that we will only briefly show how both the Givens transformations and the representation change in some cases. A more thorough study of the effect on the

structured rank after applying a Givens transformation for example can be found
in Volume I (Chapter 9). This is considered when computing the QR-factorization
of general structured rank matrices.

Let us first introduce the graphical schemes for working with Givens patterns.
These patterns were introduced in [53].

To depict graphically these Givens transformations, with regard to their order
and the rows they are acting on, we use the following scheme. Remark that in
this scheme we will only depict Givens transformations applied on the left to the
matrix. This is to illustrate how we can represent Givens transformations graphically. When working with similarity transformations, one should be aware that
these transformations are applied on two sides of a matrix.

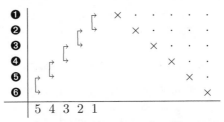

The numbered circles on the vertical axis depict the rows of the matrix. The bottom
numbers represent in some sense a time line. Let us explain this scheme in more
detail. First a Givens transformation is performed to the left of the matrix, acting
on row 5 and row 4. Secondly a Givens transformation is performed to the left
acting on row 3 and row 4 and this process continues.

First of all we will expand our graphical representation. We will graphically
depict the Givens-vector representation for the lower triangular part of a 6×6
matrix as follows. The elements $\cdot$ are not taken into consideration. We assume the
matrix to have its lower triangular part of semiseparable form.

We are interested in the interaction between the Givens transformation and our
original matrix. The elements marked with $\cdot$ are not essential in the coming analysis,
and hence we will not pay attention to them.

To illustrate more precisely what is depicted here, we will reconstruct our
semiseparable matrix from this scheme. In fact this is a graphical representation of
the reconstruction of a semiseparable matrix from its Givens-vector representation
as presented in Section 1.3.2.

We see that on the left in our graphical representation, there are still the
Givens transformations shown, whereas on the right we have just the upper triangular part of our semiseparable matrix. Applying now the first Givens transformation,
gives us the following graphical representation.

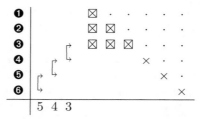

Hence, we removed one Givens transformation, and we filled up the corresponding elements in the matrix. In fact we just performed Givens transformation 1 on the matrix on the right in the graphical representation.

The elements of the upper triangular part changed, but they are not essential in our analysis, and hence we omitted them in this scheme. The element that is now in position $(2, 2)$ of the right matrix corresponds to the second element in our vector $\mathbf{v}$ of the Givens vector representation. This shows that one should be careful when directly implementing this approach, but the scheme will be very useful for deriving algorithms, as we will show in the forthcoming pages.

To conclude, we perform one more Givens transformation. This gives us the following graphical representation.

In the following section we will show how we can update our Givens-vector representation, when a transformation on the outside and on the inside is performed.

4.1.2   Interaction of Givens transformations

To be able to see how different Givens transformations interact with one another we need two operations involving Givens transformations. As already mentioned before, this is a brief summary of results discussed in Volume I.

The first lemma shows us that we can concatenate two Givens transformations acting on the same rows. The second lemma shows us that under some conditions we can rearrange the order of some Givens transformations.

Lemma 4.1. *Suppose two Givens transformations G_1 and G_2 are given:*

$$G_1 = \begin{bmatrix} c_1 & -s_1 \\ s_1 & c_1 \end{bmatrix} \ and \ G_2 = \begin{bmatrix} c_2 & -s_2 \\ s_2 & c_2 \end{bmatrix}.$$

Then we have that $G_1 G_2 = G_3$ is again a Givens transformation. We will call this the fusion of Givens transformations in the remainder of the text.

The proof is trivial. In our graphical schemes, we will depict this as follows.

$$\begin{array}{c|c} \mathbf{0} \\ \mathbf{2} \end{array} \quad \text{resulting in} \quad \begin{array}{c|c} \mathbf{0} \\ \mathbf{2} \end{array}$$

The next lemma is slightly more complicated and changes the ordering of three Givens transformations.

Lemma 4.2 (Shift through lemma). *Suppose three 3×3 Givens transformations G_1, G_2 and G_3 are given, such that the Givens transformations G_1 and G_3 act on the first two rows of a matrix, and G_2 acts on the second and third row.*
    Then we have that

$$G_1 G_2 G_3 = \hat{G}_1 \hat{G}_2 \hat{G}_3,$$

where $\hat{G}_1$ and $\hat{G}_3$ work on the second and third row and $\hat{G}_2$, works on the first two rows.

Proof. The proof is straightforward based on the factorization of a 3×3 orthogonal matrix. Suppose we have an orthogonal matrix U. We will now depict a factorization of this matrix U into two sequences of Givens transformations like described in the lemma.
The first factorization of this orthogonal matrix makes the matrix upper triangular in the traditional way. The first Givens transformation $\hat{G}_1^T$ acts on rows 2 and 3 of the matrix U, creating thereby a zero in the lower-left position:

$$\hat{G}_1^T U = \begin{bmatrix} \times & \times & \times \\ \times & \times & \times \\ 0 & \times & \times \end{bmatrix}.$$

The second Givens transformation acts on the first and second rows, to create a zero in the second position of the first column:

$$\hat{G}_2^T \hat{G}_1^T U = \begin{bmatrix} \times & \times & \times \\ 0 & \times & \times \\ 0 & \times & \times \end{bmatrix}.$$

Finally the last transformation $\hat{G}_3^T$ creates the last zero to make the matrix of upper triangular form:

$$\hat{G}_3^T \hat{G}_2^T \hat{G}_1^T U = \begin{bmatrix} \times & \times & \times \\ 0 & \times & \times \\ 0 & 0 & \times \end{bmatrix}.$$

Suppose we have chosen all Givens transformations in such a manner that the upper triangular matrix has positive diagonal elements. Because the resulting upper triangular matrix is orthogonal it has to be the identity matrix, hence we have the following factorization of the orthogonal matrix U:

$$U = \hat{G}_1 \hat{G}_2 \hat{G}_3. \tag{4.1}$$

Let us consider now a different factorization of the orthogonal matrix U. Perform a first Givens transformation to annihilate the upper-right element of the matrix U, where the Givens transformation, acts on the first and second row:

$$G_1^T U = \begin{bmatrix} \times & \times & 0 \\ \times & \times & \times \\ \times & \times & \times \end{bmatrix}.$$

Similarly as above one can continue to reduce the orthogonal matrix to lower triangular form, with positive diagonal elements. Hence one obtains a factorization of the following form:

$$U = G_1 G_2 G_3. \tag{4.2}$$

Combining Equations (4.1) and (4.2) leads to the desired result. □

Graphically we will depict the shift through lemma as follows.

resulting in

and in the other direction this becomes:

resulting in

If we cannot place the $\curvearrowright$ or $\curvearrowleft$ arrow at that specific position, then we cannot apply the shift through lemma. The reader can verify that for example in the following graphical scheme we cannot use the lemma.

To apply the shift through lemma, in some sense, we need to have some extra place to perform the action. We will now see how we can use the graphical representation and the presented interaction of the Givens transformations to retrieve information about our Givens-vector represented matrices.

Note 4.3. *Concerning the implementation of the shift through lemma we have to make some remarks. The interchanging of the order of the Givens transformations can be computed by a straightforward QR-factorization based on three Givens transformations of the involved unitary 3×3 matrix. Because the matrix is unitary, there* *is flexibility in computing this factorization into three Givens transformations. Let us clarify what is meant with this flexibility.*

Suppose a first Givens transformation is performed on the 3×3 unitary matrix U in order to annihilate the lower-left element of the matrix U. The Givens transformation G_1^T acts on the two last rows:

$$G_1^T U = \begin{bmatrix} \times & \times & \times \\ \times & \times & \times \\ 0 & \times & \times \end{bmatrix}.$$

As the final outcome of performing the three Givens transformations needs to be the identity matrix, this Givens transformation also has to make the upper-right 2×2 block of rank 1. This is necessary because the following Givens transformation acting on rows 1 and 2 needs to create zeros in positions $(2,1)$, $(1,2)$ and $(1,3)$ at once. Hence performing this first Givens transformation gives us in fact the following matrix:

$$G_1^T U = \begin{bmatrix} \times & \boxtimes & \boxtimes \\ \times & \boxtimes & \boxtimes \\ 0 & \times & \times \end{bmatrix}.$$

It is clear that the first Givens transformation could be chosen in two ways: to annihilate the lower-right element or to create a rank 1 block in the upper-right position. Similar remarks hold for the remaining Givens transformations, e.g., the Givens transformation G_2^T can be chosen to annihilate one of the following three elements marked with $\otimes$, thereby acting on the first and second row:

$$\begin{bmatrix} \times & \boxtimes & \boxtimes \\ \otimes & \boxtimes & \boxtimes \\ 0 & \times & \times \end{bmatrix} \quad \begin{bmatrix} \times & \otimes & \boxtimes \\ \times & \boxtimes & \boxtimes \\ 0 & \times & \times \end{bmatrix} \quad \begin{bmatrix} \times & \boxtimes & \otimes \\ \times & \boxtimes & \boxtimes \\ 0 & \times & \times \end{bmatrix}.$$

The outcome of either one of the Givens transformations will be theoretically identical, hence one can choose the most numerically reliable operation.

The flexibility in computing these Givens transformations should be exploited in order to make the routine as robust as possible. Details can be found in the implementation of the shift through lemma.

Finally we remark that even though we restrict ourselves to Givens rotations in the book the shift through lemma holds for general 2×2 unitary matrices. These matrices do not necessarily need to be Givens rotations. Of course the restriction to real numbers in this part of the book can also be relaxed.

4.1.3 Updating the representation

This section covers two types of transformations. Suppose we have a Givens-vector represented matrix and we perform a Givens transformation on the left of this matrix, what happens with the matrix and with its representation. Secondly we will see how we can restore a bulge in the representation by combining the correct techniques from the previous subsections.

Let us start by investigating graphically what happens with the representation if a transformation is applied on the left of the Givens-vector represented matrix. We remark that we only consider the lower triangular part of the matrix.

Suppose we apply a first Givens transformation on a semiseparable matrix. Suppose we have a matrix graphically depicted as in Subsection 4.1.1 We distinguish now between two cases: a Givens transformation performed on the last two rows or one performed elsewhere.

When applying a Givens transformation from the left on the last two rows of the structured rank matrix, one can easily update its representation by applying the fusion of Givens transformations as graphically depicted below.

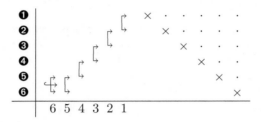

The first Givens transformation can easily fuse with the Givens transformation in position 5. Fusing the transformations in position 5 and position 6 gives us immediately an updated representation.

To illustrate the other case, we assume that we apply a Givens transformation acting on row 3 and row 4. Our first action, applying the shift through lemma, is already depicted.

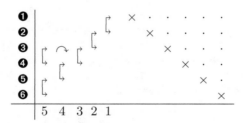

Rearranging the Givens transformations moving thereby the bottom transformation in position 5 to position 6 allows us to perform the shift through operation.

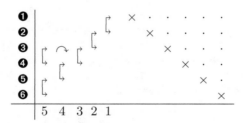

This leads to the following form

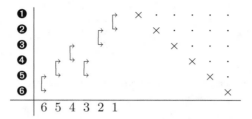

This can again be rearranged and we get the following scheme.

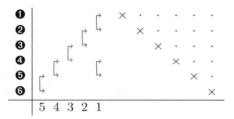

We now see that, except for the transformation acting on rows 4 and 5 in the first position, we again have a Givens-vector representation of the low rank part. Applying this Givens transformation onto the matrix gives us a kind of bulge in this matrix.

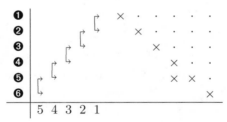

This bulge can be essential for example in bulge chasing algorithms as we will encounter them in the next chapter when steps of the QR-algorithm are performed.

The second issue we wanted to address was how to pull a Givens transformations that was located on the inside to the outside. This is in fact a similar transformation to the one above only in the reverse direction. This is however a very interesting problem. Suppose we have a bulge on the inside and we want to remove it by performing a Givens transformation on the left. The question is how to choose this Givens transformation. One can compute this Givens transformation fairly easily by removing the bulge on the inside by a Givens transformation and then dragging this Givens transformation through the representation completely to the left. The resulting Givens transformation is the inverse of the Givens transformation needed to remove the bulge on the inside. The reader can try to depict the graphical schemes themselves.

4.1.4   The reduction algorithm revisited

In this subsection we will graphically depict the reduction of a matrix to Hessenberg-like form, using thereby the Givens-vector representation. These graphical schemes can help a lot when implementing this algorithm. We choose to use the reduction to Hessenberg-like form to ease the graphical schemes. The symmetric case is however a straightforward extension.

Let us illustrate graphically the reduction procedure on a 5×5 matrix. We use the same notation as in Theorem 2.6. We start with the matrix $A_0^{(1)}$, which is of the following form (the elements marked with $\cdot$ are neglected in some sense):

$$A_0^{(1)} = \begin{bmatrix} \times & \cdot & \cdot & \cdot & \cdot \\ \times & \times & \cdot & \cdot & \cdot \\ \times & \times & \times & \cdot & \cdot \\ \times & \times & \times & \times & \cdot \\ \times & \times & \times & \times & \times \end{bmatrix}$$

The first Givens transformations or a single Householder transformation are chosen in such manner that the complete bottom row, except for the last two elements is zero. Hence we have:

$$A_3^{(1)} = \begin{bmatrix} \times & \cdot & \cdot & \cdot & \cdot \\ \times & \times & \cdot & \cdot & \cdot \\ \times & \times & \times & \cdot & \cdot \\ \times & \times & \times & \times & \cdot \\ 0 & 0 & 0 & \otimes & \times \end{bmatrix}.$$

The next Givens transformation $G_4^{(1)}$ is chosen such that applying it on the right of $A_3^{(1)}$ it annihilates the element in position $(5,4)$, marked with $\otimes$. Applying then the Givens transformation $G_4^{(1)^T}$ to the left creates the desired rank 1 block. Instead of applying this Givens transformation to the left, we store it. Hence we get the following graphical scheme for the matrix $A_0^{(2)} = A_4^{(1)} = G_4^{(1)^T} A_3^{(1)} G_4^{(1)}$.

$$\begin{array}{c|ccccc} ❶ & \times & \cdot & \cdot & \cdot & \cdot \\ ❷ & \times & \times & \cdot & \cdot & \cdot \\ ❸ & \times & \times & \times & \cdot & \cdot \\ ❹ & \otimes & \times & \times & \times & \cdot \\ ❺ & & & & & \times \\ \hline & 1 \end{array}$$

In fact this Givens transformation in position 1 can be considered as a Givens from the representation of the rank 1 block in this matrix. Based on this Givens transformation and the elements in row 4 one can reconstruct the rank 1 block in the matrix and hence also the matrix itself. We remark that this scheme reduces significantly the number of elements to be stored.[26]

---

[26] The Givens-vector representation acts in fact only on the lower triangular part. The similarity

Let us see now how the following Givens transformations interact with the cheaply stored matrix from the graphical representation. The first Givens transformation $G_1^{(2)}$ is chosen such as to annihilate the element marked with $\otimes$ in the previous graphical scheme. Performing this similarity transformation onto the matrix $A_0^{(2)}$ results in $A_1^{(2)} = G_1^{(2)^T} A_0^{(2)} G_1^{(2)}$. The Givens transformation on the right annihilates the element $\otimes$, and the Givens transformation on the left only acts on the first and second row, hence we get the following graphical scheme for the matrix $A_1^{(2)}$.

$$
\begin{array}{c|cccccc}
\mathbf{1} & \times & \cdot & \cdot & \cdot & \cdot \\
\mathbf{2} & \times & \times & \cdot & \cdot & \cdot \\
\mathbf{3} & \times & \times & \times & \cdot & \cdot \\
\mathbf{4} & & \times & \times & \times & \cdot \\
\mathbf{5} & & & & & \times \\
\hline
& 1
\end{array}
\tag{4.3}
$$

The Givens-vector representation for the last two rows remains valid. To illustrate the power of this representation we will once apply the transformation onto the matrix and compare its structure with the one from Theorem 2.6, (but adapted to the nonsymmetric case). Applying the transformation in position 1 onto the matrix (right in Scheme 4.3), we get the following matrix

$$
A_1^{(2)} = \begin{bmatrix}
\times & \cdot & \cdot & \cdot & \cdot \\
\times & \times & \cdot & \cdot & \cdot \\
\times & \times & \times & \cdot & \cdot \\
0 & \boxtimes & \boxtimes & \boxtimes & \cdot \\
0 & \boxtimes & \boxtimes & \boxtimes & \times
\end{bmatrix}.
$$

It is clear that the transformation $G_1^{(2)}$ created two zeros at once. Moreover this structure coincides entirely with the one from the results in Chapter 2, Theorem 2.6.

Continuing with the Scheme 4.3, and performing the Givens transformation $G_2^{(2)}$ onto the matrix leads to a graphical representation of the matrix $A_2^{(2)}$ of the following form.

$$
\begin{array}{c|cccccc}
\mathbf{1} & \times & \cdot & \cdot & \cdot & \cdot \\
\mathbf{2} & \times & \times & \cdot & \cdot & \cdot \\
\mathbf{3} & \times & \times & \times & \cdot & \cdot \\
\mathbf{4} & & & \times & \times & \cdot \\
\mathbf{5} & & & & & \times \\
\hline
& 1
\end{array}
$$

Now we will enter the chasing part in the reduction algorithm. In our graphical representation this will result in a new Givens-vector representation, representing the bottom three rows now. Applying the transformation $G_3^{(2)}$ results in the matrix

---

transformation performed did however also affect other elements. These elements are depicted with $\cdot$. Even though this is not visible, we assume they changed. A more detailed analysis on this interpretation can be found in Remark 8.1.

$A_3^{(2)}$, which is of the following form:

$$
\begin{array}{c|cccccc}
\mathbf{1} & & \times & \cdot & \cdot & \cdot & \cdot \\
\mathbf{2} & & \times & \times & \cdot & \cdot & \cdot \\
\mathbf{3} & & \times & \times & \times & \cdot & \cdot \\
\mathbf{4} & & & & & \times & \cdot \\
\mathbf{5} & & & & & & \times \\
\hline
 & 2 \quad 1 & & & & &
\end{array}
$$

Applying now the transformation in position 1 leads to a changed representation for the matrix $A_3^{(2)}$ (creating fill-in in the bottom row marked with $\otimes$).

$$
\begin{array}{c|cccccc}
\mathbf{1} & & \times & \cdot & \cdot & \cdot & \cdot \\
\mathbf{2} & & \times & \times & \cdot & \cdot & \cdot \\
\mathbf{3} & & \times & \times & \times & \cdot & \cdot \\
\mathbf{4} & & & & & \times & \cdot \\
\mathbf{5} & & & & & \otimes & \times \\
\hline
 & 1 & & & & &
\end{array}
$$

The Givens transformation $G_4^{(2)}$ is now chosen in order to annihilate the element in position $(5,4)$, marked with $\otimes$. Applying this transformation onto the matrix $A_3^{(2)}$ results in the matrix $A_4^{(2)} = A_0^{(3)}$.

$$
\begin{array}{c|cccccc}
\mathbf{1} & & \times & \cdot & \cdot & \cdot & \cdot \\
\mathbf{2} & & \times & \times & \cdot & \cdot & \cdot \\
\mathbf{3} & & \otimes & \times & \times & \cdot & \cdot \\
\mathbf{4} & & & & & \times & \cdot \\
\mathbf{5} & & & & & & \times \\
\hline
 & 2 \quad 1 & & & & &
\end{array}
$$

Hence we obtain due to the chasing a kind of Givens-vector representation for the bottom three lines. This procedure can easily be continued. We will depict the addition of one more row into the structure. First we annihilate the element depicted with $\otimes$ in the previous representation leading to the following scheme for $A_1^{(3)}$.

$$
\begin{array}{c|cccccc}
\mathbf{1} & & \times & \cdot & \cdot & \cdot & \cdot \\
\mathbf{2} & & \times & \times & \cdot & \cdot & \cdot \\
\mathbf{3} & & & \otimes & \times & \cdot & \cdot \\
\mathbf{4} & & & & & \times & \cdot \\
\mathbf{5} & & & & & & \times \\
\hline
 & 2 \quad 1 & & & & &
\end{array}
$$

Annihilating now the element marked with $\otimes$ creates an extra Givens on the left, which we cannot apply onto the matrix because it interferes with the Givens transformations already present in $A_2^{(3)}$. We get the following graphical representation.

$$
\begin{array}{c|ccccc}
\mathbf{1} & \times & \cdot & \cdot & \cdot & \cdot \\
\mathbf{2} & \times & \times & \cdot & \cdot & \cdot \\
\mathbf{3} & & \times & \cdot & \cdot \\
\mathbf{4} & & & \times & \cdot \\
\mathbf{5} & & & & \times \\
\hline
 & 3 & 2 & 1
\end{array}
$$

Applying the Givens transformation in position 1 leads to a slightly changed representation for $A_2^{(3)}$.

$$
\begin{array}{c|ccccc}
\mathbf{1} & \times & \cdot & \cdot & \cdot & \cdot \\
\mathbf{2} & \times & \times & \cdot & \cdot & \cdot \\
\mathbf{3} & & \times & \cdot & \cdot \\
\mathbf{4} & & \otimes & \times & \cdot \\
\mathbf{5} & & & & \times \\
\hline
 & 2 & 1
\end{array}
$$

Determining and applying the Givens transformation to annihilate $\otimes$ leads to the following form for the matrix $A_3^{(3)} = G_3^{(3)T} A_2^{(3)} G_3^{(3)}$.

$$
\begin{array}{c|ccccc}
\mathbf{1} & \times & \cdot & \cdot & \cdot & \cdot \\
\mathbf{2} & \times & \times & \cdot & \cdot & \cdot \\
\mathbf{3} & & \times & \cdot & \cdot \\
\mathbf{4} & & & \times & \cdot \\
\mathbf{5} & & & & \times \\
\hline
 & 3 & 2 & 1
\end{array}
$$

Rewriting this scheme for the matrix $A_3^{(3)}$ gives us

$$
\begin{array}{c|ccccc}
\mathbf{1} & \times & \cdot & \cdot & \cdot & \cdot \\
\mathbf{2} & \times & \times & \cdot & \cdot & \cdot \\
\mathbf{3} & & \times & \cdot & \cdot \\
\mathbf{4} & & & \times & \cdot \\
\mathbf{5} & & & \otimes & \times \\
\hline
 & 2 & 1
\end{array}
$$

This creates another element $\otimes$ that needs to be annihilated, leading to a matrix $A_4^{(3)}$ of the following form:

$$
\begin{array}{c|ccccc}
\mathbf{1} & \times & \cdot & \cdot & \cdot & \cdot \\
\mathbf{2} & \times & \times & \cdot & \cdot & \cdot \\
\mathbf{3} & & \times & \cdot & \cdot \\
\mathbf{4} & & & \times & \cdot \\
\mathbf{5} & & & & \times \\
\hline
 & 3 & 2 & 1
\end{array}
$$

This gives us clearly an updated Givens-vector representation for the bottom four rows. This procedure can easily be continued to obtain the complete Givens-vector representation of the transformed matrix. This depicts in some sense the implementation based on the Givens-vector representation as discussed in a forthcoming subsection.

Notes and references

This graphical interpretation of working with Givens transformations was introduced by Delvaux and Van Barel.

☞ S. Delvaux and M. Van Barel. A Givens-weight representation for rank structured matrices. *SIAM Journal on Matrix Analysis and Applications*, 29(4):1147–1170, 2007.

☞ S. Delvaux and M. Van Barel. Unitary rank structured matrices. *Journal of Computational and Applied Mathematics*, 215(1):268–287, March 2008.

In the article on the Givens-weight representation, the authors introduce the generalization of the Givens-vector representation. General schemes for annihilating low rank parts in matrices (of arbitrary forms) are introduced leading to the so-called Givens-weight representation. For these more general matrices, we have more Givens transformations than just one sequence. Moreover instead of a single vector we need to store now a sequence of weights that can be row vectors themselves.

In the article on unitary matrices the authors show how one can use the shift through lemma (Lemma 4.2) to obtain different patterns in the sequences of Givens transformations. This is also examined in more detail in the first volume of the book.

These different patterns of sequences of Givens transformations also give possibilities in changing the flow of an algorithm. All Givens transformations that were used in the reduction of the matrix to semiseparable form were based on zero creating Givens transformations. This means that performing the Givens transformation on a vector will create a zero in this vector. In fact this is not the only type of Givens transformation. There exists also a rank expanding Givens transformation, which can create blocks of rank 1 in a matrix. Based on these rank expanding Givens transformations one can change the reduction to semiseparable form to obtain another kind of reduction algorithm mostly based on rank expanding Givens transformations. More information on rank expanding Givens transformations can be found in the first volume (see also Subsection 9.6.3 in Volume I).

4.2   Implementation details

In this section we will provide some implementation details about the algorithms provided in Chapter 2. As the aim of the book focuses towards eigenvalue computations for semiseparable matrices, we include only details concerning the reduction algorithms of arbitrary matrices towards semiseparable or related forms. This means that we will not include details concerning the reduction towards tridiagonal, Hessenberg nor bidiagonal form. Also no details are given concerning the implementation of the algorithms for transforming sparse and structured rank matrices into each other. These details can be found in the following publications [38, 115, 77, 65, 54]. See also the notes and references following Section 2.6 in Chapter 2.

In the first subsection some details concerning the orthogonal similarity reduction towards semiseparable form are given. The second subsection briefly discusses the orthogonal similarity reduction into a similar semiseparable plus diagonal one. The last two subsections discuss, respectively, the orthogonal similarity reduction into a Hessenberg-like matrix and the orthogonal transformation into an upper triangular semiseparable matrix.

In the last subsection we briefly summarize the computational complexities of all presented reduction schemes.

4.2.1   Reduction to symmetric semiseparable form

First we describe how to implement the orthogonal similarity reduction of a symmetric matrix into semiseparable form, using the Givens-vector representation of a semiseparable matrix. This representation gives information that can directly be used in the QR-algorithm applied to the resulting semiseparable matrix, as will be shown in Chapter 8. The implementation involves detailed shuffling of indices. Therefore only the mathematical details behind the implementation are given. The MATLAB-files can be downloaded from the MASE-homepage[27].

Suppose we are at the beginning of step $k = n - j$ in the reduction algorithm this means that the rows $j + 1, j + 2, \ldots, n$ (columns $j + 1, j + 2, \ldots, n$) are already in the reduced semiseparable form. We can represent this semiseparable part by a vector $\mathbf{r}_j \in \mathbb{R}^j$, Givens transformations

$$\begin{bmatrix} c_i & -s_i \\ s_i & c_i \end{bmatrix}, \qquad i = j + 1, j + 2, \ldots, n - 1 \tag{4.4}$$

and a vector $[v_{j+1}, v_{j+2}, \ldots, v_n]^T$. At this point, the matrix $A^{(k)}$ similar to the original symmetric matrix A can be divided into four blocks:

$$\begin{bmatrix} A_k & c_{j+1}\mathbf{r}_j & c_{j+2}s_{j+1}\mathbf{r}_j & \cdots & s_{n-1} \ldots s_{j+1}\mathbf{r}_j \\ c_{j+1}\mathbf{r}_j^T & c_{j+1}v_{j+1} & c_{j+2}s_{j+1}v_{j+1} & \cdots & s_{n-1} \ldots s_{j+1}v_{j+1} \\ \hline c_{j+2}s_{j+1}\mathbf{r}_j^T & c_{j+2}s_{j+1}v_{j+1} & c_{j+2}v_{j+2} & & \vdots \\ \vdots & \vdots & & \ddots & \\ s_{n-1} \ldots s_{j+1}\mathbf{r}_j^T & s_{n-1} \ldots s_{j+1}v_{j+1} & \cdots & & v_n \end{bmatrix}, \tag{4.5}$$

with $A_k \in \mathbb{R}^{j \times j}$. In the actual implementation only the vector $[v_{j+1}, v_{j+2}, \ldots, v_n]^T$, the Givens transformations and the matrix

$$\begin{bmatrix} A_k & \mathbf{r}_j \\ \mathbf{r}_j^T & v_{j+1} \end{bmatrix}$$

are stored, as this provides all the necessary information for reconstructing the matrix $A^{(k)}$.

The first substeps in step k of the method described in the proof of Theorem 2.6, eliminate the elements $1, 2, \ldots, j - 1$ in row $j + 1$ by multiplying $A^{(k)}$ to

---
[27]http://www.cs.kuleuven.be/~mase/books/

the right by the Givens transformations $G_1^{(k)}$, $G_2^{(k)}$, ..., $G_{j-1}^{(k)}$, i.e.,

$$\mathbf{r}_j^T G_1^{(k)} G_2^{(k)} \cdots G_{j-1}^{(k)} = [0, \ldots, 0, \tilde{\alpha}_j] = \tilde{\mathbf{r}}_{j+1}^T.$$

It is clear that we can obtain a similar result by applying a Householder transformation $H^{(k)}$ on the vector $\mathbf{r}_j^T$ such that the following equation is obtained:

$$\mathbf{r}_j^T H^{(k)} = [0, \ldots, 0, \alpha_j] = \hat{\mathbf{r}}_j^T,$$

with $|\alpha_j| = |\tilde{\alpha}_j|$. Performing the embedded similarity Householder transformation on the matrix (4.5) transforms it into the following matrix:

$$\begin{bmatrix} H^{(k)^T} A_k H^{(k)} & c_{j+1}\hat{\mathbf{r}}_j & c_{j+2}s_{j+1}\hat{\mathbf{r}}_j & \cdots & s_{n-1}\ldots s_{j+1}\hat{\mathbf{r}}_j \\ c_{j+1}\hat{\mathbf{r}}_j^T & c_{j+1}v_{j+1} & c_{j+2}s_{j+1}v_{j+1} & \cdots & s_{n-1}\ldots s_{j+1}v_{j+1} \\ c_{j+2}s_{j+1}\hat{\mathbf{r}}_j^T & c_{j+2}s_{j+1}v_{j+1} & c_{j+2}v_{j+2} & & \vdots \\ \vdots & \vdots & & \ddots & \\ s_{n-1}\ldots s_{j+1}\hat{\mathbf{r}}_j^T & s_{n-1}\ldots s_{j+1}v_{j+1} & \cdots & & v_n \end{bmatrix}. \quad (4.6)$$

The upper left part $H^{(k)^T} A_k H^{(k)}$ can be written as

$$H^{(k)^T} A_k H^{(k)} = \begin{bmatrix} A_{k-1} & \mathbf{r}_{j-1} \\ \mathbf{r}_{j-1}^T & \tilde{v}_j \end{bmatrix},$$

leading to the following matrix, identical to the one in Equation (4.6)

$$\begin{bmatrix} A_{k-1} & \mathbf{r}_{j-1} & 0 & 0 & \cdots \\ \mathbf{r}_{j-1}^T & \tilde{v}_j & c_{j+1}\alpha_j & c_{j+2}s_{j+1}\alpha_j & \cdots \\ 0 & c_{j+1}\alpha_j & c_{j+1}v_{j+1} & c_{j+2}s_{j+1}v_{j+1} & \cdots \\ 0 & c_{j+2}s_{j+1}\alpha_j & c_{j+2}s_{j+1}v_{j+1} & c_{j+2}v_{j+2} & \\ \vdots & \vdots & \vdots & & \ddots \end{bmatrix}. \quad (4.7)$$

From this point the Givens transformation $G_j^{(k)}$ can be calculated, such that the following equation is satisfied[28]:

$$[\alpha_j, v_{j+1}]\, G_j^{(k)} = [0, \alpha_{j+1}] \qquad \text{with} \qquad G_j^{(k)} = \begin{bmatrix} c_j & s_j \\ -s_j & c_j \end{bmatrix}.$$

After applying the similarity Givens transformation $G_j^{(k)}$ on the Matrix (4.7) we get the following matrix:

$$\begin{bmatrix} A_{k-1} & c_j\mathbf{r}_{j-1} & s_j\mathbf{r}_{j-1} & 0 & \cdots & 0 \\ c_j\mathbf{r}_{j-1}^T & c_j\hat{v}_j & s_j\hat{v}_j & 0 & \cdots & 0 \\ s_j\mathbf{r}_{j-1}^T & s_j\hat{v}_j & c_{j+1}\alpha_{j+1} & c_{j+2}s_{j+1}\alpha_{j+1} & \cdots & s_{n-1}\ldots s_{j+1}\alpha_{j+1} \\ 0 & 0 & c_{j+2}s_{j+1}\alpha_{j+1} & c_{j+2}v_{j+2} & & \\ \vdots & \vdots & \vdots & & \ddots & \vdots \\ 0 & 0 & s_{n-1}\ldots s_{j+1}\alpha_{j+1} & \cdots & & v_n \end{bmatrix}, \quad (4.8)$$

---

[28]The Givens here is defined slightly differently, with regard to the minus sign, as the one in Equation (4.4) in order to obtain the representation Givens in the form of Equation (4.4).

with $\hat{v}_j = \tilde{v}_j c_j + s_j c_{j+1}\alpha_j$. The Givens transformation $G_j^{(k)^T}$ and the element $\hat{v}_j$ can already be stored as part of the representation of the new semiseparable part that has to be formed at the bottom of the matrix. The process explained here can now be repeated to make all rows j, $j+1$ up to n semiseparable, because the Matrix (4.8) has essentially the same structure as the Matrix (4.6). One can clearly see that the Givens transformations $G_j^{(k)}$ calculated here can immediately be stored to represent the lower right semiseparable block of the matrix.

Note 4.4. *The latter step, i.e., the reduction of the lower right part to the semiseparable structure, can also be applied directly on a semiseparable matrix. This corresponds to performing a QR-step (in fact a QL-step) without shift on this semiseparable matrix (see Chapter 3, Subsection 3.2.1).*

4.2.2 Reduction to semiseparable plus diagonal form

Carefully comparing the reduction algorithms to either semiseparable or semiseparable plus diagonal form leads to the observation that there is little difference in both methods. The only difference is that before computing and performing the Givens transformations of the chasing procedure one has to slightly change the corresponding diagonal elements. This is also the only important difference in the implementation of the reduction to semiseparable plus diagonal form. Hence we do not include a thorough study of this implementation. More details can be found for example in [172].

4.2.3 Reduction to Hessenberg-like

The implementation of this part is an easy, almost trivial extension of the implementation given in Subsection 4.2.1. Now attention is focused on the lower triangular part. All the Givens and Householder transformations are determined by the lower triangular part of the matrix.

We will describe the structure of an intermediate matrix in the reduction when we represent this matrix with the Givens-vector representation.

Suppose we are at step k of the reduction, with $j = n - k$. This means that we have a semiseparable part of dimension k and a nonsemiseparable part of dimension j. The intermediate matrix $A^{(k)}$ has the following form:

$$
\begin{bmatrix}
\begin{array}{ll}
A_k & \mathbf{q}_j \\
c_{j+1}\mathbf{r}_j^T & c_{j+1}v_{j+1}
\end{array} & B_k \\
\begin{array}{ll}
c_{j+2}s_{j+1}\mathbf{r}_j^T & c_{j+2}s_{j+1}v_{j+1} \\
\vdots & \\
\\
\\
s_{n-1}\ldots s_{j+1}\mathbf{r}_j^T &
\end{array} &
\begin{array}{lllll}
c_{j+2}v_{j+2} & a_{12} & \cdots & & a_{1,m-2} \\
& \ddots & & & \vdots \\
& & \ddots & & \\
& & & c_{n-1}v_{n-1} & a_{m-2,m-2} \\
\cdots & & & s_{n-1}v_{n-1} & v_n
\end{array}
\end{bmatrix},
$$

with $\mathbf{q}_j, \mathbf{r}_j$ as vectors of length j. The matrix B_k is a $(j+1) \times (k-1)$ block. The entries a_{ij} also have no special structure in general. One can see immediately that the structure of this matrix is completely similar to the one of Equation (4.5), except that symmetry is lost. Therefore we do not go into the details of this implementation anymore.

4.2.4   Reduction to upper triangular semiseparable form

In this section only the reduction towards an upper triangular semiseparable form is described. All the other algorithms (as described in Theorem 2.18) can be deduced in an analogous way. Again we give some explicit formulas for the matrices. Based on these formulas the reader should be able to construct the algorithm.

Suppose we are at step k of the algorithm, and we want to reduce an $m \times n$ matrix A into upper triangular semiseparable form. (Note that for this implementation we start at the top of the matrix, while the reduction in the previous section started at the end.) This means that our matrix has a part of dimension $k \times n$ already of upper triangular semiseparable form, while a $j = n - k$ dimensional part is still not reduced. At the beginning of this step our matrix $A^{(k)}$ has the following form:

$$\begin{bmatrix} v_n & s_{n-1}v_{n-1} & \cdots & & & \cdots & & s_{n-1}\cdots s_{j+1}\mathbf{r}_j^T \\ 0 & c_{n-1}v_{n-1} & & & & & & \vdots \\ \vdots & & \ddots & & & & & \\ 0 & \cdots & 0 & c_{j+2}v_{j+2} & c_{j+2}s_{j+1}v_{j+1} & & c_{j+2}s_{j+1}\mathbf{r}_j^T & \\ \hline 0 & \cdots & & 0 & c_{j+1}v_{j+1} & & c_{j+1}\mathbf{r}_j^T & \\ \vdots & & & \vdots & & & & \\ 0 & \cdots & & 0 & \mathbf{q}_j & & A_k & \end{bmatrix} \quad (4.9)$$

The matrix is partitioned in such a way that the upper left block is of dimension $(k-1) \times (k-1)$ and the lower right block is of dimension $(m - k + 1) \times (n - k + 1)$. The vectors $\mathbf{q}_j$ and $\mathbf{r}_j$ are of length $(m - k)$ and $(n - k)$, respectively. The elements v_i, c_i, s_i are defined in the same way as in the previous section. Equation (4.9) shows that the matrix $A^{(k)}$ already has the upper triangular part of the correct semiseparable form. In this step one wants to add one more row to the semiseparable structure, such that in the beginning of step $(k+1)$ we have a $(k+1) \times (k+1)$ upper triangular semiseparable matrix in the upper left $(k+1) \times (k+1)$ block. The first step consists of applying a Householder transformation or a sequence of Givens transformations to annihilate all the elements of the vector $\mathbf{r}_j^T$ except for the first one. We choose to annihilate the elements with a Householder transformation $H_r^{(k)}$. (The r denotes that we perform this transformation on the right of the matrix.):

$$\mathbf{r}_j^T H_r^{(k)} = [\alpha_j, 0, \ldots, 0] = \hat{\mathbf{r}}_j^T.$$

Embedding this transformation in the identity matrix and applying this transformation onto the matrix $A^{(k)}$ we get:

$$
\begin{bmatrix}
v_n & s_{n-1}v_{n-1} & \cdots & & & \cdots & & s_{n-1}\cdots s_{j+1}\hat{\mathbf{r}}_j^T \\
0 & c_{n-1}v_{n-1} & & & & & & \vdots \\
\vdots & & \ddots & & & & & \\
0 & \cdots & 0 & c_{j+2}v_{j+2} & c_{j+2}s_{j+1}v_{j+1} & & c_{j+2}s_{j+1}\hat{\mathbf{r}}_j^T \\
0 & \cdots & & 0 & c_{j+1}v_{j+1} & & c_{j+1}\hat{\mathbf{r}}_j^T \\
\vdots & & & \vdots & & & \\
0 & \cdots & & 0 & \mathbf{q}_j & & A_k H_r^{(k)}
\end{bmatrix}.
$$

At this stage we are ready to annihilate the element α_j by applying a Givens transformation such that

$$
[v_{j+1}, \alpha_j]\, G_{r,j}^{(k)} = [\alpha_{j+1}, 0].
$$

(Again the r denotes that the Givens transformation is performed on the right-side of the matrix.) Note that the same result can be achieved when applying a Householder transformation $\tilde{H}_r^{(k)}$ such that

$$
\left[v_{j+1}, \mathbf{r}_j^T \right] \tilde{H}_r^{(k)} = [\tilde{\alpha}_{j+1}, 0, \ldots, 0],
$$

with $|\tilde{\alpha}_{j+1}| = |\alpha_{j+1}|$. After applying the embedded Givens transformation $G_{r,j}^{(k)}$, our matrix has the following structure:

$$
\begin{bmatrix}
v_n & s_{n-1}v_{n-1} & \cdots & & s_{n-1}\cdots s_{j+1}\alpha_{j+1} & 0 \\
0 & c_{n-1}v_{n-1} & & & & \vdots \\
\vdots & & \ddots & & & \\
0 & \cdots & 0 & c_{j+2}v_{j+2} & c_{j+2}s_{j+1}\alpha_{j+1} & 0 \\
0 & \cdots & & 0 & c_{j+1}\alpha_{j+1} & 0 \\
\vdots & & & \vdots & & \\
0 & \cdots & & 0 & \hat{\mathbf{q}}_j & \hat{A}_k
\end{bmatrix}. \qquad (4.10)
$$

The vector $\mathbf{q}_j$ and the matrix $A_k H_r^{(k)}$ also changed after applying the Givens transformation. The new vector and matrix are denoted as $\hat{\mathbf{q}}_j$ and $\hat{A}_k$. The next step consists of applying another Householder transformation on the left of the Matrix (4.10) such that:

$$
H_l^{(k)^T} \hat{\mathbf{q}}_j = [\beta_{j+1}, 0, \ldots, 0]^T.
$$

(The l denotes that the transformation will be performed on the left side of the matrix.) Applying this transformation on the matrix and rewriting $H_l^{(k)^T} \hat{A}_k$ as

$$
H_l^{(k)^T} \hat{A}^{(j)} = \begin{bmatrix} v_j & \mathbf{r}_{j-1}^T \\ \mathbf{q}_{j-1} & A_{k-1} \end{bmatrix},
$$

gives us

$$
\left[
\begin{array}{cccc|ccc}
v_n & s_{n-1} & \cdots & & s_{n-1}\cdots s_{j+1}\alpha_{j+1} & 0 & 0 \\
0 & c_{n-1}v_{n-1} & & & \vdots & \vdots & \vdots \\
\vdots & & \ddots & & & & \\
0 & \cdots & 0 & c_{j+2}v_{j+2} & c_{j+2}s_{j+1}\alpha_{j+1} & 0 & 0 \\
0 & \cdots & & 0 & c_{j+1}\alpha_{j+1} & 0 & 0 \\
\vdots & & & \vdots & \beta_{j+1} & v_j & \mathbf{r}_{j-1}^T \\
0 & \cdots & & 0 & 0 & \mathbf{q}_{j-1} & A_{k-1}
\end{array}
\right]. \qquad (4.11)
$$

In the next step we will add one row to the semiseparable structure. Apply the Givens transformation $G_{l,j}^{(k)^T}$ with

$$
G_{l,j}^{(k)^T} = \left[
\begin{array}{cc}
c_j & s_j \\
-s_j & c_j
\end{array}
\right]
$$

on the rows k and $k+1$ such that $G_{l,j}^{(k)^T}[c_{j+1}\alpha_{j+1},\beta_{j+1}]^T = [\tilde{v}_{j+1},0]^T$. This will give us the following matrix:

$$
\left[
\begin{array}{cccc|cccc}
v_n & s_{n-1} & \cdots & & s_{n-1}\cdots s_{j+1}\alpha_{j+1} & & 0 & \\
0 & c_{n-1}v_{n-1} & & & \vdots & & \vdots & \\
\vdots & & \ddots & & & & & \\
0 & \cdots & 0 & c_{j+2}v_{j+2} & c_{j+2}s_{j+1}\alpha_{j+1} & & 0 & \\
0 & \cdots & & 0 & \tilde{v}_{j+1} & s_j v_j & s_j \mathbf{r}_{j-1}^T & \\
\vdots & & & \vdots & 0 & c_j v_j & c_j \mathbf{r}_{j-1}^T & \\
& & & & \vdots & & & \\
0 & \cdots & & 0 & 0 & \mathbf{q}_{j-1} & A^{(j-1)} &
\end{array}
\right].
$$

Applying the Givens transformation $G_{r,j+1}^{(k)}$ to the columns k and $k+1$ in order to annihilate the upper part of column k, we get a matrix that is essentially the same as the Matrix (4.11). One can continue and chase the complete structure upwards.

4.2.5   Computational complexities of the algorithms

We will briefly recapitulate the complexities of the proposed reduction algorithms. The complexities as presented here for the reductions to structured rank forms are based on the implementations presented in this section. Depending on the detailed implementation, the constants might slightly vary. Firstly we will discuss the complexity in case there is no deflation. Secondly we will assume that deflation is applied.

No deflation

An overview list of the complexities is presented.

- The reduction to tridiagonal form costs $4/3n^3 + \mathcal{O}(n^2)$ involving Householder transformations and $2n^3 + \mathcal{O}(n^2)$ involving Givens transformations.[29]

- The reduction to semiseparable form costs the same as the reduction to tridiagonal form, plus an extra $9n^2 + \mathcal{O}(n)$ operations.

- The reduction to semiseparable plus diagonal form costs the same as the reduction to tridiagonal form, plus an extra $10n^2 + \mathcal{O}(n)$ operations.

- The reduction to Hessenberg form costs $10/3n^3$ operations, when using Householder transformations.

- The reduction to Hessenberg-like costs an extra $3n^3 + \mathcal{O}(n^2)$ operations.

- The reduction to bidiagonal form costs $4mn^2 - 4/3n^3$ operations plus lower order terms involving Householder transformations.

- The reduction to upper triangular semiseparable form costs an extra $9n^2$ operations.

In case of deflation

In this section we will briefly compare the complexity of computing the eigenvalues of a symmetric matrix, via the reduction methods based on semiseparable and semiseparable plus diagonal matrices and the traditional method based on tridiagonal matrices. Some algorithms for computing the spectrum of semiseparable (plus diagonal) matrices are covered further in the text. Let us start first with comparing the complexities of the reduction methods.

To reduce a symmetric matrix to tridiagonal form, we only need to perform Householder transformations. The cost of performing a symmetric Householder transformation on a matrix of size n equals $4n^2 + 8n + 7$ operations, leading to a global reduction cost of $4/3n^3 + \mathcal{O}(n^2)$. This cost is shared by the reduction to semiseparable form, and one cannot get rid of this n^3 term.

The cost of the chasing steps performed in the reduction to semiseparable and semiseparable plus diagonal form equals (for a block of size n):

$$\alpha(n-1) + \beta,$$

with $\alpha = 18, \beta = 9$ for the reduction to semiseparable form, and $\alpha = 20, \beta = 13$, for the reduction to semiseparable plus diagonal form. We chose $\alpha = 0 = \beta$ in case the matrix is reduced to tridiagonal form.

During the reduction to one of the structured rank forms, deflation of blocks may occur. Deflating these blocks has its influence on the complexity of the chasing,

---

[29]The operation count might vary depending on the choice made by the authors. Sometimes operations of the form $ax + b$ are counted as one floating point operation, whereas others count this as two operations.

but it also heavily influences the complexity of the algorithms for computing the eigenvalues, as they work on smaller blocks. Let us first investigate the reduction algorithms.

Suppose we reduce a matrix A to semiseparable (plus diagonal) form. During the reduction we obtain convergence to k blocks, which we can deflate. Suppose the k blocks have the following sizes, in order of deflation: $n_1, n_2, n_3, \ldots, n_k$. Each of the blocks is deflated after $l_1, l_1 + l_2, l_1 + l_2 + l_3, \ldots$ steps.

This means that the global cost of reducing a matrix to one of the three forms is the following one. In order to deflate the first block, we need to perform

$$\sum_{i=1}^{l_1} \left(\alpha(i-1) + \beta + 4(n-i)^2 + 8(n-i) + 7 \right)$$

operations. To deflate the second block, an extra

$$\sum_{i=1}^{l_2} \left(\alpha(i-1+l_1-n_1) + \beta + 4(n-i-l_1)^2 + 8(n-i-l_1) + 7 \right)$$

operations are needed. Globally, in order to deflate block q with $1 \leq q \leq k$ we need

$$\sum_{j=1}^{q} \left(\sum_{i=1}^{l_j} \left(\alpha(i-1+\sum_{p=1}^{q-1}(l_p - n_p)) + \beta \right. \right.$$
$$\left. \left. + 4\left(n-i-\sum_{p=1}^{q-1}l_1\right)^2 + 8\left(n-i-\sum_{p=1}^{q-1}l_1\right) + 7 \right) \right)$$

operations in total. The terms in the first line are related to the chasing, and the terms in the second line are related to the Householder tridiagonalization.

We know now the number of operations needed in order to deflate blocks inside the reduction algorithms. In general, the reduction to tridiagonal form does not reveal blocks. In order to compare the eigenvalue solvers globally based on the tridiagonal reduction, and the reduction to semiseparable (plus diagonal) form, we need to have estimates on the complexity of the computation of eigenvalues of semiseparable (plus diagonal) and tridiagonal matrices.

We list here some of the complexities of algorithms for computing the eigenvalues of a matrix of size n.

- Standard QR-algorithms for tridiagonal matrices cost approximately $30n$ for one step of the QR-method, and it takes globally two steps to converge to an eigenvalue.

- Standard QR-algorithms for semiseparable (plus diagonal) matrices (see [164, 157] and also Chapter 7) cost approximately $40n$ for one step, but they converge slightly faster, at approximately 1.7 steps.

- Divide and conquer for tridiagonal matrices (see [44, 27]) takes $\mathcal{O}(n^2)$ operations.

- A divide-and-conquer method for semiseparable (plus diagonal matrices) (see [117, 40] and also Chapter 10), takes also approximately $\mathcal{O}(n^2)$ operations.

Assume now that after a block has been separated in the reduction method that we immediately compute its eigenvalues via one of the above methods. Once we have computed these eigenvalues, we continue the reduction until another block separates, of which we then compute the eigenvalues.

Experiment 4.5. *In Figure 4.1, we compare the speed of convergence of the above approaches for the computation of the eigenvalues based on the semiseparable plus diagonal and on the tridiagonal approach. On the vertical axis the number of eigenvalues computed is depicted, and on the horizontal axis, the number of flops that were needed to compute this amount of eigenvalues is presented. The algorithms used for computing the eigenvalues of both approaches are the standard QR-algorithms. We considered also the complexity with regard to the divide-and-conquer methods, but they were rather similar and hence not included.*

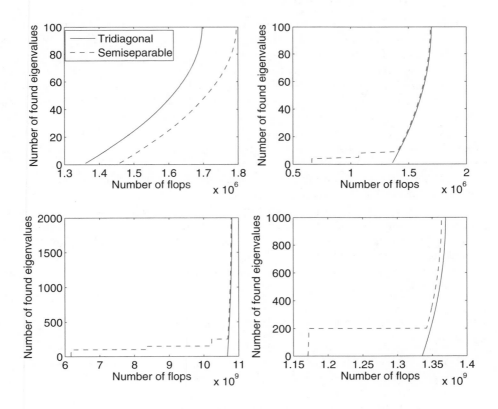

Figure 4.1. *Comparison in speed.*

In the first figure, we compared the speed, if no blocks are separated in the semiseparable plus diagonal case. In this case one can clearly see that the semiseparable case is slightly slower than the tridiagonal case. The second figure was for a matrix of size 100, separating two blocks of sizes 4 each after an extra 20 steps in the reduction procedure. The third figure is for a problem size 2000 and $l_1 = 500, l_2 = 300$ and $l_3 = 500$. The separated blocks are of sizes 100, 50 and 100, respectively. In the second and third figure, the final complexity is almost equal, but in the two cases, the semiseparable approach already revealed eigenvalues, a long time before the tridiagonal approach did. In case one is interested in only these largest eigenvalues, the semiseparable approach performs much better. In the last example only one large block is separated in the reduction of a matrix of size 1000. After 500 steps in the reduction procedure, a matrix of size 200 is separated. We see that in this case the semiseparable case outperforms the tridiagonal case and, moreover, it finds much faster these 200 dominant eigenvalues.

4.3   Numerical experiments

In this section we will illustrate, by means of numerical examples, some of the previously deduced theoretical results. The presented examples are based on the results in the following publications [159, 154, 171].

As the behavior of the reduction to Hessenberg-like and upper triangular semiseparable is closely related to the reduction to semiseparable form, we only include numerical results related to the latter reduction. Secondly we also include numerical experiments for the reduction to semiseparable plus diagonal form.

The first section deals with the reduction to semiseparable form and includes experiments illustrating the interaction between the Lanczos-Ritz values and the subspace iteration. Also an experiment is included illustrating deflation possibilities using this reduction method and an experiment showing that the diagonal elements of the resulting eigenvalues already approximate the real eigenvalues.

The second section focuses on the reduction to semiseparable plus diagonal form. It is shown how one can tune the convergence behavior using the multishift convergence theory. The interaction between the Lanczos-Ritz values and the subspace iteration is also illustrated. Finally numerical experiments are included to verify the theoretical bound on the convergence speed of the norm of the subblocks as presented in the previous chapter.

4.3.1   The reduction to semiseparable form

The following three sections are based on the algorithm, which transforms any symmetric matrix into a similar semiseparable one. In Experiment 4.6 we want to illustrate the Lanczos convergence behavior of the orthogonal similarity reduction of a symmetric matrix to semiseparable form. Experiments 4.7 up to 4.9 illustrate the Lanczos behavior and the subspace iteration convergence. In Experiment 4.7 an example is created such that the subspace iteration has a large delay and is not visible. After a clear delay in Experiment 4.8 the subspace iteration starts to convergence. In Experiment 4.9 the example is created in such a way that there is no delay in the convergence behavior of the subspace iteration. Experiment 4.10 illustrates the interaction between the Lanczos behavior and the subspace iteration. Experiment 4.11 shows that the reduction to semiseparable form orders somehow the diagonal elements such that they approximate the eigenvalues of the original matrix.

In each experiment, we obtain a symmetric matrix A by transforming the diagonal matrix $\text{diag}(\Lambda)$ containing the prescribed eigenvalues $\Lambda = \{\lambda_1, \lambda_2, \ldots, \lambda_n\}$ by an orthogonal similarity transformation $A = Q^T \text{diag}(\Lambda)Q$. The orthogonal matrix used is taken as the Q-factor in the QR factorization of a random matrix built by the MATLAB command rand(n) where n is the dimension of the matrix.

Experiment 4.6. *We choose the eigenvalues as $\lambda_i = i$ for $i = 1, 2, \ldots, 200$. (These are equidistant eigenvalues.) In Figure 4.2 the eigenvalues are located on the y-axis. In each step of the algorithm (x-axis), a cross is placed if a Ritz value approximates a real eigenvalue up to eight correct digits. This behavior is the same as the one described in [112, Section 4.2, Figure 4.1], for equally spaced eigenvalues. More information on the convergence of Ritz values towards eigenvalues can be found in [112].*

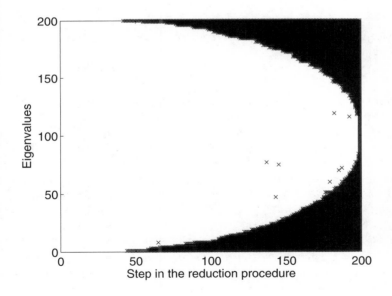

Figure 4.2. *Equally spaced eigenvalues $\{1 : 200\}$.*

Experiment 4.7. *We choose two clusters of equidistant eigenvalues, namely*[30]

---

[30]With $a : b$, where $a \leq b$ are both integers, the sequence of numbers $\{a, a+1, a+2, \ldots, b-1, b\}$

$\{1 : 100\}$ and $\{1000 : 1099\}$, each cluster has the same number of eigenvalues. The following convergence behavior of the Ritz values is computed (see Figure 4.3 left). Note however that the gap between the two intervals does not appear in the following figure (on the right). For each $i = 1 : 199$ the norm of the block $S(i + 1 : n, 1 : i)$ of the resulting semiseparable matrix is plotted. One would expect a small value for $i = 100$, because of the subspace iteration but as explained in Subsection 3.2.1, this is not the case (see Figure 4.3 right) because the Lanczos-Ritz values need to have converged first.

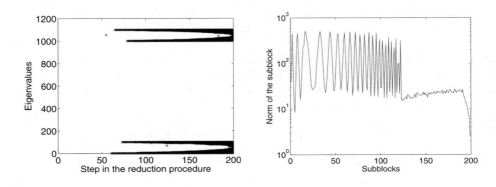

Figure 4.3. *Equally spaced eigenvalues in two clusters $\{1 : 100\}$ and $\{1000 : 1099\}$.*

Experiment 4.8. *In the previous experiment there was no sign of the influence of subspace iteration. In this experiment however we will clearly see the effect of the subspace iteration. For this example again two clusters of eigenvalues were chosen $\{1 : 100\}$ and $\{1000 : 1009\}$. One expects a clear view of the convergence of the subspace iteration in this case. Because the Lanczos-Ritz values approximate the extreme eigenvalues, it will take about 20 steps before the 10 dominant eigenvalues are approximated. After these steps one can expect to see the convergence of the subspace iteration. The first figure (left of Figure 4.4) shows for each step $j = 1, 2, \ldots, n - 1$ in the algorithm the norms of the blocks $S(i : n, 1 : i - 1)$ for $i = n - j : n$, the lines correspond to one particular submatrix, i.e., the norm of this submatrix is shown after every step in the algorithm. The second figure (right of Figure 4.4) is constructed in an analogous way as in Experiment 4.7. In Figure 4.5 it can be seen in which step the Ritz values approximate the most extreme eigenvalues well enough. This is also the point from which the convergence behavior starts in Figure 4.4.*

It is clearly seen in Figure 4.4 that the subspace iteration starts with a small delay (as soon as the Lanczos-Ritz values approximate the dominant eigenvalues well enough the convergence behavior starts).

is meant.

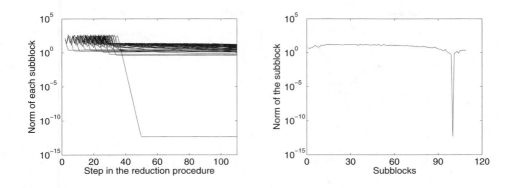

Figure 4.4. *Equally spaced eigenvalues in two clusters* $\{1 : 100\}$ *and* $\{1000 : 1009\}$.

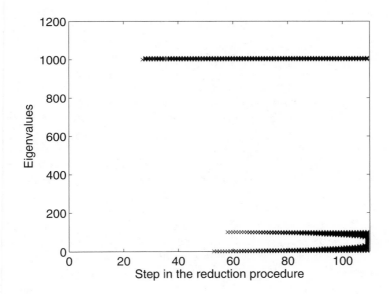

Figure 4.5. *Lanczos behavior of equally spaced eigenvalues in two clusters* $\{1 : 100\}$ *and* $\{1000 : 1009\}$.

Experiment 4.9. *The previous experiment showed that the subspace iteration started working after a delay. In this experiment the largest eigenvalues in absolute value have opposite signs, such that they will be located fast by the Lanczos algorithm and therefore the subspace iteration convergence will show up without a delay. The eigenvalues are located in three clusters* $\{-1004 : -1000, -100 : 100, 1000 : 1004\}$. *The Lanczos-Ritz values will converge fast to the dominant eigenvalues, and there-*

fore the subspace iteration convergence will start fast. Figure 4.6 shows the fast convergence after a few steps of the iteration (only the first 35 steps are shown) and also the Lanczos convergence behavior.

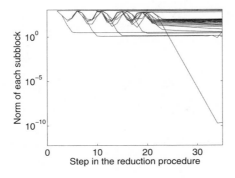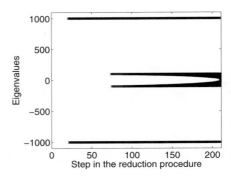

Figure 4.6.   *Equally spaced eigenvalues in three clusters* $\{-1004 : -1000, -100 : 100, 1000 : 1004\}$.

Experiment 4.10. *In the following experiment we want to illustrate the interplay between the Lanczos convergence behavior and the nested subspace iteration of the new method. We want to focus attention on the deflation possibilities of the new algorithm.*

Let us construct a symmetric matrix having the eigenvalues as shown in Figure 4.7. There are three clusters of eigenvalues where the relative gap between the first and second cluster is equal to 0.5 and the same as the relative gap between the second and third cluster. The first cluster has 20 eigenvalues geometrically distributed between 1 and 10^{-2}, the second cluster 20 eigenvalues between $5 \cdot 10^{-3}$ and $5 \cdot 10^{-5}$ and the third 60 eigenvalues between $2.5 \cdot 10^{-5}$ and $2.5 \cdot 10^{-7}$.

Let us first look at Figure 4.8 showing the behavior of the Ritz values for this example. Note that the same behavior is also obtained when using the classical tridiagonalization approach. However, the tridiagonal matrix obtained using this approach does not give a clear indication of the possible clusters or of deflation possibilities, i.e., no subdiagonal element becomes very small in magnitude. One might however choose a different initial transformation Q_0 (see Chapter 3, Section 3.1) to increase the Lanczos convergence behavior towards the last cluster, but one will not be able to recognize two clusters in the tridiagonal matrix. The magnitude of the subdiagonal elements is plotted in Figure 4.9. During the execution of the algorithm to obtain a similar semiseparable matrix, there will be a clear point where a diagonal block corresponding to the first cluster can be deflated and another point where a diagonal block corresponding to the second cluster can be deflated. Looking at Figure 4.8, we see that the eigenvalues of the first cluster are approximated well by the Lanczos-Ritz values from step 25 on. This means that at that moment the subspace iteration where the subspace has dimension 20 will converge towards the subspace

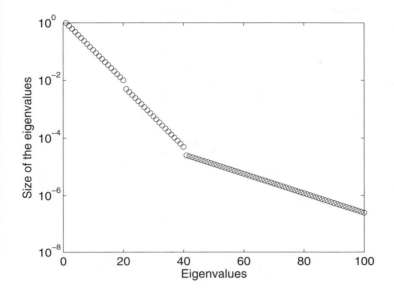

Figure 4.7. *Eigenvalues of the symmetric matrix: three clusters.*

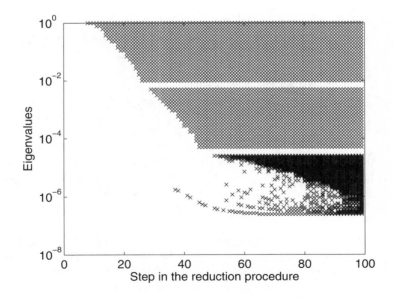

Figure 4.8. *Behavior of the Ritz values.*

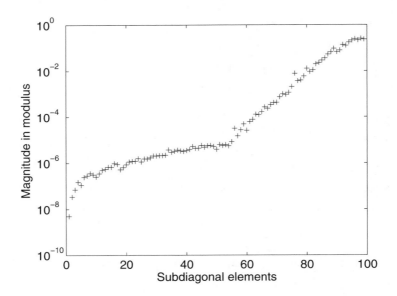

Figure 4.9. *Magnitude of subdiagonal elements of similar tridiagonal matrix.*

corresponding to the first cluster. The convergence factor will be 0.5. This can clearly be seen in Figure 4.10. This figure shows for each step $j = 1, 2, \ldots, n-1$ in the algorithm the norms of the subdiagonal blocks $S(i : n, 1 : i - 1)$ for $i = n - j : n$. Each line in the plot indicates the change of the norm at each step of the algorithm. The line indicated by "o" corresponds to the norm of the subdiagonal block $S(81 : 100, 1 : 80)$. For the eigenvalues of the second cluster a similar convergence behavior occurs around step 45. The line indicated by "" corresponds to the norm of the subdiagonal block $S(61 : 100, 1 : 60)$. The parallel lines in the figure having a smaller slope correspond to the subspace convergence behavior inside each of the clusters. Hence, at step 68 a first diagonal 20×20 block can be deflated while at step 80 the next diagonal 20×20 block can be deflated. This example shows clearly the influence of the convergence of the Lanczos-Ritz values on the convergence behavior of the nested subspace iteration.*

Experiment 4.11 (Stewart's devil's stairs for symmetric matrices).
In the following example (from [141]), we will only take a look at the diagonal elements of the semiseparable matrix and the tridiagonal matrix, after the reduction step. We compare these diagonal elements with the real eigenvalues, which are Stewart's devil's stairs. In Figure 4.11 one can see 10 stairs, with gaps of order 50 between the stairs. All the blocks are of size 10.

It can be seen that the devil's stairs are approximated much better by the diagonal elements of the semiseparable than by the diagonal elements of the tridiagonal matrix.

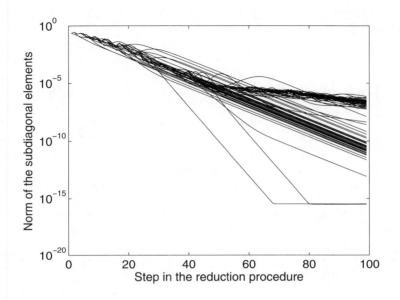

Figure 4.10. *Norms of subdiagonal matrices during the semiseparable reduction.*

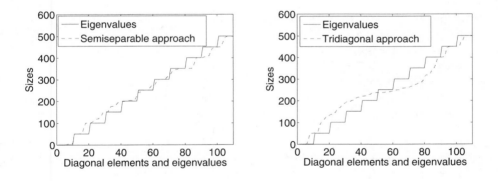

Figure 4.11. *Stewart's Devil's stairs.*

4.3.2   The reduction to semiseparable plus diagonal form

In this section, numerical experiments are given, illustrating the theoretical results presented in the previous chapter. Several types of experiments will be performed. We will investigate the delay of convergence caused by the Lanczos-Ritz values behavior. We will experimentally check the convergence speed of the subspace iteration and we will present some experiments in which the diagonal is chosen in such a way to reveal a specific part of the spectrum. All the experiments are performed in MATLAB. We use MATLAB-style notation. With zeros(i, j), we denote a zero matrix with i rows and i columns, with ones(i, j), we denote a matrix with all entries equal to 1 of dimension $i \times j$, with rand(i, j) we denote a matrix of dimension $i \times j$, with entries random chosen from a uniform distribution between 0 and 1.

Experiment 4.12 (Tuning the multishift convergence behavior). *In these first experiments we construct several matrices, with specific eigenvalues, and we choose the diagonal for the reduction in such a way that it will reveal parts of the spectrum. In the following examples the eigenvalues $\Lambda = \{\lambda_1, \ldots, \lambda_n\}$ of the matrix are given and the matrix itself is constructed as $A = Q^T \operatorname{diag}(\Lambda)Q$, where Q is the orthogonal matrix coming from the QR-factorization of a random matrix. For every example we give the eigenvalues, the diagonal and the number of Householder and Givens transformations performed before the reduction algorithm separated a block containing the desired eigenvalues. A block is separated if the norm of the off-diagonal block is relatively less than 10^{-10}. Also the maximum absolute error between the real and the computed eigenvalues is given.*

1. $\Lambda = \{\operatorname{rand}(1, 10), 100\}$ *and* $\mathbf{d} = \operatorname{zeros}(1, 11)$.
   Number of Householder transformations: 6
   Number of Givens transformations: 21
   Separated eigenvalue: 100
   Maximum absolute error: $4.2633 \cdot 10^{-14}$

2. $\Lambda = \{\operatorname{rand}(1, 100), 100\}$ *and* $\mathbf{d} = \operatorname{zeros}(1, 101)$
   Number of Householder transformations: 6
   Number of Givens transformations: 21
   Separated eigenvalue: 100
   Maximum absolute error: $1.4211 \cdot 10^{-14}$

3. $\Lambda = \{\operatorname{rand}(1, 100), 100, 101, 102\}$ *and* $\mathbf{d} = \operatorname{zeros}(1, 103)$
   Number of Householder transformations: 10
   Number of Givens transformations: 55
   Separated eigenvalues: 100, 101, 102
   Maximum absolute error: $5.6843 \cdot 10^{-14}$

4. $\Lambda = \{1, 100 + \operatorname{rand}(1, 10)\}$ *and* $\mathbf{d} = 100 \cdot \operatorname{ones}(1, 11)$
   Number of Householder transformations: 6
   Number of Givens transformations: 21
   Separated eigenvalue: 1
   Maximum absolute error: $1.4211 \cdot 10^{-14}$

5. $\Lambda = \{1, 100 + \operatorname{rand}(1, 100)\}$ *and* $\mathbf{d} = 100 \cdot \operatorname{ones}(1, 101)$
   Number of Householder transformations: 6
   Number of Givens transformations: 21
   Separated eigenvalues: 1
   Maximum absolute error: $1.4211 \cdot 10^{-14}$

6. $\Lambda = \{1, 2, 3, 100 + \mathrm{rand}(1, 100)\}$ *and* $\mathbf{d} = 100 \cdot \mathrm{ones}(1, 103)$
   Number of Householder transformations: 11
   Number of Givens transformations: 66
   Separated eigenvalue: $1, 2, 3$
   Maximum absolute error: $6.7502 \cdot 10^{-14}$

7. $\Lambda = \{\mathrm{ones}(1, 50) + \mathrm{rand}(1, 50), 100, 10000 \cdot \mathrm{ones}(1, 50) + \mathrm{rand}(1, 50)\}$ *and*
   $\mathbf{d} = [10000, 1, 10000, 1, \ldots, 10000, 1, 10000]$
   Number of Householder transformations: 12
   Number of Givens transformations: 78
   Separated eigenvalue: 100
   Maximum absolute error: $1.8190 \cdot 10^{-12}$

8. $\Lambda = \{1, 2, 3, 100 + \mathrm{rand}(1, 100), 10000, 10001, 10002\}$ *and*
   $\mathbf{d} = [\mathrm{zeros}(1, 6), \mathrm{ones}(1, 96) \cdot 100]$
   First there is convergence to the cluster with the largest eigenvalues:
   Number of Householder transformations: 10
   Number of Givens transformations: 55
   Separated eigenvalues: $1001, 1002, 1003$
   Maximum absolute error: $3.6380 \cdot 10^{-12}$
   Secondly there is convergence to the cluster with the smallest eigenvalues
   Extra number of Householder transformations: 15
   Extra number of Givens transformations: 170
   Separated eigenvalues: $1, 2, 3$
   Maximum absolute error: $1.5099 \cdot 10^{-14}$

The examples illustrate clearly that the convergence behavior can be tuned, by choosing different diagonal values, for reducing the matrix to semiseparable plus diagonal form.

In the Experiments 4.13 to 4.16, the interaction between the Lanczos and the multishift convergence behavior is shown. For each experiment two figures are given. The left figure shows the Lanczos-Ritz values behavior and the right figure shows the subspace iteration convergence. This behavior is similar to the one observed in the previous subsection.

The left figure depicts on the x-axis the iteration step of the reduction algorithm and on the y-axis the eigenvalues of the original matrix. If at step k of the reduction algorithm a Ritz value of the lower right block approximates well-enough (closer than 10^{-5}) an eigenvalue of the original matrix, a cross is placed on the intersection of this step (x-axis) and this eigenvalue (y-axis).

The right figure, shows for all off-diagonal blocks the norm (y-axis), with regard to the iteration step (x-axis).

According to the theory, one should observe decreasing norms, as soon as the Ritz values approximate well enough the dominant eigenvalues with regard to the multishift polynomial.

Experiment 4.13 (The interaction). *In this experiment (see Figure 4.12), we*

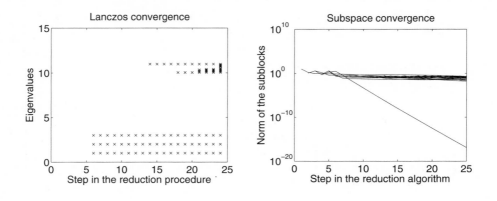

Figure 4.12. *Experiment 4.13.*

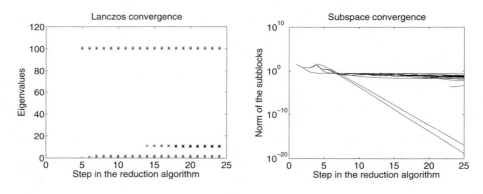

Figure 4.13. *Experiment 4.14.*

have $\Lambda = \{1, 2, 3, 10+\mathrm{rand}(1, 22)\}$, *and the diagonal used for the reduction is* $\mathbf{d} = 10 \cdot$ $\mathrm{ones}(1, 25)$. *In the left figure, we see that after six steps in the reduction algorithm, three eigenvalues are approximated up to five digits by the Lanczos convergence behavior. In the right figure, we see that after step 6 the norm of one subblock starts to decrease. This means that the subspace iteration starts separating a block with these three eigenvalues.*

Experiment 4.14 (The interaction).   *In this experiment (Figure 4.13), a matrix with three clusters in its eigenvalues, was generated:* $\Lambda = \{1, 2, 10+\mathrm{rand}(1, 21), 100, 101\}$, *and the diagonal used for the reduction* $\mathbf{d} = 10 \cdot \mathrm{ones}(1, 25)$. *As the eigenvalues are separated into three clusters and two clusters are both dominant with respect to the multishift polynomial, we would expect two clusters to be separated by the reduction to semiseparable plus diagonal form. This is exactly what we observe in Figure 4.13. The Lanczos Ritz values approximate both clusters (see the left figure), and as soon as these clusters are approximated well enough, the subspace iteration*

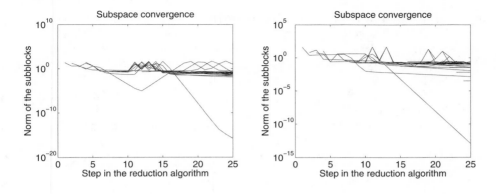

Figure 4.14. *Experiment 4.15.*

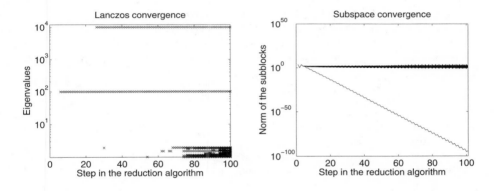

Figure 4.15. *Experiment 4.16*

starts converging. The subspace iteration converges to two clusters, and hence two subblocks show a decreasing norm.

Experiment 4.15 (The interaction). *Continuing with the previous example, but changing the diagonal* **d**, *should only influence the subspace convergence. This can clearly be seen in Figure 4.14. In the left figure we chose* **d** = [100, 101, 10 · ones(1, 23)] *and in the right figure* **d** = [100, 101, 100, 101, 100, 101, 10 · ones(1, 19)]. *For the left figure, we see that the convergence towards the small eigenvalues starts, but then the norm starts to increase again, and finally we get only convergence towards the eigenvalues* 100, 101. *For the second polynomial, however, we do get convergence to the smaller eigenvalues.*

Experiment 4.16 (The interaction). *In the last example (see Figure 4.15) convergence is forced into the middle of the spectrum of the matrix. The considered matrix has eigenvalues* Λ = {ones(1, 50) + rand(1, 50), 100, 10000 · ones(1, 50) +

rand$(1, 50)$} *and* **d** *equal to* $[10000, 1, 10000, 1, \ldots]$. *This forces convergence towards the middle of the spectrum. As soon as there is convergence of the Ritz values towards the eigenvalue* 100, *the subspace iteration starts working. We can see that the convergence rate is not as smooth as in the other cases. This is due to the changing roots in the polynomials* $p_i(\lambda)$.

In the Experiments 4.17 to 4.21 we will investigate the upper bound for the convergence speed as presented in Subsection 3.3.3.[31] In the following experiments the figures show the norm of the submatrix, which should decrease, and the upper bound for this subblock. The upper bound is dependent on the following factor:

$$\frac{\max_{1 \leq j \leq l-1} |\hat{p}_\eta(\lambda_j)|}{\min_{l \leq j \leq n} |\hat{p}_\eta(\lambda_j)|}.$$

To obtain the norm of a subblock, we need to reorder at every step of the method the eigenvalues such that $|\hat{p}_\eta(\lambda_1)| \leq |\hat{p}_\eta(\lambda_2)| \leq \ldots$. In our computation of the upper bound, we assume, however, that we know to which eigenvalues convergence will occur. Hence, we can divide the eigenvalues into two clusters, a cluster $\lambda_1, \ldots, \lambda_{l-1}$ and a cluster $\lambda_l, \ldots, \lambda_n$. We know that when there is convergence we have that

$$\max_{1 \leq j \leq l-1} |\hat{p}_\eta(\lambda_j)| \leq \min_{l \leq j \leq n} |\hat{p}_\eta(\lambda_j)|.$$

Hence our computed upper bound will be the correct one if there is convergence to the eigenvalues $\lambda_l, \ldots, \lambda_n$.

For every example we also give the constant C, which is a measure for the influence of the Lanczos convergence behavior on the subspace iteration. (See Subsection 3.3.3.)

Experiment 4.17 (The convergence speed). *The first example related to the convergence speed, is similar as in the previous two sections. A matrix is constructed with eigenvalues* $\Lambda = \{1, 2, 3, 10 + \text{rand}(1, 22)\}$, *the diagonal* **d** $= 10 \cdot \text{ones}(1, 25)$. *We expect convergence to a* 3×3 *block containing the three eigenvalues* $\{1, 2, 3\}$. *The norm of the off-diagonal block is plotted and decreases linearly as can be seen in the left figure of Figure 4.16. The size of the constant* C, *for calculating the convergence rate equals* $1.0772 \cdot 10^4$. *In the right figure, we plotted almost the same example, but the sizes of the larger eigenvalues are chosen smaller now. The eigenvalues were* $\Lambda = \{1, 2, 3, 5 + \text{rand}(1, 22)\}$, *and* **d** $= 5 \cdot \text{ones}(1, 25)$. *This clearly has an effect on the slope of the line representing the convergence speed. The size of the constant* C *equals* $1.8409 \cdot 10^3$. *In both figures we see that our upper bound predicts well the convergence behavior.*

---

[31]The implementation used in the previous sections is based on the Givens-vector representation (see [165]) and is therefore more stable than the implementation used in this section for generating the figures and calculating the constant C. That is why we get a horizontal line in these figures, once the norm of a subblock reaches the machine precision. This means that using this implementation the norm of the subblocks cannot, relatively speaking, go below the machine precision in contrast to the figures in the previous section.

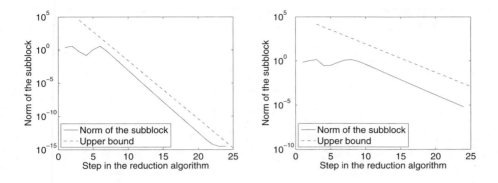

Figure 4.16. *Experiment 4.17.*

Experiment 4.18 (The convergence speed). *Also in the case of a diago-nal with varying elements, our upper bound provides an accurate estimate of the decay of the corresponding subblock. We consider the example with eigenvalues* $\Lambda = \{1, 2, 10 + \text{rand}(1, 21), 100, 101\}$. *If we choose as diagonal* $\mathbf{d} = 10 \cdot \text{ones}(1, 25)$, *we get a similar behavior as above (see the left of Figure 4.17) and convergence to the cluster containing the eigenvalues* $100, 101$. *The constant* $C = 3.4803 \cdot 10^3$. *To obtain however the cluster* $1, 2$ *we need to change our diagonal to, e.g.,* $\mathbf{d} = [100, 101, 100, 101, 100, 10 \cdot \text{ones}(1, 21)]$. *Also in this case our upper bound provides an accurate estimate of the decay (see the right of Figure 4.17). The constant* $C = 2.0396 \cdot 10^3$.

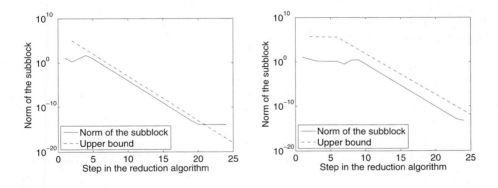

Figure 4.17. *Experiment 4.18.*

Experiment 4.19 (The convergence speed). *In the following experiment we illustrate that if convergence is also slowed down, our upper convergence bound predicts a rather accurate estimate of the convergence rate. We consider here the same matrix three times, with varying values for the diagonal, used to re-*

duce it to semiseparable plus diagonal form. The matrix has eigenvalues $\Lambda = \{1 + \text{rand}(1, 20), 100 + \text{rand}(1, 2)\}$. The diagonals considered for the reduction algorithm are as follows:

$$\mathbf{d}_1 = [\text{zeros}(1, 22)],$$
$$\mathbf{d}_2 = [100, \text{zeros}(1, 21)],$$
$$\mathbf{d}_3 = [100, 100, 100, \text{zeros}(1, 21)].$$

In the first case we expect normal convergence, in the second case a delay, and in the third, an even larger delay. This behavior is shown in Figure 4.18, where the reduction with $\mathbf{d}_1, \mathbf{d}_2$ and $\mathbf{d}_3$ is shown in this order.

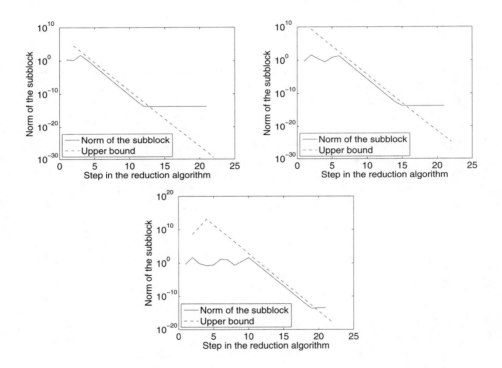

Figure 4.18. *Experiment 4.19.*

Experiment 4.20 (The convergence speed). *In the last two experiments we illustrate false convergence and how our upper bound can deal with, or predict, it. Suppose we have a matrix with eigenvalues $\Lambda = \{1 + \text{rand}(1, 2), 5 + \text{rand}(1, 36), 10 + \text{rand}(1, 2)\}$, suppose the diagonal is chosen in the following way: $\mathbf{d} = [10, 10, 10, 10, 10, 10, 5 \cdot \text{ones}(1, 38)]$. We observe in the convergence behavior (left of Figure 4.19), that first there is convergence, but then suddenly the subblock starts to diverge again. This divergence was predicted by the convergence bound.*

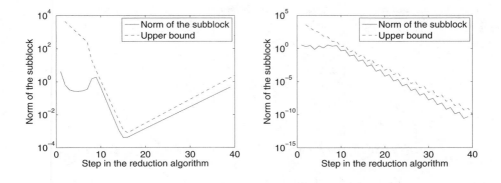

Figure 4.19. *Experiment 4.20.*

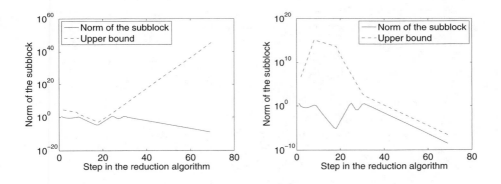

Figure 4.20. *Experiment 4.21.*

This means that the polynomial, predicting the convergence, was not yet strong enough to force convergence to the two small eigenvalues. If we however would have chosen our diagonal as $\mathbf{d} = [10, 5, 10, 5, \ldots]$, we would have been able to force convergence towards the two small eigenvalues (see right of Figure 4.19).

Experiment 4.21 (The convergence speed).   *Let us conclude with almost the same experiment, but let us increase the number of eigenvalues in the middle of the spectrum. The eigenvalues are now $\Lambda = \{1 + \mathrm{rand}(1,2), 5 + \mathrm{rand}(1,66), 10 + \mathrm{rand}(1,2)\}$, and our diagonal values are $\mathbf{d} = [10, 10, 10, 10, 10, 10, 10, 5 \cdot \mathrm{ones}(1,10), \mathrm{zeros}(1,58)]$. As in the previous example, we observe first (see left of Figure 4.20) false convergence towards the small eigenvalues. We see that our upper bound goes up very fast. In reality however we see that the norm of that subblock starts decreasing again in size, so there is convergence towards a block. This block does however not contain the small eigenvalues anymore but the largest two. This is depicted in the second figure (right of Figure 4.20), where we plotted the upper bound related to* *convergence of the largest eigenvalues. It is clear that the upper limit predicts that the separated block will contain the largest eigenvalues, instead of the smallest ones.*

These last examples illustrate that the theoretical upper bound computed for this multishift subspace iteration is rather tight. Moreover this upper bound can also be used as a theoretical device to predict the eigenvalue convergence.

4.4   Conclusions

In this chapter some final issues related to the reduction algorithms and theoretical results of the previous chapters are addressed. The mathematical details behind the implementation of the basic algorithms were given exploiting thereby already the existing structured rank part. A complexity analysis based on these implementations was presented.

Furthermore, different experiments were performed showing some of the features of the reduction algorithms. First the interaction between the two convergence behaviors as explained in the previous chapter was illustrated as well as deflation possibilities. Secondly also numerical experiments were included concerning the reduction to semiseparable plus diagonal form, showing thereby the tunability of the reduction and also comparing the theoretical convergence bound on several examples.

This chapter is the last one of this part. We are able now to efficiently reduce matrices to structured rank form. Moreover we are also aware of the tunability of the presented approach, in order to obtain deflation possibilities and so forth.

In the upcoming part QR-algorithms will be constructed. Combining the results of Parts I and II, one can compute the eigenvalue or singular value decomposition of arbitrary matrices using intermediate structured rank matrices.

Part II

QR-algorithms (eigenvalue problems)

Traditional QR-algorithms for computing the eigenvalues of (dense) matrices consist of two main parts. In a first part the matrices in question are reduced to a specific form, which will be advantageous in the second part. This reduction was considered in the first part of this book, namely the reductions to semiseparable, tridiagonal, Hessenberg, Hessenberg-like form etc. The main aim of these reductions is to reduce the computational complexity of the following QR-algorithm. These QR-algorithms are the subject of this part of the book. We have now the specific structured rank forms and we would like to develop here (implicit) QR-algorithms for these classes of matrices.

Chapter 5 (the first chapter of this part) is a brief summary of theoretical results concerning some essential tools needed to develop implicit QR-algorithms. First we show how to compute a QR-factorization and the QR-algorithm for arbitrary matrices. We will see that this is a costly algorithm. We adapt these results such that they become applicable for sparse matrices, Hessenberg matrices in our case. It is shown that both the QR-factorization and the QR-algorithm lose an order in their computational complexity. Moreover we also prove that the structure of a Hessenberg matrix is preserved under a step of the QR-method. Due to this preservation of the structure we can continue working with Hessenberg matrices throughout the complete QR-algorithm. Finally we show that we can change our QR-algorithm slightly such that we do not need to compute explicitly the factors Q and R from the QR-factorization. This approach is called the implicit QR-method, moreover a theorem, named 'the implicit Q-theorem' is needed to prove correctness of the presented approach.

In the following two chapters of this part, attention is paid to the analogue of the results above but then for the structured rank case. First some theoretical results for structured rank matrices are derived, followed by the real chasing techniques in Chapter 7.

Chapter 6 develops all the essential tools to come to an implicit QR-algorithm for the structured rank matrices as considered in this book. More precisely we will develop this method for semiseparable, Hessenberg-like and upper triangular semiseparable matrices. In a first section we will investigate the structure of the QR-factorization of a Hessenberg-like plus diagonal matrix. The 'plus diagonal' term is essential for the development of the QR-algorithm with shift. Based on this specific QR-factorization we will prove that the structure of a Hessenberg-like matrix is maintained under a step of the QR-algorithm. The second section of this chapter focuses on an implicit Q-theorem, suitable for Hessenberg-like matrices. The proof is more complicated than the corresponding simple Hessenberg case. The final section focuses briefly onto the class of Hessenberg-like plus diagonal matrices, because in Part I of the book also a reduction to this class of matrices was proposed. It is shown that the structure of such a matrix is preserved in case the diagonal has nonzero elements.

Chapter 8 discusses the implementation details of the QR-algorithms already discussed in this part. A distinction between two types of implementations will be made. First a graphical form of the implementation will be discussed thereby using the Givens-vector representation and performing the chasing techniques by using the shift through operation and the fusion. Secondly we show a more mathematical formulation of the algorithms. In the previous chapters little attention was paid to the computation of the eigenvectors. In this chapter several methods for computing the eigenvectors will be explored. Finally numerical experiments based on the previously discussed implementations will be given.

The final chapter of this part discusses some extensions of topics previously presented. A multishift version of the QR-method for structured rank matrices is given. This allows us to compute complex conjugate eigenvalues of real unsymmetric matrices using only computations in the real field. Different variants of the QR-methods for structured rank matrices will also be given. It will be shown that there are different ways to compute the QR-factorization of a structured rank matrix. Different kinds of QR-factorizations will then lead to different kinds of QR-algorithms. Based on these QR-algorithms a new type of iteration, a QH-iteration for computing the eigenvalues, will be presented. It will be shown that this iteration is cheaper to compute than the QR-iteration and moreover it will have an extra convergence behavior. This is an important algorithm as it supersedes the QR-method for structured rank matrices. Hints on how to implement a QR-method in an implicit way for quasiseparable matrices are given. The remainder of the section discusses several references related to QR-algorithms, such as for example the relation between semiseparable plus diagonal matrices and Krylov subspaces. Also results on the computation of roots of polynomials based on companion and related matrices will be given.

✎ *This part of the book focuses on the development of (implicit) QR-algorithms for different types of structured rank matrices.*

Chapter 5 recapitulates the needs and theoretical results for the development of an (implicit) QR-method. People familiar with this can omit the chapter.

In Chapter 6 the theoretical necessities for the development of an implicit QR-method are provided. Two main issues are discussed: the preservation of the rank structure under the QR-algorithm and the development of an implicit QR-algorithm for Hessenberg-like matrices. Based on the QR-factorization presented in Subsection 6.1.1, Theorem 6.8 contains the main result of the first section. Definition 6.14 formulates what is meant in our context with 'unreduced' Hessenberg-like matrices and Definition 6.21 provides us with the 'unreduced number'. These two definitions are used in formulating an implicit Q-theorem for Hessenberg-like matrices (Theorem 6.28).

The effective chasing techniques for restoring the structure in the matrices, a tool needed for implicit QR-algorithms, are deduced in Chapter 7. Based on an unreduced semiseparable matrix, the chasing technique is explored in Subsection 7.1.3. The important structure restoring transformations for Hessenberg-like matrices are provided in Section 7.3. How to perform an implicit chasing technique on an upper triangular semiseparable matrix, in order to compute its singular values, is explained in Subsections 7.4.2 and 7.4.3.

Section 8.1 in Chapter 8 is important as it discusses a graphical viewpoint for interpreting the chasing techniques used in the QR-algorithms. This chasing tech-

nique will be used in the multishift algorithm discussed in Chapter 9, Section 9.4.

Chapter 9 contains miscellaneous topics of which the alternative fast QH-algorithm in Section 9.5 and the eigenvalue methods for computing eigenvalues of polynomials in Section 9.6, are certainly worth investigating.

Chapter 5

Introduction: traditional sparse QR-algorithms

This chapter is merely a repetition of the theory of QR-algorithms for sparse matrices. All the ingredients needed for designing such algorithms will be investigated.

In a first section we will reconsider what is meant by a QR-factorization and a QR-algorithm. It turns out that applying the QR-algorithm directly onto dense matrices is very expensive.

We know, however, from part I, that we can reduce matrices via orthogonal similarity transformations to sparse form, such as Hessenberg or tridiagonal form. Therefore we investigate in Section 5.2 how to efficiently compute the QR-factorization of a Hessenberg matrix. Moreover it will also be shown that the structure of a Hessenberg matrix is preserved after performing a step of the QR-method. This means that we can apply the QR-algorithm on Hessenberg matrices, with a reduced computational complexity.

Rewriting the formulas determining the QR-algorithm we see that we can also apply a similarity transformation onto the matrix, without computing the upper triangular factor R. In Section 5.3 we will see that based on the first Givens transformation of the QR-factorization that we will be able to perform a complete step of the QR-method without having to compute the QR-factorization. This technique is called implicit. To be able to prove the correctness of the presented approach an implicit Q-theorem is needed.

The last section of this chapter pays attention to the QR-algorithm for computing the singular values. The algorithm is an adaptation of the QR-method for tridiagonal matrices. The global algorithm also consists of two steps. Firstly a matrix is reduced to upper bidiagonal form, to reduce the complexity of the second step. In the second step an adaptation of the traditional QR-algorithm is applied to the upper bidiagonal matrices to let them converge to a diagonal matrix containing the singular values on its diagonal.

✎ *This chapter repeats well-known results related to QR-algorithms (for sparse matrices). The sections and subsections in this chapter are ordered somehow* *similarly as the ones in the forthcoming two chapters. This is done to clearly see the similarities and differences between the approaches. This chapter is unessential for the reader familiar with the notions, QR-algorithm, implicit QR-algorithm, QR-factorization, chasing technique, implicit Q-theorem, unreducedness, maintaining of the Hessenberg structure under a step of the QR-method and the implicit QR-algorithm for bidiagonal matrices to compute the singular values.*

For those interested in a brief repetition of these results let us present a concise summary of the most important results. The computation of the QR-factorization is described in Subsection 5.1.1. The QR-algorithm itself is presented in Subsection 5.1.2. Both these subsections illustrate the heavy computational complexity of the QR-algorithm in case no structure of the matrices is involved. Adaptations towards sparse matrices are discussed in Section 5.2. Successively the QR-factorization and the preservation of the structure are discussed. How to obtain an implicit QR-algorithm is discussed in Section 5.3. The idea behind an implicit QR-algorithm, the chasing technique and the implicit Q-theorem, proving the correctness of the followed approach are discussed in this order. Especially the definition of unreducedness, discussed in Subsection 5.3.2 and the implicit Q-theorem (Theorem 5.5) are important. To conclude, also the ideas behind the QR-algorithm for computing the singular values are included (Section 5.4).

5.1   On the QR-algorithm

In this section we will briefly explain the QR-factorization and the QR-algorithm (with shift).

5.1.1   The QR-factorization

Let us start by shortly repeating what is meant with a QR-factorization.

Suppose we have a matrix A and a factorization of the form:

$$A = QR,$$

with Q a matrix having orthonormal columns and R an upper triangular matrix. If Q is a square matrix, and hence orthogonal, this is called the full QR-factorization (in this case R is rectangular). Otherwise, if Q has the same number of columns as A this is called the thin (or reduced) QR-factorization (in this case R is square).

In fact this means that the first k columns of A span the same space as the first k columns of Q. This leads to the following equation, which we already met before in Part I:

$$A\langle \mathbf{e}_1, \mathbf{e}_2, \ldots, \mathbf{e}_k \rangle = \langle \mathbf{q}_1, \mathbf{q}_2, \ldots, \mathbf{q}_k \rangle, \tag{5.1}$$

where $Q = [\mathbf{q}_1, \mathbf{q}_2, \ldots, \mathbf{q}_k]$. Remark that in this case A performs a step of subspace iteration onto the space $\langle \mathbf{e}_1, \mathbf{e}_2, \ldots, \mathbf{e}_k \rangle$, we will come back to this in the next subsection.

Computing the QR-factorization of a matrix can be done in theory, by applying for example the Gram-Schmidt orthogonalization procedure to the columns of the matrix A.

One can formulate and prove the following two theorems, related to the QR-decomposition of a matrix A.

Theorem 5.1. *For every matrix A there exists a full (and hence also a reduced) QR-factorization.*

Concerning the unicity of the decomposition we have the following result.

Theorem 5.2. *Each matrix A of full column rank has an essentially[1] unique reduced QR-factorization.*

Proofs of both of the above theorems can be found in, e.g., [145, 58, 140, 94]. The QR-factorization, as already mentioned, above can be computed by the classical Gram-Schmidt procedure. However, the modified Gram-Schmidt procedure gives better numerical results. Most techniques, used in practice, involve Householder/Givens transformations for making the matrix A upper triangular.

Computing the QR-factorization of a standard dense matrix (not symmetric) involves $4/3n^3 + \mathcal{O}(n^2)$ operations when using Householder transformations and $2/3n^3 + \mathcal{O}(n^2)$ operations in case the matrix is made upper triangular by Givens transformations. We will graphically depict here the computation of the QR-factorization[2] of a dense matrix by means of Householder transformations.

The technique is illustrated on a 5×5 example. Remark that we do not perform similarity transformations anymore, we will only act on the left-hand side of the matrix A. Let $A_0 = A$. Determine a first Householder transformation H_1^T which will annihilate the elements $\otimes$ placed in the first column:

$$
\begin{bmatrix}
\times & \times & \times & \times & \times \\
\otimes & \times & \times & \times & \times \\
\otimes & \times & \times & \times & \times \\
\otimes & \times & \times & \times & \times \\
\otimes & \times & \times & \times & \times
\end{bmatrix}
\longrightarrow
\begin{bmatrix}
\times & \times & \times & \times & \times \\
0 & \times & \times & \times & \times \\
0 & \times & \times & \times & \times \\
0 & \times & \times & \times & \times \\
0 & \times & \times & \times & \times
\end{bmatrix}
$$

$$\Updownarrow$$

$$A_0 \xrightarrow{H_1^T A_0} A_1.$$

We see that clearly the first column is already in upper triangular form. Applying a second Householder transformation, chosen in order to annihilate the elements

---

[1]With essentially we mean, that up to the signs, the columns are identical. A more formal definition will be given in Section 6.2, Definition 6.20.

[2]Even though we speak about 'the' QR-factorization, the factorization is not always unique. Further in the text examples different QR-factorizations will be given, which will not be suitable for applying the QR-algorithm on sparse or structured rank matrices. These QR-factorizations will not preserve the structure, as we will show later on.

in the second column just below the diagonal gives us the following situation:

$$
\begin{bmatrix}
\times & \times & \times & \times & \times \\
0 & \times & \times & \times & \times \\
0 & \otimes & \times & \times & \times \\
0 & \otimes & \times & \times & \times \\
0 & \otimes & \times & \times & \times
\end{bmatrix}
\longrightarrow
\begin{bmatrix}
\times & \times & \times & \times & \times \\
0 & \times & \times & \times & \times \\
0 & 0 & \times & \times & \times \\
0 & 0 & \times & \times & \times \\
0 & 0 & \times & \times & \times
\end{bmatrix}
$$

$$
\Updownarrow
$$

$$
A_1 \xrightarrow{H_2^T A_1} A_2.
$$

Applying two more Householder transformations completes the job. We depict here the application of these transformations:

$$
\begin{bmatrix}
\times & \times & \times & \times & \times \\
0 & \times & \times & \times & \times \\
0 & 0 & \times & \times & \times \\
0 & 0 & \otimes & \times & \times \\
0 & 0 & \otimes & \times & \times
\end{bmatrix}
\longrightarrow
\begin{bmatrix}
\times & \times & \times & \times & \times \\
0 & \times & \times & \times & \times \\
0 & 0 & \times & \times & \times \\
0 & 0 & 0 & \times & \times \\
0 & 0 & 0 & \times & \times
\end{bmatrix}
$$

$$
\Updownarrow
$$

$$
A_2 \xrightarrow{H_3^T A_2} A_3.
$$

One last element needs to be annihilated:

$$
\begin{bmatrix}
\times & \times & \times & \times & \times \\
0 & \times & \times & \times & \times \\
0 & 0 & \times & \times & \times \\
0 & 0 & 0 & \times & \times \\
0 & 0 & 0 & \otimes & \times
\end{bmatrix}
\longrightarrow
\begin{bmatrix}
\times & \times & \times & \times & \times \\
0 & \times & \times & \times & \times \\
0 & 0 & \times & \times & \times \\
0 & 0 & 0 & \times & \times \\
0 & 0 & 0 & 0 & \times
\end{bmatrix}
$$

$$
\Updownarrow
$$

$$
A_3 \xrightarrow{H_4^T A_3} A_4.
$$

It is clear that this method involves a heavy computational effort as $\mathcal{O}(n^2)$ elements need to be annihilated from the lower triangular part.

Computing however the QR-factorization of, e.g., a Hessenberg matrix is much cheaper. Instead of annihilating $\mathcal{O}(n^2)$ elements, only $\mathcal{O}(n)$ elements need to be removed. In the next section we will see how we can exploit this property for reducing the overall complexity when computing the eigenvalues via the QR-algorithm. Also when computing the QR-factorization of structured rank matrices one can decrease the computational effort by exploiting the low rank properties of the involved matrices. An elaborate study on computing QR-factorizations of general structured rank matrices can be found in the first volume and in [56].

5.1.2 The QR-algorithm

The QR-algorithm is a straightforward tool for computing the eigenvalues and eigenvectors of a given matrix. The idea is to compute a sequence of similar matrices converging to a diagonal matrix. We start with a matrix $A = A^{(0)}$ and compute the following sequence:

$$A^{(0)} \rightarrow A^{(1)} \rightarrow A^{(2)} \rightarrow \cdots,$$

where the matrices $A^{(i)}$ are computed as follows[3]:

$$A^{(i)} - \mu_i I = Q^{(i)} R^{(i)}$$
$$A^{(i+1)} = R^{(i)} Q^{(i)} + \mu_i I,$$

where μ_i is a cleverly chosen shift,[4] to speed up convergence A step to go from $A^{(i)}$ to $A^{(i+1)}$ is called a step of the QR-method or a QR step. So, first the QR-factorization of the matrix $A^{(i)}$ minus a shift is computed, then these factors are used for computing the next iterate.

In fact one can interpret this QR-algorithm also as a kind of subspace iteration as we did in the first part. Namely a subspace iteration with a basis transformation after each step, such that one works with changing matrices instead of with changing subspaces. This is clear when considering Equation (5.1). This results in convergence of the eigenvalues closest to μ_i. These eigenvalues will appear at the bottom right of the matrices $A^{(i)}$. More information on how to interpret this subspace iteration and on convergence speed of shifted subspace iterations can be found in the first part and in [182, 179, 177, 178, 191].

After an eigenvalue has converged, one can deflate this value and continue by choosing another μ_i to force convergence to another eigenvalue and so forth. The orthogonal matrices $Q^{(i)}$ need to be stored in order to compute the eigenvectors, or one can also compute the eigenvectors via inverse iteration, but this might result in loss of orthogonality.

It is clear that the QR-algorithm as presented here needs to compute in every step the QR-factorization of the matrix $A^{(i)}$. In case the matrix A is dense the cost of one step of the QR-method is huge, namely $\mathcal{O}(n^3)$ in every step. Hence the total complexity of computing the eigenvalues via this method is $\mathcal{O}(n^4)$ (assuming a fixed number of iterations to find each eigenvalue).

In the following section we will reconsider what we have discussed here, but we will try to reduce the complexity of the method presented here by exploiting the fact that one can more easily (cheaper) compute the QR-factorization of a Hessenberg matrix.

---

[3]We assume in this book that all matrices taken into consideration are real. In general, when working with complex matrices and complex eigenvalues, one also needs to change the transpose to the hermitian conjugate.

[4]Also the shift can be chosen complex in some cases. This can happen when all eigenvalues are not real.

Notes and references

The QR-factorization is said to be a basic linear algebra tool, having its impact in different topics of linear algebra. The factorization is therefore already covered in many basic linear algebra textbooks such as Trefethen and Bau [145], Demmel [58], Golub and Van Loan [94] and Stewart.

> ☞ G. W. Stewart. *Matrix Algorithms, Volume I: Basic Decompositions.* SIAM, Philadelphia, Pennsylvania, USA, 1998.

These books discuss the interpretation via Gram-Schmidt and derive the modified Gram-Schmidt procedure. Finally also other triangularization methods, via Householder and/or Givens transformations are discussed.

The QR-algorithm is closely related to the QR-factorization and is covered in detail the following textbooks: Stewart [142], Parlett [129] and Wilkinson.

> ☞ J. H. Wilkinson. *The Algebraic Eigenvalue Problem.* Numerical Mathematics and Scientific Computation. Oxford University Press, New York, USA, 1999.

The books by Stewart, Parlett, Wilkinson and Watkins are entirely dedicated to eigenvalue problems. Of course information can also be found in the books mentioned above on the QR-factorization, such as [58, 94, 145, 180].

These textbooks also contain all the information that will be mentioned in the forthcoming two sections. They deal with QR-algorithms including the reduction to sparse form and all theoretical results needed for developing implicit QR-algorithms for sparse matrices.

5.2   A QR-algorithm for sparse matrices

In the previous section we briefly discussed the setting of the QR-algorithm. The algorithm as it was discussed there was expensive. $\mathcal{O}(n^4)$ operations were necessary to compute all the eigenvalues of a matrix.

In the first part of this book we discussed an orthogonal similarity transformation for reducing a matrix to sparse format, preserving thereby the eigenvalues. We showed how to reduce a matrix to Hessenberg form or to tridiagonal form in case the original matrix was symmetric.

Let us investigate the possibility of using these sparse matrices for computing the eigenvalues instead of the dense matrices considered in the previous section. We will consider only Hessenberg matrices in this section. The approach for symmetric matrices, using tridiagonal matrices instead of Hessenberg matrices is similar.

5.2.1   The QR-factorization

We assume in this section that we are working with a Hessenberg matrix H, for which we would like to compute the eigenvalues. A Hessenberg matrix has already many more zeros than the original similar matrix A, hence the number of operations involved for computing the QR-factorization will be significantly reduced.

Suppose H to be an $n \times n$ Hessenberg matrix, then there exist $n-1$ Givens transformations such that

$$G_{n-1}^T G_{n-2}^T \ldots G_1^T H = R,$$

with R an upper triangular matrix. The Givens transformations G_i^T are constructed such that applying them onto the matrix H annihilates the element in column i, row $i+1$. Hence every Givens transformation is chosen to annihilate a subdiagonal element starting with the bottom one. There are $n-1$ subdiagonal elements. Hence, we need to perform $n-1$ Givens transformations from top to bottom. Computing this QR-factorization involves only $3n^2 + \mathcal{O}(n)$ operations, in contrast to the $\mathcal{O}(n^3)$ operations from the previous chapter.

In order to be able to use this factorization in every step of the QR-algorithm, we would like to obtain again a Hessenberg matrix after a step of the QR-algorithm. Let us prove that the structure of a Hessenberg matrix is maintained during the QR-algorithm.

5.2.2   Maintaining the Hessenberg structure

In this subsection we will prove that applying one step of the QR-algorithm on the Hessenberg matrix H with shift μ[5], namely:

$$H - \mu I = QR$$
$$\hat{H} = RQ + \mu I,$$

results in a matrix $\hat{H}$, which will also be of Hessenberg form.

The matrix Q^T is constructed in such a way, that applying it onto the matrix $H - \mu I$ makes the matrix upper triangular, moreover this matrix consists of a sequence of $n-1$ Givens transformations from bottom to the top.[6]

Hence the second equation, in which we apply the matrix Q to the right of the matrix R, fills this matrix up with an extra subdiagonal. Let us briefly depict what happens. Assume we have the intermediate matrix R and apply the matrix $Q = G_1 G_2 \cdots G_{n-1}$ to its right. We demonstrate this for a 5×5 example. First we apply the Givens transformation G_1 to the right of the matrix R, the Givens transformation acts on the first two columns of this matrix, this results in a matrix R_1:

---

[5]The shift can also be complex in this case, and hence one has to switch to complex computations.

[6]It is essential that this type of QR-factorization is used, otherwise maintaining the structure is not necessarily true as will be shown in the forthcoming example.

$$\begin{bmatrix} \times & \times & \times & \times & \times \\ & \times & \times & \times & \times \\ & & \times & \times & \times \\ & & & \times & \times \\ & & & & \times \end{bmatrix} \longrightarrow \begin{bmatrix} \times & \times & \times & \times & \times \\ \times & \times & \times & \times & \times \\ & & \times & \times & \times \\ & & & \times & \times \\ & & & & \times \end{bmatrix}$$

$$\Updownarrow$$

$$R \xrightarrow{RG_1} R_1.$$

The second Givens transformation G_2 acts on the second and third column and creates the following intermediate matrix R_2:

$$\begin{bmatrix} \times & \times & \times & \times & \times \\ \times & \times & \times & \times & \times \\ & & \times & \times & \times \\ & & & \times & \times \\ & & & & \times \end{bmatrix} \longrightarrow \begin{bmatrix} \times & \times & \times & \times & \times \\ \times & \times & \times & \times & \times \\ & \times & \times & \times & \times \\ & & & \times & \times \\ & & & & \times \end{bmatrix}$$

$$\Updownarrow$$

$$R_1 \xrightarrow{R_1 G_2} R_2.$$

One can easily continue with this procedure and the resulting matrix will again be a Hessenberg matrix. In the following example we show that the choice of the QR-factorization is essential for the preservation of the structure. In case one does not use the QR-factorization as presented here, based on the sequence of Givens transformations from bottom to top, one cannot guarantee the preservation of the structure in general.

Example 5.3 For Hessenberg matrices, the type of QR-factorization is important to prove that the Hessenberg structure is maintained under one step of the QR-algorithm, as the example will show. Suppose we have a Hessenberg matrix of the following form:

$$H = \begin{bmatrix} \times & \times & \times & \times & \times \\ \otimes & \times & \times & \times & \times \\ 0 & 0 & 0 & \times & \times \\ 0 & 0 & 0 & \times & \times \\ 0 & 0 & 0 & \otimes & \times \end{bmatrix}. \tag{5.2}$$

We perform a sequence of Givens transformations on the Hessenberg matrix such that it becomes upper triangular $G_2^T G_1^T H = R$, for which G_1 and G_2 are Givens transformations and R is an upper triangular matrix. In fact only the two elements marked with $\otimes$ in matrix (5.2) need to be annihilated. Therefore G_1^T performs a Givens transformation on rows 3 and 5 to annihilate the element in position $(5, 4)$ and G_2^T annihilates the element in position $(2, 1)$ by performing a Givens transformation on the first and second row. We can calculate now the matrix $RG_1 G_2$. This corresponds to performing one step of QR without shift to the Hessenberg matrix H. The matrix RG_1G_2 is of the following form:

$$RG_1G_2 = \begin{bmatrix} \times & \times & \times & \times & \times \\ \times & \times & \times & \times & \times \\ 0 & 0 & \times & \times & \times \\ 0 & 0 & \times & \times & \times \\ 0 & 0 & \times & 0 & \times \end{bmatrix}.$$

One can clearly see that the resulting matrix is not Hessenberg anymore. Therefore some assumptions have to be placed on the type of QR-factorization such that the resulting matrix after a step of the QR-algorithm is again Hessenberg. The QR-factorization presented in [94] and in the previous subsection, consists of annihilating all the subdiagonal elements by a Givens transformation on the subdiagonal element and the diagonal element above it. When we use this type of QR-factorization to compute the QR-steps, we will always have again a Hessenberg matrix. ∎

The preservation of the structure implies that instead of applying the QR-algorithm directly on a dense matrix it is better to first reduce it to Hessenberg form. The reduction itself does cost $\mathcal{O}(n^3)$ operations, but the following QR-steps are of the order $\mathcal{O}(n^2)$. This leads to an overall complexity of $\mathcal{O}(n^3)$ instead of $\mathcal{O}(n^4)$ when applying the QR-algorithm to dense matrices.

The QR-algorithm as it is designed here, with the formation of the matrices Q and R and performing the multiplication RQ is called an explicit QR-algorithm. In the following section we will propose a variant of this method namely an implicit QR-algorithm. In an implicit QR-algorithm the matrices Q and R are not formed explicitly.

5.3 An implicit QR-method for sparse matrices

In the previous section we briefly mentioned that the method proposed above was an explicit QR-method, involving the explicit computation of the matrices Q and R. Let us see here what the ingredients are to compute a step of the QR-algorithm implicitly, without explicitly computing the QR-factorization.

5.3.1 An implicit QR-algorithm

Rewriting the equations that determine a step of the QR-algorithm as developed in the previous section, we obtain that the matrix $A^{(i+1)}$ is similar to the matrix $A^{(i)}$ in the following fashion:

$$A^{(i+1)} = Q^{(i)^T} A^{(i)} Q^{(i)},$$

with

$$A^{(i)} - \mu_i I = Q^{(i)} R^{(i)}.$$

The idea of an implicit QR-algorithm is to compute the matrix $A^{(i+1)}$, using only the first formula by exploiting some information from the second equation.

The implicit QR-algorithm as it will be developed in the remainder of this section is only suitable for sparse matrices. Remember that with sparse matrices we mean Hessenberg or tridiagonal matrices.

Let us first develop the bulge chasing technique that will give us the implicit QR-algorithm.

5.3.2 Bulge chasing

We will now develop a structure restoring algorithm. More precisely, assume that our Hessenberg structure has been disturbed in the first two columns by a similarity Givens transformation acting on the first two rows and columns. This means that the Hessenberg matrix has a nonzero element in the first column on the third position. Starting from that matrix we will bring the matrix back to Hessenberg form. It will be shown that the initial matrix needs to obey certain conditions, called the unreducedness of the Hessenberg matrix. In the following section we will see that if we choose this disturbing Givens transformation in a good way, that the resulting Hessenberg matrix is identical (up to the sign) to the one resulting from a step of the QR-algorithm.

Let us construct the bulge chasing technique. Assume we start working on a 5×5 Hessenberg matrix on which we perform an initial similarity Givens transformation G_1, acting on the first two rows and columns. We assume moreover that all considered subdiagonal elements are different from zero. This gives us the following figure, with $H_0 = H$ (the elements not shown in the figures below are assumed to be zero):

$$
\begin{bmatrix}
\times & \times & \times & \times & \times \\
\times & \times & \times & \times & \times \\
 & \times & \times & \times & \times \\
 & & \times & \times & \times \\
 & & & \times & \times
\end{bmatrix}
\longrightarrow
\begin{bmatrix}
\times & \times & \times & \times & \times \\
\times & \times & \times & \times & \times \\
\times & \times & \times & \times & \times \\
 & & \times & \times & \times \\
 & & & \times & \times
\end{bmatrix}
$$

$$\Updownarrow$$

$$H_0 \xrightarrow{\;G_1^T H_0 G_1\;} H_1.$$

The Hessenberg structure in the matrix H_1 is clearly disturbed in the first column. In the next step we perform a similarity Givens transformation in order to remove the element in position $(3, 1)$. We determine Givens transformation G_2, such that applying it onto the second and third row of the matrix H_1 annihilates this disturbing element. Performing this similarity transformation creates the following situation, the element that will be annihilated is again marked with $\otimes$:

$$
\begin{bmatrix}
\times & \times & \times & \times & \times \\
\times & \times & \times & \times & \times \\
\otimes & \times & \times & \times & \times \\
 & & \times & \times & \times \\
 & & & \times & \times
\end{bmatrix}
\longrightarrow
\begin{bmatrix}
\times & \times & \times & \times & \times \\
\times & \times & \times & \times & \times \\
 & \times & \times & \times & \times \\
 & \times & \times & \times & \times \\
 & & & \times & \times
\end{bmatrix}
$$

$$
\Updownarrow
$$

$$
H_1 \xrightarrow{G_2^T H_1 G_2} H_2 .
$$

Unfortunately, removing the first bulge creates another one in position $(4, 2)$. Removing this bulge by a good chosen Givens transformation G_3 gives us the following result:

$$
\begin{bmatrix}
\times & \times & \times & \times & \times \\
\times & \times & \times & \times & \times \\
 & \times & \times & \times & \times \\
 & \otimes & \times & \times & \times \\
 & & & \times & \times
\end{bmatrix}
\longrightarrow
\begin{bmatrix}
\times & \times & \times & \times & \times \\
\times & \times & \times & \times & \times \\
 & \times & \times & \times & \times \\
 & & \times & \times & \times \\
 & & \times & \times & \times
\end{bmatrix}
$$

$$
\Updownarrow
$$

$$
H_2 \xrightarrow{G_3^T H_2 G_3} H_3 .
$$

A last transformation is necessary to remove the final bulge in the last row:

$$
\begin{bmatrix}
\times & \times & \times & \times & \times \\
\times & \times & \times & \times & \times \\
 & \times & \times & \times & \times \\
 & & \times & \times & \times \\
 & & \otimes & \times & \times
\end{bmatrix}
\longrightarrow
\begin{bmatrix}
\times & \times & \times & \times & \times \\
\times & \times & \times & \times & \times \\
 & \times & \times & \times & \times \\
 & & \times & \times & \times \\
 & & & \times & \times
\end{bmatrix}
$$

$$
\Updownarrow
$$

$$
H_3 \xrightarrow{G_4^T H_3 G_4} H_4 .
$$

We see that this technique gives us an easy way to transform a disturbed Hessenberg matrix via similarity transformations back to Hessenberg form. This technique is often called bulge chasing because a bulge, not satisfying the Hessenberg structure, is moved downwards in order to remove it.

We assumed however when starting this reduction procedure that our initial Hessenberg matrix had no subdiagonal elements equal to zero. One can clearly check in the above reduction procedure that a single zero subdiagonal element causes the algorithm to break down in an early stage in the middle of the matrix.

Therefore we assume that we are working with an unreduced Hessenberg matrix. A matrix H is said to be unreduced or irreducible if all the subdiagonal elements are different from zero. Remark that when computing eigenvalues there is no loss of generality when assuming this. A matrix which is not unreduced can be split into several unreduced Hessenberg matrices, such that the eigenvalues of the global matrix are the union of the eigenvalues of the unreduced Hessenberg matrices. Splitting up such a reducible Hessenberg matrix is sometimes referred to as deflation. Assume for example that we have the following Hessenberg matrix H with one subdiagonal element zero:

$$H = \left[\begin{array}{cc} H_{11} & H_{12} \\ 0 & H_{22} \end{array} \right],$$

such that the matrix H is of size $n \times n$, and the matrices H_{11} and H_{22} of sizes $n_1 \times n_1$ and $n_2 \times n_2$, respectively, with $n = n_1 + n_2$. The matrix H is clearly reducible as the subdiagonal element in position $(n_1 + 1, n_1)$ is zero. This matrix can be decomposed into two unreduced Hessenberg matrices H_1 and H_{22}. Moreover the eigenvalues of H_{11} together with the ones of H_{22} give the eigenvalues of H.

Note 5.4. *We remark that the definition of unreducedness for a symmetric tridiagonal matrix is similar to the one for Hessenberg matrices, namely all subdiagonal elements should be different from zero.*

Another feature of an unreduced Hessenberg matrix H is that it has an essentially unique QR-factorization. Moreover also $H - \mu I$, with μ a shift has an essentially unique QR-factorization. This is straightforward because of the zero structure below the subdiagonal. Only the last column can be dependent of the first $n - 1$ columns. Essentially unique means that the sign of the columns of the orthogonal matrices Q can differ as well as the signs of the elements in R.[7] Because the previous $n - 1$ columns are linearly independent, the first $n - 1$ columns of the matrix Q in the QR-factorization of $H = QR$, are linearly independent. Q is an orthogonal matrix, and the first $n - 1$ columns are essentially unique. As the dimension is n and we already have $n - 1$ orthogonal columns, the last column is uniquely defined as being orthogonal to the first $n - 1$ columns. This means that the QR-factorization is essentially unique. Summarizing, this means that the QR-factorization of an unreduced Hessenberg matrix H has the following form:

$$H = Q \left[\begin{array}{c|c} R & \mathbf{w} \\ \hline 0 & \alpha \end{array} \right],$$

for which R is a nonsingular upper triangular matrix of dimension $(n-1) \times (n-1)$, $\mathbf{w}$ is a column vector of length $(n - 1)$, and α is an element in $\mathbb{R}$. We have that $\alpha = 0$ if and only if H is singular. This means that the last column of H depends on the previous $n - 1$.

In the next subsection we will prove that a cleverly chosen G_1, which is performed first on the unreduced Hessenberg matrix H, followed by the bulge chasing technique, can be interpreted as performing a step of QR on the matrix H.

---

[7]A formal definition of essentially uniqueness will be given in Chapter 6, Subsection 6.2.1.

5.3.3 The implicit Q-theorem

Assume we choose the initial transformation G_1 simply as the Givens transformation annihilating the first subdiagonal element of the matrix $H - \mu I$. Choosing the Givens transformation in this way guarantees that performing the chasing technique is equivalent to performing a step of the QR method on the matrix H with shift μ. Let us prove this in this section. To do so we need the implicit Q-theorem.

The implicit Q-theorem for Hessenberg matrices is completely similar to the one for symmetric tridiagonal matrices. We formulate only the Hessenberg case here.

Theorem 5.5 (Implicit Q-theorem, from [94, p. 347]). *Suppose $Q = [q_1, \ldots, q_n]$ and $V = [v_1, \ldots, v_n]$ are orthogonal matrices with the property that both $Q^T A Q = H$ and $V^T A V = G$ are Hessenberg matrices where $A \in \mathbb{R}^{n \times n}$. Let k denote the small positive integer for which $h_{k+1,k} = 0$ with the convention that $k = n$ if H is unreduced. If $v_1 = q_1$, then $v_i = \pm q_i$ and $|h_{i,i-1}| = |g_{i,i-1}|$ for $i = 2, \ldots, k$. Moreover, if $k < n$, then $g_{k+1,k} = 0$.*

The implicit Q-theorem for symmetric tridiagonal matrices as well as the proofs for both the theorems can be found in [94].

Assume the matrix $\hat{A} = Q^T A Q$, the result of applying a step of the explicit QR-method with shift μ onto the matrix A. We remark that the matrix $A - \mu I = QR$. By construction we know that the orthogonal matrix Q consists of $n - 1$ Givens transformations $Q = G_1 G_2 \ldots G_{n-1}$. Choosing as an initial transformation for the bulge chasing technique the matrix G_1, i.e., that we apply our bulge chasing technique on the matrix $G_1^T A G_1$, this results in a matrix $\tilde{A} = \tilde{Q}^T A \tilde{Q}$. Due to the construction of the chasing technique one can see that $Q\mathbf{e}_1 = \tilde{Q}\mathbf{e}_1$, hence we know that both matrices are essentially unique, under the assumption that both $\tilde{A}$ and $\hat{A}$ are unreduced.[8]

5.4 On computing the singular values

In this book we will also discuss an alternative method for computing the singular values. Instead of using bidiagonal matrices as intermediate matrices we will use now upper triangular semiseparable matrices.

The idea is to use all the tools available for computing eigenvalues for symmetric matrices and to translate these tools to the singular value case. In fact we will adapt the algorithm for computing the eigenvalues of the symmetric matrix $A^T A$ in order to compute the singular values of the matrix A. Let us describe the main ideas in this method. This section is mainly based on [94, pp. 448–460].

Let $A \in \mathbb{R}^{m \times n}$. Without loss of generality, assume that $m \geq n$. In a first phase the matrix A is transformed into a bidiagonal one by orthogonal transformations

---

[8]This can be assumed without loss of generality as otherwise the problem decouples.

U_B and V_B to the left and to the right, respectively, i.e.,

$$U_B^T A V_B = \begin{bmatrix} B \\ 0 \end{bmatrix}, \text{ with } B = \begin{bmatrix} d_1 & f_1 & 0 & \cdots & & 0 \\ 0 & d_2 & \ddots & \ddots & & \vdots \\ & & \ddots & \ddots & \ddots & 0 \\ \vdots & & & \ddots & \ddots & f_{n-1} \\ 0 & \cdots & & & 0 & d_n \end{bmatrix}.$$

This transformation was already discussed extensively in the first part of this book.

Without loss of generality we can assume that all the elements f_i and d_i in the matrix B are different from zero (see [94, p. 454]), otherwise we can apply deflation, split the matrix into two parts or easily chase away the zero. These demands correspond to the fact that the matrix $B^T B$ is unreduced.

Knowing B we will apply now the QR-method to it. This method will generate a sequence of bidiagonal matrices converging to a block diagonal one.

Here a short description of one iteration of the QR-method for computing the singular values is considered. More details can be found in [94, pp. 452–456].

Let $\tilde{V}_1$ be a Givens rotation such that the following equality is satisfied:

$$\tilde{V}_1^T \begin{bmatrix} d_1^2 - \mu \\ d_1 f_1 \end{bmatrix} = \begin{bmatrix} \times \\ 0 \end{bmatrix},$$

where μ is a suitably chosen shift. We remark that this Givens transformation is essentially equal to the Givens transformation chosen in order to annihilate the first subdiagonal element of the matrix $B^T B - \mu I$.

Let V_1 be the matrix in which $\tilde{V}_1$ is embedded, i.e., $V_1 = \text{diag}(\tilde{V}_1, I_{n-2})$, with $I_k, k \geq 1$, the identity matrix of order k. The matrix B is multiplied to the right by V_1, introducing the bulge denoted by $\otimes$ in the matrix BV_1,

$$\begin{bmatrix} \times & \times & 0 & \cdots & & \\ \otimes & \times & \times & 0 & \cdots & \\ 0 & 0 & \times & \times & 0 & \cdots \\ & & & \ddots & \ddots & \end{bmatrix}.$$

The remaining part of the iteration consists of hopping the bulge around until it is eventually removed. We will illustrate this hopping technique on a 5×5 example. A Givens transformation U_1^T is applied to the left of the matrix BV_1 in order to remove the element marked with $\otimes$ in the previous equation. The Givens transformation acts on the first two rows:

$$U_1^T B V_1 = \begin{bmatrix} \times & \times & \otimes & 0 & 0 \\ 0 & \times & \times & 0 & 0 \\ 0 & 0 & \times & \times & 0 \\ 0 & 0 & 0 & \times & \times \\ 0 & 0 & 0 & 0 & \times \end{bmatrix}.$$

Applying this transformation causes the bulge to hop around from the subdiagonal to the second superdiagonal. A Givens transformation V_2 is performed to the right acting on column 2 and 3 to remove the marked element:

$$U_1^T B V_1 V_2 = \begin{bmatrix} \times & \times & 0 & 0 & 0 \\ 0 & \times & \times & 0 & 0 \\ 0 & \otimes & \times & \times & 0 \\ 0 & 0 & 0 & \times & \times \\ 0 & 0 & 0 & 0 & \times \end{bmatrix}.$$

In the next step we remove the element $\otimes$ by a Givens transformation U_2^T acting on rows 2 and 3. This gives us the following equality:

$$U_2^T U_1^T B V_1 V_2 = \begin{bmatrix} \times & \times & 0 & 0 & 0 \\ 0 & \times & \times & \otimes & 0 \\ 0 & 0 & \times & \times & 0 \\ 0 & 0 & 0 & \times & \times \\ 0 & 0 & 0 & 0 & \times \end{bmatrix}.$$

This procedure can be repeated to chase away the bulge. In general one needs to apply $2n-3$ Givens rotations, $U_1, \ldots, U_{n-1}, V_2, \ldots, V_{n-1}$ moving the bulge along the bidiagonal and eventually removing it,

$$U^T B V = (U_{n-1}^T \cdots U_1^T) B (V_1 V_2 \cdots V_{n-1}) = \begin{bmatrix} \times & \times & 0 & \cdots & \\ 0 & \times & \times & 0 & \cdots \\ 0 & 0 & \times & \times & 0 & \cdots \\ & & & \ddots & \ddots & \end{bmatrix}.$$

Hence, the bidiagonal structure is preserved after one iteration of the QR-method. It can be seen that $V\mathbf{e}_1 = V_1 \mathbf{e}_1$, where $\mathbf{e}_1$ is the first vector of the canonical basis.

We will now relate the chasing algorithm presented above to the QR-factorization of the related tridiagonal matrix $B^T B$, because in one of the forthcoming chapters, we will describe a similar algorithm for upper triangular semiseparable matrices S_u and we will base this algorithm on a QR-step applied to the semiseparable matrix $S = S_u^T S_u$, just like here.

Suppose we have the bidiagonal matrix B and we want to compute its singular values. A naive way to do this consists of calculating the eigenvalues of the symmetric tridiagonal matrix $T = B^T B$. In fact this is not a good idea, because it will square the condition number. Let us apply now one step of the QR-algorithm to this tridiagonal matrix T:

$$\begin{cases} B^T B - \mu I &=& QR \\ \tilde{T} &=& RQ + \mu I. \end{cases}$$

Let us compare this with the transformations applied to the matrix B as described above. B is transformed into another upper bidiagonal matrix $\tilde{B} = U^T B V$. This means that we have the following two equations:

$$\begin{cases} \tilde{T} &=& Q^T (B^T B) Q \\ \tilde{B}^T \tilde{B} &=& V^T (B^T B) V. \end{cases}$$

We know that $Q\mathbf{e}_1 = V\mathbf{e}_1$. Using the implicit Q-theorem for tridiagonal matrices, we get the desired result, namely the matrices $\tilde{T}$ and $\tilde{B}^T\tilde{B}$ are essentially the same. This means that if $\tilde{T}$ converged to an eigenvalue, then $\tilde{B}$ converged to a singular value.[9] Moreover, using the technique as described in the beginning of this section we can calculate the singular values, without explicitly forming the product B^TB, which would have squared the condition number of the matrix.

Summarizing, the QR-method for computing the singular values of a matrix $A \in \mathbb{R}^{m\times n}$ can be divided into two phases.

1. Reduction of A into a bidiagonal matrix B by means of orthogonal transformations.

2. The QR-method to compute the singular values is applied to B.

In the new method based on structured rank matrices, the role of the bidiagonal matrices is played by upper triangular semiseparable matrices, which are the inverses of bidiagonal matrices, as proved in Theorem 1.10.

Notes and references

Computing the singular values via this QR-method is a well-known tool. It is covered in basic textbooks such as the Golub and Van Loan [94], Trefethen and Bau [145] and many others.

The use of the traditional QR-algorithm towards the computation of the singular values was presented by Golub and Kahan [93] and was discussed in Subsection 2.4.1 in Chapter 2.

5.5   Conclusions

In this chapter some known concepts related to QR-algorithms for sparse matrices were refreshed. It was mentioned what is meant with a QR-factorization and a QR-algorithm. It was shown how both algorithms can be suitably adapted to low cost algorithms in case one is working with sparse matrices. A bulge chasing technique was developed for Hessenberg matrices, leading to an alternative QR-algorithm (implicit) for computing the eigenvalues. Finally we also showed the relation between the QR-algorithm for tridiagonal matrices and the QR-algorithm for computing the singular values.

Now we know which tools are necessary for developing QR-algorithms we will adapt all this material towards the structured rank case. In the next chapter first some theoretical results will be derived, followed by the chasing techniques in Chapter 7.

---

[9]Again there is no loss in assuming both matrices to be unreduced, otherwise deflation could be applied.

Chapter 6

Theoretical results for structured rank QR-algorithms

In Chapter 7 of this book we will design different types of implicit QR-algorithms for matrices of semiseparable form. More precisely the chasing techniques will be developed in that chapter. Before we are able to construct these algorithms some theoretical results are needed. In this chapter we will provide answers to the following problems for Hessenberg-like matrices. How can we calculate the QR-factorization of a Hessenberg-like plus diagonal matrix? What is the structure of an unreduced Hessenberg-like matrix and how can we transform a matrix to the unreduced form? Is the Hessenberg-like structure maintained after a step of QR with shift? Does some kind of analogue of the implicit Q-theorem exist for Hessenberg-like matrices?

The theorems, definitions and results in this chapter are applied to Hessenberg-like matrices, because this is the most general structure. In the following chapter, we will adapt, if it is necessary, the theorems to the other cases, namely the symmetric case and the upper triangular semiseparable case, for computing the singular values. As these structures, semiseparable and upper triangular semiseparable, have more structure with regard to the Hessenberg-like case, we will often be able to simplify some definitions or to obtain stronger results.

This chapter consists of three sections. In the first two sections results related to Hessenberg-like matrices will be shown. The preservation of the Hessenberg-like structure after a step of the QR-algorithm is the subject of the first section and the second section focuses on the development of an implicit Q-theorem. The last section briefly discusses the Hessenberg-like plus diagonal case.

In the first section of this chapter we focus towards the QR-factorization and the structure after performing a QR-step. In a first subsection we will investigate "a type" of QR-factorization of Hessenberg-like plus diagonal matrices. We say a type of because we will show in the next section, that the QR-factorization is not necessarily unique. The factorization we propose is based on the paper [156] and was also presented in Volume I, Chapter 9 of this book [169]. In the factorization presented here, the orthogonal matrix Q is constructed as a product of $2n - 2$

Givens transformations, with n the dimension of the matrix. Suppose we have a Hessenberg-like plus diagonal matrix $Z + D$. It will be shown that a first sequence of $n - 1$ Givens transformations Q_1 will transform the matrix Z into an upper triangular matrix $Q_1^T Z$. These Givens transformations will transform the matrix D into a Hessenberg matrix. Our resulting matrix $Q_1^T(Z + D)$ will therefore be Hessenberg. The second sequence of Givens transformations will reduce the Hessenberg matrix to upper triangular form. In the second and third subsection we will investigate the structure of a Hessenberg-like matrix after having applied a step of the QR-algorithm with shift. First we show that the choice of the QR-factorization is essential to obtain again a Hessenberg-like matrix after a QR-step. This fact is illustrated by an example. Using the QR-factorization for Hessenberg-like matrices as explained in Subsection 6.1.1, we will prove the following theorems: The structure of a Hessenberg-like matrix is maintained after a step of QR without shift; the strictly lower triangular rank structure of a matrix is maintained after a QR-step with shift; the Hessenberg-like structure is maintained after a QR-step with shift. Note that for these theorems no assumptions concerning the singularity of the matrices are made.

In the second section we will develop an implicit Q-theorem for Hessenberg-like matrices. In order to do so, we define what is meant with an unreduced Hessenberg-like matrix. It will be shown later on that this unreducedness demand plays a very important role in the implicit Q-theorem and in the construction of the implicit QR-algorithm. Moreover, using the definition of an unreduced Hessenberg-like matrix, it is rather easy to prove the essential uniqueness of the QR-factorization for Hessenberg-like matrices. By essentially unique we mean that two QR-factorizations only differ for the sign of the elements. In the first subsection we will define unreduced Hessenberg-like matrices, and in the third Subsection 6.2.3 we will prove an implicit Q-theorem for Hessenberg-like matrices. This implicit Q-theorem strongly depends on the unreducedness of the Hessenberg-like matrix. Therefore we will provide in Subsection 6.2.2 an easy way of transforming a Hessenberg-like matrix to unreduced form. The algorithm provided performs in fact a special QR-step without shift on the Hessenberg-like matrix. This results in an algorithm, revealing immediately the singularities of the corresponding Hessenberg-like matrix. In the third subsection (Subsection 6.2.3) we will provide an implicit Q-theorem for Hessenberg-like matrices. The theorem is quite similar to the theorem for Hessenberg matrices. It will prove to be very powerful in the next chapter when we start with the construction of the different QR-algorithms.

In the final section of this chapter we will briefly discuss the case of Hessenberg-like plus diagonal matrices. The development of an implicit QR-algorithm for Hessenberg-like plus diagonal matrices is similar to the Hessenberg-like case. Unfortunately the diagonal makes the matter more complicated, involving a lot of theoretical details. Hence we will not cover the QR-algorithm for Hessenberg-like plus diagonal matrices in detail. We will only prove here, that in case the diagonal has all diagonal elements different from zero, that the rank structure of the matrix is maintained after a step of the QR-algorithm. Moreover also the diagonal is maintained. In the upcoming two chapters the issues related to an implicit QR-

algorithm for semiseparable plus diagonal matrices will be discussed only briefly. A graphical interpretation of the connected chasing algorithm is given in Section 8.2, Subsection 8.1.3. The notes and references will point to articles, containing more detailed information for computing the eigenvalues of Hessenberg-like plus diagonal matrices. Also a divide-and-conquer method for computing the eigenvalues of semiseparable plus diagonal matrices will be discussed in a forthcoming chapter (see Chapter 10).

✎ *This chapter is concerned with providing the theoretical tools for the development of an (implicit) QR-algorithm for structured rank matrices.*

Initially we start with the preservation of the Hessenberg-like structure under a step of the QR-algorithm. Important in this context is the type of QR-factorization used (Subsection 6.1.1). Some special cases in which the structure is preserved are presented in Subsection 6.1.2, but the main theorem is Theorem 6.8 proving the preservation of the structure of a Hessenberg-like matrix under a step of the QR-algorithm. It is worth reading the proof of the theorem as it uses elementary matrix properties. Example 6.12 is important as it shows that the type of QR-factorization is essential for preserving the structure.

An implicit Q-theorem is based on the definition of unreducedness (Definition 6.14). The reduction to unreduced form is necessary before being able to start with an implicit QR-algorithm. Subsection 6.2.2 depicts graphically this reduction procedure. The implicit Q-theorem is proved in Subsection 6.2.3, and involves a lot of theoretical details. The theorem itself (Theorem 6.28) is however very powerful and looks a lot like the one of the sparse case. Necessary for the understanding of the theorem are the notions "essentially the same" (Definition 6.20) and "the unreduced number" (Definition 6.21). The power of this implicit Q-theorem also extends towards the reductions to structured rank form as mentioned in Note 6.30, consider also the Examples 6.31 and 6.32.

The proof that the structure of a Hessenberg-like plus diagonal matrix is maintained under inversion is not trivial. A simplified proof if the diagonal elements are different from zero is provided in Theorem 6.33. The general case is considered in Theorem 6.35.

6.1 Preserving the structure under a QR-step

The main goal of this section is to prove that the structure of a Hessenberg-like (plus diagonal) matrix is preserved under a step of the QR-algorithm. This is essential for the development of an effective implicit QR-algorithm. This section is divided into three subsections. The first subsection briefly repeats how to compute the QR-factorization of a Hessenberg-like plus diagonal matrix. In the second subsection we discuss, respectively, the preservation of the Hessenberg-like structure, when performing a step of the QR-method. Results on the preservation of the structure if one is working with a Hessenberg-like plus diagonal matrix are presented in the last section of this chapter (Section 6.3).

6.1.1   The QR-factorization

Before QR-algorithms can be introduced it is important to retrieve more information about the QR-factorization of Hessenberg-like plus diagonal matrices. We consider the class of Hessenberg-like plus diagonal matrices instead of the Hessenberg-like class, because we will deduce implicit QR-algorithms with a shift.

The QR-factorization we will propose here is based on the QR-factorization of semiseparable plus diagonal matrices as described in [156] and in the first volume of this book [169]. First we will explain the algorithm for computing the QR-factorization of Hessenberg-like matrices. Then we will specify in more detail the QR-factorization of Hessenberg-like plus diagonal matrices. The computation of the QR-factorization of a Hessenberg-like matrix Z is straightforward. Due to the structure of the matrix involved, applying $n-1$ Givens rotations G_i^T, $i = 1, \ldots, n-1$, from bottom to top (G_1^T acts on the last two rows of the matrix, annihilating all the elements in the last row of Z below the main diagonal, G_2^T acts on the rows $n-2$ and $n-1$ of $G_1^T Z$, annihilating all the elements in row $n-1$ below the main diagonal, G_{n-1}^T acts on the first two rows of $G_{n-2}^T \cdots G_1^T Z$, annihilating the element in position $(2,1)$) reduces the Hessenberg-like matrix Z to an upper triangular matrix $\tilde{R}$:

$$G_{n-1}^T \ldots G_2^T G_1^T Z = \tilde{R}.$$

Note 6.1. *These Givens transformations are the transpose of the Givens transformations stored in the Givens-vector representation of the Hessenberg-like matrix. This gives an essential advantage when implementing the QR-algorithm for Hessenberg-like matrices as they are known a priori when working with the Givens-vector representation.*

Let us now calculate the QR-factorization of a Hessenberg-like plus diagonal matrix: $Z + D$. The Q factor of this reduction consists of $2n - 2$ Givens transformations. The first $n - 1$ Givens transformations are performed on the rows of the Hessenberg-like matrix from bottom to top. These Givens transformations are exactly the same as the ones described in the previous paragraph. These Givens transformations transform the Hessenberg-like matrix into an upper triangular matrix. When taking into consideration the diagonal, one can see that these $n - 1$ Givens transformations transform the diagonal part into a new Hessenberg matrix $\tilde{H}$:

$$G_{n-1}^T \ldots G_2^T G_1^T (Z + D) = \tilde{R} + \tilde{H} = H.$$

Recombining the upper triangular matrix $\tilde{R}$ and the Hessenberg matrix $\tilde{H}$ gives another Hessenberg matrix H. This matrix will then be made upper triangular by another $n - 1$ Givens transformations G_i^T, $i = n, \ldots, 2n - 2$, from top to bottom:

$$\begin{aligned}
G_{2n-2}^T &\ldots G_2^T G_1^T (Z + D) \\
&= G_{2n-2}^T \ldots G_n^T \left(\tilde{R} + \tilde{H} \right) \\
&= R.
\end{aligned}$$

Note 6.2. *If one of the Givens transformations is not uniquely determined (this means that the Givens transformation had to be calculated based on the vector $[0,0]^T$) then we choose the Givens transformation equal to the identity matrix.*

Note 6.3. *The number of Givens transformations $2n-2$ can be reduced to $2n-3$, because the last Givens transformation of the first sequence and the first Givens transformation of the second sequence can be combined into one single Givens transformation.*

Note 6.4. *In the previous part in Chapter 4 some techniques were presented for working efficiently with Givens transformations. Also the shift through lemma was presented. Combining these techniques, one can construct different patterns of Givens transformations for computing the QR-factorization (more details can be found in Volume I [169] and [57]). These different patterns of annihilation can also be used for constructing different types of QR-algorithms. More information is given in the notes and references section.*

When applying this technique to a semiseparable plus diagonal matrix, more general theorems concerning the structure, e.g., the structure of R for general rank structures, can be derived. These theorems can be found in [56, 156, 169]. This QR-factorization can also be used to compute eigenvectors via inverse iteration.

6.1.2   A QR-step maintains the rank structure

In this section we will prove that a QR-step applied to a Hessenberg-like matrix is again a Hessenberg-like matrix. Mathematically speaking this means that for a Hessenberg-like matrix Z, with μ a shift and

$$Z - \mu I = QR,$$

the QR-decomposition of $Z - \mu I$ (with the QR-decomposition as described in Subsection 6.1.1[10]), the matrix $\tilde{Z}$ satisfying

$$\tilde{Z} = RQ + \mu I$$

is again a Hessenberg-like matrix. Because we want to place this in a general framework, we derive a theorem proving this without putting any constraints on the Hessenberg-like matrix. However, we do need to place demands on the type of QR-factorization used. This is illustrated by the following example for Hessenberg matrices.

The type of QR-factorization we will use to compute the decomposition for Hessenberg-like matrices is the one as presented in Subsection 6.1.1 of this chapter. The Q factor of the QR-decomposition consists of two sequences of $n-1$ Givens

---

[10]Further in the text we will show, by an example, that it is essential that this specific type of QR-factorization is needed. Otherwise one cannot guarantee in general that the structure will be preserved. This remark is also true in the case of the preservation of the structure of Hessenberg matrices as was shown in the previous chapter.

transformations, let us define $Q_1 = G_1 \ldots G_{n-1}$ and $Q_2 = G_n \ldots G_{2n-2}$. This means that the QR-decomposition of a Hessenberg-like matrix minus a shift matrix has the following form:

$$Q_2^T Q_1^T (Z - \mu I) = R,$$

leading to

$$(Z - \mu I) = Q_1 Q_2 R = QR.$$

Completing the step of the QR-algorithm leads to the following equations

$$\tilde{Z} = R Q_1 Q_2 + \mu I$$
$$= Q_2^T Q_1^T Z Q_1 Q_2.$$

Note 6.5. *If the matrix R is invertible, which is not always the case, then we also have the following equation:*
$$\tilde{Z} = R Z R^{-1}.$$

Using this equation one can easily prove that the Hessenberg-like structure is maintained. This is true, because multiplying a Hessenberg-like matrix on the right or on the left with an upper triangular matrix, does not destroy its semiseparable rank structure in the lower triangular part (see, e.g., [23] in which this fact is exploited). Moreover if the matrix Z is invertible and the shift equals zero, one can directly switch, via inversion, to the Hessenberg case, which provides us with another easy proof for maintaining the Hessenberg-like structure. If the shift is not equal to zero, the QR-factorization is computed for the matrix $Z - \mu I$, inverting this matrix gives again a Hessenberg-like plus diagonal matrix and not a Hessenberg matrix. Hence in the general case, $\mu \neq 0$, one cannot use the argument of inversion.

In the remaining part of this section we will first prove that the Hessenberg-like structure is maintained when a step of QR without shift is performed. Then we will prove theorems for a step of QR with shift applied on matrices having the strictly lower triangular part of semiseparable form, Hessenberg-like matrices.

Theorem 6.6. *Suppose we have a Hessenberg-like matrix Z and a step of the QR-algorithm (with the QR-decomposition as in Subsection 6.1.1) without shift is performed on the matrix Z:*
$$\begin{cases} Z & = & QR \\ \tilde{Z} & = & RQ. \end{cases}$$
Then the matrix Z will be a Hessenberg-like matrix.

Proof. Because no shift is involved, only the first $n-1$ Givens transformations need to be performed on the Hessenberg-like matrix Z to retrieve the upper triangular matrix R.

$$R = Q_1^T Z$$
$$= G_{n-1}^T \ldots G_1^T Z.$$

The Givens transformation G_k^T performs an operation on rows $(n-k)$ and $(n-k+1)$. Therefore one can clearly see that the transformations $RG_1 \ldots G_{n-1}$ will create a matrix $\tilde{Z}$ whose lower triangular part is semiseparable. As an illustration we perform the transformations on a matrix of dimension 4, the elements marked with $\boxtimes$ denote the part of the matrix satisfying the semiseparable structure.

$$
R = \begin{bmatrix} \times & \times & \times & \times \\ 0 & \times & \times & \times \\ 0 & 0 & \times & \times \\ 0 & 0 & 0 & \times \end{bmatrix}
\xrightarrow{RG_1}
\begin{bmatrix} \times & \times & \times & \times \\ 0 & \times & \times & \times \\ 0 & 0 & \boxtimes & \times \\ 0 & 0 & \boxtimes & \boxtimes \end{bmatrix}
$$

$$
\xrightarrow{RG_1G_2}
\begin{bmatrix} \times & \times & \times & \times \\ 0 & \boxtimes & \times & \times \\ 0 & \boxtimes & \boxtimes & \times \\ 0 & \boxtimes & \boxtimes & \boxtimes \end{bmatrix}
\xrightarrow{RG_1G_2G_3}
\begin{bmatrix} \boxtimes & \times & \times & \times \\ \boxtimes & \boxtimes & \times & \times \\ \boxtimes & \boxtimes & \boxtimes & \times \\ \boxtimes & \boxtimes & \boxtimes & \boxtimes \end{bmatrix}
$$

$\square$

Using this theorem, and the knowledge about the Givens transformations involved in the QR-factorization we can easily formulate a theorem which states that the structure of a matrix with the strictly lower triangular part of semiseparable form is maintained under one step of the QR-algorithm with shift. (Note that for matrices A with a strictly lower triangular semiseparable part the shift is not essential, because the matrix A plus a diagonal still has the strictly lower triangular part of semiseparable form.)

Theorem 6.7. *Suppose we have a matrix A whose strictly lower triangular part is semiseparable and a step of QR with shift (with the QR-decomposition as in Subsection 6.1.1) is performed on this matrix:*

$$
\begin{cases} A - \mu I &= QR \\ \tilde{A} &= RQ + \mu I. \end{cases}
$$

The matrix $\tilde{A}$ will have the strictly lower triangular part of semiseparable form.

Proof. The transformation Q consists of two sequences of Givens transformations Q_1 and Q_2: $Q = Q_1 Q_2$. The first sequence of transformations Q_1^T will make use of the structured rank of the matrix A to reduce A to a Hessenberg matrix. This transformation Q_1^T will also transform the shift matrix μI into a Hessenberg matrix. We have

$$
Q_1^T A - \mu Q_1^T I = H_1 + H_2 = H,
$$

for which H_1, H_2 and H are all Hessenberg matrices. The second sequence of Givens transformations Q_2^T will be applied from top to bottom, to transform H into an upper triangular matrix R:

$$
Q_2^T Q_1^T (A - \mu I) = Q_2^T H = R.
$$

The matrix $\tilde{A}$ satisfies now the following equation:

$$\tilde{A} = RQ_1Q_2 + \mu I.$$

Because of the structure of the orthogonal matrices Q_1 and Q_2 which consist of a sequence of Givens transformations in a specific order we have that RQ_1 is a matrix with lower triangular rank of semiseparable form. Performing the transformation Q_2 to the right of RQ_1, RQ_1Q_2 will transform the lower triangular semiseparable part to a strictly lower triangular semiseparable part. Therefore our matrix $\tilde{A}$ will have the desired rank structure.   □

The theorem above already explains that for a Hessenberg-like plus diagonal matrix the structure of the strictly lower triangular part is preserved. Unfortunately we know from Volume I, that not every matrix having the strictly lower triangular part of semiseparable form, can be written as a Hessenberg-like plus diagonal matrix. Hence we do not yet know that after performing a step of the QR-algorithm on a Hessenberg-like plus diagonal matrix we obtain again a Hessenberg-like plus diagonal matrix. This is proved in the next theorem.

Theorem 6.8.   *Suppose we have a Hessenberg-like matrix Z, and we apply one step of the shifted QR-algorithm (with the QR-decomposition as in Subsection 6.1.1) on this matrix:*

$$\begin{cases} Z - \mu I &=& QR \\ \tilde{Z} &=& RQ + \mu I. \end{cases}$$

Then the matrix $\tilde{Z}$ will be a Hessenberg-like matrix.

Proof. Without loss of generality, we may assume that the element in the lower left corner is different from zero, otherwise there are zero blocks below the diagonal (a theorem similar to Proposition 1.15 can be derived to state this fact), and we can split the matrix into different blocks. Suppose we perform now a QR-step on the Hessenberg-like matrix Z, then we have also the following two equalities:

$$Q_1^T Z = R_1 \tag{6.1}$$

$$Q_2^T (R_1 - \mu Q_1^T I) = R_2 \tag{6.2}$$

with R_1 and R_2 upper triangular, and μ as a shift. Assume μ to be different from zero, otherwise we can refer to Theorem 6.6. Using the Equations (6.1) and (6.2) we have that

$$Q_1^T (-\mu I) = H_1$$

where H_1 is a Hessenberg matrix. The matrix H_1 has all the subdiagonal elements different from zero if all Givens transformations involved are different from the identity matrix. Because the element in the lower left corner of the matrix Z is different from zero, we know that the sequence of Givens transformations will not contain Givens transformations equal to the identity matrix. Hence our matrix H_1 has all subdiagonal elements different from zero.

This means that the Hessenberg matrix H_1 and therefore also $R_1 + H_1$ is unreduced and has an essentially unique QR-decomposition (see Subsection 5.3.2 from Chapter 5):

$$Q_2^T(R_1 + H_1) = R,$$

where R has two possible structures: all the diagonal elements are different from zero or, all the diagonal elements are different from zero, except for the last one (see Subsection 5.3.2 from Chapter 5). We have to distinguish between the two cases.

- **Case 1.** Suppose all the diagonal elements are different from zero, this means that the matrix R is invertible, this gives us:

$$\tilde{Z} = RQ + \mu I$$
$$= RZR^{-1}.$$

  We can see that the multiplication with the upper triangular matrices will not change the rank of the submatrices below the diagonal. Therefore $\tilde{Z}$ is again a Hessenberg-like matrix.[11]

- **Case 2.** Suppose now that our matrix R has the lower right element equal to zero. We partition the matrices Q and R in the following way:

$$R = \left[\begin{array}{c|c} \tilde{R} & \mathbf{w} \\ \hline 0 & 0 \end{array} \right]$$
$$Q = \left[\begin{array}{c|c} \tilde{Q} & \mathbf{q} \end{array} \right],$$

  with $\tilde{R}$ an $(n-1) \times (n-1)$ matrix, $\mathbf{w}$ a column vector of length $n-1$, $\tilde{Q}$ a matrix of dimension $n \times (n-1)$ and $\mathbf{q}$ a vector of length n. This gives us the following equalities:

$$Z - \mu I = \left[\begin{array}{c|c} \tilde{Q} & \mathbf{q} \end{array} \right] \left[\begin{array}{c|c} \tilde{R} & \mathbf{w} \\ \hline 0 & 0 \end{array} \right]$$
$$= \left[\begin{array}{c|c} \tilde{Q}\tilde{R} & \tilde{Q}\mathbf{w} \end{array} \right].$$

Denote with P an $n \times (n-1)$ projection matrix of the following form $[I_{n-1} \ 0]^T$, then we get (because $\tilde{R}$ is invertible):

$$(Z - \mu I)P = \tilde{Q}\tilde{R}$$
$$(Z - \mu I)P\tilde{R}^{-1} = \tilde{Q}$$
$$(Z - \mu I)\left[\begin{array}{c} \tilde{R}^{-1} \\ 0 \end{array} \right] = \tilde{Q}.$$

---

[11]This is easily verified when writing down explicitly the product of the matrix Z with upper triangular matrices.

Using these equalities we get:

$$\tilde{Z} = RQ + \mu I_n$$

$$= \left[\begin{array}{c|c} \tilde{R} & \mathbf{w} \\ \hline 0 & 0 \end{array} \right] Q + \mu I_n$$

$$= \left[\begin{array}{c|c} \tilde{R} & \mathbf{w} \\ \hline 0 & 0 \end{array} \right] \left[\; [Z - \mu I] \left[\begin{array}{c} \tilde{R}^{-1} \\ 0 \end{array} \right] \; \middle| \; \mathbf{q} \; \right] + \mu I_n$$

$$= \left[\begin{array}{c|c} \hat{Z} - \mu I_{n-1} & \mathbf{z} \\ \hline 0 & 0 \end{array} \right] + \mu I$$

$$= \left[\begin{array}{c|c} \hat{Z} & \mathbf{z} \\ \hline 0 & \mu \end{array} \right]$$

where $\hat{Z}$ is Hessenberg-like, and therefore also $\tilde{Z}$ is a Hessenberg-like matrix, which we had to prove.   □

Note 6.9. *It is clear from the previous proof that the matrix $\tilde{Z}$ (from Case 2) immediately reveals the eigenvalue μ. This μ is called the perfect shift.*

Note 6.10. *In the previous theorem we spoke about maintaining the structure of Hessenberg-like matrices after a step of QR. The proof can however be adapted very easily for generator representable Hessenberg-like matrices. After a step of QR performed on a generator representable Hessenberg-like matrix we get again a generator representable Hessenberg-like matrix, or we will have convergence to one eigenvalue in the lower right corner, while the remaining part in the upper left corner will again be a generator representable Hessenberg-like matrix.*

Note 6.11. *Once more we state that it is important that the sequence of Givens transformations are performed in this special order. Otherwise it is possible to construct a QR-factorization which will generally not give again a Hessenberg-like matrix as the following example shows.*

Example 6.12 Suppose we have the following Hessenberg-like matrix:

$$Z = \left[\begin{array}{cccc} 1 & 0 & 0 & 0 \\ 1 & 1 & 1 & 0 \\ 0 & 0 & 0 & 1 \\ 1 & 1 & 1 & 1 \end{array} \right].$$

Annihilating the complete last row with a Givens transformation applied on the second and fourth row, and then annihilating the first element of the second row, one gets a QR-factorization of the matrix Z. Applying the transformations again on the right side of the matrix, will not give as result a Hessenberg-like matrix, because there will be a block of rank 2 in the lower triangular part.   ∎

Essential for the development of an implicit QR-algorithm, was the preservation of the structure, proved here, and the chasing technique. We proved in the previous chapter that the chasing technique for Hessenberg matrices was correct, based on an implicit Q-theorem for Hessenberg matrices. As the chasing technique for Hessenberg-like matrices is more complicated than in the sparse case, we solve this problem in the next chapter. In the current chapter we will prove the implicit Q-theorem for Hessenberg-like matrices and we will also define unreduced Hessenberg-like matrices and so forth.

Notes and references

In this section we briefly showed how to compute the QR-factorization of a Hessenberg-like matrix and moreover we showed that the structure of such a Hessenberg-like matrix is preserved under a step of the QR-method. The preservation of the structure of this method is already discussed before in several articles.

The following article by Van Camp, Mastronardi and Van Barel investigates the QR-decomposition of semiseparable plus diagonal matrices (as used in this book) to solve systems of equations. Also the first volume of this book [169] covers the QR-factorization in detail, including higher order structured rank matrices. Moreover Volume I contains an extensive list of references related to the subject.

☞ E. Van Camp, N. Mastronardi, and M. Van Barel. Two fast algorithms for solving diagonal-plus-semiseparable linear systems. *Journal of Computational and Applied Mathematics*, 164–165:731–747, 2004.

The QR-factorization was also studied in a more general form in the book by Dewilde and van der Veen. More precisely, the authors present a new type of representation for sequentially semiseparable matrices or low Hankel rank matrices (quasiseparable matrices also belong to this class). Also a QR-factorization and inversion formulas for this class of matrices are presented.

☞ P. Dewilde and A.-J. van der Veen. *Time-Varying Systems and Computations*. Kluwer Academic Publishers, Boston, Massachusetts, USA, 1998.

The authors Delvaux and Van Barel proposed a general scheme for computing the QR-factorization of structured rank matrices. The matrices considered in this article have low rank blocks (with or without shift), starting from the lower left corner. Also a solver based on this representation is given. The method also exploits the rank properties of the upper triangular factor R for solving the system of equations. The proposed method assumes that the Givens-weight representation of the matrices involved is precomputed. The Givens-weight representation is a generalization of the Givens-vector representation.

☞ S. Delvaux and M. Van Barel. A QR-based solver for rank structured matrices. *SIAM Journal on Matrix Analysis and Applications*, 30(2):464–490, 2008.

The authors Vandebril, Van Barel and Mastronardi prove in the following article that the QR-factorization of a structured rank matrix inherits also rank properties of the original matrix. The proof of the theorem is based on a the nullity theorem.

☞ R. Vandebril and M. Van Barel. A note on the nullity theorem. *Journal of Computational and Applied Mathematics*, 189:179–190, 2006.

Maintaining the structure under a step of the QR-algorithm was extensively studied in the following articles by Delvaux and Van Barel. Matrices satisfying a so-called polynomial structure as well as a rank structure are considered here. The singular case is slightly more complicated and hence included in a second article.

☞ S. Delvaux and M. Van Barel. Rank structures preserved by the QR-algorithm: the singular case. *Journal of Computational and Applied Mathematics*, 189:157–178, 2006.

☞ S. Delvaux and M. Van Barel. Structures preserved by the QR-algorithm. *Journal of Computational and Applied Mathematics*, 187(1):29–40, March 2006.

Also the authors Bini, Gemignani and Pan investigate the preservation of the structure of structured rank matrices under a step of the QR-algorithm. Moreover, they derive an algorithm for performing a step of QR on a generalized semiseparable matrix. The lower part of this generalized matrix is in fact lower triangular semiseparable. The authors present an alternative representation, consisting of three vectors, to represent this part.

☞ D. A. Bini, L. Gemignani, and V. Y. Pan. Fast and stable QR eigenvalue algorithms for generalized companion matrices and secular equations. *Numerische Mathematik*, 100(3):373–408, 2005.

In the article by Bevilacqua and Del Corso [18] also a proof is included, stating that the structure of a nonsingular generator representable semiseparable matrix is maintained under a step of the QR-method. This article was discussed before in the notes and references of Section 2.2 in Chapter 2.

6.2 An implicit Q-theorem

In this section we will derive an implicit Q-theorem for Hessenberg-like matrices. Closely related to an implicit Q-theorem we have to study the unreducedness of the Hessenberg-like matrices and hence also the reduction to this unreduced form.

These details are discussed in this section. Firstly the definition of unreduced Hessenberg-like matrices, followed by the reduction to unreduced Hessenberg-like form is given and finally we present the implicit Q-theorem.

Note 6.13. *We remark that in this chapter the definitions and theorems are always developed for the case of Hessenberg-like matrices. This is logical because this is the most general class. Results applicable for these matrices have a natural extension to for example semiseparable matrices, which are the symmetric version of Hessenberg-like matrices. The symmetric case, due to its extra structure, quite often leads to stronger results or to more easy definitions. When developing in the next section the chasing technique for semiseparable matrices, we reconsider some of the results presented here and adapt them to the symmetric case.*

6.2.1 Unreduced Hessenberg-like matrices

Unreduced Hessenberg matrices are very important for the development of an implicit QR-algorithm for Hessenberg matrices. It is essential for the application of an implicit QR-algorithm that the corresponding Hessenberg matrix is unreduced, otherwise the algorithm could break down. More information concerning unreduced Hessenberg matrices can be found in textbooks [94, 129] and in Chapter 5.

Let us define now an unreduced Hessenberg-like matrix, in the following way:

Definition 6.14. *A Hessenberg-like matrix Z is said to be unreduced if*

1. *all the blocks $Z(i+1:n, 1:i)$ (for $i = 1, \ldots, n-1$) have rank equal to 1; this corresponds to the fact that there are no zero blocks below the diagonal;*

2. *all the blocks $Z(i:n, 1:i+1)$ (for $i = 1, \ldots, n-1$) have rank strictly higher than 1, this means that on the superdiagonal, no elements are includable in the semiseparable structure.*

Note 6.15. *If a Hessenberg-like matrix Z is unreduced, it is also nonsingular. This can be seen by calculating the QR-factorization of Z as described in Subsection 6.1.1 of this chapter. Because none of the elements above the diagonal is includable in the semiseparable structure below the diagonal, all the diagonal elements of the upper triangular matrix R in the QR-factorization of Z will be different from zero, implying the nonsingularity of the Hessenberg-like matrix Z.*

When these demands are placed on the Hessenberg-like matrix, and they can be checked rather easily, we have the following theorem connected to the QR-factorization:

Theorem 6.16. *Suppose Z is an unreduced Hessenberg-like matrix. Then the matrix $Z - \mu I$, with μ as a shift has an essentially unique QR-factorization.*

Proof. If $\mu = 0$, the theorem is true because Z is nonsingular. Suppose $\mu \neq 0$. Because the Hessenberg-like matrix Z is unreduced, the first $n-1$ Givens transformations G_i of Subsection 6.1.1 of this chapter are nontrivial. Applying them to the left of the matrix $Z - \mu I$ results therefore in an unreduced Hessenberg matrix. As this Hessenberg matrix has the first $n-1$ columns independent of each other also the matrix $Z - \mu I$ has the first $n-1$ columns independent. Hence, the QR-factorization of $Z - \mu I$ is essentially unique.   □

Note 6.17. *An unreduced Hessenberg-like matrix Z is always nonsingular and has, therefore, always an essentially unique QR-factorization. If one however admits that the block $Z(n-1:n, 1:n)$ is of rank 1, the matrix Z also will have an essentially unique QR-factorization.*

Note 6.18. *The unreduced Hessenberg-like matrix Z is always nonsingular, but the matrix $Z - \mu I$ can be singular.*

This demand of unreducedness will play an important role in the proof of the implicit Q-theorem for semiseparable matrices. As in the sparse case, before starting an implicit QR-step on a matrix, one reduces the matrix to unreduced form. Also in the next section we will always assume we are working with unreduced matrices. Let us see how we can easily transform via orthogonal similarity transformations a matrix to unreduced form.

6.2.2   The reduction to unreduced Hessenberg-like form

The implicit QR-algorithm for Hessenberg matrices is based on the unreducedness of the corresponding Hessenberg matrix. For Hessenberg matrices it is straightforward how to split the matrix into two or more Hessenberg matrices that are unreduced. (See Subsection 7.5.1 in [94].)

In this subsection we will reduce a Hessenberg-like matrix to unreduced form as presented in Definition 6.14. The first demand that there are no zero blocks in the lower left corner of the matrix can be satisfied by dividing the matrix into different blocks similar to the Hessenberg case. This corresponds to deflation of the matrix.

With the second demand, the fact that the semiseparable structure does not extend above the diagonal is not solved in such an easy way. It can be seen that matrices having this special structure are singular. In the solution we propose, we will chase in fact the dependent rows completely to the lower right where they will form zero rows that can be removed. This chasing technique is achieved by performing a special QR-step without shift on the matrix. We will demonstrate this technique on a 5×5 matrix. The $\boxtimes$ denote the dependent elements. One can clearly see that the semiseparable structure of the matrix extends above the diagonal:

$$Z = \begin{bmatrix} \boxtimes & \times & \times & \times & \times \\ \boxtimes & \boxtimes & \boxtimes & \times & \times \\ \boxtimes & \boxtimes & \boxtimes & \times & \times \\ \boxtimes & \boxtimes & \boxtimes & \boxtimes & \times \\ \boxtimes & \boxtimes & \boxtimes & \boxtimes & \boxtimes \end{bmatrix}.$$

We start the annihilation with the traditional orthogonal matrix $Q_1 = G_1 G_2 G_3 G_4$. This results in the following matrix $\tilde{R} = G_4^T G_3^T G_2^T G_1^T Z$:

$$\tilde{R} = \begin{bmatrix} \times & \times & \times & \times & \times \\ 0 & \times & \times & \times & \times \\ 0 & 0 & 0 & \times & \times \\ 0 & 0 & 0 & \times & \times \\ 0 & 0 & 0 & 0 & \times \end{bmatrix}.$$

In normal circumstances one would apply the transformation Q_1 now on the right of the above matrix to complete one step of QR without shift. Instead of applying now this transformation we continue and annihilate the elements marked with $\otimes$ with the transformations G_5 and G_6. G_5 is performed on rows 3 and 4, while G_6 is performed on rows 4 and 5:

$$
\begin{bmatrix}
\times & \times & \times & \times & \times \\
0 & \times & \times & \times & \times \\
0 & 0 & 0 & \times & \times \\
0 & 0 & 0 & \otimes & \times \\
0 & 0 & 0 & 0 & \times
\end{bmatrix}
\xrightarrow{G_5^T \tilde{R}}
\begin{bmatrix}
\times & \times & \times & \times & \times \\
0 & \times & \times & \times & \times \\
0 & 0 & 0 & \times & \times \\
0 & 0 & 0 & 0 & \times \\
0 & 0 & 0 & 0 & \otimes
\end{bmatrix}
$$

$$
\xrightarrow{G_6^T G_5^T \tilde{R}}
\begin{bmatrix}
\times & \times & \times & \times & \times \\
0 & \times & \times & \times & \times \\
0 & 0 & 0 & \times & \times \\
0 & 0 & 0 & 0 & \times \\
0 & 0 & 0 & 0 & 0
\end{bmatrix}.
$$

We have now finished performing the transformations on the left side of the matrix. Denote $R = G_6^T G_5^T \tilde{R}$. To complete the QR-step applied to the matrix Z, we have to perform the transformations on the right side of the matrix:

$$
R =
\begin{bmatrix}
\times & \times & \times & \times & \times \\
0 & \times & \times & \times & \times \\
0 & 0 & 0 & \times & \times \\
0 & 0 & 0 & 0 & \times \\
0 & 0 & 0 & 0 & 0
\end{bmatrix}
\xrightarrow{RG_1}
\begin{bmatrix}
\times & \times & \times & \times & \times \\
0 & \times & \times & \times & \times \\
0 & 0 & 0 & \boxtimes & \times \\
0 & 0 & 0 & \boxtimes & \boxtimes \\
0 & 0 & 0 & 0 & 0
\end{bmatrix}
$$

$$
\xrightarrow{RG_1G_2}
\begin{bmatrix}
\times & \times & \times & \times & \times \\
0 & \times & \times & \times & \times \\
0 & 0 & \boxtimes & \boxtimes & \times \\
0 & 0 & \boxtimes & \boxtimes & \boxtimes \\
0 & 0 & 0 & 0 & 0
\end{bmatrix}
\xrightarrow{RG_1G_2G_3}
\begin{bmatrix}
\times & \times & \times & \times & \times \\
0 & \boxtimes & \times & \times & \times \\
0 & \boxtimes & \boxtimes & \boxtimes & \times \\
0 & \boxtimes & \boxtimes & \boxtimes & \boxtimes \\
0 & 0 & 0 & 0 & 0
\end{bmatrix}
$$

$$
\xrightarrow{RQ_1}
\begin{bmatrix}
\boxtimes & \times & \times & \times & \times \\
\boxtimes & \boxtimes & \times & \times & \times \\
\boxtimes & \boxtimes & \boxtimes & \boxtimes & \times \\
\boxtimes & \boxtimes & \boxtimes & \boxtimes & \boxtimes \\
0 & 0 & 0 & 0 & 0
\end{bmatrix}
\xrightarrow{RQ_1G_5}
\begin{bmatrix}
\boxtimes & \times & \times & \times & \times \\
\boxtimes & \boxtimes & \times & \times & \times \\
\boxtimes & \boxtimes & \boxtimes & \boxtimes & \times \\
\boxtimes & \boxtimes & \boxtimes & \boxtimes & \boxtimes \\
0 & 0 & 0 & 0 & 0
\end{bmatrix}
$$

$$
\xrightarrow{RQ_1G_5G_6}
\left[
\begin{array}{cccc|c}
\boxtimes & \times & \times & \times & \times \\
\boxtimes & \boxtimes & \times & \times & \times \\
\boxtimes & \boxtimes & \boxtimes & \times & \times \\
\boxtimes & \boxtimes & \boxtimes & \boxtimes & \times \\
\hline
0 & 0 & 0 & 0 & 0
\end{array}
\right].
$$

It is clear that the last matrix can be deflated such that we get an unreduced matrix and one eigenvalue that has already converged.

The same technique can be applied for matrices with larger blocks crossing the diagonal or more blocks crossing the diagonal.

Note 6.19. *Suppose we have a symmetric semiseparable matrix coming from the reduction algorithm as presented in Part I. Then we know that this matrix can* *be written as a block diagonal matrix, for which all the blocks are generator representable semiseparable matrices. Moreover, for all these blocks condition (2) in Definition 6.14 is satisfied in a natural way. This can be seen by combining the knowledge that the reduction algorithm performs steps of QR without shift, the fact that if condition (2) is not satisfied the matrix has to be singular and the note following Theorem 6.8. For these matrices the scheme as presented here does not need to be performed. The same is true for Hessenberg-like and upper triangular semiseparable matrices coming from one of these reduction algorithms.*

6.2.3   An implicit Q-theorem for Hessenberg-like matrices

We will now prove an implicit Q-theorem for Hessenberg-like matrices. Combined with the theoretical results provided in the previous sections, we will be able to derive implicit QR-algorithms for semiseparable and Hessenberg-like matrices in the next chapter. The theorem is proved in a similar way as the implicit Q-theorem for Hessenberg matrices as formulated in [94, Subsection 7.4.5]. First two definitions and some propositions are needed.

Definition 6.20. *Two matrices Z_1 and Z_2 are called essentially the same if there exists a matrix $W = \mathrm{diag}([\pm 1, \pm 1, \dots, \pm 1])$ such that the following equation holds:*

$$Z_1 = W Z_2 W^T.$$

Before proving the implicit Q-theorem, we define the so-called unreduced number of a Hessenberg-like matrix.

Definition 6.21.   *Suppose Z to be a Hessenberg-like matrix. The unreduced number k of Z is the smallest integer such that one of the following two conditions is satisfied:*

  1. The submatrix $S(k + 1 : n; 1 : k) = 0$.

  2. The element $S(k + 1, k + 2)$ with $k < n - 1$ is includable in the lower semiseparable structure.

If the matrix is unreduced, $k = n$.

To prove an analogue of the implicit Q-theorem for Hessenberg matrices for the Hessenberg-like case, some propositions are needed.

Proposition 6.22. *Suppose we have a Hessenberg-like matrix Z which is not unreduced, and does not have any zero blocks below the diagonal, i.e., whose bottom left element is nonzero. Then this matrix can be transformed via similarity transformations to a Hessenberg-like matrix, for which the upper left $(n - l) \times (n - l)$ matrix is of unreduced form and the last l rows equal to zero, where l equals the nullity of the matrix. Moreover if k is the unreduced number of the matrix Z then the orthogonal* *transformation can be chosen in such a way that the upper left $k \times k$ block of this orthogonal transformation equals the identity matrix.*

Proof. We will explicitly construct the transformation matrix. We illustrate this technique on a matrix of dimension 5×5 of the following form

$$
Z = \begin{bmatrix}
\boxtimes & \times & \times & \times & \times \\
\boxtimes & \boxtimes & \times & \times & \times \\
\boxtimes & \boxtimes & \boxtimes & \boxtimes & \times \\
\boxtimes & \boxtimes & \boxtimes & \boxtimes & \times \\
\boxtimes & \boxtimes & \boxtimes & \boxtimes & \boxtimes
\end{bmatrix}.
$$

We annihilate first the elements of the fifth and fourth row, with Givens transformations G_1 and G_2. This results in the matrix

$$
G_2^T G_1^T Z = \begin{bmatrix}
\boxtimes & \times & \times & \times & \times \\
\boxtimes & \boxtimes & \times & \times & \times \\
\boxtimes & \boxtimes & \boxtimes & \boxtimes & \times \\
0 & 0 & 0 & 0 & \times \\
0 & 0 & 0 & 0 & \otimes
\end{bmatrix}.
$$

The Givens transformation G_3 is constructed to annihilate the element $\otimes$:

$$
G_3^T G_2^T G_1^T Z = \begin{bmatrix}
\boxtimes & \times & \times & \times & \times \\
\boxtimes & \boxtimes & \times & \times & \times \\
\boxtimes & \boxtimes & \boxtimes & \boxtimes & \times \\
0 & 0 & 0 & 0 & \times \\
0 & 0 & 0 & 0 & 0
\end{bmatrix}.
$$

Completing the similarity transformation by applying the transformations G_1, G_2 and G_3 on the right of $G_3^T G_2^T G_1^T Z$ gives us the following matrix:

$$
\tilde{Z} = G_3^T G_2^T G_1^T Z G_1 G_2 G_3 = \begin{bmatrix}
\boxtimes & \times & \times & \times & \times \\
\boxtimes & \boxtimes & \times & \times & \times \\
\boxtimes & \boxtimes & \boxtimes & \times & \times \\
0 & 0 & \times & \times & \times \\
0 & 0 & 0 & 0 & 0
\end{bmatrix}.
$$

The matrix does not yet satisfy the desired structure, as we want the upper left 4×4 block to be of Hessenberg-like form. To do so, we remake the matrix semiseparable, by similar techniques as in Part I. Let us perform the transformation G_4 on the right of the matrix $\tilde{Z}$ to annihilate the marked element. Completing this similarity transformation gives us:

$$
\tilde{Z} = \begin{bmatrix} \boxtimes & \times & \times & \times & \times \\ \boxtimes & \boxtimes & \times & \times & \times \\ \boxtimes & \boxtimes & \boxtimes & \times & \times \\ 0 & 0 & \otimes & \times & \times \\ 0 & 0 & 0 & 0 & 0 \end{bmatrix} \quad \xrightarrow{\tilde{Z}G_4} \quad \begin{bmatrix} \boxtimes & \times & \times & \times & \times \\ \boxtimes & \boxtimes & \times & \times & \times \\ \boxtimes & \boxtimes & \boxtimes & \times & \times \\ 0 & 0 & 0 & \times & \times \\ 0 & 0 & 0 & 0 & 0 \end{bmatrix}
$$

$$
\xrightarrow{G_4^T \tilde{Z} G_4} \quad \left[\begin{array}{cccc|c} \boxtimes & \times & \times & \times & \times \\ \boxtimes & \boxtimes & \times & \times & \times \\ \boxtimes & \boxtimes & \boxtimes & \times & \times \\ \boxtimes & \boxtimes & \boxtimes & \boxtimes & \times \\ \hline 0 & 0 & 0 & 0 & 0 \end{array} \right]
$$

The resulting matrix has the upper left 4×4 block of unreduced Hessenberg-like form, while the last row is zero. The transformations involved did not change the upper left 2×2 block as desired.   □

Note 6.23. *The resulting matrices from Proposition 6.22 are sometimes called zero-tailed matrices (see [18]).*

Note 6.24. *Proposition 6.22 can be generalized to Hessenberg-like matrices with zero blocks below the diagonal. These matrices can be transformed to a Hessenberg-like matrix with zero rows on the bottom and all the Hessenberg-like blocks on the diagonal of unreduced form.*

Proposition 6.25. *Suppose Z is an unreduced Hessenberg-like matrix. Then Z can be written as the sum of a rank 1 matrix and a strictly upper triangular matrix, where the superdiagonal elements of this matrix are different from zero. We have that*

$$
Z = \mathbf{u}\mathbf{v}^T + R,
$$

and u_n and v_1 are different from zero.

Proof. Straightforward, because of the fact that the Hessenberg-like matrix is unreduced and hence is of generator representable form. Moreover, the superdiagonal elements are not includable in the rank structure of the lower triangular part.   □

Proposition 6.26. *Suppose we have the following equality:*

$$
WZ = XW, \tag{6.3}
$$

where W is an orthogonal matrix with the first column equal to $\mathbf{e}_1$ and Z and X

are two Hessenberg-like matrices of the following form:

$$Z = \begin{bmatrix} Z_1 & \times & \times \\ 0 & Z_2 & \times \\ 0 & 0 & 0 \end{bmatrix} \quad X = \begin{bmatrix} X_1 & \times & \times \\ 0 & X_2 & \times \\ 0 & 0 & 0 \end{bmatrix},$$

both having l zero rows at the bottom and the matrices Z_1, Z_2, X_1 and X_2 of unreduced form. If we denote the dimension of the upper left block of Z with n_{Z_1} and the dimension of the upper left block of X with n_{X_1}. Then we have that $n_{X_1} = n_{Z_1}$ and W has the lower left $(n - n_{X_1}) \times n_{X_1}$ block equal to zero.

Proof. Without loss of generality we assume $n_{Z_1} \geq n_{X_1}$. When considering the first n_{Z_1} columns of Equation (6.3) we have

$$W(1 : n, 1 : n_{Z_1}) Z_1 = V, \tag{6.4}$$

where V is of dimension $n \times n_{Z_1}$, with the last l rows of V equal to zero. Because Z_1 is invertible we know that $W(1 : n, 1 : n_{Z_1}) = V Z_1^{-1}$ has the last l rows equal to zero. We will prove by induction that all the columns $\mathbf{w}_k$ ($W = [\mathbf{w}_1, \mathbf{w}_2, \ldots, \mathbf{w}_n]$) with $1 \leq k \leq n_{X_1}$ have the components with index higher than n_{X_1} equal to zero.

- **Step 1.** $k = 1$. Because $W\mathbf{e}_1 = \mathbf{e}_1$, we know already that this is true for $k = 1$. Let us write the matrices Z_1 and X_1 as (using Proposition 6.25):

$$Z_1 = \mathbf{u}^{(1)}\mathbf{v}^{(1)T} + R^{(1)},$$
$$X_1 = \mathbf{u}^{(2)}\mathbf{v}^{(2)T} + R^{(2)},$$

with $v_1^{(1)} = v_1^{(2)} = 1$. Multiplying Equation (6.3) to the right with $\mathbf{e}_1$ gives us

$$W \begin{bmatrix} \mathbf{u}^{(1)} \\ 0 \\ \vdots \\ 0 \end{bmatrix} = \begin{bmatrix} \mathbf{u}^{(2)} \\ 0 \\ \vdots \\ 0 \end{bmatrix}.$$

Note that $\mathbf{u}^{(1)}$ is of length n_{Z_1} and $\mathbf{u}^{(2)}$ is of length n_{X_1}.

- **Step k.** Suppose $2 \leq k \leq n_{X_1}$. We prove that $\mathbf{w}_k$ has the components $n_{X_1} + 1, \ldots, n$ equal to zero. We know by induction that this is true for the columns $\mathbf{w}_i$ with $i < k$. Multiply both sides of Equation (6.3) with $\mathbf{e}_k$, this gives us:

$$W \begin{bmatrix} \begin{bmatrix} \mathbf{u}^{(1)} \\ 0 \\ \vdots \\ 0 \end{bmatrix} v_k^{(1)} + \begin{bmatrix} r_{1,k}^{(1)} \\ \vdots \\ r_{k-1,k}^{(1)} \\ 0 \\ \vdots \\ 0 \end{bmatrix} \end{bmatrix} = X\mathbf{w}_k.$$

This can be rewritten as:

$$\mathbf{u}^{(2)} v_k^{(1)} + W \begin{bmatrix} r_{1,k}^{(1)} \\ \vdots \\ r_{k-1,k}^{(1)} \\ 0 \\ \vdots \\ 0 \end{bmatrix} = X\mathbf{w}_k.$$

Because of the induction procedure, we know that the left-hand side of the former equation has the components below element n_{X_1} equal to zero. The vector $\mathbf{w}_k$ only has the first $n - l$ components different from zero. Because the matrices X_1 and X_2 are nonsingular we know that $\mathbf{w}_k$ has only the first n_{X_1} components different from zero, because $X\mathbf{w}_k$ can only have the first n_{X_1} components different from zero. This proves the induction step.

This means that the matrix W has in the lower left position a $(n - n_{X_1}) \times n_{X_1}$ block of zeros. A combination of Equation (6.3) and the zero structure of W leads to the fact that WZ needs to have a zero block in the lower left position of dimension $(n - n_{X_1}) \times n_{X_1}$. Therefore the matrix Z_1 has to be of dimension n_{X_1} which proves the proposition.   □

Note 6.27. *The proposition above can be formulated in a similar way for more zero blocks below the diagonal.*

Using this property, we can prove the following implicit Q-theorem for Hessenberg-like matrices.

Theorem 6.28 (implicit Q-theorem for Hessenberg-like matrices).
Suppose the following equations hold:

$$Q_1^T A Q_1 = Z \tag{6.5}$$
$$Q_2^T A Q_2 = X \tag{6.6}$$

with $Q_1 \mathbf{e}_1 = Q_2 \mathbf{e}_1$, where Z and X are two Hessenberg-like matrices, with unreduced numbers k_1 and k_2, respectively, and Q_1 and Q_2 are orthogonal matrices. Let us denote $k = \min(k_1, k_2)$. Then we have that the first k columns of Q_1 and Q_2 are the same, up to the sign, and the upper left $k \times k$ submatrices of Z and of X are essentially the same. More precisely, there exists a matrix $V = \mathrm{diag}([1, \pm 1, \pm 1, \ldots, \pm 1])$, of size $k \times k$, such that we have the following two equations:

$$Q_1(1:k, 1:k)\, V = Q_2(1:k, 1:k),$$
$$Z(1:k, 1:k)\, V = V\, X(1:k, 1:k).$$

Proof. Using Proposition 6.22 and the note following the proposition we can assume that the matrices Z and X in (6.5) and (6.6) are of the following form (for simplicity, we assume the number of blocks to be equal to two):

$$
Z = \begin{bmatrix} Z_1 & \times & \times \\ 0 & Z_2 & \times \\ 0 & 0 & 0 \end{bmatrix} \quad X = \begin{bmatrix} X_1 & \times & \times \\ 0 & X_2 & \times \\ 0 & 0 & 0 \end{bmatrix},
$$

with Z_1, Z_2, X_1 and X_2 unreduced Hessenberg-like matrices. This does not affect our statements from the theorem, as the performed transformations from Proposition 6.22 do not affect the upper left $k \times k$ block of the matrices Z and/or X. Denoting $W = Q_1^T Q_2$ and using the Equations (6.5) and (6.6) the following equality holds:

$$
ZW = WX. \tag{6.7}
$$

Moreover, we know by Proposition 6.26 that the matrix W has the lower left block of dimension $(n - n_1) \times n_1$ equal to zero, where n_1 is the dimension of the block Z_1. If we can prove now that for the first n_1 columns of $W = [\mathbf{w}_1, \mathbf{w}_2, \ldots, \mathbf{w}_n]$ the following equality holds: $\mathbf{w}_k = \pm \mathbf{e}_k$, then we have that (as $n_1 \geq k$ and by Proposition 6.22) the theorem holds. According to Proposition 6.25 we can write the upper rows of the matrix Z in the following form:

$$
Z(1 : n_1; 1 : n) = \mathbf{u}^{(1)} \left[\mathbf{v}^{(1)T}, \tilde{\mathbf{v}}^{(1)T} \right] + R^{(1)},
$$

where $\mathbf{u}^{(1)}, \mathbf{v}^{(1)}$ have length n_1. The vector $\tilde{\mathbf{v}}^{(1)}$ is of length $n - n_1$ and is chosen in such a way that the matrix $R^{(1)} \in \mathbb{R}^{n_1 \times n}$ has the last row equal to zero. Moreover the matrix $R^{(1)}$ has the left $n_1 \times n_1$ block strictly upper triangular. Also the left part of X can be written as

$$
X(1 : n; 1 : n_1) = \begin{bmatrix} \mathbf{u}^{(2)} \\ \tilde{\mathbf{u}}^{(2)} \end{bmatrix} \mathbf{v}^{(2)T} + R^{(2)}
$$

with $\mathbf{u}^{(2)}$ and $\mathbf{v}^{(2)}$ of length n_1, $\tilde{\mathbf{u}}^{(2)}$ is of length $n - n_1$ and $R^{(2)}$ a strictly upper triangular matrix of dimension $n \times n_1$. Both of the strictly upper triangular parts $R^{(1)}$ and $R^{(2)}$ have nonzero elements on the superdiagonals. The couples $\mathbf{u}^{(1)}, \mathbf{v}^{(1)}$ and $\mathbf{u}^{(2)}, \mathbf{v}^{(2)}$ are the generators of the semiseparable matrices Z_1 and X_1, respectively. They are normalized in such a way that $v_1^{(1)} = v_1^{(2)} = 1$. Denoting with P the projection operator $P = [I_{n_1}, 0]$ we can calculate the upper left $n_1 \times n_1$ block of the matrices in Equation (6.7):

$$
PZWP^T = PWXP^T
$$
$$
\left(\mathbf{u}^{(1)} \left[\mathbf{v}^{(1)T}, \tilde{\mathbf{v}}^{(1)T} \right] + R^{(1)} \right) WP^T = PW \left(\begin{bmatrix} \mathbf{u}^{(2)} \\ \tilde{\mathbf{u}}^{(2)} \end{bmatrix} \mathbf{v}^{(2)T} + R^{(2)} \right). \tag{6.8}
$$

The fact that $Q_1 \mathbf{e}_1 = Q_2 \mathbf{e}_1$ leads to the fact that $\mathbf{w}_1 = \mathbf{e}_1$. Hence, also the first row of W equals $\mathbf{e}_1^T$. Multiplying (6.8) to the right by $\mathbf{e}_1$, gives

$$
\mathbf{u}^{(1)} = PW \begin{bmatrix} \mathbf{u}^{(2)} \\ \tilde{\mathbf{u}}^{(2)} \end{bmatrix}. \tag{6.9}
$$

Because of the structure of P and W, multiplying (6.9) to the left by $\mathbf{e}_1^T$ gives us that $u_1^{(1)} = u_1^{(2)}$. Using Equation (6.9), we will prove now by induction that for $i \leq n_1 : \mathbf{w}_i = \pm \mathbf{e}_i$ and

$$\left[\mathbf{v}^{(1)^T}, \tilde{\mathbf{v}}^{(1)^T} \right] W P^T = \left[\mathbf{v}^{(1)^T}, \tilde{\mathbf{v}}^{(1)^T} \right] [\mathbf{w}_1, \mathbf{w}_2, \ldots, \mathbf{w}_{n_1}] = \mathbf{v}^{(2)^T}, \qquad (6.10)$$

which will prove the theorem.

- **Step 1.** $l = 1$. This is a trivial step, because $v_1^{(1)} = v_1^{(2)} = 1$ and $\mathbf{w}_1 = \mathbf{e}_1$.

- **Step l.** $1 < l \leq n_1$. By induction we show that $\mathbf{w}_i = \pm \mathbf{e}_i \quad \forall 1 \leq i \leq l - 1$ and (6.10) holds for the first $(l-1)$ columns. This means that:

$$\left[\mathbf{v}^{(1)^T}, \tilde{\mathbf{v}}^{(1)^T} \right] [\mathbf{w}_1, \mathbf{w}_2, \ldots, \mathbf{w}_{l-1}] = \left[v_1^{(2)}, v_2^{(2)}, \ldots, v_{l-1}^{(2)} \right].$$

Taking (6.9) into account, (6.8) becomes

$$\mathbf{u}^{(1)} \left(\left[\mathbf{v}^{(1)^T}, \tilde{\mathbf{v}}^{(1)^T} \right] W P^T - \mathbf{v}^{(2)^T} \right) = \left(P W R^{(2)} - R^{(1)} W P^T \right). \qquad (6.11)$$

Multiplying (6.11) to the right by $\mathbf{e}_l$, we have

$$\mathbf{u}^{(1)} \left(\left[\mathbf{v}^{(1)^T}, \tilde{\mathbf{v}}^{(1)^T} \right] W P^T - \mathbf{v}^{(2)^T} \right) \mathbf{e}_l = \left(P W R^{(2)} - R^{(1)} W P^T \right) \mathbf{e}_l. \qquad (6.12)$$

Because of the special structure of $P, W, R^{(1)}$ and $R^{(2)}$ the element in the n_1-th position in the vector on the right-hand side of (6.12) is equal to zero. We know that $u_{n_1}^{(1)}$ is different from zero, because of the unreducedness assumption. Therefore the following equation holds:

$$u_{n_1}^{(1)} \left(\left[\mathbf{v}^{(1)^T}, \tilde{\mathbf{v}}^{(1)^T} \right] W P^T - \mathbf{v}^{(2)^T} \right) \mathbf{e}_l = 0.$$

This means that

$$\left[\mathbf{v}^{(1)^T}, \tilde{\mathbf{v}}^{(1)^T} \right] \mathbf{w}_l - v_l^{(2)} = 0.$$

Therefore Equation (6.10) is already satisfied up to element l:

$$\left[\mathbf{v}^{(1)^T}, \tilde{\mathbf{v}}^{(1)^T} \right] [\mathbf{w}_1, \mathbf{w}_2, \ldots, \mathbf{w}_l] = \left[v_1^{(2)}, v_2^{(2)}, \ldots, v_l^{(2)} \right]. \qquad (6.13)$$

Using Equation (6.13) together with Equation (6.12) leads to the fact that the complete right-hand side of Equation (6.12) has to be zero. This gives the following equation:

$$R^{(1)} W P^T \mathbf{e}_l = P W R^{(2)} \mathbf{e}_l$$

leading to

$$R^{(1)} W \mathbf{e}_l = \sum_{j=1}^{l-1} r_{j,l}^{(2)} \mathbf{w}_j \qquad (6.14)$$

with the $r_{j,l}^{(2)}$ as the elements of column l of matrix $R^{(2)}$. Hence, the right-hand side can only have the first $l-1$ components different from zero. Because the superdiagonal elements of the left square block of $R^{(1)}$ are nonzero and because the first n_1 columns of W have the last $(n-n_1)$ elements equal to zero, only the first l elements of $\mathbf{w}_l$ can be different from zero. This together with the fact that W is orthogonal means that $\mathbf{w}_l = \pm\mathbf{e}_l$, which proves the induction step.

$\square$

Note 6.29. *In the definition of the unreduced number we assumed that also the element $Z(1,2)$ was not includable in the semiseparable structure. The reader can verify that one can admit this element to be includable in the semiseparable structure, and the theorem and the proof still remain true for this newly defined unreduced number.*

Note 6.30. *This theorem can also be applied to the reduction algorithms which transform matrices to Hessenberg-like, symmetric semiseparable and upper triangular semiseparable form, thereby stating the uniqueness of these reduction algorithms in case the outcome is in unreduced form. If the outcome is not in unreduced form, one knows to which extent the matrices are essentially unique.*

Finally we give some examples connected with this theorem. (See also [18].) The first condition in Definition 6.21 is quite logical, and we will not give any examples connected with this condition.

Example 6.31 Suppose we have the matrices:

$$A = \begin{bmatrix} 1 & 1 & 1 \\ 1 & 0 & 0 \\ 1 & 0 & 0 \end{bmatrix} \quad Q_1 = \begin{bmatrix} 1 & 0 & 0 \\ 0 & \frac{1}{\sqrt{2}} & \frac{-1}{\sqrt{2}} \\ 0 & \frac{1}{\sqrt{2}} & \frac{1}{\sqrt{2}} \end{bmatrix}$$

and $Q_2 = I$. Then we have that $X = A$ and

$$Z = \begin{bmatrix} 1 & \sqrt{2} & 0 \\ \sqrt{2} & 0 & 0 \\ 0 & 0 & 0 \end{bmatrix}.$$

One can see that the unreduced number equals 1 because the element $X(2,2)$ is includable in the lower semiseparable structure of the matrix X. Thus we know by the theorem that equality is only guaranteed for the upper left element of the matrices X and Z. $\blacksquare$

Example 6.32 Suppose we have the matrices:

$$A = \begin{bmatrix} 1 & 0 & 1 \\ 0 & 0 & 0 \\ 1 & 0 & 0 \end{bmatrix} \quad Q_1 = \begin{bmatrix} 1 & 0 & 0 \\ 0 & 0 & 1 \\ 0 & 1 & 0 \end{bmatrix}$$

and $Q_2 = I$. Then we have that $X = A$ and

$$
Z = \begin{bmatrix} 1 & 1 & 0 \\ 1 & 0 & 0 \\ 0 & 0 & 0 \end{bmatrix}.
$$

One can see that the unreduced number equals 0 because the element $X(1,2)$ is includable in the lower semiseparable structure of the matrix X. But using the definition of the unreduced number connected to the note following Theorem 6.28, we know that the unreduced number equals 1 and the equality holds for the upper left element of Z and X.   ∎

Notes and references

The main focus of this section was the development of an implicit Q-theorem for Hessenberg-like matrices, without linking these matrices to their inverse, the Hessenberg matrices. The theorem presented in this section is very general, stating to which extent the matrices are essentially unique. In most of the other articles assumptions were made concerning nonsingularity of the matrices working with. The nonsingularity implied that one could work via the inverse of the matrix R, or that one could switch to the sparse matrix case via inversion.

The results of this section are based on the results in the articles

☞ R. Vandebril, M. Van Barel, and N. Mastronardi. An implicit Q theorem for Hessenberg-like matrices. *Mediterranean Journal of Mathematics*, 2:59–275, 2005.

☞ R. Vandebril, M. Van Barel, and N. Mastronardi. An implicit QR-algorithm for symmetric semiseparable matrices. *Numerical Linear Algebra with Applications*, 12(7):625–658, 2005.

The implicit Q-theorem as proved in this section is the one formulated in the article with the identical title. The second article focuses on the development of the implicit QR-algorithm for semiseparable matrices. In this article several definitions such as the unreduced number and a manner to reduce symmetric matrices to unreduced form are presented. Also the chasing technique, which is the subject of the forthcoming chapter, is discussed in this article.

In the article by Bevilacqua and Del Corso [18] (discussed before in Section 2.2 in Chapter 2), a weaker formulation of the implicit Q-theorem as proposed in this chapter was also presented. Their formulation of the theorem is valid for generator representable semiseparable matrices and is based on inverting these matrices and using the implicit Q-theorem as formulated for tridiagonal matrices. Using this theorem the authors provide results on the uniqueness concerning the reduction to semiseparable form as presented in the first part of this book.

6.3   On Hessenberg-like plus diagonal matrices

In the first part of this book attention was also paid to the special orthogonal similarity transformation of a matrix to semiseparable plus diagonal form. (This is the symmetric version of the reduction to Hessenberg-like plus diagonal form.)

In this chapter however we did not yet pay attention to the structure of semi-separable plus diagonal matrices. The deduction of an implicit QR-algorithm for semiseparable plus diagonal matrices is rather similar to the one for semiseparable matrices, just like the reduction of a matrix to semiseparable and semiseparable plus diagonal form were closely related. Unfortunately the admittance of the diagonal creates a lot of details that obscure the algorithm. Also the definition of unreducedness and the reduction to unreduced form, just like the development of an implicit Q-theorem, become complicated. Hence we will not derive all steps for developing an implicit QR-algorithm for semiseparable plus diagonal matrices. We will however present an algorithm based on a divide-and-conquer approach for computing the eigenvalues of semiseparable plus diagonal matrices (see Chapter 10).

In this section we will present a result, interesting in our opinion, related to the structure of a semiseparable plus diagonal matrix (or Hessenberg-like plus diagonal), when performing steps of the QR-algorithm. Not all details are shown or included as they obscure the global picture. References will be provided at the end of this section.

Theorem 6.7 stated that the strictly lower triangular structure of a Hessenberg-like matrix plus diagonal was maintained under the QR-algorithm with shift.

In the following proof, we show that also the diagonal term of the resulting Hessenberg-like plus diagonal matrix is the same as the original Hessenberg-like plus diagonal matrix. So in fact not only the rank structure but even the diagonal is maintained. The general proof of this statement can be found in [52, 50], investigating all kinds of structures. Here we will only prove a simple formulation of this statement, i.e., the case in which the diagonal does not have zeros.

Theorem 6.33. *Suppose Z is a Hessenberg-like matrix and D is a diagonal matrix, such that all diagonal elements are different from zero. Applying a step of QR (with the QR-decomposition as in Subsection 6.1.1) without shift on the matrix $Z + D$ results in a Hessenberg-like plus diagonal matrix, which can always be written as a Hessenberg-like plus diagonal matrix with the same diagonal D: $\tilde{Z} + D$.*

Note 6.34. *We remark that this theorem is also valid for QR-steps with a shift μ, the only demand is that the shift is different from any of the diagonal elements of the matrix D. In this case the matrix $\tilde{Z} = Z - \mu I$ is again a Hessenberg-like plus diagonal matrix, with another (shifted) diagonal.*

Proof. We will assume that we perform a QR-step without shift on the Hessenberg-like plus diagonal matrix $Z + D$. We have

$$\begin{cases} Q_1^T Z &= R_1 \\ Q_2^T (R_1 + Q_1^T D) &= R, \end{cases}$$

with R_1 and R as two upper triangular matrices. Let us denote with H the Hessenberg matrix $R_1 + Q_1^T D$. We will distinguish between different cases:

- Suppose R is invertible. This is the same as stating that $Z + D$ is nonsingular. We know that the matrix after one QR-step without shift has the following form:

$$R(Z+D)R^{-1} = RZR^{-1} + RDR^{-1}$$
$$= \hat{Z} + \hat{R},$$

where $\hat{Z}$ is clearly a Hessenberg-like matrix, and $\hat{R}$ an upper triangular matrix with as diagonal elements, the diagonal elements of D. This can clearly be rewritten as:

$$\tilde{Z} + D,$$

which proves one case of the theorem.

- Suppose the Hessenberg-like matrix Z has the element in the lower left corner different from zero. This means (like in the regular case) that the first sequence of Givens transformations does not have Givens transformations equal to the identity matrix. As all elements in the diagonal of the Hessenberg-like plus diagonal matrix are different from zero, we know that our resulting Hessenberg matrix H will be unreduced. This means that

$$Q_2^T H = \left[\begin{array}{c|c} \tilde{R} & \mathbf{w} \\ \hline 0 & \alpha \end{array} \right],$$

with $\tilde{R}$ invertible. If α is different from zero, the matrix R is invertible, and we are in case 1 of the proof. So we assume α to be zero. Similar to the proof of Theorem 6.8 we can write:

$$RQ = \left[\begin{array}{c|c} \tilde{R} & \mathbf{w} \\ \hline 0 & 0 \end{array} \right] Q$$
$$= \left[\begin{array}{c|c} \tilde{R} & \mathbf{w} \\ \hline 0 & 0 \end{array} \right] \left[\; [Z+D] \left[\begin{array}{c} \tilde{R}^{-1} \\ 0 \end{array} \right] \; \middle| \; \mathbf{q} \; \right]$$
$$= \left[\begin{array}{c|c} \hat{Z} + \hat{D} & \mathbf{z} \\ \hline 0 & 0 \end{array} \right],$$

where $\hat{Z}$ is a Hessenberg-like matrix of dimension $(n-1) \times (n-1)$, if the dimension of Z is n and $\hat{D}$ is a diagonal matrix with as elements the first $n-1$ elements of the matrix D. Denoting the last element of D with d_n, we can rewrite this as:

$$\left[\begin{array}{c|c} \hat{Z} + \hat{D} & \mathbf{z} \\ \hline 0 & 0 \end{array} \right] = \left[\begin{array}{c|c} \hat{Z} & \mathbf{z} \\ \hline 0 & -d_n \end{array} \right] + D$$
$$= \tilde{Z} + D$$

for which $\tilde{Z}$ is a Hessenberg-like and D the diagonal, which proves the theorem if H is unreduced.

- Suppose our Hessenberg-like matrix Z has the element in the lower left corner equal to zero. We know now that we can decompose our matrix Z as a block diagonal matrix, having all the blocks of Hessenberg-like plus diagonal form. For each of these blocks separately one can apply the reasoning in the previous bullet. Hence we get a block diagonal matrix with all blocks of Hessenberg-like plus diagonal form. Combining them gives globally a Hessenberg-like matrix with the same diagonal.

□

We remark that the construction above is not valid for Hessenberg-like plus diagonal matrices, for which some of the diagonal elements equal zero. But in theory one can generically prove that for any kind of Hessenberg-like plus diagonal matrix the structure and the diagonal are maintained after a step of the QR-algorithm.

Theorem 6.35. *Suppose Z is a Hessenberg-like matrix and D is a diagonal matrix. Applying a step of QR (with the QR-decomposition as in Subsection 6.1.1) without shift on the matrix $Z+D$ results in a Hessenberg-like plus diagonal matrix, which can always be written as a Hessenberg-like plus diagonal matrix with the same diagonal D: $\tilde{Z} + D$.*

Proof. One can prove the theorem by slightly perturbing the diagonal, and then taking the limit for the perturbation going to zero. We will not go into the details, but a general proof of this theorem and the preservation of more general structures can be found in [52, 50].   □

When defining unreduced Hessenberg-like matrices (in order to develop chasing technique as we will do in the next chapter), one has again to be very careful, because the extra diagonal creates several theoretical difficulties. As mentioned before, we do not go into the details of developing an implicit QR-algorithm for semiseparable plus diagonal matrices. The details for doing so can be found in some articles mentioned in the notes and references. We will however show in Chapter 10 how one can effectively compute the eigenvalues and eigenvectors of a Hessenberg-like plus diagonal matrix based on a divide-and-conquer method.

Notes and references

Even though we will not discuss QR-algorithms for Hessenberg-like plus diagonal or semiseparable plus diagonal matrices, we will present some references in which details on how to do so are given.

A technical report by Van Camp, Van Barel, Mastronardi and Vandebril describes all details on how to change the implicit QR-algorithm for semiseparable matrices in order to obtain an algorithm suitable for semiseparable plus diagonal matrices.

☞ E. Van Camp, M. Van Barel, R. Vandebril, and N. Mastronardi. An implicit QR-algorithm for symmetric diagonal-plus-semiseparable matrices. Technical Report TW419, Department of Computer Science, Katholieke Universiteit Leuven, Celestijnenlaan 200A, 3000 Leuven (Heverlee), Belgium, March 2005.

In the following article by Delvaux and Van Barel the authors present a method for performing the explicit QR-algorithm on structured rank matrices. The idea is based on the preservation of the underlying structured rank. The algorithm is based on the Givens-weight or unitary weight representation.

☞ S. Delvaux and M. Van Barel. The explicit QR-algorithm for rank structured matrices. Technical Report TW459, Department of Computer Science, Katholieke Universiteit Leuven, Celestijnenlaan 200A, 3000 Leuven (Heverlee), Belgium, May 2006.

In the following article Eidelman, Gohberg and Olshevsky develop an explicit QR-algorithm for hermitian quasiseparable matrices of any order. Explicit formulas are given to represent the change of structure and the quasiseparable representation of the intermediate matrices in the algorithm.

☞ Y. Eidelman, I. C. Gohberg, and V. Olshevsky. The QR iteration method for Hermitian quasiseparable matrices of an arbitrary order. *Linear Algebra and its Applications*, 404:305–324, July 2005.

6.4   Conclusions

In this chapter we investigated theoretical results in order to design implicit QR-algorithms. All ingredients, except for the chasing itself, were be provided here. The chasing is the subject of the next chapter.

We showed that a special kind of QR-factorization is needed in order to have the structure maintained under a QR-step. Using this QR-factorization, which consists of $2n - 1$ Givens transformations, we proved that the structure of a Hessenberg-like matrix is maintained. As in the sparse case we defined an unreduced Hessenberg-like matrix and a method to transform any Hessenberg-like matrix to unreduced form. Based on the unreduced number, an implicit Q-theorem for Hessenberg-like matrices was derived. In the final section we proved that the structure of a Hessenberg-like plus diagonal matrix was maintained under a step of the QR-method, in case the diagonal had all elements different from zero.

Chapter 7

Implicit QR-methods for semiseparable matrices

In the previous chapters some of the well-known results for QR-algorithms related to sparse matrices, such as tridiagonal and Hessenberg matrices, were recapitulated. Also interesting theorems connected to QR-algorithms for Hessenberg-like matrices were provided. Starting with a Hessenberg-like matrix we know how to transform it to unreduced form. The structure of a Hessenberg-like matrix after an explicit step of the QR-algorithm is known and an implicit Q-theorem for Hessenberg-like matrices was proved. To construct an implicit algorithm, similarly as for the sparse matrix case, only the implicit step of the QR-method is missing. In this chapter we will provide this step for Hessenberg-like matrices and for symmetric semiseparable matrices. Moreover, we will translate this implicit QR-step such that we can use it to calculate the singular values of an upper triangular semiseparable matrix. The results provided in this chapter can be combined in a straightforward way with the reduction algorithms presented in Part I. Using this combination we can compute the eigenvalues of symmetric and unsymmetric matrices via intermediate semiseparable and Hessenberg-like matrices and we can also compute the singular values of arbitrary matrices via intermediate upper triangular semiseparable matrices.

In the first section an implicit QR-algorithm for semiseparable matrices will be designed. Brief discussions about the unreducedness and the type of shift are included. One should make use of the symmetry of the matrix when transforming it to unreduced form. We will consider the Wilkinson shift. The actual implicit QR-algorithm consists of two main parts. In a first part a step of QR without shift will be performed on the semiseparable matrix. The resulting matrix is again semiseparable. In a second part a similarity Givens transformation will be applied on the first two columns and first two rows of the new semiseparable matrix. This will disturb the semiseparable structure. From now on we switch to the implicit approach to restore the semiseparable structure. We will prove that the combination of these two steps corresponds to performing one step of the explicit QR-algorithm.

In the third section we focus on the development of an implicit QR-algorithm for Hessenberg-like matrices. The approach followed is similar to the symmetric case. First a step of QR-step without shift is performed. Afterwards a similarity Givens transformation is applied. The resulting disturbed Hessenberg-like matrix will then be reduced back to Hessenberg-like form. Using the implicit Q-theorem one can easily prove that it corresponds to performing an explicit step of QR. In this section we do not yet cover the double shift approach as in Chapter 9 the more general multishift strategy will be discussed.

The last section focuses on the computation of singular values of upper triangular semiseparable matrices S_u. The implicit approach will be deduced by looking at the implicit QR-algorithm applied to the symmetric matrix $S = S_u^T S_u$. We will explain what the structure of an unreduced upper triangular semiseparable matrix is and how we can transform it to this form. It will be shown that one step of the QR-method applied to the upper triangular semiseparable matrix S_u corresponds to four main steps: transforming the upper triangular semiseparable matrix to lower triangular form; creating a disturbance in the lower triangular semiseparable matrix; transforming the matrix back to lower triangular semiseparable form; finally transforming the resulting semiseparable matrix back to upper triangular semiseparable form.

✎ *In this chapter the chasing techniques for restoring the rank structure are developed.*

Initially we start with symmetric semiseparable matrices. Extra information on unreduced symmetric semiseparable matrices is provided in Subsection 7.1.1. This is however not essential for understanding the remainder of the book. The essential ideas on this algorithm are contained in Subsection 7.1.3. It explains how to perform the two sequences of Givens transformations, the computation of the shift and the different transformations needed for restoring the structure in the matrix. Even though the proof of Theorem 7.6 is though it gives insight into the inner structured rank relations. A brief summary of the algorithm and a proof of correctness are presented in Subsection 7.1.4 for the interested reader.

Similar ideas as for the symmetric case can be applied to the Hessenberg-like case. The most important propositions on structure restoration are presented in Subsection 7.3.3.

The translation of the QR-algorithm for singular values has to start in some sense from scratch. Unreduced upper triangular semiseparable matrices are defined in Definition 7.15. The general concept of performing a QR-step is explained in Subsection 7.4.2 followed by the chasing techniques in Subsection 7.4.3. A summary of the method and the proof of correctness are presented for the interested reader in Subsection 7.4.4.

7.1   An implicit QR-algorithm for symmetric semiseparable matrices

In this section we will derive an implicit QR-algorithm for symmetric semiseparable matrices. We will make extensive use of the theorems and tools provided in the previous chapter for Hessenberg-like matrices. Because the class of matrices we are working with is symmetric, we will sometimes adapt the tools slightly, taking into account the symmetry.

Suppose we have a symmetric semiseparable matrix S, and a shift μ then we want to calculate the semiseparable matrix $\tilde{S}$ for which:

$$\left\{ \begin{array}{rcl} S - \mu I & = & QR \\ \tilde{S} & = & RQ + \mu I. \end{array} \right. \tag{7.1}$$

(The QR-factorization is defined as in Chapter 6 Subsection 6.1.1.) Performing the transformations of (7.1) explicitly leads to an explicit QR-algorithm for semiseparable matrices. This means first calculating the QR-factorization and then performing the product RQ. Explicit QR-algorithms can be found in the literature, see, e.g., [23]. We can however also write the semiseparable matrix $\tilde{S}$ as

$$\tilde{S} = Q^T S Q.$$

In this section we will see that we calculate the matrix $\tilde{S}$ based on these formulas without explicitly forming the orthogonal matrix Q. Because Q consists of a sequence of $2n - 2$ Givens transformations, we will apply these Givens transformations one after the other on both sides of the matrix. In fact we will calculate a semiseparable matrix $\hat{S}$ satisfying

$$\hat{S} = \hat{Q}^T S \hat{Q},$$

such that $Q\mathbf{e}_1 = \hat{Q}\mathbf{e}_1$. The matrix $\hat{Q}$ will be constructed implicitly, without considering the complete QR-factorization of the matrix S. Using the implicit Q-theorem (see Theorem 6.28) we will then prove that the semiseparable matrices $\hat{S}$ and $\tilde{S}$ are essentially the same. Hence the computed matrix $\hat{S}$ is the result of applying a QR-step on the matrix S, with shift μ. We will now consider the different steps to obtain an implicit QR-algorithm for symmetric semiseparable matrices.

7.1.1 Unreduced symmetric semiseparable matrices

In Subsection 6.2.1 of the previous chapter we have developed a method to transform a Hessenberg-like matrix to unreduced form. This technique is also naturally applicable to symmetric matrices.

However, because the matrix we are working with is symmetric, it is sometimes easier to reduce such a matrix to unreduced form. In the remainder of this section we will assume that the matrices in question do not have a zero block $S(i : n, 1 : i - 1)$ (with $i = 2 : n$) below the diagonal. In case there is a zero block, we can divide the matrix in different blocks, all of symmetric semiseparable form. Moreover, we know (see Proposition 1.15), that this matrix is therefore of generator representable semiseparable form. The fact that this matrix is of generator representable semiseparable form will make the reduction to unreduced form computationally less expensive.

Suppose for example we have the following symmetric semiseparable matrix:

$$\begin{bmatrix} \boxtimes & \times & \times & \times & \times \\ \boxtimes & \boxtimes & \boxtimes & \times & \times \\ \boxtimes & \boxtimes & \boxtimes & \times & \times \\ \boxtimes & \boxtimes & \boxtimes & \boxtimes & \times \\ \boxtimes & \boxtimes & \boxtimes & \boxtimes & \boxtimes \end{bmatrix}.$$

And assume more precisely that we have the following situation, in which a lot of zeros are present:

$$\begin{bmatrix} \boxtimes & \times & 0 & \times & \times \\ \boxtimes & \boxtimes & 0 & \times & \times \\ 0 & 0 & 0 & 0 & 0 \\ \boxtimes & \boxtimes & 0 & \boxtimes & \times \\ \boxtimes & \boxtimes & 0 & \boxtimes & \boxtimes \end{bmatrix}.$$

It is clear that simply applying a suitable permutation reduces this matrix to a similar unreduced semiseparable matrix. Or for example the following situation can also be transformed to unreduced form quite easily.

$$\begin{bmatrix} \boxtimes & 0 & 0 & \times & \times \\ 0 & 0 & 0 & \times & \times \\ 0 & 0 & 0 & \times & \times \\ \boxtimes & \boxtimes & \boxtimes & \boxtimes & \times \\ \boxtimes & \boxtimes & \boxtimes & \boxtimes & \boxtimes \end{bmatrix}.$$

Due to the symmetry and the semiseparable structure, we know that the 2×2 block which is the intersection between rows 2 and 3 and the columns 4 and 5 is of rank 1. Therefore a well chosen Givens transformation acting on rows 2 and 3 can immediately create zeros in the entire row 3. This brings us back to the previous case in which a simple permutation solved the problem.

Suppose now that the 2×2 block, which is the intersection between rows 2 and 3 and columns 2 and 3 is different from zero and still has rank 1, then it is obvious that the complete rows 2 and 3 are dependent of each other. Hence a single Givens transformation can zero out the complete row 3. This leads hence to the first case we considered. It seems that in fact in all cases we can perform a much cheaper reduction towards unreduced form. Globally we can make the following statement concerning the rank structure of a symmetric semiseparable matrix.

Proposition 7.1. *Suppose we have a symmetric generator representable semiseparable matrix S for which the element in position $(2, 3)$ is includable in the semiseparable structure. Then row 2 and row 3 are linearly dependent.*

Proof. The proof consists in checking all possible cases. To prove this statement we distinguish between two simple cases.

- Assume the 2×2 block $S(2 : 3, 2 : 3)$ different from zero. This symmetric 2×2 block is contained in the rank 1 part to the left of it and also to the right of it. This means that globally the completely row needs to share this block its rank 1 property and hence the complete row is of rank 1. (All columns in these two rows need to be multiples of a nonzero column out of this 2×2 block.) To illustrate it more clearly we depict the following matrix:

$$\begin{bmatrix} & & \ddots & & & & & & \\ \boxtimes & \cdots & \boxtimes & \boxdot & \boxdot & \boxplus & \cdots & \boxplus \\ \boxtimes & \cdots & \boxtimes & \boxdot & \boxdot & \boxplus & \cdots & \boxplus \\ & & & & & & \ddots & \end{bmatrix}.$$

The $\ddots$ elements depict the diagonal in the matrix. Assume the block depicted with the elements $\boxdot$ is the rank 1 block crossing the diagonal (which is not entirely zero). Due to the demands we know that this block is includable in the structure to the left of it consisting of the elements $\boxtimes$ and to the right of it consisting of the elements $\boxplus$. Due to the nonzeroness and the rank one assumptions posed, we know that the complete row needs to be of rank 1.

- Assume the 2×2 block $S(2:3, 2:3)$ equals zero. The statement from the previous item is not valid anymore. Assume for example we have the following rows:

$$\begin{bmatrix} 1 & 0 & 0 & 1 & 1 \\ 1 & 0 & 0 & 2 & 2 \end{bmatrix}.$$

It is clear that both submatrices

$$\begin{bmatrix} 1 & 0 & 0 \\ 1 & 0 & 0 \end{bmatrix} \text{ and } \begin{bmatrix} 0 & 1 & 1 \\ 0 & 2 & 2 \end{bmatrix}$$

are of rank 1, whereas the combination of both rows is not of rank 1. But, in this case we have to take the symmetry of the matrices into consideration! We have the following situation:

$$\begin{bmatrix} \boxtimes & \times & \times & \times & \times \\ \boxtimes & 0 & 0 & \times & \times \\ \boxtimes & 0 & 0 & \times & \times \\ \boxtimes & \boxtimes & \boxtimes & \boxtimes & \times \\ \boxtimes & \boxtimes & \boxtimes & \boxtimes & \boxtimes \end{bmatrix}.$$

From Theorem 1.14, we know that every zero (in a generator representable semiseparable matrix) needs to be contained in a zero row or a zero column (included in the semiseparable structure of course). This means that we have a situation in which at least the marked elements are already zero. We remark that we already exploited the symmetry to construct extra zero elements to construct the following two possibilities[12]:

$$
\begin{bmatrix}
\boxtimes & \times & \times & \times & \times \\
\boxtimes & 0 & 0 & 0 & 0 \\
\boxtimes & 0 & 0 & 0 & 0 \\
\boxtimes & 0 & 0 & \boxtimes & \times \\
\boxtimes & 0 & 0 & \boxtimes & \boxtimes
\end{bmatrix}
\quad \text{or} \quad
\begin{bmatrix}
\boxtimes & 0 & 0 & \times & \times \\
0 & 0 & 0 & \times & \times \\
0 & 0 & 0 & \times & \times \\
\boxtimes & \boxtimes & \boxtimes & \boxtimes & \times \\
\boxtimes & \boxtimes & \boxtimes & \boxtimes & \boxtimes
\end{bmatrix}
$$

We remark once more, that there are possibly more zeros than the ones shown. It is trivial to say that these two matrices have rows 2 and 3 dependent of each other (for the figure on the right one has to take the semiseparable structure of the upper triangular part into consideration).

Hence we can conclude, that if an element above the diagonal is includable in the structure, that this creates two dependent rows and columns. Moreover this also implies that the matrix is singular. □

The previous proposition shows that it is very easy to remove these troublesome elements from our symmetric semiseparable matrix. Having a symmetric semiseparable matrix of the following structure:

$$
S =
\begin{bmatrix}
\boxtimes & \times & \times & \times & \times \\
\boxtimes & \boxtimes & \boxtimes & \times & \times \\
\boxtimes & \boxtimes & \boxtimes & \times & \times \\
\boxtimes & \boxtimes & \boxtimes & \boxtimes & \times \\
\boxtimes & \boxtimes & \boxtimes & \boxtimes & \boxtimes
\end{bmatrix},
$$

one can apply a simple similarity Givens transformation G in order to zero out an entire row and column. This creates the following situation:

$$
G^T S G =
\begin{bmatrix}
\boxtimes & \times & 0 & \times & \times \\
\boxtimes & \boxtimes & 0 & \times & \times \\
0 & 0 & 0 & 0 & 0 \\
\boxtimes & \boxtimes & 0 & \boxtimes & \times \\
\boxtimes & \boxtimes & 0 & \boxtimes & \boxtimes
\end{bmatrix}.
$$

The procedure can be completed by performing one more permutation P, moving the zero row and column to the right and bottom. This creates a so-called zero-tailed matrix of the following form:

$$
P^T G^T S G P =
\begin{bmatrix}
\boxtimes & \times & \times & \times & 0 \\
\boxtimes & \boxtimes & \times & \times & 0 \\
\boxtimes & \boxtimes & \boxtimes & \times & 0 \\
\boxtimes & \boxtimes & \boxtimes & \boxtimes & 0 \\
0 & 0 & 0 & 0 & 0
\end{bmatrix}.
$$

---

[12]The reader can try to construct other possible cases in which the zeros are distributed along one row and one column. But one can see that combining one row and one column to capture the four zeros in the middle block is not possible. One row combined with one column leaves one zero uncaptured. Hence an extra row or column is always necessary to include this zero in a row or column.

This last zero row can immediately be removed. It corresponds to the eigenvalue 0.

We can conclude that in the symmetric case one can more easily transform a matrix to unreduced form, with regard to the Hessenberg-like case.

7.1.2   The shift μ

The choice of the shift is very important to achieve fast convergence to one of the eigenvalues. The shifts we consider here are discussed in [94]. Suppose we have the following symmetric semiseparable matrix represented with the Givens-vector representation:

$$
\begin{bmatrix}
c_1 v_1 \\
c_2 s_1 v_1 & c_2 v_2 \\
c_3 s_2 s_1 v_1 & c_3 s_2 v_2 & c_3 v_3 \\
\vdots & & & \ddots \\
c_{n-1} s_{n-2} \cdots s_1 v_1 & \cdots & & c_{n-1} v_{n-1} \\
s_{n-1} s_{n-2} \cdots s_1 v_1 & \cdots & & s_{n-1} v_{n-1} & v_n
\end{bmatrix}
$$

One can choose v_n as a shift, the so-called Rayleigh shift, or one can consider as a shift the eigenvalue of

$$
\begin{bmatrix}
c_{n-1} v_{n-1} & s_{n-1} v_{n-1} \\
s_{n-1} v_{n-1} & v_n
\end{bmatrix},
$$

that is closest to v_n: the Wilkinson shift [184]. Using this shift in the tridiagonal case will give cubic convergence. The numerical results provided in Chapter 8 will experimentally prove the same rate of convergence for the symmetric semiseparable case.

7.1.3   An implicit QR-step

The first $n-1$ Givens transformations are in fact completely determined by the semiseparable matrix. To perform these transformations, the shift μ is not yet needed. When applying these Givens transformations to the left of the semiseparable matrix $S_0 = S$, this matrix becomes upper triangular:

$$
G_{n-1}^T \dots G_1^T S = R.
$$

Directly applying the Givens transformations $G_1 \dots G_n$ on the right of the matrix R will construct a matrix $RG_1 \dots G_{n-1}$ whose lower triangular part is semiseparable. (See the proof of Theorem 6.6 to see that one can apply both transformations at the same time. The proof is for nonsymmetric matrices.) Because of symmetry reasons the resulting matrix $RG_1 \dots G_{n-1}$ is a symmetric semiseparable matrix. The application of the different Givens transformations can be done at the same time, i.e., instead of first applying all the transformations on the left, we apply them to the left and to the right at the same time. Implementation details about this step can be found in the next chapter. We obtain successively:

$$
S_1 = G_1^T S_0 G_1
$$

followed by

$$S_2 = G_2^T S_1 G_2$$

and so on.

As stated before, this step corresponds to applying a QR-step without shift to the semiseparable matrix.

The application of the second sequence of Givens transformations is the hardest step of the two and requires some theoretical results. To initialize this step the knowledge of the Givens transformation G_n^T is crucial. G_n^T is the Givens transformation that will start to reduce the Hessenberg matrix $G_{n-1}^T \ldots G_1^T (S - \mu I)$ to upper triangular form. The algorithm however did not calculate $G_{n-1}^T \ldots G_1^T (S - \mu I)$ but a semiseparable matrix $S_{n-1} = G_{n-1}^T \ldots G_1^T S G_1 \ldots G_{n-1}$. However, because we use the Givens-vector representation as mentioned in the first section, we know that the upper left element of the matrix $G_{n-1}^T \ldots G_1^T S$ is the first element in the vector $\mathbf{v}$ from the Givens-vector representation of the matrix S. It can be seen that the elements in the upper left positions $(1, 1)$ and $(2, 1)$ of the matrix $G_{n-1}^T \ldots G_1^T \mu I$ equal

$$G_{n-1}^T \begin{bmatrix} \mu \\ 0 \end{bmatrix} .$$

The Givens transformation G_n^T is constructed to annihilate the first subdiagonal element in the Hessenberg matrix $G_{n-1}^T \ldots G_1^T (S - \mu I)$. Due to the Givens-vector representation and the structure of $G_{n-1}^T \ldots G_1^T \mu I$, we know that the first two elements of the first column of the considered Hessenberg matrix are the following ones:

$$\begin{bmatrix} v_1 \\ 0 \end{bmatrix} - G_{n-1}^T \begin{bmatrix} \mu \\ 0 \end{bmatrix} .$$

Hence one can easily compute the transformation G_n^T. This means that the Givens transformation G_n^T already can be applied to the matrix S_{n-1}. Hence, we obtain $S_n = G_n^T S_{n-1} G_n$. From this point we will work directly on the matrix S_n and therefore we switch to the implicit approach.

The matrix S_{n-1} was a semiseparable matrix and the output of a complete step of the implicit QR-algorithm also has to be a semiseparable matrix. However after having applied the similarity transformation $G_n^T S_{n-1} G_n$, the semiseparable structure is disturbed. The following sequence of Givens transformations that will be applied to the matrix S_n will restore the semiseparable structure. Even more: the resulting matrix will essentially be the same as the matrix coming from one step of the QR-algorithm with shift μ. We will show that it is possible to rebuild a semiseparable matrix out of S_n without changing the first row and column. To do so we use some kind of chasing technique. More information about general chasing techniques can be found in, e.g., [181, 183]. To prove the two main theorems, two propositions are needed.

Proposition 7.2. *Suppose the following symmetric 3×3 matrix is given:*

$$A = \begin{bmatrix} x & a & d \\ a & b & e \\ d & e & f \end{bmatrix} , \tag{7.2}$$

which is not yet semiseparable. Then there exists a Givens transformation

$$G = \begin{bmatrix} 1 & 0 \\ 0 & \hat{G} \end{bmatrix} \quad with \quad \hat{G} = \frac{1}{\sqrt{1+t^2}} \begin{bmatrix} t & -1 \\ 1 & t \end{bmatrix}$$

such that the following matrix

$$G^T A G$$

is a symmetric semiseparable matrix.

Proof. We can assume that $ae - bd$ is different from zero, otherwise, the matrix would already be semiseparable. The proof is constructive, i.e., the matrix G will be constructed such that the matrix $G^T A G$ indeed is a semiseparable matrix. Calculating explicitly the product $G^T A G$ gives the following matrix (with $s = 1 + t^2$):

$$\frac{1}{s} \begin{bmatrix} sx & \sqrt{s}(at+d) & \sqrt{s}(-a+dt) \\ \sqrt{s}(ta+d) & (tb+e)t+(te+f) & (-1)(tb+e)+(te+f)t \\ \sqrt{s}(-a+td) & (-b+te)t+(-e+tf) & (-1)(-b+te)+(-e+tf)t \end{bmatrix}.$$

When this matrix has to be semiseparable the determinant of the lower left 2×2 block has to be zero, i.e.,

$$\big(ta+d\big)\big((-b+te)t+(-e+tf)\big) - \big((tb+e)t+(te+f)\big)\big(-a+td\big) = 0.$$

Solving this equation for t gives the following result:

$$t = \frac{de - af}{ae - db},$$

which is properly defined because $ae - db \neq 0$. $\qquad\square$

Proposition 7.3. *Suppose the following symmetric 4×4 matrix is given,*

$$A = \begin{bmatrix} x & a & 0 & d \\ a & b & 0 & e \\ 0 & 0 & 0 & f \\ d & e & f & y \end{bmatrix}, \tag{7.3}$$

which is not yet semiseparable. Then there exists a Givens transformation

$$G = \begin{bmatrix} 1 & 0 & 0 \\ 0 & \hat{G} & 0 \\ 0 & 0 & 1 \end{bmatrix} \quad with \quad \hat{G} = \frac{1}{\sqrt{1+t^2}} \begin{bmatrix} t & -1 \\ 1 & t \end{bmatrix}$$

such that the upper 3×3 block of the following matrix

$$G^T A G$$

is a symmetric semiseparable matrix. And the lower left 3×2 block is of rank one.

Proof. The proof is straightforward. Calculating the product $G^T A G$ shows that the block

$$\begin{bmatrix} -a & -tb \\ d & te + f \end{bmatrix}$$

has to be of rank 1. This corresponds with $t = -af/(ae - bd)$, where $(ae - bd)$ is different from zero because the matrix is not yet semiseparable.   □

Proposition 7.4. *Suppose the following symmetric 4×4 matrix is given,*

$$A = \begin{bmatrix} x & y & a & d \\ y & 0 & 0 & 0 \\ a & 0 & b & e \\ d & 0 & e & z \end{bmatrix}, \tag{7.4}$$

which has the 2×2 matrix

$$\begin{bmatrix} a & b \\ d & e \end{bmatrix},$$

of rank 2. This means that the matrix is not yet semiseparable. Then there exists a Givens transformation

$$G = \begin{bmatrix} I_2 & 0 \\ 0 & \hat{G} \end{bmatrix},$$

such that the complete 4×4 matrix $G^T A G$ will be of semiseparable form.

Proof. The proof is straightforward. Deleting the second column and row gives us a 3×3 matrix as in Proposition 7.2.   □

One might wonder how both theorems can be used for larger matrices. This is shown in the following theorem:

Theorem 7.5. *Suppose an unreduced symmetric semiseparable matrix S is given. On this matrix a similarity Givens transformation $\tilde{G}$ is performed, involving the first two columns and rows:*

$$\tilde{S} = \tilde{G}^T S \tilde{G}.$$

Then there exists a Givens transformation G:

$$G = \begin{bmatrix} 1 & 0 & 0 \\ 0 & \hat{G} & 0 \\ 0 & 0 & I \end{bmatrix} \quad \text{with} \quad \hat{G} = \frac{1}{\sqrt{1+t^2}} \begin{bmatrix} t & -1 \\ 1 & t \end{bmatrix}$$

such that the upper left 3×3 block of the matrix $G^T \tilde{S} G$ is semiseparable.

Proof. The proof is divided in different cases. We will see that, due to the restrictions posed by the unreducedness, only two possibilities remain. These possibilities coincide with Proposition 7.2 and 7.3. Before being able to distinguish between the different cases we can pose some extra constraints.

- **Condition 1.** $\tilde{G} \neq I$. We can clearly assume that the initial disturbing Givens transformation $\tilde{G}$ is different from zero, otherwise we can take $\hat{G}$ also equal to the identity and the theorem is proved. Assume therefore in the remainder of the proof that $\tilde{G} \neq I$.

- **Condition 2.** Suppose the upper left 2×2 block of the matrix $\tilde{S}$ is of the following form:

$$\tilde{S}(1:2, 1:2) = \begin{bmatrix} x & a \\ a & b \end{bmatrix}. \tag{7.5}$$

Condition 2 states that a and b cannot be zero at the same time. This means that either a or b or both are different from zero. Let us briefly prove this statement. Suppose both a and b to be zero at the same time. This will lead to a contradiction. This means that we have:

$$\tilde{S}(1:2, 1:2) = \begin{bmatrix} x & 0 \\ 0 & 0 \end{bmatrix}.$$

As a Givens transformation does not change the rank we hence obtain that the rank of $S(1:2, 1:2)$ with

$$S(1:2, 1:2) = \tilde{G}(1:2, 1:2)\, \tilde{S}(1:2, 1:2)\, \tilde{G}(1:2, 1:2)^T,$$

also needs to be equal to 1. This is in contradiction with the initial assumption of unreducedness. Hence we know that at least one of the two elements a or b needs to be different from zero.

- **Condition 3.** Suppose our matrix S has the upper left 3×3 block of the following form:

$$\tilde{S}(1:3, 1:3) = \begin{bmatrix} x & a & d \\ a & b & e \\ d & e & f \end{bmatrix}. \tag{7.6}$$

We show that if d and/or e are different from zero, then $ae \neq bd$. Again we will prove this statement by contradiction. Assume namely that $ae = bd$, this means that the lower right 2×2 block in the Matrix (7.6) is of rank strict equal to 1 (strict because d and/or e are different from zero and a and/or b are different from zero, hence the two rows are a multiple of each other). As in Condition 2 we will see that this leads to a contradiction with the unreducedness of the original matrix S. Depict the matrix $\tilde{S} = (1:3, 1:3)$ as follows:

$$\tilde{S}(1:3, 1:3) = \begin{bmatrix} \times & \times & \times \\ \boxtimes & \boxtimes & \times \\ \boxtimes & \boxtimes & \times \end{bmatrix}. \tag{7.7}$$

The matrix is symmetric, but we only depict the lower right structured rank block to overcome an overloaded notation. The matrix S satisfies the following relation:

$$S(1:3, 1:3) = \tilde{G}(1:3, 1:3) \, \tilde{S}(1:3, 1:3) \, \tilde{G}(1:3, 1:3)^T.$$

Applying the Givens transformation to the right of the matrix $\tilde{S}(1:3, 1:3)$ creates the following structure:

$$\tilde{S}(1:3, 1:3) \, \tilde{G}(1:3, 1:3)^T = \begin{bmatrix} \times & \times & \times \\ \boxtimes & \boxtimes & \times \\ \boxtimes & \boxtimes & \times \end{bmatrix}. \qquad (7.8)$$

Nothing changes because the Givens transformation replaces the first two columns by a linear combination of these first two columns. We still have to perform the Givens transformation to the left of this matrix. This Givens transformation involves the first two rows. Due to Condition 1 we also know that this Givens transformation is different from the identity transformation. We also know that the resulting matrix $S(1:3, 1:3)$, which is the original semiseparable matrix, has the lower right 2×2 block of rank 1. Hence the Givens transformation acting on the upper two rows cannot change the rank of the lower 2×2 block. Due to the constraints posed, this is only possible if the complete left two columns are of rank 1. This means that we have the following equation:

$$S(1:3, 1:3) = \begin{bmatrix} \boxtimes & \boxtimes & \times \\ \boxtimes & \boxtimes & \times \\ \boxtimes & \boxtimes & \times \end{bmatrix}. \qquad (7.9)$$

This is clearly in contradiction with the unreducedness assumption of the matrix.

- **Condition** 4. Assume the upper left 4×4 matrix follows form: (This corresponds to $d = e = f = 0$ in Equation (7.6); $f = 0$ is naturally implied as $f \neq 0$ implies the matrix to be of block diagonal form, which is in contradiction with the unreducedness.)

$$\begin{bmatrix} x & a & 0 & \hat{d} \\ a & b & 0 & \hat{e} \\ 0 & 0 & 0 & \hat{f} \\ \hat{d} & \hat{e} & \hat{f} & y \end{bmatrix}. \qquad (7.10)$$

Then $\hat{d}$ and/or $\hat{e}$ have to be different from zero and moreover $a\hat{e} \neq b\hat{d}$.

  - Assume $\hat{d} = \hat{e} = 0$ and $\hat{f} \neq 0$. This would mean that the complete columns underneath e and d would be zero, because the original matrix S is semiseparable. This is in contradiction with the unreducedness of the original matrix S, which can be written in this case as a block diagonal matrix. This is not possible.

- Suppose now $\hat{d} = \hat{e} = \hat{f} = 0$, this would mean that we have the following upper 4×4 matrix:

$$\begin{bmatrix} x & a & 0 & 0 \\ a & b & 0 & 0 \\ 0 & 0 & 0 & 0 \\ 0 & 0 & 0 & y \end{bmatrix}$$

  For this matrix the element in position $(3, 4)$ can be included in the semiseparable structure of the original matrix S. This contradicts the unreducedness assumption.

- Assume $\hat{a}\hat{e} = \hat{b}\hat{d}$. This is not possible. A similar reasoning as the one used in Condition 3 can be applied here.

These conditions make it possible to prove that in fact only two types of Givens transformations exist. Suppose we have the upper left 3×3 submatrix $\tilde{S}$ as in Equation (7.6). Depending on the values of d and e we distinguish between two cases.

- **Case 1.** Assume d and/or e are different from zero. Using Condition 3 from above, we see that the matrix is equal to the one from Equation (7.2) and hence we can apply Proposition 7.2.

- **Case 2.** Assume $d = e = 0$. To solve this case we have to consider the structure of the entire upper 4×4 subblock of the matrix $\tilde{S}$. Suppose the matrix $\tilde{S}$ has the upper left 4×4 block equal to the matrix in Equation 7.10.

Because of the conditions posed above, there are only two possible cases, due to the initial unreducedness assumption posed on the matrix S. $\square$

Now we are ready to reduce a semiseparable matrix S which is disturbed in the upper left two by two block $\tilde{S} = \tilde{G}^T S \tilde{G}$, back to semiseparable form. The idea is to interpret the structure restoring Givens transformation applied on the second row and column as a disturbing Givens transformation on the semiseparable matrix starting from row 2 and column 2 up to the end. To interpret the algorithm in this way some extra rank 1 blocks are needed. This is the subject of the next theorem.

Theorem 7.6. *Suppose we have an $n \times n$ unreduced symmetric semiseparable matrix S, which will be disturbed in the first two rows and columns by means of a similarity Givens transformation G. Then there exists an orthogonal transformation U with $U\mathbf{e}_1 = \mathbf{e}_1$ such that $U^T G^T S G U$ is again a symmetric semiseparable matrix.*

Proof. In this theorem a 5×5 matrix is considered, and we will use the special Givens transformation from Theorem 7.5. Some more notation is needed to simplify the construction of U: Denote with G_i the orthogonal transformation that performs a Givens transformation on the columns i and $i + 1$ of the matrix S. To prove that the algorithm gives the desired result, several figures are included. Starting with the matrix S each figure shows all the dependencies in the matrix. The elements

placed in a box satisfy the semiseparable structural constraints. The elements ·
form a rank 1 part. Hence an element ⊡ forms a rank 1 part and is includable in
a semiseparable structure. In the following figures we denote the lower triangular
part and the upper triangular part of the involved matrices separately. This is done
because in the beginning we perform the similarity transformations at the same
time, but to clearly understand what exactly is going on, we need to depict them
sometimes separately.

With S_l we denote the structure of the lower triangular part, with S_u the structure
of the upper triangular part. Elements not shown are not considered at that time.
Initially our matrix has the following structure, with both the upper and lower
triangular structure having a 3×3 block of rank 1:

$$
S_l = (S_0)_l =
\begin{bmatrix}
\boxtimes & & & & \\
\boxtimes & \boxtimes & & & \\
\boxdot & \boxdot & \boxdot & & \\
\boxdot & \boxdot & \boxdot & \boxtimes & \\
\boxdot & \boxdot & \boxdot & \boxtimes & \boxtimes
\end{bmatrix}
\qquad
S_u = (S_0)_u =
\begin{bmatrix}
\boxtimes & \boxtimes & \boxdot & \boxdot & \boxdot \\
 & \boxtimes & \boxdot & \boxdot & \boxdot \\
 & & \boxdot & \boxdot & \boxdot \\
 & & & \boxtimes & \boxtimes \\
 & & & & \boxtimes
\end{bmatrix}
$$

Applying the first disturbing Givens transformation, acting on the first row and
column, destroys the semiseparable structure in the first row and column. We get
the following situation (Let $S_1 = G_1^T S_0 G_1$):

$$
(S_1)_l =
\begin{bmatrix}
\boxplus & & & & \\
\boxplus & \boxplus & & & \\
\cdot & \boxdot & \boxdot & & \\
\cdot & \boxdot & \boxdot & \boxtimes & \\
\cdot & \boxdot & \boxdot & \boxtimes & \boxtimes
\end{bmatrix}
\qquad
(S_1)_u =
\begin{bmatrix}
\boxplus & \boxplus & \cdot & \cdot & \cdot \\
 & \boxplus & \boxdot & \boxdot & \boxdot \\
 & & \boxdot & \boxdot & \boxdot \\
 & & & \boxtimes & \boxtimes \\
 & & & & \boxtimes
\end{bmatrix}.
$$

We see that the resulting matrix $G_1^T S_0 G_1$ has a new 2×2 semiseparable matrix in
the upper left corner, has a remaining 4×4 semiseparable matrix in the lower right
corner and has a block of rank 1 in the lower left and upper right corner.

Applying the structure restoring Givens transformation onto the matrix can be
interpreted as disturbing the semiseparable structure in the lower right 4×4 semiseparable matrix. This clearly illustrates the inductive procedure because one can
again determine a structure restoring Givens transformation, based on the this disturbed 4×4 lower right semiseparable matrix. However in order to prove that the
structure of the created upper left semiseparable matrix grows, an additional condition needs to be satisfied. *After having applied the first structure restoring Givens
transformation, the first two columns of the matrix have to be of rank 1 (except for
the top elements).* This means that the created 2×2 rank 1 block in the upper
left 3×3 matrix extends in fact throughout the complete first two columns. We
will prove first this statement, and we will then see that we can continue with the
inductive procedure for restoring the structure.

Depending on the kind of structure restoring Givens transformation applied, the
proof slightly changes.

- **Case 2 of Theorem 7.5.** We start with Case 2, as this is the most simple one.
  The disturbing Givens transformation has created a matrix of the following

form (illustrated on a 5×5 example):

$$(S_1)_l = \begin{bmatrix} \boxplus & & & & \\ \boxplus & \boxplus & & & \\ 0 & 0 & 0 & & \\ \cdot & \boxdot & \boxdot & \boxtimes & \\ \cdot & \boxdot & \boxdot & \boxtimes & \boxtimes \end{bmatrix} \qquad (S_1)_u = \begin{bmatrix} \boxplus & \boxplus & 0 & \cdot & \cdot \\ & \boxplus & 0 & \boxdot & \boxdot \\ & & 0 & \boxdot & \boxdot \\ & & & \boxtimes & \boxtimes \\ & & & & \boxtimes \end{bmatrix}$$

Applying the structure restoring Givens transformation G_2 acting on row 2 and 3 and on column 2 and column 3, creates a matrix of the following form:

$$(S_2)_l = \begin{bmatrix} \boxplus & & & & \\ \boxdot & \boxdot & & & \\ \boxdot & \boxdot & \boxplus & & \\ \cdot & \cdot & \boxtimes & \boxtimes & \\ \cdot & \cdot & \boxtimes & \boxtimes & \boxtimes \end{bmatrix} \qquad (S_2)_u = \begin{bmatrix} \boxplus & \boxdot & \boxdot & \cdot & \cdot \\ & \boxdot & \boxdot & \cdot & \cdot \\ & & \boxplus & \boxtimes & \boxtimes \\ & & & \boxtimes & \boxtimes \\ & & & & \boxtimes \end{bmatrix}.$$

$$(7.11)$$

Hence we have a grown semiseparable block in the upper left corner, a reduced semiseparable block in the lower right corner, and we also have the structural constraints posed on the first two columns and rows.

- **Case 1 of Theorem 7.5.** This is the more difficult case. When applying the Givens transformation of the first kind, we take a closer look to see how the dependencies will change. First we apply the transformation $G_2{}^T$ only on the left of the matrix. We remark that we depict now two times the same structure, namely the structure of the upper triangular part of the matrix $G_2^T S_1$:

$$\left(G_2^T S_1\right)_u = \begin{bmatrix} \boxplus & \boxdot & \cdot & & \\ & \boxdot & \boxdot & \boxtimes & \boxtimes \\ & & \boxtimes & \boxtimes & \boxtimes \\ & & & \boxtimes & \boxtimes \\ & & & & \boxtimes \end{bmatrix}$$

$$\left(G_2^T S_1\right)_u = \begin{bmatrix} \boxplus & \boxplus & \cdot & \cdot & \cdot \\ & \boxplus & \boxdot & \boxdot & \boxdot \\ & & \boxdot & \boxdot & \boxdot \\ & & & \boxtimes & \boxtimes \\ & & & & \boxtimes \end{bmatrix}.$$

There are now two rank 1 parts, the small 2×2 matrix (left figure) and the larger 3×3 matrix, which was already present before (right figure). The small block has to be of rank 1 because the next Givens transformation G_2 applied to the right will not change the rank of this block, and after this Givens transformation that block is part of the semiseparable matrix, and therefore of rank 1. The 3×3 block remains of rank 1 after the Givens transformation is performed. This means that after applying the Givens transformation on the left we have a 2×4 matrix of rank 1 as depicted below:

$$
\left(G_2^T S_1\right)_u =
\begin{bmatrix}
\boxplus & \boxdot & \cdot & \cdot & \cdot \\
 & \boxdot & \boxdot & \boxdot & \boxdot \\
 & & \boxtimes & \boxtimes & \boxtimes \\
 & & \boxtimes & \boxtimes & \\
 & & & \boxtimes &
\end{bmatrix}.
$$

This is only the case because the first three elements of column 3 of the matrix S_1 are not all equal to zero. If they were all zero, we would have been in Case 2 of Theorem 7.5.

Applying the transformation G_2 to the right of the matrix gives the matrix $S_2 = G_2^T S_1 G_2$, which because of symmetry reasons has the following structure:

$$
(S_2)_l =
\begin{bmatrix}
\boxplus & & & & \\
\boxdot & \boxdot & & & \\
\boxdot & \boxdot & \boxplus & & \\
\cdot & \cdot & \boxtimes & \boxtimes & \\
\cdot & \cdot & \boxtimes & \boxtimes & \boxtimes
\end{bmatrix}
\qquad
(S_2)_u =
\begin{bmatrix}
\boxplus & \boxdot & \boxdot & \cdot & \cdot \\
 & \boxdot & \boxdot & \cdot & \cdot \\
 & & \boxplus & \boxtimes & \boxtimes \\
 & & & \boxtimes & \boxtimes \\
 & & & & \boxtimes
\end{bmatrix}. \qquad (7.12)
$$

Hence we obtain a matrix that contains an upper left 3×3 semiseparable matrix, a lower right 3×3 semiseparable matrix and moreover a large 4×2 block of rank 1 in the first two columns (a similar rank 1 block is present in the first two rows).

The figure shows that the upper semiseparable part has increased, while the lower semiseparable part is reduced in dimension. Very important are the remaining rank 1 parts.

We have proved now that in either case, the rank one blocks in the first two columns and rows are created. The dependency in these blocks makes sure that the next Givens transformation indeed creates a semiseparable matrix of dimension 4. This is in fact the inductive procedure. Let us see how it works.

The next Givens transformation G_3 is calculated by using Theorem 7.5 applied to the matrix S_2 without the first row and column. In fact we will apply now a structure restoring Givens transformation on the lower right 4×4 disturbed semiseparable matrix. Applying the transformation G_3, this means calculating $S_3 = G_3^T S_2 G_3$, will create a semiseparable block in the middle of the matrix. However because of the specific rank 1 parts in the first two columns, the complete upper left 4 by 4 block will become dependent. When exploiting the rank one parts as shown in this equation:

$$
(S_2)_l =
\begin{bmatrix}
\boxplus & & & & \\
\boxplus & \boxplus & & & \\
\boxdot & \boxdot & \boxplus & & \\
\cdot & \cdot & \boxtimes & \boxtimes & \\
 & & \boxtimes & \boxtimes & \boxtimes
\end{bmatrix}
\qquad
(S_2)_u =
\begin{bmatrix}
\boxplus & \boxplus & \boxdot & \cdot & \\
 & \boxplus & \boxdot & \cdot & \\
 & & \boxplus & \boxtimes & \boxtimes \\
 & & & \boxtimes & \boxtimes \\
 & & & & \boxtimes
\end{bmatrix},
$$

and then performing the following Givens similarity transformation on the rows 3 and 4 and columns 3 and 4, we obtain the following situation:

$$(S_3)_l = \begin{bmatrix} \boxplus & & & & \\ \boxplus & \boxplus & & & \\ \boxplus & \boxplus & \boxplus & & \\ \boxplus & \boxplus & \boxplus & \boxplus & \\ & & & \boxtimes & \boxtimes \end{bmatrix} \qquad (S_3)_u = \begin{bmatrix} \boxplus & \boxplus & \boxplus & \boxplus & \\ & \boxplus & \boxplus & \boxplus & \\ & & \boxplus & \boxplus & \\ & & & \boxplus & \boxtimes \\ & & & & \boxtimes \end{bmatrix}. \qquad (7.13)$$

This means that our resulting matrix S_3 has the upper left 4×4 block of semiseparable structure and a remaining 2×2 block in the bottom right. The reasoning applied above is not valid if the second column consists entirely of zeros except for the first element. In this case apply the Givens transformation from Proposition 7.4, which then provides the desired structure.

Before performing the final Givens transformation, one can also search the rank 1 blocks (extending throughout the first three columns now) such that the last Givens transformation will transform the matrix into a complete semiseparable one.

Even though we proved the theorem here for a semiseparable matrix of dimension 5, one can easily see that the statements remain true, also for higher dimensions of the matrix. □

Note 7.7. *The theorem provided here is valid, regardless of the kind of representation used. In practice, however, a representation is chosen in order to minimize the storage cost for the structured rank parts. Starting from a specific representation, the different cases of transformations often collapse, all the information concerning zeros, etc., is simply stored inherently in the representation, and there is no need to distinguish between these cases. In Section 8.2 this will be illustrated for the Givens-vector representation. Based on this representation we will see that all the different cases of Givens transformation are incorporated in the method.*

This final theorem produces an algorithm to transform the semiseparable matrix with a disturbance in the upper left part back to an orthogonal similar semiseparable matrix. In the next subsection it will be proven that the constructed semiseparable matrix will be essentially the same as the semiseparable matrix coming directly from the QR-algorithm.

7.1.4 Proof of the correctness of the implicit approach

In this final part of Section 7.1 we will prove that the matrix arising from the theorem above is essentially the same as a matrix coming from an explicit step of QR to an unreduced semiseparable matrix. First we will briefly combine all the essential information provided in this section to come to an implicit QR-algorithm for semiseparable matrices. We will recapitulate all the essential steps. Suppose we have a given semiseparable matrix S. The matrix is unreduced when the following conditions are satisfied:

1. All the blocks $S(i+1:n, 1:i)$ (for $i = 1, \ldots, n-1$) have rank equal to 1. This means that there are no zero blocks below the diagonal.

2. All the blocks $S(i:n, 1:i+1)$ (for $i = 1, \ldots, n-1$) have rank strictly higher than 1. This means that on the superdiagonal, no elements are includable in the semiseparable structure.

While executing the algorithm, the second demand only needs to be checked once, because this corresponds to singularities, which cannot be created afterwards. Therefore, before performing a QR-step, we transform the matrix S to unreduced form as described in Subsection 7.1.1. All the singularities are now removed. We have an unreduced matrix S_0. After a step of QR with or without shift, it is possible that the matrix has almost a block diagonal structure. If this is the case, one should apply deflation and continue with the smaller blocks. More on the deflation criterion used can be found in Chapter 8, Subsection 8.2.4.

A brief description of one step of the implicit QR-algorithm:

1. Calculate the shift μ.

2. Apply the first orthogonal transformation Q_1 to the matrix S: $\tilde{S} = Q_1^T S Q_1$. (This corresponds to performing a step of the QR-algorithm without shift on the matrix S.)

3. Apply deflation to the matrix $\tilde{S}$, if there are numerically zero blocks created in the matrix $\tilde{S}$.

4. Apply the second step of the implicit QR method to the blocks, deflated from the matrix $\tilde{S}$.

5. Again apply deflation to the blocks resulting from the implicit step applied onto $\tilde{S}$.

In fact the deflation in the middle of the algorithm is not necessary. If there are zero blocks created in the first orthogonal similarity, the chasing step will simply break down. This is not dramatic, as one can then immediately proceed with two blocks. One can also incorporate this check into the chasing part.

Because of the implicit Q-theorem 6.28, we know that we performed a step of the QR-algorithm similar to the original matrix. Let us briefly explain this in more detail. Suppose our matrix S is transformed via the explicit QR-algorithm into the matrix $\tilde{S}$ and denote with $\hat{S}$ the matrix resulting from the implicit QR-algorithm. We have the following equalities:

$$\tilde{S} = \tilde{Q}^T S \tilde{Q}$$
$$\hat{S} = \hat{Q}^T S \hat{Q}.$$

We know that the first n Givens transformations performed in the implicit approach are identical to the ones performed in the explicit approach, namely $G_n^T \ldots G_1^T$. The remaining Givens transformations in the explicit approach do not act on the first row anymore, hence we have:

$$\tilde{Q}\mathbf{e}_1 = G_1 \ldots G_n \mathbf{e}_1.$$

In the implicit approach, the chasing technique was constructed such that it did not act on the first row, hence also for the implicit approach we obtain:

$$\hat{Q}\mathbf{e}_1 = G_1 \ldots G_n \mathbf{e}_1.$$

This implies that the first columns of the orthogonal transformations applied to obtain $\tilde{S}$ and $\hat{S}$ are identical. Due to the implicit Q-theorem, we now know that our matrices are essentially unique, thereby justifying our implicit approach for performing a step of QR on a semiseparable matrix S.

Notes and references

The results presented in this section are based on the ones presented in [164], in which the implicit QR-algorithm for symmetric semiseparable matrices was developed.

In some articles (e.g., [18]) one considers also the following definition of unreducedness if one is working with a symmetric semiseparable matrix S.

Definition 7.8. *A symmetric semiseparable matrix S is considered unreduced if there are no zero blocks below the diagonal and the following condition is satisfied:*

$$\text{rank}\,(S(i:i+1, i:i+1)) > 1 \quad (\forall i = 1, \ldots, n-1)$$

One might consider this definition to be equivalent to the one posed on semiseparable matrices. But in fact this is not at all the case. Consider for example the following matrix, which is symmetric:

$$A = \begin{bmatrix} 1 & 2 & 3 & 0 & 4 \\ 2 & 2 & 3 & 0 & 4 \\ 3 & 3 & 3 & 0 & 4 \\ 0 & 0 & 0 & 0 & 4 \\ 4 & 4 & 4 & 4 & 4 \end{bmatrix}.$$

Now considering the element in position $(3, 4)$, this element is not includable in the semiseparable structure according to this book's definition of unreducedness, but corresponding to the alternative definition it is includable. According to the new definition this matrix would not be considered unreduced and there would be the need for transforming this matrix to unreduced form. Hence it seems that the definition considered in this book is more detailed. Being not unreduced according to the definition used in the book always implies being not unreduced according to this new definition, but not vice versa. But this new definition is in some sense stronger than the one we considered, so if a matrix is unreduced according to this new definition, it will surely be unreduced with respect to the old definition. Hence in practical implementations, it is natural to use the latter definition. It is more severe but easier to check however.

The example above is not at all singular. In the definition used throughout the book however singularity is necessary if an element in the superdiagonal is includable in the semiseparable structure.

Another interesting theoretical observation related to the duality between tridiagonal and semiseparable matrices is the following. Assume for simplicity that we are working with nonsingular matrices, hence we can invert them. Assume we have a tridiagonal matrix T and a semiseparable matrix $S = T^{-1}$. Applying the standard QR-algorithm without shift to the matrix S, causes the matrix to converge to its dominant eigenvalue (in normal circumstances). Assume now the matrix $\hat{S}$ has converged to its dominant eigenvalue after a few steps of the QR-method. This means that $\hat{S} = Q^H S Q$. We now have the following equations:

$$T = S^{-1}$$
$$Q^H T Q = Q^H S^{-1} Q = \left(Q^H S Q \right)^{-1}$$
$$\left[\begin{array}{c|c} \hat{T} & \hat{\epsilon} \\ \hline \hat{\epsilon}^T & \lambda_1 \end{array} \right] = \left[\begin{array}{c|c} \hat{S} & \epsilon \\ \hline \epsilon^T & 1/\lambda_1 \end{array} \right]^{-1},$$

in which ϵ and $\hat{\epsilon}$ denote vectors of small elements. This idea can be exploited in different forms. For example one can determine the first Givens transformation from the semiseparable QR-step accurately, apply it onto the matrix T and then continue with the implicit approach for tridiagonal matrices to let it converge to its smallest eigenvalue. To obtain accurate results after every step of the QR-method one can also try to invert the resulting semiseparable matrix (using $\mathcal{O}(n)$ operations) in order to determine the first Givens transformation as accurately as possible.

We would like to conclude with a simple remark. The analysis above might result in some questions. Is it really necessary to use two sequences of Givens transformations in order to obtain a step of the QR-method, whereas only one sequence is needed in the tridiagonal case. This is because a tridiagonal matrix minus a shift matrix remains a tridiagonal matrix, and therefore its QR-factorization can be computed by $n-1$ Givens transformations. For semiseparable matrices the structure is however not preserved when subtracting a diagonal matrix. This leads to a semiseparable plus diagonal matrix for which two sequences of Givens transformations each involving $n-1$ transformations are necessary.

7.2   A QR-algorithm for semiseparable plus diagonal

As mentioned before, we will not cover all details for developing an implicit QR-algorithm for semiseparable plus diagonal matrices. The changes with regard to the method presented above are limited. The idea is similar to the one in the reduction method. The reduction of a symmetric matrix to semiseparable form changed only slightly in order to obtain a semiseparable plus diagonal matrix. The idea is to subtract in the reduction procedure some values from the diagonal and to construct the Givens transformations based on this slightly changed information. Here, one can do exactly the same. Slightly changing the diagonal elements before computing the structure restoring Givens transformations results in an implicit QR-algorithm for semiseparable plus diagonal matrices. Unfortunately this creates an overhead of tracking for nonzeroness and rank deficiency, problems that are more complicated than in the traditional implicit QR-method for semiseparable matrices. Already in the simple semiseparable case it was shown in Theorem 7.6, that one has to distinguish between many cases. More information on this implicit QR-method for semiseparable plus diagonal matrices can be found in [157].

Nevertheless, we will present another approach for developing the implicit QR-algorithm for semiseparable plus diagonal matrices in the next chapter. There we will discuss a graphical interpretation of the chasing based on the Givens-vector representation. This approach is more appealing, which makes it easier to find troublesome cases (almost zero and so forth) and moreover, it makes it easy to write the implementation.

In order to devise this graphical implementation, we will present here some aspects of the QR-method for semiseparable plus diagonal matrices, which were not discussed before.

Lemma 7.9. *When applying the orthogonal similarity transformation based on Q_1 to a symmetric $n \times n$ semiseparable plus diagonal matrix $D + S$, the resulting matrix is semiseparable plus diagonal again:*

$$Q_1^T(S + D)Q_1 = \tilde{S} + \tilde{D},$$

and the diagonal is of the following form $\tilde{D} = \mathrm{diag}([0, d_1, d_2, \ldots, d_{n-1}])$. The matrix Q_1, consists of the n Givens transformations $G_1, \ldots, G_n$ to make the semiseparable matrix S, upper triangular (see Subsection 6.1.1).

Proof. By induction. The first Givens transformation G_1^T working on the last two rows and its transpose on the last two columns transforms $S + D$ into

$$
\begin{bmatrix}
\boxtimes & \cdots & \boxtimes & \boxtimes & 0 \\
\vdots & \ddots & \vdots & \vdots & \vdots \\
\boxtimes & \cdots & \boxtimes & \boxtimes & 0 \\
\boxtimes & \cdots & \boxtimes & \times & \times \\
0 & \cdots & 0 & \times & \times
\end{bmatrix}
+
\begin{bmatrix}
d_1 & & & & \\
& \ddots & & & \\
& & d_{n-2} & & \\
\hline
& & & \times & \times \\
& & & \times & \times
\end{bmatrix},
$$

where $\boxtimes$ are elements belonging to the symmetric semiseparable structure. The 2×2 blocks in the lower-right corners can always be written as

$$
\begin{bmatrix}
0 & \\
& d_{n-1}
\end{bmatrix}
+
\begin{bmatrix}
\boxplus & \boxplus \\
\boxplus & \boxplus
\end{bmatrix},
$$

where $\boxplus$ denote an element belonging to the new semiseparable structure constructed in the lower-right corner, because any 2×2 block is a semiseparable block.

Suppose that after $n - i$ steps, the semiseparable plus diagonal matrix $S + D$ is transformed into

$$
\begin{array}{c}
\hspace{3.5cm} \overset{i-1}{\downarrow} \;\; \overset{i}{\downarrow} \\[-2pt]
\begin{bmatrix}
\boxtimes & \cdots & \boxtimes & \boxtimes & \boxtimes & 0 & \cdots & 0 \\
\vdots & \ddots & \vdots & \vdots & \vdots & \vdots & & \vdots \\
\boxtimes & \cdots & \boxtimes & \boxtimes & \boxtimes & 0 & \cdots & 0 \\
\boxtimes & \cdots & \boxtimes & \boxtimes & \boxtimes & 0 & \cdots & 0 \\
\boxtimes & \cdots & \boxtimes & \boxtimes & \boxdot & \boxplus & \cdots & \boxplus \\
0 & \cdots & 0 & 0 & \boxplus & \boxplus & \cdots & \boxplus \\
\vdots & & \vdots & \vdots & \vdots & \vdots & \ddots & \vdots \\
0 & \cdots & 0 & 0 & \boxplus & \boxplus & \cdots & \boxplus
\end{bmatrix} +
\end{array}
$$

$$
\begin{array}{c}
\hspace{0.8cm} \overset{i-1}{\downarrow} \;\; \overset{i}{\downarrow} \\[-2pt]
\begin{bmatrix}
d_1 & & & & & & & \\
& \ddots & & & & & & \\
& & d_{i-2} & & & & & \\
& & & d_{i-1} & & & & \\
& & & & 0 & & & \\
& & & & & d_i & & \\
& & & & & & \ddots & \\
& & & & & & & d_{n-1}
\end{bmatrix} ,
\end{array}
$$

where $\boxtimes$ denote an element belonging to the semiseparable block in the upper-left corner and $\boxplus$ an element of the new constructed semiseparable block in the lower-right corner. The elements $\boxdot$ belongs to both semiseparable blocks.

When applying a Givens transformation on the $(i-1)$-th and i-th row and column, the diagonal part will not remain diagonal any more. This can be avoided by changing the 0 on the i-th position of the diagonal part into d_{i-1}. Hence, both diagonal elements, the $(i-1)$-th and i-th, are equal and any orthogonal similarity transformation working on these two rows and columns will never change the diagonal part.

Of course, we also need to adapt the corresponding element in the semiseparable part. But this element is $\boxdot$ and whatever its value it will always belong to both semiseparable parts because the lower-right element and the upper-left element of a semiseparable matrix can be chosen arbitrarily.

The Givens transformation G_{n-i+1}^T is constructed such that it annihilates the i-th row up to the main diagonal. Hence, after the application of G_{n-i+1} and G_{n-i+1}^T, we get the following:

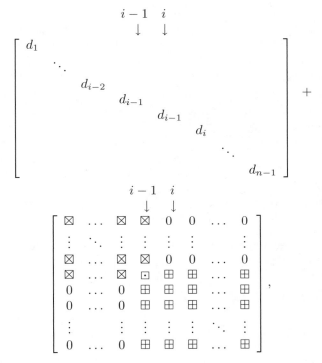

where we have enlarged the semiseparable block in the lower-right corner by one row and column.

Only the $(i-1)$-th diagonal element d_{i-1} should be zero in order to prove the lemma, but, as before, we can change that element into 0 and adapt the corresponding element $\boxdot$ in the semiseparable part. The adaptation of $\boxdot$ will not change the existing semiseparable structures.

This ends the proof of Lemma 7.9. □

We know that after the first sequence of Givens transformations the diagonal shifts down one position. Therefore the second sequence of Givens transformations must shift up the diagonal one position, as finally the structure of the semiseparable plus diagonal matrix is maintained after one step of the QR-method (see Section 6.3).

As seen in the previous section, the preservation of the semiseparable plus diagonal structure (see Section 6.3) is not enough to develop an implicit method. As the semiseparable plus diagonal case is quite similar to the semiseparable case, we will not discuss all these issues in detail. Below you can find a list of important topics, which should not be forgotten when developing an implicit method.

- **Unreducedness:** As a semiseparable plus diagonal matrix $S+D$ consists of a complete semiseparable matrix, one can apply the existing deflation techniques onto the semiseparable matrix. The diagonal is of no influence here.

- **Implicit Q-theorem for semiseparable plus diagonal matrices:** An implicit Q-theorem for semiseparable plus diagonal matrices does not exist to our knowledge. Nevertheless, one can prove this the way the semiseparable case is proved, based on the unreducedness of the involved matrices, and thereby using the fact that an unreduced matrix is generator representable.

The implicit Q-theorem is in some sense only needed for proving that one did perform a step of the QR-method in the implicit case. In fact the implicit Q-theorem is not needed in this case. If one can prove that one performs essentially the same Givens transformations in the implicit as in the classical manner the statement holds. Because the first sequence of Givens transformations is uniquely determined this is not so difficult to prove (see [157, 55]). Also based on an inverse eigenvalue problem one can state essential uniqueness of the matrices resulting from an implicit or an explicit QR-step, see [72] and the discussion of this article in Section 9.7.

The chasing technique for semiseparable plus diagonal matrices will be discussed in the forthcoming chapter.

Notes and references

As mentioned before, a more detailed analysis of the QR-method (without the graphical scheme) for semiseparable plus diagonal matrices can be found in [157] (see Section 6.3). Instead of an implicit Q-theorem one can use two approaches.

☞ S. Delvaux and M. Van Barel. Eigenvalue computation for unitary rank structured matrices. *Journal of Computational and Applied Mathematics*, 213(1):268–287, March 2008.

This article by Delvaux and Van Barel discusses the construction of an implicit method for computing the eigenvalues of unitary Hessenberg matrices.

Another approach is discussed in more detail in [72] (see also Section 9.7).

7.3   An implicit QR-algorithm for Hessenberg-like matrices

In the previous section we developed an implicit QR-algorithm for symmetric semiseparable matrices. This was useful because every symmetric matrix can be transformed into a similar semiseparable one. As nonsymmetric matrices can be transformed via similarity transformations into Hessenberg-like matrices, we will deduce in this section an implicit QR-algorithm for this class of matrices.

7.3.1   The shift μ

For Hessenberg-like matrices one can also consider the Rayleigh and Wilkinson shift as for the symmetric case. However, one important remark has to be made. Our Hessenberg-like matrix is not necessarily symmetric anymore and therefore does not necessarily contain all real eigenvalues. It is perfectly possible to have complex conjugate eigenvalues in the Hessenberg-like matrix. To obtain convergence to these eigenvalues, one can try the following, as presented in [94, p. 355], which is called the double shift strategy or the performance of a Francis QR-step. We will present this for Hessenberg matrices. Suppose we have a Hessenberg matrix $H^{(1)} = H$, for which the lower right 2×2 block contains two complex conjugate eigenvalues μ_1 and μ_2. The double shift strategy performs in fact two QR-steps with different shifts:

$$H^{(1)} - \mu_1 I = Q^{(1)} R^{(1)}$$
$$H^{(2)} = R^{(1)} Q^{(1)} + \mu_1 I$$
$$H^{(2)} - \mu_2 I = Q^{(2)} R^{(2)}$$
$$H^{(3)} = R^{(2)} Q^{(2)} + \mu_2 I.$$

The resulting Hessenberg matrix $H^{(3)}$ should be real but numerical rounding errors can prevent that. Therefore one switches to the implicit approach, in which two steps of QR are applied simultaneously. In this section we will however assume that all our eigenvalues of the Hessenberg-like matrix are real and that we therefore do not need this implicit double shift strategy[13].

The technique of using two shifts is often referred to as the double shift strategy. Instead of presenting this technique here, we will immediately present the multishift strategy in Chapter 9, Section 9.4, which enables the performance of more than two steps of the QR-algorithm at once. This part is not yet included here as the presented algorithm makes extensive use of the graphical schemes for operating with Givens transformations, which will be presented in Chapter 8.

7.3.2 An implicit QR-step on the Hessenberg-like matrix

Without loss of generality we assume our Hessenberg-like matrix Z to be unreduced. In the previous chapter a method was proposed for bringing Hessenberg-like matrices back to unreduced form. To execute an implicit QR-step, we know that we have to perform two sequences of Givens transformations on the matrix Z. The first sequence of Givens transformations Q_1 can be performed immediately on both sides of the matrix at the same time. This can be seen from the proof of Theorem 6.6:

$$Z_1 = Q_1^T Z Q_1.$$

By Theorem 6.6, the matrix Z is again a Hessenberg-like matrix. The second step is performed in an implicit way, just like in the symmetric case. To start with the implicit approach, we need an initialization step. We know that H_1 is a Hessenberg matrix:

$$H_1 = Q_1^T Z - \mu Q_1^T$$
$$= R_1 - \mu Q_1^T.$$

---

[13]One can also choose a complex shift and continue working with complex shifts and complex matrices. One should work with the complex conjugate instead of with the transpose. In this book everything is based on the transpose, which is why we make this assumption.

This makes it easy to calculate the first Givens G_n^T transformation of the matrix Q_2. This transformation G_n^T will annihilate the element $(2,1)$ in the matrix H_1. Switching to the implicit approach, we perform the transformation G_n^T on the Hessenberg-like matrix Z_1:

$$Z_2 = G_n^T Z_1 G_n.$$

Note that the matrix Z_2 is not Hessenberg-like anymore but has the structure disturbed in the upper left 2×2 block, with the remaining part of the matrix still of the correct form. We will now design a chasing technique that will remove the disturbance in the matrix.

7.3.3 Chasing the disturbance

We will now derive three propositions that determine Givens transformations that can be used to reestablish the structure of the Hessenberg-like matrix. In fact the last proposition is the most general one, incorporating also the previous two propositions. It is included to compare it with the QR-method for symmetric semiseparable matrices.

Proposition 7.10. *Suppose the following 3×3 matrix is given,*

$$A = \begin{bmatrix} a & g & h \\ b & c & i \\ d & e & f \end{bmatrix},$$

for which $be - dc \neq 0$. Then there exists a Givens transformation

$$G = \begin{bmatrix} 1 & 0 \\ 0 & \hat{G} \end{bmatrix} \quad with \quad \hat{G} = \frac{1}{\sqrt{1+t^2}} \begin{bmatrix} t & -1 \\ 1 & t \end{bmatrix}$$

such that the following matrix

$$G^T A G$$

has the lower left 2×2 block of rank 1.

Proof. The proof is similar to the one of Theorem 7.2. A simple calculation reveals that the parameter t of the Givens transformation

$$t = \frac{id - fb}{eb - dc}$$

is well defined because $eb - dc \neq 0$. □

This is however not enough to restore the complete lower triangular semiseparable structure. Just like in the symmetric case another type of Givens transformation is needed.

Proposition 7.11. *Suppose the following 4×4 matrix is given,*

$$A = \begin{bmatrix} a & g & h & j \\ b & c & i & k \\ 0 & 0 & 0 & l \\ d & e & f & m \end{bmatrix},$$

which is not yet Hessenberg-like. Then there exists a Givens transformation

$$G = \begin{bmatrix} 1 & 0 & 0 \\ 0 & \hat{G} & 0 \\ 0 & 0 & 1 \end{bmatrix} \quad with \quad \hat{G} = \frac{1}{\sqrt{1+t^2}} \begin{bmatrix} t & -1 \\ 1 & t \end{bmatrix}$$

such that the lower left 3×3 block of the following matrix

$$G^T A G$$

is Hessenberg-like. And the lower left 3×2 block is of rank 1.

Proof. A straightforward calculation reveals

$$t = \frac{id - bf}{eb - cd},$$

which is well defined.   □

In fact the structure given in Proposition 7.11 is not the most general. The following proposition will capture all the possible transformations.

Proposition 7.12. *Suppose the following $n \times n$ matrix is given,*

$$A = \begin{bmatrix} a & g & \cdots & & & \\ b & c & i & \cdots & & \\ 0 & 0 & 0 & l & \cdots & \\ \vdots & & \vdots & \ddots & \ddots & \\ 0 & \cdots & 0 & \cdots & 0 & j \\ d & e & f & \cdots & & m \end{bmatrix},$$

which is not yet Hessenberg-like. Then there exists a Givens transformation

$$G = \begin{bmatrix} 1 & 0 & 0 & \cdots \\ 0 & \hat{G} & 0 & \\ 0 & 0 & 1 & \\ \vdots & & & \ddots \end{bmatrix} \quad with \quad \hat{G} = \frac{1}{\sqrt{1+t^2}} \begin{bmatrix} t & -1 \\ 1 & t \end{bmatrix}$$

such that the lower left $(n-1)$ block of the following matrix

$$G^T A G$$

is of Hessenberg-like form. And the first two columns are dependent, except for the first element.

Proof. A calculation reveals:

$$t = \frac{id - bf}{eb - cd},$$

which is well defined.   □

Note 7.13.   *The transformation from Proposition 7.12 is not necessary in the symmetric case. In the symmetric case only two different types of Givens transformations are needed (see proof of Theorem 7.5).*

Note 7.14. *In fact one should also consider transformations similar to the one in Proposition 7.4, but for simplicity reasons we omit them here.*

Graphically we have also a bulge chasing algorithm in order to restore the structure. Let us depict it with some figures. Suppose we start with the following unreduced Hessenberg-like matrix Z_{n-1}, on which already the orthogonal similarity transformation $Q_1 = G_1 \ldots G_{n-1}$ has been performed:

$$Z_{n-1} = \begin{bmatrix} \boxtimes & \times & \times & \times & \times \\ \boxtimes & \boxtimes & \times & \times & \times \\ \boxtimes & \boxtimes & \boxtimes & \times & \times \\ \boxtimes & \boxtimes & \boxtimes & \boxtimes & \times \\ \boxtimes & \boxtimes & \boxtimes & \boxtimes & \boxtimes \end{bmatrix}.$$

The first disturbing Givens transformation G_n is applied to the matrix Z_{n-1} leading to the following structure:

$$Z_n = G_n^T Z_{n-1} G_n = \begin{bmatrix} \times & \times & \times & \times & \times \\ \times & \times & \times & \times & \times \\ \boxtimes & \boxtimes & \boxtimes & \times & \times \\ \boxtimes & \boxtimes & \boxtimes & \boxtimes & \times \\ \boxtimes & \boxtimes & \boxtimes & \boxtimes & \boxtimes \end{bmatrix}.$$

Let us apply now the first structure restoring Givens transformation onto this matrix. We know what the result should be, a 2×2 block of rank 1 should be created in the matrix. Let us apply the transformations on both sides of the matrix separately, to clearly depict what will happen with the structure of the matrix. First we apply the transformation to the right side of the matrix. This gives us the following structural properties of the matrix:

$$Z_n G_{n+1} = \begin{bmatrix} \times & \times & \times & \times & \times \\ \times & \times & \times & \times & \times \\ \boxtimes & \boxtimes & \boxtimes & \times & \times \\ \boxtimes & \boxtimes & \boxtimes & \boxtimes & \times \\ \boxtimes & \boxtimes & \boxtimes & \boxtimes & \boxtimes \end{bmatrix}. \tag{7.14}$$

The Givens transformation G_{n+1} acts on columns 2 and 3 and makes therefore a linear combination of these columns. This results in the fact that the remaining semiseparable part is still intact. Applying now the Givens transformation G_{n+1}^T to the left of this matrix creates a rank 1 block, unfortunately it destroys the semiseparable structure denoted by ⊠ in the element $(3,3)$. In the following figure the newly created semiseparable structure is denoted by ⊞ and the vanishing semiseparable structure by ⊠:

$$
G_{n+1}^T Z_n G_{n+1} = \begin{bmatrix} ⊞ & \times & \times & \times & \times \\ ⊞ & ⊞ & \times & \times & \times \\ ⊞ & ⊞ & \times & \times & \times \\ ⊠ & ⊠ & ⊠ & ⊠ & \times \\ ⊠ & ⊠ & ⊠ & ⊠ & ⊠ \end{bmatrix}.
$$

In this matrix the elements ⊞ denote the newly created semiseparable structure.

Before we can continue with our structure restoring procedure we have to investigate more closely the underlying rank structure of this newly created matrix. Even though it might seem that the new upper structured rank part is completely disconnected from the elder lower structured rank part this is not at all the case. Let us reconsider the matrix from Equation (7.14). On this matrix only one more transformation needs to be applied, namely on the left. This transformation combines rows 2 and 3 and after this transformation has been performed we know that the block $Z_{n+1}(2:3, 1:2)$ has to be of rank 1. The Givens transformation applied to the left of the matrix in Equation (7.14) cannot change the rank of this block in the matrix $Z_n G_{n+1}$. This is because it makes just a linear combination of the two rows involved. This means that in the matrix $Z_n G_{n+1}$ this block has to be rank 1. Denoting the elements of this rank 1 part with $\cdot$ we get the following situation:

$$
Z_n G_{n+1} = \begin{bmatrix} \times & \times & \times & \times & \times \\ \cdot & \cdot & \times & \times & \times \\ \boxdot & \boxdot & ⊠ & \times & \times \\ ⊠ & ⊠ & ⊠ & ⊠ & \times \\ ⊠ & ⊠ & ⊠ & ⊠ & ⊠ \end{bmatrix}. \tag{7.15}
$$

(Without loss of generality we assume that we are always working with the Givens transformation as proposed in Proposition 7.10. In the other cases the analysis becomes simpler as one can skip most of the analysis presented here.) Because also in this matrix the lower three rows satisfy a structured rank constraint we obtain that the matrix $Z_n G_{n+1}$ satisfies the following structured rank conditions (due to the assumption that we use the transformation proposed in Proposition 7.10, we know that the elements in the third row are nonzero, and therefore we can expand the rank 1 block):

$$
Z_n G_{n+1} = \begin{bmatrix} \times & \times & \times & \times & \times \\ \cdot & \cdot & \times & \times & \times \\ \boxdot & \boxdot & ⊠ & \times & \times \\ \boxdot & \boxdot & ⊠ & ⊠ & \times \\ \boxdot & \boxdot & ⊠ & ⊠ & ⊠ \end{bmatrix}. \tag{7.16}
$$

Completing the transformation of this matrix by performing the transformation G_{n+1} also on the left side creates a matrix of the form:

$$G_{n+1}^T Z_n G_{n+1} = \begin{bmatrix} \boxplus & \times & \times & \times & \times \\ \boxdot & \boxdot & \times & \times & \times \\ \boxdot & \boxdot & \times & \times & \times \\ \boxdot & \boxdot & \boxtimes & \boxtimes & \times \\ \boxdot & \boxdot & \boxtimes & \boxtimes & \boxtimes \end{bmatrix}.$$

It is clear that both newly created structured rank parts are still connected tightly via the rank 1 block intersecting both structures. This rank 1 block is essential for the continuity of the chasing procedure.

To continue the analysis we will show the effect of one more Givens transformation for chasing the disturbance. First we apply a Givens transformation on the columns 3 and 4[14]:

$$Z_{n+1} G_{n+2} = \begin{bmatrix} \boxplus & \times & \times & \times & \times \\ \boxdot & \boxdot & \times & \times & \times \\ \boxdot & \boxdot & \times & \times & \times \\ \boxdot & \boxdot & \boxtimes & \boxtimes & \times \\ \boxdot & \boxdot & \boxtimes & \boxtimes & \boxtimes \end{bmatrix}.$$

Applying this Givens transformation, does not change anything essential but due to a similar remark as in the previous transformation, we know that the 2×2 block which is the intersection between rows 3 and 4 and columns 2 and 3 needs to be of rank 1. Due to the existing rank 1 structure in the first two columns the whole block needs to be of rank 1. Therefore we get the following situation (only the essential rank 1 block is depicted now):

$$Z_{n+1} G_{n+2} = \begin{bmatrix} \boxplus & \times & \times & \times & \times \\ \boxplus & \boxplus & \times & \times & \times \\ \boxdot & \boxdot & \boxdot & \times & \times \\ \boxdot & \boxdot & \boxdot & \boxtimes & \times \\ \boxtimes & \boxtimes & \boxtimes & \boxtimes & \boxtimes \end{bmatrix}.$$

Completing the similarity transformation with an operation on the left creates therefore the following structured rank matrix (we omitted now the rank 1 parts):

$$G_{n+2}^T Z_{n+1} G_{n+2} = \begin{bmatrix} \boxplus & \times & \times & \times & \times \\ \boxplus & \boxplus & \times & \times & \times \\ \boxplus & \boxplus & \boxplus & \times & \times \\ \boxtimes & \boxtimes & \boxtimes & \times & \times \\ \boxtimes & \boxtimes & \boxtimes & \boxtimes & \boxtimes \end{bmatrix}.$$

---

[14]In case the elements in the second column are zero, one uses the elements of a column left of it for defining the Givens transformation proposed in Proposition 7.10. We do not discuss this in detail as it makes the analysis unnecessarily complicated. Moreover in a practical implementation one is not confronted with this problem, as one stores the structured rank part efficiently and one works with one nonzero column defining all the others up to a multiple. This nonzero column is used for determining this Givens transformation.

It is clear that the new structured rank part gradually grows. To continue this analysis, again a new rank 1 part extending across the structures has to be searched for.

7.3.4 Proof of correctness

A similar analysis as in the symmetric case ensures that the matrix resulting from this reduction procedure is essentially identical to a Hessenberg-like matrix on which an explicit step of the QR-algorithm was performed. One can again assume all Hessenberg-like matrices involved to be of unreduced form as otherwise deflation occurs.

Notes, references and open problems

The results in this section are based on the results presented in [159].

The main open problem of this section is the incorporation of the double shift strategy for sticking to real computations instead of switching to complex numbers. The difficulty lies in the fact that before we can compute the special Givens transformation G_n we first need to perform the first sequence of Givens transformations from bottom to top. Hence when one wants to perform two steps of the QR-method at once, this becomes difficult as this second special Givens transformation is heavily dependent of the previously performed Givens transformation. This problem will be adressed in Chapter 9. Before we can derive the multishift QR-algorithm we need more knowledge on working effectively with the Givens-vector representation.

7.4 An implicit QR-algorithm for computing the singular values

In this section an adaptation of the QR-algorithm of Section 7.1 will be made such that it becomes suitable for computing the singular values of matrices via upper triangular semiseparable matrices. The adaptations are achieved in a completely similar way as the translation of the QR-algorithm for tridiagonal matrices towards the QR-algorithm for bidiagonal matrices.

In this section we will derive, what is meant by an unreduced upper triangular semiseparable matrix S_u. We will prove that one iteration of the implicit QR-method applied to an upper triangular semiseparable matrix S_u is uniquely determined by the first column of the orthogonal factor Q of the QR-factorization of

$$S_u{}^T S_u - \mu I, \tag{7.17}$$

where μ is a suitable shift. Exploiting the particular structure of Matrix (7.17), we will show how the first column of Q can be computed without explicitly computing $S_u{}^T S_u$. Finally we will make sure that the resulting matrix after one step of the implicit QR-method is again an upper triangular semiseparable matrix.

7.4.1   Unreduced upper triangular semiseparable matrices

Let us first define what is meant by an unreduced upper triangular semiseparable matrix.

Definition 7.15. *Suppose S_u is an upper triangular semiseparable matrix. S_u is called unreduced if*

1. *The ranks of the blocks $S_u(1 : i, i + 1 : n)$, for $i = 1, \ldots, n - 1$ are 1. This means that there are no zero blocks above the diagonal.*

2. *All the diagonal elements are different from zero.*

Note 7.16. *An unreduced lower triangular semiseparable matrix is defined in a similar way.*

Assume that we have an upper triangular matrix S_u, and we want to transform it into an unreduced one. Condition 1 can be satisfied rather easily by dividing the matrix into different blocks. This means that the resulting block-matrices are generator representable. From now on, we will consider each of these blocks separately. Assume the generators of such an upper triangular block matrix to be $\mathbf{u}$ and $\mathbf{v}$. If Condition 2 is not satisfied, i.e., a diagonal element $d_i = u_i v_i$ is zero, then either u_i or v_i equals zero. Assume the matrix S_u to be of the following form:

$$
S_u = \begin{bmatrix}
u_1 v_1 & u_2 v_1 & u_3 v_1 & \cdots & u_n v_1 \\
0 & u_2 v_2 & u_3 v_2 & \cdots & u_n v_2 \\
 & 0 & u_3 v_3 & \cdots & u_n v_3 \\
\vdots & & \ddots & \ddots & \vdots \\
0 & \cdots & & 0 & u_n v_n
\end{bmatrix}. \tag{7.18}
$$

Assume $u_i = 0$. This means that we have a matrix of the following form (let us take for example a matrix of size 5×5, with $u_3 = 0$):

$$
\begin{bmatrix}
\boxtimes & \boxtimes & 0 & \boxtimes & \boxtimes \\
0 & \boxtimes & 0 & \boxtimes & \boxtimes \\
0 & 0 & 0 & \boxtimes & \boxtimes \\
0 & 0 & 0 & \boxtimes & \boxtimes \\
0 & 0 & 0 & 0 & \boxtimes
\end{bmatrix}.
$$

We want to transform this matrix to unreduced form. In this case we do not need similarity transformations anymore and we can use an orthogonal transform on the right that is different from the one on the left. Let us apply a Givens transformation G^T on rows 3 and 4 of the matrix, which will annihilate the complete row 4, because these rows are dependent. Thus we get the following matrix:

$$
\begin{bmatrix}
\boxtimes & \boxtimes & 0 & \boxtimes & \boxtimes \\
0 & \boxtimes & 0 & \boxtimes & \boxtimes \\
0 & 0 & 0 & \boxtimes & \boxtimes \\
0 & 0 & 0 & \boxtimes & \boxtimes \\
0 & 0 & 0 & 0 & \boxtimes
\end{bmatrix}
\xrightarrow{G^T S_u}
\begin{bmatrix}
\boxtimes & \boxtimes & 0 & \boxtimes & \boxtimes \\
0 & \boxtimes & 0 & \boxtimes & \boxtimes \\
0 & 0 & 0 & \boxtimes & \boxtimes \\
0 & 0 & 0 & 0 & 0 \\
0 & 0 & 0 & 0 & \boxtimes
\end{bmatrix}.
$$

In the matrix on the right we can clearly remove the zero row and column to get an unreduced upper triangular semiseparable matrix when all other diagonal elements are different from zero.

The results applied here can be generalized to $v_i = 0$, by annihilating a column via a Givens transformation on the right. Also the extension towards more zeros on the diagonal is straightforward.

7.4.2 An implicit QR-step

In this section we will apply all the necessary transformations to the upper triangular matrix S_u such that the first column of the final orthogonal transformation applied to the right of S_u will be the same as the first column of the matrix Q in the QR-factorization in $S_u^T S_u - \mu I$. We can assume that the matrix S_u is unreduced, and therefore this matrix is representable with two generators $\mathbf{u}$ and $\mathbf{v}$. Moreover, the matrix will be nonsingular. Let S_u be as in Equation (7.18). Let $\tau_i = \sum_{k=1}^{i} v_k^2$, $i = 1, \ldots, n$. An easy calculation reveals that

$$
S = S_u^{\ T} S_u =
\begin{bmatrix}
u_1\, u_1\tau_1 & u_2\, u_1\tau_1 & u_3\, u_1\tau_1 & \cdots & u_n\, u_1\tau_1 \\
u_2\, u_1\tau_1 & u_2\, u_2\tau_2 & u_3\, u_2\tau_2 & & u_n\, u_2\tau_2 \\
u_3\, u_1\tau_1 & u_3\, u_2\tau_2 & u_3\, u_3\tau_3 & & \vdots \\
\vdots & & & \ddots & \\
u_n\, u_1\tau_1 & u_n\, u_2\tau_2 & \cdots & & u_n\, v_n\tau_n
\end{bmatrix}
\tag{7.19}
$$

is a symmetric semiseparable matrix. Let us denote the generators for this symmetric semiseparable matrix S by $\hat{\mathbf{u}}$ and $\hat{\mathbf{v}}$,

$$
\hat{\mathbf{u}} = \mathbf{u} = [u_1, u_2, u_3, \cdots, u_n]
$$
$$
\hat{\mathbf{v}} = [u_1\tau_1, u_2\tau_2, u_3\tau_3, \cdots, u_n\tau_n].
$$

Therefore $S - \mu I$ is a semiseparable plus diagonal matrix.

Theorem 7.17. *Assume S_u is an unreduced upper triangular semiseparable matrix, then the matrix $S = S_u^T S_u$ will be an unreduced semiseparable matrix. Moreover all the elements in the generators $\mathbf{u}$ and $\mathbf{v}$ are different from zero.*[15]

Proof. By Equation (7.19) and knowing that because of the unreducedness of S_u the elements u_i and v_i are all different from zero, we have no zero blocks in the

---
[15]The idea is of course to choose the definition of unreducedness for the upper triangular semiseparable matrix, such that this theorem holds.

matrix S. Assume however that the semiseparable structure below the diagonal expands above the diagonal. This will lead to a contradiction. Assume for example that the element $u_3 u_2 \tau_2$ can be incorporated in the semiseparable structure below the diagonal. This implies that the 2×2 matrix

$$\begin{bmatrix} u_2\, u_2\tau_2 & u_3\, u_2\tau_2 \\ u_3\, u_2\tau_2 & u_3\, u_3\tau_3 \end{bmatrix}$$

is of rank 1[16]. This means that the matrix is singular and

$$u_2\, u_2\tau_2 u_3\, u_3\tau_3 - u_3\, u_2\tau_2 u_3\, u_2\tau_2 = 0$$

this can be simplified, leading to:

$$\tau_3 - \tau_2 = 0$$

which means, that $v_3 = 0$, which is a contradiction. $\square$

We will investigate how the application of one step of the QR-algorithm on the matrix $S - \mu I$ interacts with the upper triangular matrix S_u. Once the first column of the orthogonal transformation applied to the right is completely determined, we switch to the implicit approach. We make use of the structure of the QR-factorization as presented in Chapter 6, Subsection 6.1.1. We know that the Q factor of the QR-factorization of a diagonal plus semiseparable matrix of order n can be given by the product of $2n - 2$ Givens transformations. The first $n - 1$ Givens transformations, $G_i, i = 1, \ldots, n - 1$, applied from bottom to top transform the semiseparable matrix S into an upper triangular matrix and the diagonal matrix μI into a Hessenberg one. Moreover if $\tilde{Q} = G_1 G_2 \cdots G_{n-1}$, then $S_l = S_u Q_1$ is a lower triangular semiseparable matrix.[17] We have now already performed the first sequence of transformations on the matrix $S - \mu I$ which corresponds to transforming the upper triangular matrix S_u to lower triangular form S_l. We now calculate the first Givens transformation G_n which will annihilate the element in position $(2,1)$ of the Hessenberg matrix $H = Q_1^T (S - \mu I)$. The knowledge of this Givens transformation will determine the first column of the transformation performed on the right of the matrix S_u.

Taking into account that S_l is a lower triangular matrix, and

$$\begin{aligned} H &= Q_1^T (S - \mu I) \\ &= Q_1^T S_u^T S_u - \mu Q_1^T \\ &= S_l^T S_u - \mu Q_1^T, \end{aligned}$$

when denoting the Givens transformation embedded in G_{n-1} as

$$\begin{bmatrix} c_{n-1} & -s_{n-1} \\ s_{n-1} & c_{n-1} \end{bmatrix},$$

---

[16] In the notes and references at the end of the previous section some remarks were made concerning this 2×2 block with regard to the unreducedness.

[17] This transformation can be omitted. See the notes and references section for more information hereabout.

we show that the first two elements of the first column of H are

$$\left[\begin{array}{c} (S_l)_{11}(S_u)_{11} - \mu c_{n-1} \\ \mu s_{n-1} \end{array} \right].$$

This means that we can determine the Givens transformation G_n rather easily.

Moreover, taking a closer look at the first column of $Q = Q_1 Q_2$ we have

$$\begin{aligned} Q\mathbf{e}_1 &= Q_1 Q_2 \mathbf{e}_1 \\ &= G_1 G_2 \cdots G_{n-1} G_n G_{n+1} \cdots G_{2n-2} \mathbf{e}_1 \\ &= G_1 G_2 \cdots G_{n-1} G_n \mathbf{e}_1 \end{aligned}$$

since $G_i \mathbf{e}_1 = \mathbf{e}_1$ for $i = n + 1, \ldots, 2n - 2$. This means that the first column of the Q factor of the QR-factorization of $S - \mu I$ depends only on the product $G_1 G_2 \cdots G_{n-1} G_n$.

Furthermore, let

$$\tilde{S}_l = S_u G_1 G_2 \cdots G_{n-1} G_n = S_l G_n. \tag{7.20}$$

If we do not perform any transformations to the right of the matrix anymore involving the first column, we will be able to apply the implicit Q-theorem. From now on we will start with the implicit approach. The bulge, just created by the Givens transformation G_n, will be chased by a sequence of Givens transformations performed on the left and the right of the matrix.

7.4.3   Chasing the bulge

The Matrix (7.20) differs from a lower triangular semiseparable matrix, because there is a bulge in position $(1, 2)$. In order to retrieve the lower triangular semiseparable structure an algorithm is presented in this subsection. At each step of the algorithm the bulge is chased down one position along the superdiagonal, by applying orthogonal transformations to $\tilde{S}_l$. Only orthogonal transformations with the first column equal to $\mathbf{e}_1$ are applied to the right of $\tilde{S}_l$. Before describing the algorithm, we consider the following proposition.

Proposition 7.18. *Let*

$$C = \left[\begin{array}{ccc} u_1 v_1 & \alpha & 0 \\ u_2 v_1 & u_2 v_2 & u_2 v_3 \end{array} \right], \tag{7.21}$$

with u_1, u_2, v_1, v_2, v_3 all different from zero. Then there exists a Givens transformation G such that

$$\tilde{C} = C \left[\begin{array}{cc} 1 & 0 \\ 0 & G \end{array} \right] \tag{7.22}$$

has a linear dependency between the first two columns of $\tilde{C}$.

Proof. The theorem is proven by explicitly constructing the Givens transformation G

$$G = \frac{1}{\sqrt{1 + t^2}} \left[\begin{array}{cc} t & -1 \\ 1 & t \end{array} \right]. \tag{7.23}$$

Taking (7.21) and (7.23) into account, (7.22) can be written in the following way:

$$\tilde{C} = \frac{1}{\sqrt{1+t^2}} \left[\begin{array}{ccc} u_1 v_1 \sqrt{1+t^2} & t\alpha & -\alpha \\ u_2 v_1 \sqrt{1+t^2} & tu_2 v_2 + u_2 v_3 & -u_2 v_2 + tu_2 v_3 \end{array} \right].$$

Dependency between the first two columns leads to the following condition on the coefficients of the previous matrix:

$$t\alpha \left(\sqrt{1+t^2}\, u_2 v_1 \right) = \left(\sqrt{1+t^2}\, u_1 v_1 \right) (tu_2 v_2 + u_2 v_3).$$

Simplification leads to

$$t\alpha u_2 = u_1 \left(tu_2 v_2 + u_2 v_3 \right).$$

Extracting the factor t out of the previous equation proves the existence of the Givens transformation G:

$$t = \frac{v_3}{\left(\frac{\alpha}{u_1} - v_2 \right)}.$$

The denominator is clearly different from zero. Otherwise the left 2×2 block of the matrix in Equation (7.21) would already have been of rank 1 and we would have chosen the Givens transformation equal to the identity matrix. $\square$

The next theorem yields the algorithm that transforms $\tilde{S}_l$ into a lower triangular semiseparable matrix.

Theorem 7.19. *Let $\tilde{S}_l$ be an unreduced lower triangular semiseparable matrix. For which the strictly upper triangular part is zero except for the entry $(1,2)$. Then there exist two orthogonal matrices $\tilde{U}$ and $\tilde{V}$ such that*

$$\hat{S}_l = \tilde{U}^T \tilde{S}_l \tilde{V}$$

is a lower triangular semiseparable matrix and $\tilde{V}\mathbf{e}_1 = \mathbf{e}_1$.

Proof. The theorem is proven by constructing an algorithm that transforms $\tilde{S}_l$ into a lower triangular semiseparable matrix, in which the orthogonal transformations applied to the right have the first column equal to $\mathbf{e}_1$. Without loss of generality we assume $\tilde{S}_l \in \mathbb{R}^{5 \times 5}$. Let

$$\tilde{S}_l = \left[\begin{array}{ccccc} \boxtimes & \otimes & 0 & 0 & 0 \\ \boxtimes & \boxtimes & 0 & 0 & 0 \\ \boxtimes & \boxtimes & \boxtimes & 0 & 0 \\ \boxtimes & \boxtimes & \boxtimes & \boxtimes & 0 \\ \boxtimes & \boxtimes & \boxtimes & \boxtimes & \boxtimes \end{array} \right],$$

where $\otimes$ denotes the bulge to be chased. Moreover, the entries of the matrix satisfying the semiseparable structure are denoted by $\boxtimes$. At the first step a Givens transformation $\tilde{U}_1^T$ is applied to the left of $\tilde{S}_l$ in order to annihilate the bulge,

$$
\tilde{U}_1^T \tilde{S}_l =
\begin{bmatrix}
\times & 0 & 0 & 0 & 0 \\
\times & \times & 0 & 0 & 0 \\
\boxtimes & \boxtimes & \boxtimes & 0 & 0 \\
\boxtimes & \boxtimes & \boxtimes & \boxtimes & 0 \\
\boxtimes & \boxtimes & \boxtimes & \boxtimes & \boxtimes
\end{bmatrix}.
\tag{7.24}
$$

Although $\tilde{U}_1^T \tilde{S}_l$ is a lower triangular, the semiseparable structure is lost in its first two rows. In order to retrieve it, a Givens transformation $\tilde{V}_1$, constructed according to Theorem 7.18 and acting to the second and the third column of $\tilde{U}_1^T \tilde{S}_l$, is applied to the right, in order to make the first two columns in the lower triangular part proportional:

$$
\tilde{U}_1^T \tilde{S}_l =
\begin{bmatrix}
\boxtimes & 0 & 0 & 0 & 0 \\
\times & \times & 0 & 0 & 0 \\
\boxtimes & \boxtimes & \boxtimes & 0 & 0 \\
\boxtimes & \boxtimes & \boxtimes & \boxtimes & 0 \\
\boxtimes & \boxtimes & \boxtimes & \boxtimes & \boxtimes
\end{bmatrix}
\xrightarrow{\tilde{U}_1^T \tilde{S}_l \tilde{V}_1}
\begin{bmatrix}
\boxtimes & 0 & 0 & 0 & 0 \\
\boxtimes & \boxtimes & \otimes & 0 & 0 \\
\boxtimes & \boxtimes & \boxtimes & 0 & 0 \\
\boxtimes & \boxtimes & \boxtimes & \boxtimes & 0 \\
\boxtimes & \boxtimes & \boxtimes & \boxtimes & \boxtimes
\end{bmatrix}.
$$

Hence, applying $\tilde{U}_1^T$ and $\tilde{V}_1$ to the right and to the left of $\tilde{S}_l$, respectively, the bulge is moved one position down the superdiagonal, retrieving the semiseparable structure in the lower triangular part. Recursively applying the latter procedure the matrix $\tilde{U}_4^T \cdots \tilde{U}_1^T \tilde{S}_l \tilde{V}_1 \cdots \tilde{V}_3$ will be a lower triangular semiseparable matrix. Then the theorem holds choosing $\tilde{U}^T = \tilde{U}_4^T \cdots \tilde{U}_1^T$ and $\tilde{V} = \tilde{V}_1 \cdots \tilde{V}_3$. $\qquad\square$

7.4.4 Proof of correctness

In this subsection we describe one iteration of the proposed algorithm for computing the singular values of an upper triangular semiseparable matrix. Moreover we prove the equivalence between the latter iteration and one iteration of the QR-method applied to an upper triangular semiseparable matrix. Before we start with the QR-iteration the corresponding upper triangular semiseparable matrix needs to be in unreduced form. We have to apply deflation. After having made the matrix unreduced, one iteration of the proposed method consists of the following four steps.

- **Step 1.** $n-1$ Givens transformations $G_1, \ldots, G_{n-1}$ are performed to the right of S_u swapping it into a lower triangular semiseparable matrix S_l.

- **Step 2.** One more Givens transformation G_n is computed in order to introduce the shift. As seen in Subsection 7.4.3, the application of this Givens transformation to the right of the lower triangular semiseparable matrix creates a bulge.

- **Step 3.** The bulge is chased by $n-2$ Givens transformations, $G_{n+1}, \ldots, G_{2n-2}$ on the left of the matrix, and also $n-2$ Givens transformations on the right retrieving the lower triangular semiseparable structure.

- **Step** 4. The latter matrix is swapped back to upper triangular form. This is done by applying $n - 1$ more Givens transformations to the left, without destroying the semiseparable structure.

- **Step** 5. Apply deflation to the resulting upper triangular semiseparable matrix.

Note 7.20. *Actually, the Givens transformations of Step 1 and Step 4 do not need to be explicitly applied to the matrix. The semiseparable structure allows us to calculate the resulting semiseparable matrix (the representation of this matrix) in a very cheap way with $\mathcal{O}(n)$ flops.*

Note 7.21. *If the shift is not considered, only Step 1 and Step 4 are performed at each iteration of the method.*

We still have to prove that one iteration of the latter method is equivalent to one iteration of the QR-method applied to an upper triangular semiseparable matrix S_u, i.e., if $\tilde{Q}$ is the orthogonal factor of the QR-factorization of $S_u^T S_u - \mu I = S - \mu I$, and Q is the matrix of the product of the orthogonal matrices applied to the right of S_u during one iteration of the proposed method, then $\tilde{Q}$ and Q have the same columns up to the sign, i.e.,

$$\tilde{Q} = Q \operatorname{diag}([\pm 1, \ldots, \pm 1]).$$

Assume a QR-step is performed on the matrix $S = S_u^T S_u$. This can be written in the following form:

$$\tilde{Q}^T S \tilde{Q} = \tilde{S}. \tag{7.25}$$

Now we show how one iteration of the new method performed on the matrix S_u can also be rewritten in terms of the matrix $S = S_u{}^T S_u$. This is achieved in the following way. (All the steps mentioned in the beginning of the subsection are now applied to the matrix S_u.) First the $n - 1$ Givens transformations are performed on the matrix S_u, transforming the matrix into a lower triangular semiseparable form. In the following equations, all the transformations performed on the matrix S_u are fitted in the equation $S = S_u{}^T S_u$ to see what happens to the matrix S:

$$S_l^T S_l = \left(G_{n-1}^T \cdots G_1^T S_u{}^T \right) \left(S_u G_1 \cdots G_{n-1} \right)$$
$$= G_{n-1}^T \cdots G_1^T S G_1 \cdots G_{n-1}.$$

One more Givens transformation G_n is now applied, introducing a bulge in the matrix S_l:

$$G_n^T S_l^T S_l G_n = G_n^T G_{n-1}^T \cdots G_1^T S G_1 \cdots G_{n-1} G_n.$$

Taking Theorem 7.18 into account, there exist two orthogonal matrices U and V, with $V \mathbf{e}_1 = \mathbf{e}_1$, such that $U^T S_l G_n V$ is again a lower triangular semiseparable matrix. This leads to the equation

$$V^T G_n^T S_l^T U U^T S_l G_n V = V^T G_n^T G_{n-1}^T \cdots G_1^T S G_1 \cdots G_{n-1} G_n V \tag{7.26}$$

The final step consists of swapping the lower triangular matrix $U^T S_l G_n V^T$ back into upper triangular semiseparable form. This is accomplished by applying $n - 1$ more Givens transformations to the left of $U^T S_l G_n V$. Thus, let $Q = G_1 \cdots G_{n-1} G_n V$ and taking (7.19) into account, we have

$$Q^T S_u^T S_u Q = \hat{S} \tag{7.27}$$

with Q orthogonal and $\hat{S}$ a semiseparable matrix.

We observe that $\tilde{Q}\mathbf{e}_1 = Q\mathbf{e}_1$. This holds because the first n Givens transformations are the same Givens transformations as performed in a QR-step on S, and the fact that $V\mathbf{e}_1 = \mathbf{e}_1$.

One can apply the implicit Q theorem now (Theorem 6.28), which justifies our approach. Hence we developed a QR-variant for upper triangular semiseparable matrices. Combined with the results of Part II, we can now compute the singular values of not only upper triangular semiseparable matrices but also of arbitrary matrices.

Notes and references

The results in this section on the implicit QR-method for the singular value decomposition were presented in [162]. As mentioned before this book discusses the reduction to upper triangular semiseparable form, the inherited convergence properties as well as the new chasing technique.

The complexity of the proposed method can still be reduced slightly. Reconsidering the different individual steps when performing one step of the implicit QR-method we see that we can omit in some sense Step 4 in this procedure. In Step 4 the lower triangular semiseparable form is swapped back to upper triangular form in order to maintain the upper triangular structure. The transformations however to reduce the matrix back to upper triangular form are performed on the left of the matrix, considering Equations (7.26) and (7.27) we see that they do not have an impact on the proof of the correctness of the implicit approach. The resulting lower triangular semiseparable matrix $\hat{S}_l = U^T S_l G_n V$ can be used immediately for performing a new step of the QR-method. Instead of applying the QR-method on the matrix product $\hat{S}_l^T \hat{S}_l$ one can do it on the matrix product $\hat{S}_l \hat{S}_l^T$, which has exactly the same eigenvalues and moreover $\hat{S}_l^T$ is upper triangular. So one can simply omit Step 4 in the procedure and continue working with its transpose. The first swapping is however essential to prove correctness.

Of course there are various ways to continue without working with the transpose. One can also develop an implicit technique for $\hat{S}_l^T \hat{S}_l$ (which is also a semiseparable matrix) or one can immediately try to reduce during the chasing procedure to upper triangular semiseparable form.

7.5 Conclusions

In this chapter we developed implicit QR-algorithms for computing the eigenvalues and/or singular values. In order to compute the eigenvalues, we explored the method for symmetric semiseparable and for Hessenberg-like matrices. The singular values were computed for upper triangular semiseparable matrices.

Combined with the reduction algorithms provided in the previous part we now have algorithms to compute the eigenvalues/singular values of arbitrary matrices via semiseparable matrices.

In order to obtain these algorithms, some results from the previous chapter such as unreducedness and the implicit Q-theorem were used extensively.

Chapter 8

Implementation and numerical experiments

In the previous chapters of this part, different theoretical results were combined and developed in order to build implicit QR-algorithms for semiseparable matrices. We know how such an algorithm should be designed. Also details on the implementation of the designed algorithms will be given. In the previous chapters, nothing was mentioned about the deflation criterion, i.e., when do we assume that an eigenvalue is accurate enough, or even more, when should we deflate an almost block diagonal matrix into several blocks. This problem, together with a fast computation for the norms of the off-diagonal blocks, is solved in this chapter too. The final important problem related to the eigenvalue decomposition is the calculation of the eigenvectors. Some methods for calculating the eigenvectors will be discussed. Finally this chapter is closed by showing some numerical examples related to the implementation of the implicit QR-methods presented in this chapter.

Implementing the QR-methods as presented in the previous chapter can be done in two different ways. First one can translate all algorithms to formulas and use all the theorems designed in Chapter 7 to obtain a fine grained algorithm. This approach is presented in Section 8.2. The second approach is more rigid and more easy to implement. In this approach one uses the Givens-vector representation and one uses implementations for the shift through lemma and the fusion. Based on these two basic operations the global implementation becomes easy. This approach is shown in the first section. This approach will also be exploited in the next chapter in which the multishift algorithm and the QH-algorithm for structured rank matrices will be developed.

In the first section of the chapter we will exploit some of the techniques presented in Section 4.1 to interpret the QR-method in a different way. The technique is based on the Givens-vector representation. Both the top bottom and the right left representations are considered for developing a chasing algorithm. We will also translate the fusion and the shift through lemma towards Givens operations performed on the right. First the QR-method is discussed in a graphical manner for the semiseparable case. Secondly all details are presented on how to implement the

QR-method for semiseparable plus diagonal matrices. For both algorithms a step of the QR-method is divided into two parts. A first part, which is performed rather easily, corresponds to a step of the QR-method without shift for the semiseparable case. The second part will be the chasing part. This part is more difficult for the semiseparable plus diagonal case, but it is much easier to understand than a technical treatment such as presented in the previous chapter for the semiseparable case.

In the second section, we will design the implementation of a QR-step for a symmetric semiseparable matrix. This implementation is subdivided in different subsections. First the implementation of the first part in the QR-method is discussed, namely the QR-step without shift. This corresponds to performing a similarity transformation, with the first sequence of Givens transformations as designed in Subsection 6.1.1. Instead of performing first all the transformations on the left side of the semiseparable matrix, and then performing all the transformations on the right side of the resulting matrix, we will perform the transformations simultaneously. This means that at the same time as performing the transpose of a Givens transformation on the left, we will perform the Givens transformation also on the right. The details of this implementation are presented in this section.

Secondly we discuss in Subsection 8.2.2 the orthogonal similarity transformation of a matrix to unreduced form. Attention is also paid to a criterion stating whether or not the semiseparable structure extends above the diagonal.

Subsection 8.2.3 details the implementation of an implicit QR-step on a symmetric semiseparable matrix with shift. As the QR-step with shift initially performs a QR-step without shift, we assume that the matrix we start with has the structure of the resulting matrix from a step of the QR-method without shift. The QR-step with shift is initialized by performing the similarity Givens transformation G_n, which will divide the matrix in two semiseparable matrices: one semiseparable matrix of dimension 2×2 at the upper left position, and one semiseparable matrix of dimension $n - 1$ by $n - 1$ at the lower right position. Both of the semiseparable matrices will be represented by the Givens-vector representation. The consecutive Givens transformations performed on the matrix will increase the dimension of the upper left semiseparable matrix by one at each transformation, and they will decrease the dimension of the lower right semiseparable matrix by one at each step. At each step therefore the Givens-vector representation of the upper left and lower right semiseparable matrix have to be updated.

Deflation is the subject of the following subsection. It is investigated when we should divide the symmetric semiseparable matrix in different blocks, such that all the eigenvalues of these blocks approximate very well the eigenvalues of the original matrix. In the tridiagonal approach the original matrix is divided if the corresponding subdiagonal elements are close enough to zero. This is logical because the subdiagonal elements contain all the information of the corresponding lower left block, which only has the upper right element, which is the subdiagonal element, different from zero. In our semiseparable case however the lower left blocks are full. Calculating the norms of all of the lower right blocks to see whether they are close enough to zero is a very costly operation. Fortunately we can compute all the norms of the off-diagonal blocks in $\mathcal{O}(n)$ operations. Two different criterions to decide whether an off-diagonal block is small enough are discussed.

In several cases, not only are the eigenvalues desired but also sometimes some or all of the eigenvectors are needed. Section 8.3 considers some possibilities for calculating the eigenvectors of the symmetric semiseparable matrix. E.g., inverse iteration and a technique in the case of clustered eigenvalues are explored. It will be shown that the latter technique is well-suited for the class of structured rank matrices.

In Subsection 8.4.1 different numerical experiments are performed on the eigenvalue solver based on semiseparable matrices. Arbitrary matrices are transformed to a similar symmetric semiseparable one and afterwards the implicit QR-algorithm is applied to the semiseparable matrix in order to calculate its eigenvalues. Four different experiments are reported. A first experiment investigates an almost block matrix and compares the tridiagonal approach with the semiseparable one. In a second experiment a matrix is constructed, for which the eigenvalues form a stair. After the reduction to tridiagonal and the corresponding reduction to the semiseparable form, the diagonal elements of both the tridiagonal and the semiseparable matrix are compared. In a last experiment the accuracy of the eigenvalues, computed by two different deflation criterions, using the semiseparable approach is compared.

In Section 8.4 several numerical experiments are performed using the algorithm to compute the singular values via intermediate upper triangular semiseparable matrices. Two types of experiments are performed. In one type of experiment the accuracy between the bidiagonal and the upper triangular semiseparable approach are compared. In the second experiment the number of QR-steps needed to compute all the singular values, via both the algorithms are compared.

✎ *This chapter contains more detailed material related to the implicit QR-algorithms proposed in this part. In our opinion Section 8.1 is interesting for people working with Givens-vector represented structured rank matrices. The chasing technique simplifies extremely when using this representation. For those implementing the method Section 8.1 and Section 8.2 are essential. Computing eigenvectors is briefly discussed in Section 8.3.*

8.1 Working with Givens transformations

In a previous section (Chapter 4, Section 4.1) it was shown how we can graphically depict Givens transformations acting on the rows of matrices. Moreover also the Givens-vector representation from top to bottom was interpreted in this fashion. If one wants to represent however the QR-algorithm transformations to the right of the matrix have to be applied also. Therefore we will introduce here also the graphical representation for Givens transformations performed to the right of a matrix. Also the representation from right to left will be considered. Then in Subsection 8.1.2 we will use this graphical representation to depict a step of the QR-method on a Givens-vector represented Hessenberg-like matrix and in Subsection 8.1.3 we will depict the QR-method for semiseparable plus diagonal matrices.

A graphical scheme of Givens transformations applied onto a matrix from the right has the following form.

$$
\begin{array}{c|cccccccc}
4 & & & & \llcorner\!\rightarrow & & & & \\
3 & & \llcorner\!\rightarrow & & & & \llcorner\!\rightarrow & & \\
2 & \llcorner\!\rightarrow & & & & & & \llcorner\!\rightarrow & \\
1 & \llcorner\!\rightarrow & & & & & & & \\
\hline
& ❶ & ❷ & ❸ & ❹ & ❺ & ❻ & ❼ & ❽
\end{array}
$$

Let us interpret this scheme. The elements ❶, ❷, ... on the horizontal axis denote the columns of the matrix the Givens transformations act on; the elements $1, 2, 3, \ldots$, on the vertical axis denote the order in which the Givens transformations are applied onto the matrix.

The first Givens transformation, presented in the row following 1, acts on the first and second column of the matrix. The second row of transformations consists of two Givens transformations, one transformation acting on the first two columns and a second one acting on columns 7 and 8. These transformations do not interfere with each other. The third row also contains two Givens transformations acting, respectively, on columns 2 and 3 and on columns 6 and 7. The final Givens transformation in row 4 acts on columns 4 and 5.

Being familiar with the row based representation (from top to bottom), we will now explain the second (swapped) representation, from right to left or the column representation. The row based representation, for a Hessenberg-like matrix, is as presented in Section 4.1 of the following form:

$$
\begin{array}{c|ccccccc}
❶ & & & & \times & \cdot & \cdot & \cdot & \cdot & \cdot \\
❷ & & & & & \times & \cdot & \cdot & \cdot & \cdot \\
❸ & & & & & & \times & \cdot & \cdot & \cdot \\
❹ & & & & & & & \times & \cdot & \cdot \\
❺ & & & & & & & & \times & \cdot \\
❻ & & & & & & & & & \times \\
\hline
& 5 & 4 & 3 & 2 & 1 &
\end{array}
\tag{8.1}
$$

Gradually the matrix presented on the right is filled up with a low rank part. The column based representation for a 5×5 Hessenberg-like matrix is of the following form:

$$
\begin{array}{c|ccccc}
4 & \llcorner\!\rightarrow & & & & \\
3 & & \llcorner\!\rightarrow & & & \\
2 & & & \llcorner\!\rightarrow & & \\
1 & & & & \llcorner\!\rightarrow & \\
& & & & & \\
& \times & \cdot & \cdot & \cdot & \cdot \\
& & \times & \cdot & \cdot & \cdot \\
& & & \times & \cdot & \cdot \\
& & & & \times & \cdot \\
& & & & & \times \\
\hline
& ❶ & ❷ & ❸ & ❹ & ❺
\end{array}
$$

The Givens transformations are applied from right to left and they gradually fill up the lower triangular part of the matrix with a part being of semiseparable form.

Let us now translate the shift through lemma and the fusion of Givens transformations towards this new scheme.

8.1.1   Interaction of Givens transformations

Lemma 4.1 on the fusion of Givens transformations as well as the shift through lemma (Lemma 4.2) stay of course valid.

Graphically the fusion of two Givens transformations acting on the right is depicted as follows:

resulting in

The shift through lemma is depicted as follows:

resulting in

and in the other direction this becomes

resulting in

Based on these graphical schemes, we will illustrate in the following subsection how we can more easily interpret a QR-step if one is working with a Hessenberg-like matrix.

8.1.2   Graphical interpretation of a QR-step

In this section we will see that distinguishing between the different cases (see all the propositions to determine the Givens transformations in the Hessenberg-like case in Subsection 7.3.3) to chase the disturbance in the Hessenberg-like matrix away is not at all necessary if one works with the Givens-vector representation and the shift through lemma. In fact the explanation here is much more simple than the one presented in Chapter 7. The chasing technique as discussed here can easily be modified to suit also the symmetric semiseparable and the bidiagonal case. We choose the Hessenberg-like case because it consisted of the many different propositions for determining the Givens transformations in order to restore the structure in the lower triangular part.

We will see that when exploiting the Givens-vector representation and the shift through technique this can all be combined into one simple case. Hence there is no need to distinguish between the different types.

Assume we start with a Hessenberg-like matrix, represented with the Givens-vector representation from top to bottom. This means that graphically our matrix has the structure as shown in (8.1). We would like to perform an implicit step of the QR-method onto this matrix. From the results in Chapter 7, Section 7.3 we know that we have to perform two orthogonal similarity transformations Q_1 and Q_2 onto our matrix Z. Initially the matrix Z is transformed into the matrix

$$Z_1 = Q_1^T Z Q_1.$$

Then an initial disturbing transformation G_n is applied acting on the first two rows and columns, followed by a chasing technique leading to a new Hessenberg-like matrix. Remark that the chasing technique needs to be constructed in such a fashion, that it does not involve the first row and first column of the disturbed matrix.

Let us see what happens with the representation if we apply the first sequence of Givens transformations denoted by the orthogonal matrix Q_1. We remark that Q_1 is fully determined by the Givens-vector representation of the Hessenberg-like matrix (see Subsection 8.2.1). Starting with the graphical representation as in (8.1) and performing the orthogonal similarity transformation involving only the matrix Q_1 results in the fact that the Givens transformations on the left of the Givens-vector representation are fully annihilated. However the matrix Q_1 applied to the right cannot be neglected and results in fact in a representation from right to left. Hence we have again a Hessenberg-like matrix Z_1, as predicted before, but now represented with a Givens-vector representation from right to left. This step can be seen in the proof of Theorem 6.6.

We will illustrate now the chasing procedure on a $5{\times}5$ matrix. The Hessenberg-like matrix Z_1 is now of the following form:

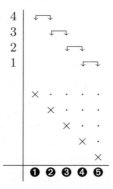

In order to distinguish between a theoretical and implementational viewpoint of the scheme above we have the following remark.

Note 8.1. *Below we will present some remarks concerning the interpretation of the scheme above, with regard to actual implementations based on this scheme.*

- *Remember that the Givens-vector representation for a semiseparable part consists of a sequence of Givens transformations and of a vector* **v**. *To construct*

*the semiseparable part of a Givens representation from right to left, in every
step an element from the vector was added and the Givens transformation G_i
was applied on it. Denote the Givens transformations as G_i based on the cosine c_i and the sine s_i, acting on the columns i and $i+1$. In fact one can then
also write the semiseparable part of the Hessenberg-like matrix Z_1 constructed
with the Givens-vector representation as*

$$\operatorname{tril}(Z_1) = \operatorname{tril}(\operatorname{diag}(\hat{\mathbf{v}})Q), \qquad (8.2)$$

where $Q = G_{n-1}G_{n-2}\ldots G_2G_1$ and the vector $\hat{\mathbf{v}}$ equals

$$\hat{\mathbf{v}} = \left[\frac{v_1}{c_1}, \frac{v_2}{c_2}, \ldots, \frac{v_{n-1}}{c_{n-1}}, v_n\right].$$

*This slightly changed vector is due to the definition of the representation,
namely that first a transformation is applied and then a new vector element
(weight) is added before applying the next transformation. In Equation (8.2)
the vector elements are however already present and affected by more Givens
transformations than in the definition of the representation.*

- *Reconsidering the graphical scheme and the remark above, we have in fact that
the elements marked with $\cdot$ are stored, and not affected by the depicted Givens
transformations coming from the representation. Hence the scheme should be
interpreted as follows*

$$Z_1 = \operatorname{tril}(\operatorname{diag}(\hat{\mathbf{v}})Q) + R_u, \qquad (8.3)$$

  *where Z_1 denotes the Hessenberg-like matrix, $\hat{\mathbf{v}}$ the vector from above (and
related to the elements $\times$ shown in the graphical representation), Q is the
orthogonal transformation consisting of the Givens transformations depicted
and R_u is a strictly upper triangular matrix consisting of the elements $\cdot$ from
above.*

- *These two remarks above are mostly important for those people wanting to
implement this presented method. When implementing the method, one should
consider the splitting from the previous item. Therefore one should store the
elements $\cdot$, the vector $\mathbf{v}$ and the Givens transformations G_i.*

- *Theoretically, it is however not easy to see the impact of a Givens transformation performed to the left or to the right of Equation (8.3). It involves
carefulness with respect to the operator $\operatorname{tril}(\cdot)$ and more. These details are left
for the software developers as mentioned in the previous item. For our pur-
pose we will use a slightly different interpretation of the scheme above, which
will give us the opportunity to neglect this detailed and careful index shuffling
and so forth.*

*Reconsidering Equation (8.3) we will rewrite it to give another interpretation
to the graphical scheme. The matrix $\operatorname{diag}(\hat{\mathbf{v}})Q$ in Equation (8.3) is not lower
triangular, therefore the $\operatorname{tril}(\cdot)$ operator was involved. The matrix $\operatorname{diag}(\hat{\mathbf{v}})Q$*

is however a lower Hessenberg matrix. This is due to the special order of the Givens transformations. Therefore if a matrix is chosen especially to annihilate the superdiagonal elements of the matrix, we can remove the $\mathrm{tril}(\cdot)$ *operator:*

$$\mathrm{tril}(\mathrm{diag}(\hat{\mathbf{v}})Q) = \mathrm{diag}(\hat{\mathbf{v}})Q - \hat{R}_u,$$

where $\hat{R}_u$ *is a strictly upper triangular matrix with only the superdiagonal elements different from zero, such that the superdiagonal elements of* $\mathrm{diag}(\hat{\mathbf{v}})Q$ *are annihilated.*

Combining all this into Equation (8.3) leads to the following form for the matrix Z_1:

$$
\begin{aligned}
Z_1 &= \mathrm{tril}(\mathrm{diag}(\hat{\mathbf{v}})Q) + R_u \\
&= \mathrm{diag}(\hat{\mathbf{v}})Q - \hat{R}_u + R_u \\
&= \mathrm{diag}(\hat{\mathbf{v}})Q + \left(\left(-\hat{R}_u + R_u \right) Q^T \right) Q \\
&= \mathrm{diag}(\hat{\mathbf{v}})Q + \tilde{R}Q
\end{aligned}
$$

The strictly upper triangular matrix $\left(-\hat{R}_u + R_u \right)$, *changed by an orthogonal transformation* Q^T *on the right, is transformed into an upper triangular matrix* $\tilde{R}$ *(not strictly upper triangular anymore). Denoting* $R = \mathrm{diag}(\hat{\mathbf{v}}) + \tilde{R}$, *gives us the following equation for the matrix* Z_1:

$$Z_1 = RQ.$$

The matrix Q *is exactly the same as the matrix representing the Givens transformations from the Givens-vector representation. Hence we can interpret the graphical scheme as just an RQ-factorization, instead of the decoupled form in Equation (8.3). This is a much simpler interpretation which we can use without loss of generality as the elements in the representation are not specified. For the implementation however, it is more appropriate to use the decoupled form and the formulas presented here already denote details one has to take into consideration.*

- *To conclude we will show the effect of a Givens transformation performed to the right and to the left of the graphical scheme. Performing a Givens transformation* $\tilde{G}$ *to the left of the representation corresponds to*

$$
\begin{aligned}
\tilde{G}Z_1 &= \tilde{G}RQ \\
&= \left(\tilde{G}R \right) Q.
\end{aligned}
$$

Hence this transformation only affects the upper triangular part in the representation. The depicted Givens transformations do not change. Applying a transformation to the right results in

$$
\begin{aligned}
Z_1\tilde{G} &= RQ\tilde{G} \\
&= R \left(Q\tilde{G} \right).
\end{aligned}
$$

This means that a transformation applied to the right only affects the Givens transformations and not the upper triangular part in the graphical interpretation.

Let us continue with the QR-step started before the remark. To complete the QR-step a disturbing Givens transformation has to be performed on the matrix Z_1. This is a similarity transformation. Performing it on the left side of the matrix creates a bulge marked with $\otimes$. The Givens transformation performed on the right has its effect on the Givens transformations and is added to the top of the sequence of Givens transformations, namely on position 5. It is already depicted that we can concatenate the Givens transformation in positions 4 and 5. Hence the matrix $Z_2 = G_n^T Z_1 G_n$ can be represented as follows:

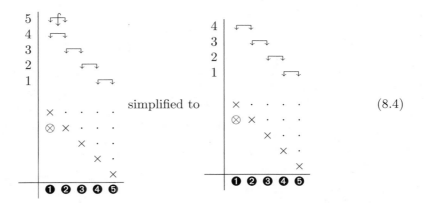

$$(8.4)$$

Note 8.2. *We remark that instead of working with the right left representation, one can also use the top bottom representation. In the next chapter we will exploit the top bottom representation for the QR-method for semiseparable plus diagonal matrices. To change the right left representation to a top bottom representation one has to use a swapping procedure (see Subsection 1.3.2).*

Now we have to determine a similarity Givens transformation, acting on rows 2 and 3 and on columns 2 and 3, such that the present bulge (marked with $\otimes$) is chased down one position. In order to determine this transformation, let us rewrite the matrix on the right of (8.4). We annihilate the bulge by performing a Givens transformation acting on columns 1 and 2. This transformation is of course placed before transformation 1. This Givens transformation, does not interact with the transformations in positions 2 and 3 (see the left of (8.5)) hence the scheme can be rewritten (see right of (8.5)).

$$\text{simplified to} \tag{8.5}$$

After having determined this transformation, the shift through lemma is applied as depicted in the following scheme. Remark that this is just another representation of exactly the same matrix represented in (8.4), namely the matrix Z_2.

$$\text{leads to} \tag{8.6}$$

The rewritten matrix Z_2 is now of Hessenberg-like form, except for one transformation to the right, acting on columns 2 and 3.

Now it is time to apply an orthogonal similarity transformation, namely a Givens transformation, to restore this disturbed structure. This Givens transformation is chosen to be the transpose (inverse) of the Givens transformation present in the fifth position, acting on columns 2 and 3. Hence applying this transformation will annihilate the Givens transformation in position 5. Unfortunately we have to perform a similarity transformation and therefore also a Givens transformation has to be performed on the left side of the matrix Z_2. This transformation creates a bulge. The resulting matrix $Z_3 = G_{n+1}^T Z_2 G_{n+1}$ is hence of the following form.

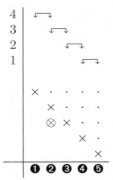

We can clearly see that the bulge has moved down one position. The construction of the next structure restoring Givens transformation proceeds similarly. Let us conclude by showing this graphically. In order to determine this Givens transformation we first have to rewrite our matrix Z_3 again, by annihilating the bulge with a Givens transformation on columns 2 and 3 and applying the shift through lemma.

The structure restoring Givens transformation is now determined in order to annihilate the Givens transformation present in position 5, acting on column 3 and 4. The Givens transformation therefore acts on rows 3 and 4 and columns 3 and 4. Performing this orthogonal similarity transformation however creates a bulge again.

It is clear that this procedure can be continued. Finally one obtains a Hessenberg-like matrix with the representation from right to left. In order to obtain the representation again from top to bottom, one needs to swap the representation (see Chapter 1).

It is interesting to see that no special cases have to be considered. All information is captured in the Givens-vector representation. Moreover the structure restoring Givens transformations do not involve the first column or row and hence one can easily prove that the result of this procedure is equivalent to performing a step of the explicit QR-algorithm.

8.1.3 A QR-step for semiseparable plus diagonal matrices

In the previous section only some hints were given on how to construct the QR-method for semiseparable plus diagonal matrices. The algorithm was not discussed in full detail as it involved similar operations to the QR-method for semiseparable matrices, but in every step few changes were made, making it quite detailed and complex.

Nevertheless this method can be found in the technical report [157]. The method discussed there can also be written in the graphical form as done in the previous subsection. This graphical form will make it much easier to understand the QR-method for semiseparable plus diagonal matrices, without the burden of all exceptional cases. Similar as in the previous subsection we will divide the QR-step into two parts. A first part will discuss performing the first sequence of Givens transformations, and the second part will discuss the chasing.

Performing the first sequence of Givens transformations

Only a graphical form of the implementation will be presented. Based on the presented schemes the reader should be able to design an implementation based on the Givens-vector representation, the shift through lemma and the fusion.

In the previous section we assumed to be working with a right left Givens-vector representation. Here we will work again with the top bottom form. There is no loss in generality in assuming our matrix to be of dimension 5×5. Our matrix $S + D$ is hence of the following form:

where clearly the diagonal matrix D is of the form $D = \mathrm{diag}([d_1, d_2, d_3, d_4, d_5])$, and S is represented already in the graphical form. Important to remark is that the diagonal D is not included in the graphical scheme, i.e., the matrix D is not included in the representation of the matrix S!

We know (see Section 7.2) that after performing the first similarity transformation $Q_1^T(S+D)Q_1$ we obtain again a semiseparable plus diagonal matrix S_1+D_1, in which S_1 is of semiseparable form and the diagonal $D_1 = \text{diag}([0, d_1, d_2, d_3, d_4])$. This means that the diagonal has shifted down one position. Moreover, the factor Q_1^T consists of four Givens transformations, which annihilate the Givens transformations from the Givens-vector representation of the matrix S. We will use a similar notation as before: $S + D = S_0 + D_0$. After performing a similarity transformation with a Givens transformation, the superscript goes up.

Note 8.3. *In the real implementation, one needs to use the fact that the semiseparable matrix is symmetric. For simplicity we will not take this into consideration. For symmetric matrices, one can also use the Givens-vector representation of the upper triangular part and use the shift through lemma and the fusion for easy updating this representation.*

Let us start now by performing the orthogonal similarity transformation onto the semiseparable plus diagonal matrix. Before performing the first Givens transformation G_1, we will rewrite the matrix above slightly. We will subtract some information from the element in the lower right corner of the matrix S and add it to the diagonal such that the diagonal becomes $\hat{D}_0 = \text{diag}([d_1, d_2, d_3, d_4, d_4])$. Important to remark is that this does not at all change the Givens transformations of the Givens-vector representation. Only the element marked with $\boxplus$ has changed, as well as the lower right element of D_0. We obtain the following graphical form of $S_0 + D_0$, which we have rewritten as $S_0 + D_0 = \hat{S}_0 + \hat{D}_0$:

Performing now the similarity transformation leads to the following equations:

$$G_1^T \left(S_0 + D_0 \right) G_1 = G_1^T \left(\hat{S}_0 + \hat{D}_0 \right) G_1$$
$$= G_1^T \hat{S}_0 G_1 + \hat{D}_0$$
$$= S_1 + D_1.$$

Obviously $G_1^T \hat{D}_0 G_1 = \hat{D}_0$, as the Givens transformations act on the two bottom rows and the last two columns and the diagonal entries of the matrix $\hat{D}_0$ in these positions are both equal to d_4.

The transformation $G_1^T \hat{S}_0$, will remove the Givens transformation from the Givens-vector representation in position 4. The transformation G_1 performed to the right of $G_1^T \hat{S}_0$, will then create fill-in in position $(5, 4)$. Hence we obtain as a graphical representation for $S_1 + D_1$:

$$
\begin{array}{l}
\textbf{1} \\ \textbf{2} \\ \textbf{3} \\ \textbf{4} \\ \textbf{5}
\end{array}
\left|
\begin{array}{ccccc}
 & \times & \cdot & \cdot & \cdot \\
 & & \times & \cdot & \cdot \\
 & & & \times & \cdot \\
 & & & \boxtimes & \cdot \\
 & & & \boxtimes & \boxtimes
\end{array}
\right.
\;+\;
\left[
\begin{array}{ccccc}
d_1 & & & & \\
 & d_2 & & & \\
 & & d_3 & & \\
 & & & d_4 & \\
 & & & & d_4
\end{array}
\right].
$$

$$\overline{\quad 3 \;\; 2 \;\; 1 \quad}$$

In the lower right corner, some elements (including the fill-in) are marked with ⊠ as they make up a semiseparable part in the matrix. We will show now that in the following steps, this part will gradually grow to create a new semiseparable matrix.

Applying the second similarity transformation proceeds in a similar way. First we rewrite the matrix $S_1 + D_1$ as $\hat{S}_1 + \hat{D}_1$:

$$
\begin{array}{l}
\textbf{1} \\ \textbf{2} \\ \textbf{3} \\ \textbf{4} \\ \textbf{5}
\end{array}
\left|
\begin{array}{ccccc}
 & \times & \cdot & \cdot & \cdot \\
 & & \times & \cdot & \cdot \\
 & & & \times & \cdot \\
 & & & \boxplus & \cdot \\
 & & & \boxtimes & \boxtimes
\end{array}
\right.
\;+\;
\left[
\begin{array}{ccccc}
d_1 & & & & \\
 & d_2 & & & \\
 & & d_3 & & \\
 & & & d_3 & \\
 & & & & d_4
\end{array}
\right].
$$

$$\overline{\quad 3 \;\; 2 \;\; 1 \quad}$$

The element ⊞ has changed, as well as the fourth diagonal entry of the matrix D_1. Even though the element ⊞ changed, it is still includable in the semiseparable structure, which consists now of the elements ⊞ and ⊠. We can write similar equations for performing the second Givens transformation G_2:

$$
\begin{aligned}
G_2^T \left(S_1 + D_1 \right) G_2 &= G_2^T \left(\hat{S}_1 + \hat{D}_1 \right) G_2 \\
&= G_2^T \hat{S}_1 G_2 + \hat{D}_1 \\
&= S_2 + D_2.
\end{aligned}
$$

The orthogonal similarity transformation does not change the matrix $\hat{D}_1$. The transformation G_2^T performed on the left of $\hat{S}_1$ removes another Givens transformation from the Givens-vector representation of the matrix $\hat{S}_1$ and the transformation G_2 performed on the right of $G_2^T \hat{S}_1$ expands the existing semiseparable structure. We obtain therefore:

$$
\begin{array}{l}
\textbf{1} \\ \textbf{2} \\ \textbf{3} \\ \textbf{4} \\ \textbf{5}
\end{array}
\left|
\begin{array}{ccccc}
 & \times & \cdot & \cdot & \cdot \\
 & & \times & \cdot & \cdot \\
 & & \boxtimes & \cdot & \cdot \\
 & & \boxtimes & \boxtimes & \cdot \\
 & & \boxtimes & \boxtimes & \boxtimes
\end{array}
\right.
\;+\;
\left[
\begin{array}{ccccc}
d_1 & & & & \\
 & d_2 & & & \\
 & & d_3 & & \\
 & & & d_3 & \\
 & & & & d_4
\end{array}
\right].
$$

$$\overline{\quad 2 \;\; 1 \quad}$$

Note 8.4. *In the depicted schemes it seems that we do fill up the lower triangular part of the matrix completely to obtain the semiseparable structure. Of course, this is not done in a real implementation. Simply store the Givens transformations* *performed onto the right as they form the new right left representation of the newly created semiseparable matrix.*

Continuing this process of slightly changing the diagonal elements before performing the similarity transformation results finally in the following scheme for the matrix $S_4 + D_4$:

$$
\begin{bmatrix}
\boxtimes & \cdot & \cdot & \cdot & \cdot \\
\boxtimes & \boxtimes & \cdot & \cdot & \cdot \\
\boxtimes & \boxtimes & \boxtimes & \cdot & \cdot \\
\boxtimes & \boxtimes & \boxtimes & \boxtimes & \cdot \\
\boxtimes & \boxtimes & \boxtimes & \boxtimes & \boxtimes
\end{bmatrix}
+
\begin{bmatrix}
d_1 & & & & \\
& d_1 & & & \\
& & d_2 & & \\
& & & d_3 & \\
& & & & d_4
\end{bmatrix}.
$$

The element in the upper left position can be chosen freely without disturbing the semiseparable structure. Hence, the matrix can easily be rewritten as $\hat{S}_4 + \hat{D}_4$:

$$
\begin{bmatrix}
\boxplus & \cdot & \cdot & \cdot & \cdot \\
\boxtimes & \boxtimes & \cdot & \cdot & \cdot \\
\boxtimes & \boxtimes & \boxtimes & \cdot & \cdot \\
\boxtimes & \boxtimes & \boxtimes & \boxtimes & \cdot \\
\boxtimes & \boxtimes & \boxtimes & \boxtimes & \boxtimes
\end{bmatrix}
+
\begin{bmatrix}
0 & & & & \\
& d_1 & & & \\
& & d_2 & & \\
& & & d_3 & \\
& & & & d_4
\end{bmatrix},
$$

which gives us the desired structure.

The second sequence of Givens transformations

In the previous subsection, in which we discussed the QR-method for semiseparable matrices, we assumed our matrix to be represented with the right left representation. To give also another viewpoint onto the chasing we will assume here our intermediate matrix to be represented again with the top bottom representation. This can easily be obtained by the swapping procedure (see Subsection 1.3.2).

Assume we know the initial disturbing Givens transformation G_6. Based on the knowledge that the resulting matrix will again be of semiseparable plus diagonal form with the diagonal as $\mathrm{diag}([d_1, d_2, d_3, d_4, d_5])$, we will now design the chasing technique.

Initially we have the following situation for the matrix $\hat{S}_4 + \hat{D}_4$:

$$
\begin{array}{l}
\textbf{1} \\
\textbf{2} \\
\textbf{3} \\
\textbf{4} \\
\textbf{5}
\end{array}
\left|
\begin{array}{ccccc}
\times & \cdot & \cdot & \cdot & \cdot \\
 & \times & \cdot & \cdot & \cdot \\
 & & \times & \cdot & \cdot \\
 & & & \times & \cdot \\
 & & & & \times
\end{array}
\right.
+
\begin{bmatrix}
0 & & & & \\
& d_1 & & & \\
& & d_2 & & \\
& & & d_3 & \\
& & & & d_4
\end{bmatrix}.
\qquad (8.7)
$$

$$4 \ 3 \ 2 \ 1$$

We start by performing the initial disturbing orthogonal similarity transformation. Directly applying the transformation would result in a full 2×2 block in the diagonal matrix $\hat{D}_4$. Therefore we would like to change the upper left element into d_1. Unfortunately it is not as easy as in the first paragraph to change the upper left diagonal element of $\hat{D}_4$ into d_1, due to the Givens-vector representation from bottom to top. In order to change the upper left diagonal element, we have to take into consideration the representation. Let us illustrate how the procedure works.

First we rewrite $\hat{D}_4$ as follows:

$$
\begin{bmatrix} 0 & & & & \\ & d_1 & & & \\ & & d_2 & & \\ & & & d_3 & \\ & & & & d_4 \end{bmatrix} = \begin{bmatrix} -d_1 & & & & \\ & 0 & & & \\ & & 0 & & \\ & & & 0 & \\ & & & & 0 \end{bmatrix} + \begin{bmatrix} d_1 & & & & \\ & d_1 & & & \\ & & d_2 & & \\ & & & d_3 & \\ & & & & d_4 \end{bmatrix}.
$$

The diagonal term on the right is of the good form, but now we need to incorporate the matrix $\mathrm{diag}([-d_1,0,0,0,0])$ into the Givens-vector representation of $\hat{S}_4$. Assume that the Givens-vector representation of $\hat{S}_4$ is written as $\hat{S}_4 = Q_4^{(r)} R_4^{(r)}$, where $Q_4^{(r)}$ consists of the Givens transformations and $R_4^{(r)}$ is the upper triangular matrix[18]. We will perform the following operations:

$$
\hat{S}_4 + \mathrm{diag}([-d_1,0,0,0,0]) = Q_4^{(r)} R_4^{(r)} + \mathrm{diag}([-d_1,0,0,0,0])
$$
$$
= Q_4^{(r)} \left(R_4^{(r)} + Q_4^{(r)^T} \mathrm{diag}([-d_1,0,0,0,0]) \right).
$$

We are interested in the structure of the matrix

$$
\tilde{R}_4^{(r)} = \left(R_4^{(r)} + Q_4^{(r)^T} \mathrm{diag}([-d_1,0,0,0,0]) \right). \tag{8.8}
$$

Let us depict graphically this structure. Due to the transpose we obtain now an ascending sequence of Givens transformations instead of a descending one, we obtain for $Q_4^{(r)^T} \mathrm{diag}([-d_1,0,0,0,0])$:

Applying the Givens transformations leads to a matrix of the following form,

$$
\begin{bmatrix} \times & & & & \\ \times & 0 & & & \\ & & 0 & & \\ & & & 0 & \\ & & & & 0 \end{bmatrix}, \tag{8.9}
$$

which needs to be added to the upper triangular matrix $R_4^{(r)}$ (see Equation (8.8)).

---

[18]The superscript (r) is only added to stress that these matrices come from the representation of the matrix considered.

Going back to the original Scheme 8.7, we see that we can rewrite it obtaining the following form $S_4 + D_4 = \hat{S}_4 + \hat{D}_4 = \tilde{S}_4 + \tilde{D}_4$:

$$
\begin{array}{c}
❶ \\ ❷ \\ ❸ \\ ❹ \\ ❺
\end{array}
\left|
\begin{array}{ccccc}
 & + & \cdot & \cdot & \cdot \\
 & + & \times & \cdot & \cdot \\
 & & \times & \cdot & \cdot \\
 & & & \times & \cdot \\
 & & & & \times
\end{array}
\right.
+
\left[
\begin{array}{ccccc}
d_1 & & & & \\
 & d_1 & & & \\
 & & d_2 & & \cdot \\
 & & & d_3 & \cdot \\
 & & & & d_4
\end{array}
\right]
\qquad (8.10)
$$
$$
\overline{ 4\ 3\ 2\ 1 }
$$

In this scheme the two elements marked with $+$ have changed due to the adding of the Matrix (8.9) to the existing upper triangular structure. This new matrix $\tilde{S}_4$ is represented as $\tilde{S}_4 = Q_4^{(r)} \tilde{R}_4^{(r)}$. The matrix $Q_4^{(r)}$ did not change as it contains still the same Givens transformations from the representation of the matrix $\hat{S}_4$. The matrix $R_4^{(r)}$ has changed into $\tilde{R}_4^{(r)}$ and it has a bulge now. In Scheme 8.10, $Q_4^{(r)}$ corresponds to the Givens transformations shown in positions 1 up to 4, and $\tilde{R}_4^{(r)}$ represents the matrix on the right of this scheme.

We are now ready to perform the disturbing initial transformation G_5. We obtain:

$$
\begin{aligned}
G_5^T \left(S_4 + D_4 \right) G_5 &= G_5^T \left(\tilde{S}_4 + \tilde{D}_4 \right) G_5 \\
&= G_5^T \tilde{S}_4 G_5 + \tilde{D}_4 \\
&= S_5 + D_5.
\end{aligned}
$$

The matrix $\tilde{D}_4$ remains unaltered by the similarity transformation. Let us see more clearly the effect of the similarity transformation onto the matrix $\tilde{S}_4$. Applying the transformation G_5 onto the right of the matrix $\tilde{S}_4 = Q_4^{(r)} \tilde{R}_4^{(r)}$ only affects the matrix $\tilde{R}_4^{(r)}$ but will change nothing essential in the structure of the first two columns of this matrix (see Scheme 8.10). The Givens transformation G_5^T applied on the left is depicted in position 5 in Scheme 8.11 (since $\tilde{D}_4$ remains unaltered, the matrix is not shown in the next scheme):

$$
\begin{array}{c}
❶ \\ ❷ \\ ❸ \\ ❹ \\ ❺
\end{array}
\left|
\begin{array}{ccccc}
 & + & \cdot & \cdot & \cdot \\
 & \oplus & \times & \cdot & \cdot \\
 & & \times & \cdot & \cdot \\
 & & & \times & \cdot \\
 & & & & \times
\end{array}
\right.
\qquad (8.11)
$$
$$
\overline{ 5\ 4\ 3\ 2\ 1 }
$$

Before using the Givens transformation in position 5, we will remove the newly created bulge (marked with $\oplus$ in Scheme 8.11) in position $(2,1)$. The annihilation is followed by a fusion. This gives us the following scheme.

Combining the Givens transformations in position 1 and 2 gives us.

Now we can reorder the Givens transformations. The position where to apply the shift through lemma is already depicted.

Applying then the shift through lemma leads to the following situation:

Applying now the Givens transformation in position 1 onto the upper triangular matrix results in the following scheme, which completes the performance of the step related to the initial disturbing Givens transformation.

We remark that all the reshuffling with the Givens transformations did not affect the diagonal matrix D_5, only the Givens-vector representation of the matrix S_5 has changed.

We are now ready to start the chasing procedure for restoring the structure of the semiseparable plus diagonal matrix. To restore the structure, we need to remove the newly created bulge in position $(3, 2)$, similar to the traditional semiseparable case. Before determining the Givens transformation, which will remove the bulge, we will change the diagonal matrix D_5. Completely similar as done before, we rewrite D_5 as follows:

$$D_5 = \begin{bmatrix} 0 & & & & \\ & -d_2 + d_1 & & & \\ & & 0 & & \\ & & & 0 & \\ & & & & 0 \end{bmatrix} + \begin{bmatrix} d_1 & & & & \\ & d_2 & & & \\ & & d_2 & & \\ & & & d_3 & \\ & & & & d_4 \end{bmatrix}.$$

Again we want to incorporate the second term $\mathrm{diag}([0, -d_2 + d_1, 0, 0, 0])$ into the representation of the matrix $S(5) = Q_5^{(r)} R_5^{(r)}$. A similar construction as in the previous case leads to the following form:

$$Q_5^{(r)^T} \mathrm{diag}([0, -d_2 + d_1, 0, 0, 0]) = \begin{bmatrix} 0 & \times & & & \\ & \times & & & \\ & \times & 0 & & \\ & & & 0 & \\ & & & & 0 \end{bmatrix}.$$

Hence rewriting $S_5 + D_5$ into $\hat{S}_5 + \hat{D}_5$ gives us the following scheme:

We choose now the next Givens transformation G_6 such that applying it onto the right of $\hat{S}_5$ annihilates the bulge in position $(3, 2)$, shown in the graphical representation of $\hat{S}_5$. Hence we obtain the following equations:

$$G_6^T (S_5 + D_5) G_6 = G_6^T \left(\hat{S}_5 + \hat{D}_5 \right) G_6$$
$$= G_6^T \hat{S}_5 G_6 + \hat{D}_5$$
$$= S_6 + D_6.$$

Again the diagonal $\hat{D}_5$ does not change. Applying the transformation G_6 onto the right of the matrix $\hat{S}_5$ will remove the bulge. In the following scheme the Givens transformation G_6 applied to the left can be found in position 5:

$$
\begin{array}{c}
\mathbf{1} \\ \mathbf{2} \\ \mathbf{3} \\ \mathbf{4} \\ \mathbf{5}
\end{array}
\left|
\begin{array}{cccccc}
 & & \times & \cdot & \cdot & \cdot & \cdot \\
 & & & \times & \cdot & \cdot & \cdot \\
 & & & & \times & \cdot & \cdot \\
 & & & & & \times & \cdot \\
 & & & & & & \times
\end{array}
\right.
+
\left[
\begin{array}{ccccc}
d_1 & & & & \\
 & d_2 & & & \\
 & & d_2 & & \\
 & & & d_3 & \\
 & & & & d_4
\end{array}
\right].
$$

$$\overline{\quad 5\ \ 4\ \ 3\ \ 2\ \ 1 \quad}$$

The procedure above can now be continued. Reshuffling the Givens transformations makes it possible to apply the shift through lemma, which results in a new bulge created in position $(4, 3)$. To remove the bulge, we first have to rewrite the diagonal matrix D_6, such that the elements in positions 3 and 4 on the diagonal are equal to d_3.

After having applied the similarity transformation G_7 and after rewriting the diagonal and so forth we obtain the following matrix $\hat{S}_7 + \hat{D}_7$ which is graphically represented as follows (even though it is not depicted in the following scheme, all elements in the fourth column change):

$$
\begin{array}{c}
\mathbf{1} \\ \mathbf{2} \\ \mathbf{3} \\ \mathbf{4} \\ \mathbf{5}
\end{array}
\left|
\begin{array}{cccccc}
 & & \times & \cdot & \cdot & \cdot & \cdot \\
 & & & \times & \cdot & \cdot & \cdot \\
 & & & & \times & \cdot & \cdot \\
 & & & & & + & \cdot \\
 & & & & & + & \times
\end{array}
\right.
+
\left[
\begin{array}{ccccc}
d_1 & & & & \\
 & d_2 & & & \\
 & & d_3 & & \\
 & & & d_4 & \\
 & & & & d_4
\end{array}
\right].
$$

$$\overline{\quad 4\ \ 3\ \ 2\ \ 1 \quad}$$

The last Givens transformation G_8 is determined to annihilate the bulge in position $(5, 4)$, when applying it to the right of the matrix $\hat{S}_7$. This gives us the following scheme for the matrix $S_8 + D_8 = G_8^T \left(\hat{S}_7 + \hat{D}_7 \right) G_8$:

$$
\begin{array}{c}
\mathbf{1} \\ \mathbf{2} \\ \mathbf{3} \\ \mathbf{4} \\ \mathbf{5}
\end{array}
\left|
\begin{array}{cccccc}
 & & \times & \cdot & \cdot & \cdot & \cdot \\
 & & & \times & \cdot & \cdot & \cdot \\
 & & & & \times & \cdot & \cdot \\
 & & & & & \times & \cdot \\
 & & & & & & \times
\end{array}
\right.
+
\left[
\begin{array}{ccccc}
d_1 & & & & \\
 & d_2 & & & \\
 & & d_3 & & \\
 & & & d_4 & \\
 & & & & d_4
\end{array}
\right].
$$

$$\overline{\quad 5\ \ 4\ \ 3\ \ 2\ \ 1 \quad}$$

It is already depicted how we can get rid of the Givens transformation G_8^T show in the fifth position. Applying the fusion and rewriting the lower right element of the semiseparable matrix gives us: $\hat{S}_8 + \hat{D}_8$:

$$
\begin{array}{c}
\mathbf{1} \\ \mathbf{2} \\ \mathbf{3} \\ \mathbf{4} \\ \mathbf{5}
\end{array}
\left|
\begin{array}{cccccc}
 & & \times & \cdot & \cdot & \cdot & \cdot \\
 & & & \times & \cdot & \cdot & \cdot \\
 & & & & \times & \cdot & \cdot \\
 & & & & & \times & \cdot \\
 & & & & & & +
\end{array}
\right.
+
\left[
\begin{array}{ccccc}
d_1 & & & & \\
 & d_2 & & & \\
 & & d_3 & & \\
 & & & d_4 & \\
 & & & & d_5
\end{array}
\right].
$$

$$\overline{\quad 4\ \ 3\ \ 2\ \ 1 \quad}$$

The resulting matrix has the desired structure and is the result of performing one step of the implicitly shifted QR-method onto a semiseparable plus diagonal matrix.

Notes and references

The graphical scheme, as mentioned before, was introduced in [53]. An explicit QR-algorithm for general structured rank matrices was considered by Delvaux and Van Barel in [49]. The idea on using the shift through lemma for restoring the disturbed structure, was also exploited in [55], which discusses an implicit QR-algorithm for unitary matrices.

More information on the QR-algorithm for semiseparable plus diagonal matrices can be found in Chapter 7, Section 7.2 and in the article [157].

8.2 Implementation of the QR-algorithm for semiseparable matrices

In this section a mathematical description of the implementation of the implicit QR-algorithm for semiseparable matrices is proposed. The implementation is based on the Givens-vector representation for semiseparable matrices. We will see that this Givens vector representation is useful for developing such QR-methods.

This section contains several subsections. First the QR-method without shift is examined, followed by a numerical decision criterion for rank 1 blocks. Next, the real implementation of the QR-method with shift is discussed, after which again a numerical criterion for deflation is derived.

In the next section the computation of the eigenvectors will be discussed. Moreover, one should be aware that the combination of this algorithm together with the reduction to a symmetric semiseparable matrix gives a tool to compute the eigendecomposition of arbitrary symmetric matrices.

8.2.1 The QR-algorithm without shift

To perform a step of the QR-algorithm, one has to perform first a sequence of Givens transformations from bottom to top on the rows, and at the same time from right to left on the columns. This corresponds to performing a QR-step without shift on the original symmetric semiseparable matrix.

Suppose our semiseparable matrix S is built up with the Givens transformations G and the vector $\mathbf{v}$. We will now perform the first $n-1$ Givens transformations on both sides of the matrix, and we will retrieve the representation of the resulting semiseparable matrix. The matrix S has the following structure:

$$\begin{bmatrix} S^{(n-2)} & c_{n-1}\mathbf{r}_{n-1} & s_{n-1}\mathbf{r}_{n-1} \\ c_{n-1}\mathbf{r}_{n-1}^T & c_{n-1}v_{n-1} & s_{n-1}v_{n-1} \\ s_{n-1}\mathbf{r}_{n-1}^T & s_{n-1}v_{n-1} & v_n \end{bmatrix}, \tag{8.12}$$

where $S^{(n-2)}$ represents a semiseparable matrix of order $n-2$ and the Givens transformations are denoted in the usual manner. We denote the Givens transformations used in the QR-factorization with $\tilde{G}_i$. We remark that the Givens transformations needed for the first step of the QR-algorithm are exactly the same Givens transformations from the representation, more precisely we have the following equivalences $\tilde{G}_1 = G_{n-1}, \ldots, \tilde{G}_{n-1} = G_1$. This is an advantage, because these Givens transformations do not need to be calculated anymore.

Applying the first transformation $\tilde{G}_1^T = G_{n-1}^T$ on the left of Matrix (8.12) gives us the following equations

$$\hat{v}_1 = -s_{n-1}^2 v_{n-1} + c_{n-1} v_n$$

and the matrix looks like:

$$\begin{bmatrix} S^{(n-2)} & c_{n-1}\mathbf{r}_{n-1} & s_{n-1}\mathbf{r}_{n-1} \\ \mathbf{r}_{n-1}^T & v_{n-1} & s_{n-1}(c_{n-1}v_{n-1} + v_n) \\ 0 & 0 & \hat{v}_1 \end{bmatrix}.$$

Applying the transformation on the right gives the following equations:

$$\begin{bmatrix} S^{(n-2)} & \mathbf{r}_{n-1} & 0 \\ \mathbf{r}_{n-1}^T & \tilde{d}_{n-1} & s_{n-1}\hat{v}_1 \\ 0 & s_{n-1}\hat{v}_1 & c_{n-1}\hat{v}_1 \end{bmatrix},$$

with $\tilde{d}_{n-1} = \left(1 + s_{n-1}^2\right) c_{n-1} v_{n-1} + s_{n-1}^2 v_n$. When denoting the new representation from right to left with $\hat{G}$ and $\hat{\mathbf{v}}$, we get:

$$\hat{G}_1 = \begin{bmatrix} c_{n-1} & -s_{n-1} \\ s_{n-1} & c_{n-1} \end{bmatrix}$$

and $\hat{v}_1$. This procedure can be continued to find all the Givens transformations $\hat{G}_i$ and the vector $\hat{\mathbf{v}}$. Note once more that this representation is constructed from right to left and only the new diagonal elements need to be calculated.

8.2.2   The reduction to unreduced form

With the mathematical details given in the previous section, the reader should be able to derive the numerical implementation of the algorithm to transform the matrix into unreduced form. Therefore, we do not include further details of this implementation.

A remaining question that we want to address in detail is the following: when do we assume an element above the diagonal to be includable in the lower semiseparable structure? We will design a proper numerical criterion. Suppose we have a $2 \times n$ matrix. We will assume that this matrix is of rank 1 if the following relation holds, between the smallest σ_2 and the largest singular value σ_1:

$$\sigma_2 \leq \epsilon_M \sigma_1, \tag{8.13}$$

where ϵ_M denotes the machine precision.

Suppose we are performing the reduction as described in Subsection 6.2.2 of Chapter 6 and we perform the following Givens transformation G_{k+1}^T on row k and

$k + 1$ to annihilate row k up to the diagonal: (Only the first $k + 1$ elements are shown.)

$$A = G_{k+1}^T \begin{bmatrix} c_k \mathbf{r}_k^T & c_k v_k & e \\ s_k \mathbf{r}_k^T & s_k v_k & v_{k+1} \end{bmatrix} = \begin{bmatrix} \mathbf{r}_k^T & v_k & b \\ 0 & 0 & \delta \end{bmatrix}.$$

We will assume these two rows to be dependent if the criterion from (8.13) is satisfied. Calculating the singular values of this last $2 \times (k + 1)$ submatrix A, via calculating the eigenvalues of AA^T gives (with $a^2 = v_k^2 + \|\mathbf{r}_k\|_2^2$):

$$\lambda_1 = \frac{a^2 + b^2}{2} + \frac{\delta^2}{2} + \frac{1}{2}\sqrt{a^4 + 2a^2 b^2 - 2a^2 \delta^2 + b^4 + 2b^2 \delta^2 + \delta^4}$$

$$\lambda_2 = \frac{a^2 + b^2}{2} + \frac{\delta^2}{2} - \frac{1}{2}\sqrt{a^4 + 2a^2 b^2 - 2a^2 \delta^2 + b^4 + 2b^2 \delta^2 + \delta^4}.$$

(The efficient calculation of the norm $\|\mathbf{r}_k\|_2^2$ will be addressed in Subsection 8.2.4.) We assume δ to be small (i.e. $\delta^2 \ll a^2 + b^2$), and then λ_1 and λ_2 can be approximated as

$$\lambda_1 \approx a^2 + b^2$$

$$\lambda_2 \approx \frac{\delta^2 a^2}{a^2 + b^2}.$$

In this way we get

$$\sigma_1 \approx \sqrt{a^2 + b^2}$$

$$\sigma_2 \approx \frac{|\delta||a|}{\sqrt{a^2 + b^2}}.$$

Therefore we assume that this matrix is of rank 1 if

$$|\delta||a| \leq \epsilon_M \left(a^2 + b^2 \right).$$

One might consider checking whether the matrix is unreduced rather costly. In fact it is not, because one can check this condition on the fly, i.e., during the performance of a step of the QR-method without shift. As this is always the first step in the performance of a QR-step with shift, this operation is not costly at all. Moreover this has to be checked only once, because this condition is linked to the singularity of the matrix. Therefore only in the first QR-step with shift, during the performance of the initial step without shift one has to check this condition.

8.2.3 The QR-algorithm with shift

As the implicit QR-algorithm with shift also starts with performing a step of the QR-method without shift, we assume that we start with a matrix as given at the end of Subsection 8.2.1. The following part in the algorithm in fact performs the next $n - 1$ Givens transformations on the matrix. It starts with the special Givens transformation G_n. Because the following sequence of Givens transformations will divide the matrix into two semiseparable parts, we have to store twice a symmetric semiseparable matrix. The decreasing lower right semiseparable matrix will be stored in the Givens-vector representation $G, \mathbf{v}$. While the growing upper left part will also be stored in the Givens-vector representation $\hat{G}, \hat{\mathbf{v}}$.

Suppose we first perform the special Givens transformation G_n (see Subsection 7.1.3 Chapter 7 for more details concerning the transformation G_n) on the matrix S_1, which looks like (this is different from the matrix in the previous section, because the representation is from right to left now):

$$
\begin{bmatrix}
v_n & s_{n-1}v_{n-1} & s_{n-1}\mathbf{r}_{n-1} \\
s_{n-1}v_{n-1} & c_{n-1}v_{n-1} & c_{n-1}\mathbf{r}_{n-1} \\
s_{n-1}\mathbf{r}_{n-1}^T & c_{n-1}\mathbf{r}_{n-1}^T & S_1^{(n-2)}
\end{bmatrix}.
\tag{8.14}
$$

Applying the first special Givens transformation G_n^T on the left of the Matrix (8.14), we get:

$$
\begin{bmatrix}
c_nv_n + s_ns_{n-1}v_{n-1} & (c_ns_{n-1} + s_nc_{n-1})\,v_{n-1} & (c_ns_{n-1} + s_nc_{n-1})\,\mathbf{r}_{n-1} \\
-s_nv_n + c_ns_{n-1}v_{n-1} & (-s_ns_{n-1} + c_nc_{n-1})\,v_{n-1} & (-s_ns_{n-1} + c_nc_{n-1})\,\mathbf{r}_{n-1} \\
s_{n-1}\mathbf{r}_{n-1}^T & c_{n-1}\mathbf{r}_{n-1}^T & S_1^{(n-2)}
\end{bmatrix}.
$$

Applying the Givens transformation G_n on the right gives us

$$
\begin{bmatrix}
\hat{v}_1 & \alpha_1 & f_1\mathbf{r}_{n-1} \\
\alpha_1 & \tilde{v}_{n-1} & f_2\mathbf{r}_{n-1} \\
f_1\mathbf{r}_{n-1}^T & f_2\mathbf{r}_{n-1}^T & S_1^{(n-2)}
\end{bmatrix},
$$

with

$$
\begin{aligned}
\hat{v}_1 &= c_n^2v_n + s_n^2c_{n-1}v_{n-1} + 2c_ns_nc_{n-1}v_{n-1}, \\
f_1 &= (c_ns_{n-1} + s_nc_{n-1}), \\
f_2 &= (-s_ns_{n-1} + c_nc_{n-1}), \\
\alpha_1 &= -c_ns_nv_n + \left((c_n^2 - s_n^2)\,s_{n-1} - s_nc_nc_{n-1}\right)v_{n-1}, \\
\tilde{v}_{n-1} &= s_n^2v_n + \left(-2c_ns_ns_{n-1} + c_n^2c_{n-1}\right)v_{n-1}.
\end{aligned}
$$

The lower right reduced semiseparable matrix can be constructed by the old representation and the knowledge of $\tilde{v}_{n-1}$ and the factor f_2.

The upper left 3×3 block can now be used to construct the next Givens transformation according to Theorem 7.2. One can clearly see that this procedure can be repeated to find the new diagonal element $\hat{v}_2$ and the second subdiagonal element α_2, and so on.

Also the new representation is built up at the same time, by storing extra information concerning the values of α_1 and $\tilde{v}_{n-1}$.

Note 8.5. *The implementation as presented here presumes the ideal situation in which the factors f_i are all different from zero. Little changes have to be made if zeros might occur.*

8.2.4 Deflation after a step of the QR-algorithm

An important, yet uncovered topic, is the deflation or cutting criterion. When should we divide the semiseparable matrix into smaller blocks, without losing too much information. For semiseparable matrices two things have to be considered.

The first point of difference with the tridiagonal approach is the fact that an off-diagonal element in the tridiagonal matrix has all the information corresponding to the nondiagonal block in which the element appears. This is straightforward, because all the other elements are zero. This is however not the case for semiseparable matrices; in fact they are dense matrices. This means that we should derive a way to calculate the norms of the off-diagonal blocks in a fast way. Moreover comparing the norms of all the off-diagonal blocks towards the cutting criterion should in total cost not more than $\mathcal{O}(n)$ operations. Otherwise this would be the slowest step in the algorithm, which is unacceptable.

The second issue is whether the norm of the block is small enough to divide the problem into two subproblems. This is a difficult problem and in fact we will test two different cutting criteria and see what the difference in accuracy is. The two cutting criteria, which will be compared in the numerical experiments section, are the aggressive and the normal cutting criterion [94]. The aggressive criterion allows deflation when the norm of the block is relatively smaller than the square root of the machine precision. The normal criterion allows deflation when the norm is relatively smaller then the machine precision. Denoting the machine precision with ϵ_M, we consider the following two deflation criteria: The aggressive:

$$\|S(i+1:n,1:i)\|_F \leq \sqrt{|v_i v_{i+1}|}\sqrt{\epsilon_M}$$

or the normal deflation criterion:

$$\|S(i+1:n,1:i)\|_F \leq \sqrt{|v_i v_{i+1}|}\epsilon_M,$$

where the v_i denote the diagonal elements of the matrix. When the deflation criterion is satisfied, deflation is allowed and the matrix S is divided into two matrices $S(1:i,1:i)$ and $S(i+1:n,i+1:n)$, thereby neglecting the 'almost zero' block $S(i+1:n,1:i)$.

In the remaining part of this section we will derive an order n algorithm to compute the norms of the off-diagonal blocks and to use them in the current cutting criterion. The semiseparable structure should be exploited when calculating these norms. An easy calculation shows that for a semiseparable matrix S with the Givens-vector representation the following equations are satisfied:

$$\|S(2:n,1:1)\|_F = \sqrt{(s_1 v_1)^2}$$

$$\|S(3:n,1:2)\|_F = \sqrt{(s_2 s_1 v_1)^2 + (s_2 v_2)^2}$$

$$= |s_2|\sqrt{(s_1 v_1)^2 + v_2^2}$$

$$= |s_2|\sqrt{\|S(2:n,1:1)\|_F^2 + v_2^2}.$$

This process can be continued and in general we get:

$$\|S(i+1:n,1:i)\|_F = |s_i|\sqrt{\|S(i:n,1:i-1)\|_F^2 + v_i^2}.$$

This formula allows us to derive an $\mathcal{O}(n)$ algorithm to compute and use the norms of these blocks in the actual cutting criterion.

Notes and references

The implementation as it is discussed here can be downloaded from the website related to the book. Even though we did not discuss the implementation for the nonsymmetric eigenvalue problem, nor the implementation for the singular value computation, their implementation is available as well. The implementations for these last two algorithms are based on similar derivations as presented in this section.

8.3   Computing the eigenvectors

Up to now we did not at all discuss the computation of the eigenvectors of semi-separable and related matrices. Here, we will present some methods for computing these. There exist of course various ways to compute the eigenvectors of matrices. We will only discuss some of these methods.

We will distinguish between several cases. If one desires all the eigenvectors or if one only needs a few eigenvectors.

In this section we will discuss the computation of eigenvalues of symmetric semiseparable and related matrices. The ideas on how to compute the eigenvectors for Hessenberg-like and singular vectors for upper triangular semiseparable matrices are similar. We denote the involved matrices in this part with S, but the presented results apply for all structures mentioned above. We will therefore name the matrices structured rank matrices in this part.

8.3.1   Computing all the eigenvectors

If one desires all the eigenvectors of the structured rank matrix S, one can store all the orthogonal transformations performed in the implicit QR-algorithm. In this way we construct the matrix Q, such that:

$$Q^T S Q = \Delta,$$

where Δ is a diagonal matrix containing all the eigenvalues $\Lambda = \{\lambda_1, \ldots, \lambda_n\}$ of the matrix S. This can be rewritten as

$$SQ = Q\Delta.$$

Hence, the columns of the orthogonal matrix Q are the eigenvectors of the matrix S.

8.3.2   Selected eigenvectors

Suppose not all the eigenvectors but just a few of them are desired. Performing or storing all the transformations as described above is too expensive. Instead, we can compute these eigenvectors via inverse iteration. Moreover via this technique we can even enhance the accuracy of the computed eigenvalues.

Suppose we already have a good approximation of the eigenvalues. With inverse iteration applied to the structured rank matrix S and these approximations, the eigenvectors can be computed efficiently.

In fact, if one wants to calculate the eigenvector corresponding to the eigenvalue λ_i, when one knows an approximation $\tilde{\lambda}$ for λ_i, the following system of equations needs to be solved several times:

$$(S - \tilde{\lambda}I)\mathbf{x} = \mathbf{b},$$

with $\tilde{\lambda}$ close to the eigenvalue λ_i. The problem above corresponds to solving a system with, e.g., a semiseparable plus diagonal matrix. In case of a semiseparable plus diagonal matrix, this can be achieved in order $\mathcal{O}(n)$ with the QR-factorization as explained in Chapter 6, Subsection 6.1.1. In case of other structured rank matrices one can also solve this system of equations in an efficient way. Various other methods exist for solving such systems of equations. They are discussed in Volume I of this book.

We can solve the system mentioned above now in a stable and accurate way, and therefore we are able to calculate the eigenvectors of the structured rank matrix via inverse iteration. In the numerical tests the following algorithm from [142] is used. This algorithm is often called the Rayleigh quotient iteration.

Algorithm 8.6 (Inverse iteration).
Input: *A structured rank matrix S, a random vector $\mathbf{v}$ and an approximate eigenvalue λ. Output: Normalized eigenvector $\mathbf{v}$, corresponding to the updated eigenvalue λ.*

While (not good_approximation) do

1. *Solve the system $(S - \lambda I)\mathbf{y} = \mathbf{v}$;*

2. $\hat{\mathbf{v}} = \mathbf{y}/\|\mathbf{y}\|_2$;

3. $\mathbf{w} = \mathbf{v}/\|\mathbf{y}\|_2$;

4. $\rho = \hat{\mathbf{v}}^T\mathbf{w}$; *(Rayleigh quotient)*

5. $\mu = \lambda + \rho$;

6. $\lambda = \mu$;

7. $\mathbf{r} = \mathbf{w} - \rho\hat{\mathbf{v}}$;

8. $\mathbf{v} = \hat{\mathbf{v}}$;

9. *good_approximation$= (\|\mathbf{r}\|_2/\|S\|_F \leq \epsilon)$;*

endwhile;

Starting with a random vector $\mathbf{v}$ and the shift λ equal to an approximation of an eigenvalue coming from the implicit QR-algorithm, the eigenvectors can be calculated in a fast and accurate way.

It is obvious that computing the eigenvectors via inverse iteration is much faster than via the first technique. Unfortunately, computing the eigenvectors one by one using this technique, might lead to loss of orthogonality, when the eigenvalues are clustered. In the next section a technique is discussed to overcome this problem.

8.3.3   Preventing the loss of orthogonality

In principle, the whole set of eigenvectors of the semiseparable, semiseparable plus diagonal, tridiagonal and other matrices can be computed by means of inverse iteration in $O(n^2)$ operations. The disadvantage in this approach is that the computed eigenvectors may not be numerically orthogonal if clusters are present in the spectrum. To enforce orthogonality the Gram-Schmidt procedure can be used, requiring $O(n^3)$ operations in the worst case.

In this section we describe a numerical method to compute the set of the eigenvectors numerically orthogonal. The method requires $O(n^2)$ floating point operations if few clusters of small size are present in the spectrum. The method is based on the fact that the orthogonal matrices generated by one step of the QR method applied to symmetric tridiagonal matrices and symmetric semiseparable ones are highly structured. For instance, if the matrix is symmetric tridiagonal, the orthogonal matrix generated by one step of the QR method is an upper Hessenberg semiseparable one. Therefore, the product of the latter matrices by a vector can be accomplished in $O(n)$ floating point operations, where n is the size of the involved matrix.

Some notation

Suppose we would like to compute the eigenvalue decomposition of a matrix S, which can be either semiseparable, semiseparable plus diagonal, tridiagonal and so forth, as long as one can easily solve the system of equations involving this matrix it is fine. At each step of the algorithm we obtain a matrix satisfying the same structural constraints. Suppose a step of the QR-method is performed onto the matrix $S^{(i)}$, with shift μ, using the orthogonal matrix $Q^{(i)}$:

$$S^{(i+1)} = Q^{(i)^T} S^{(i)} Q^{(i)}.$$

Of key importance in the forthcoming algorithm is the structure of the matrix $Q^{(i)}$. E.g., in the case of a tridiagonal matrix T, the matrix $Q^{(i)}$ will be of Hessenberg form, having the upper triangular part of semiseparable form. More precisely the matrix $Q^{(i)}$ will be an orthogonal Hessenberg matrix.

Exploiting the structure of $Q^{(i)}$ the product by a vector can be accomplished in $\mathcal{O}(n)$ flops, for all considered cases such as semiseparable, semiseparable plus diagonal and so forth. Of course, $Q^{(i)}$ does not need to be explicitly computed. It is sufficient to store the sine and cosine coefficients s_i and c_i, $i = 1, \ldots, n-1$, of the performed Givens transformations into two vectors, respectively.

From a theoretical point of view, it turns out that the off-diagonal norm $S^{(1)}(n, 1 : n-1) = 0$ if the chosen shift $\mu = \lambda_i$, i.e., if μ is one of the eigenvalues of $S^{(0)}$. Such a μ is called a perfect shift [129]. Moreover, the last column of the matrix $Q^{(0)}$ is the normalized eigenvector of $S^{(0)}$ associated to λ_i. Therefore, if the eigenvalue λ_i of $S^{(0)}$ is known the corresponding eigenvector can be computed applying one step of the implicitly shifted QR method, choosing the shift equal to λ_i. Exploiting then the structure of $Q^{(0)}$ the corresponding eigenvector is computed in $\mathcal{O}(n)$ floating point operations.

Unfortunately, if the last entries of the eigenvector are very tiny, then the off-diagonal norm $S^{(1)}(n, 1 : n-1)$ could not be small (see [129]) and the last column of $Q^{(0)}$ could not be parallel to the eigenvector associated to λ_i.

By means of iterative refinement as described in the previous subsection, one can overcome this problem.

Iterative refinement

Let us show how to compute the eigenvector of $S^{(0)}$ associated to λ_i by using the implicitly shifted QR approach.

The idea is to iterate the QR method using λ_i as shift. Due to the cubic convergence of the implicitly shifted QR method, after few iterations, let's say l iterations, the off-diagonal norm becomes smaller than a given tolerance τ_1: $\|S^{(l)}(n, 1 : n-1)\| < \tau_1$, where τ_1 is in fact the deflation criterion.

The corresponding eigenvector of $S^{(0)}$ associated to λ_i is then given by

$$Q^{(1)}Q^{(2)} \ldots Q^{(l-1)}Q^{(l)}\mathbf{e}_n. \tag{8.15}$$

We observe that the product $Q^{(1)}Q^{(2)} \ldots Q^{(l-1)}Q^{(l)}$ is not explicitly computed. In fact, it is necessary only to store the sine and cosine coefficients generated at each step of the method.

Furthermore, exploiting the structure of $Q^{(j)}$, $j = 1, \ldots, l$, the corresponding eigenvector is computed in $\mathcal{O}(ln)$ floating point operations. On average, very few steps of iterative refinement are sufficient to compute the eigenvector up to the considered tolerance.

Numerical multiple eigenvalues

At first sight, the method just presented also suffers from the loss of orthogonality in the presence of clusters in the spectrum. In this section we show that the latter drawback is easily overcome by an adaptation of the previous method. Suppose the eigenvalues are ordered such that

$$\lambda_i \le \lambda_j, \ i < j.$$

The eigenvalue λ does not belong to a cluster, if the following condition is satisfied [62]: The eigenvalue λ should have an adequate separation from the rest of the spectrum, i.e.,

$$\text{gap}(\lambda) = \min_{\lambda \neq \mu} |\lambda - \mu| \geq \tau_2,$$

where μ ranges over the other eigenvalues of $S^{(0)}$.

A good choice of τ_2 is $10^{-3} \|S^{(0)}\|_\infty$, [62], [185, p. 322]. Therefore, we will say that two eigenvalues λ and μ of $S^{(0)}$ belong to the same cluster if

$$|\lambda - \mu| < \tau_2.$$

Without loss of generality, we suppose that a cluster of two eigenvalues is present in the spectrum, i.e., $\lambda_i \approx \lambda_{i+1}$, $1 \leq i < n$.

The algorithm to compute the eigenvectors corresponding to the eigenvalues belonging to the cluster is depicted in Algorithm 8.7. The algorithm can be easily extended to a cluster of larger size.

We will briefly explain the flow of the method. We assume below that a cluster of two eigenvalues is present, hence we start iterating with one of the two eigenvalues, say λ_{i+1}, until convergence occurs. Because $\lambda_i \approx \lambda_{i+1}$, we obtain either convergence to a single eigenvalue in the lower right corner or to a block of size 2×2, containing both eigenvalues λ_i and λ_{i+1}. Depending on the case two possible actions are performed. In case of convergence to a single eigenvalue, we perform deflation and continue performing the QR-method onto the deflated matrix. Important to remark is that for computing the eigenvector for the eigenvalue λ_i, we will make use now of the information gathered from eigenvalue λ_{i+1} and we will update this eigenvector. In case of convergence to a 2×2 block we compute the eigenvalue decomposition of this block, which we use for accurate computations of the final eigenvectors.

Algorithm 8.7 (Compute all eigenvectors).
Input: *A matrix S and its eigenvalues λ_i.*
Output: *Matrix of eigenvectors V.*

While (not all eigenvectors are computed) do

1. *Iterate the QR-method with implicit shift λ_{i+1} until $|\beta_j^{(l)}| < \tau_1$, for some l and $n - 2 \leq j \leq n - 1$, and $\beta_j^{(l)} = \|S^{(l)}(j + 1 : n, 1 : j)\|$.*

2. *Let $Q_{l:1} = Q^{(l)} Q^{(l-1)} \cdots Q^{(1)}$.*

3. *If $j = n - 1$, then $Q_{l:1} \mathbf{e}_n$ is a good approximation for the eigenvector of $S^{(0)}$ corresponding to λ_{i+1}.*

   (a) *Deflate $S^{(l)}$ (delete the last row and the last column).*

   (b) *Let $\tilde{S}^{(0)} = S^{(l)}(1 : n - 1, 1 : n - 1)$ and iterate the QR-method with implicit shift λ_i onto this matrix. Repeat this until $|\tilde{\beta}_{n-2}^{(k)}| < \tau$, for some k.*

(c) Let $\tilde{Q}_{k:1} = \tilde{Q}^{(k)} \tilde{Q}^{(k-1)} \cdots \tilde{Q}^{(1)}$ and embed the matrix $\tilde{Q}_{k:1}$ into one of size $n \times n$:

$$\hat{Q}_{k:1} = \begin{bmatrix} \tilde{Q}_{k:1} & \\ & 1 \end{bmatrix},$$

(d) $\hat{Q}_{k:1} Q_{l:1} \mathbf{e}_{n-1}$ is the eigenvector of $S^{(0)}$ corresponding to λ_i.

4. Else (if $j = n - 2$)

(a) $Q_{l:1}[\mathbf{e}_{n-1}, \mathbf{e}_n]$ is the subspace spanned by the eigenvectors of $S^{(0)}$ associated to the eigenvalues λ_i and λ_{i+1}.

(b) Let

$$\check{V}^T \begin{bmatrix} \lambda_i & \\ & \lambda_{i+1} \end{bmatrix} \check{V}$$

be the spectral decomposition of $S^{(0)}(n-1:n, n-1:n)$.

(c) Let

$$V = \begin{bmatrix} I_{n-2} & \\ & \check{V} \end{bmatrix}.$$

(d) $V Q_{l:1} \mathbf{e}_{n-1}$ and $V Q_{l:1} \mathbf{e}_n$ are the eigenvectors of $S^{(0)}$ associated to the eigenvalues λ_i and λ_{i+1}.

endwhile;

Note 8.8. *If the size of the cluster is k, approximately $O(k^3 n)$ floating point operations are needed to compute the eigenvectors. In many cases, however, more than one row and column can be deflated from the matrix reducing considerably the computational complexity.*

It is clear that Algorithm 8.7 is suitable for computing orthogonal eigenvectors in case of tridiagonal, semiseparable and semiseparable plus diagonal matrices, due to the efficient multiplication between the matrices $Q^{(i)}$ and a vector.

8.3.4 The eigenvectors of an arbitrary symmetric matrix

Suppose our structured rank matrix S is the result of the orthogonal similarity transformations as described in Part I, applied to an arbitrary matrix A and one is interested in the eigenvectors of the original matrix. We compute via one of the techniques described above the eigenvectors of the matrix S. These eigenvectors need to be transformed to the original eigenvectors.

The most natural way to achieve this goal is to store the orthogonal transformations performed while reducing the matrix A into the matrix S (e.g., the reduction to tridiagonal, band, Hessenberg and so forth). In case of the reduction to semiseparable form, this corresponds to keeping either a sequence of $n-1$ Householder transformations plus $n(n-1)/2$ Givens transformations, or keeping (when using Givens transformations instead of Householder transformations) $n(n-1)$ Givens transformations.

Notes and references

A more elaborate study, focused on tridiagonal matrices on how to apply inverse iteration for clusters of eigenvalues can be found in the following article.

☞ N. Mastronardi, M. Van Barel, E. Van Camp, and R. Vandebril. On computing the eigenvectors of a class of structured matrices. *Journal of Computational and Applied Mathematics*, 189:580–591, 2006.

The article contains several numerical examples illustrating the effectiveness of this approach.

More information can also be found in the following articles by Ipsen and Dhillon.

☞ I. S. Dhillon. Current inverse iteration software can fail. *BIT*, 38:685–704, 1998.

☞ I. C. F. Ipsen. Computing an eigenvector with inverse iteration. *SIAM Review*, 39(2):254–291, 1997.

8.4   Numerical experiments

In this section several numerical experiments related to the algorithms developed in this part will be given. In several of the experiments the algorithms from the first part of the book are combined with the methods from this part to obtain algorithms, able to compute eigenvalues and/or singular values of arbitrary matrices.

First we will discuss some examples concerning the computation of eigenvalues, followed by numerical experiments discussing the singular value computations.

8.4.1   On the symmetric eigenvalue solver

In this section several numerical tests are performed to compare the traditional algorithm for finding all the eigenvalues with the new semiseparable approach. The algorithm is based on the QR-step as described in Subsection 8.2.1 and Subsection 8.2.3 and implemented in a recursive[19] way: if division in blocks is possible (i.e., that the deflation criterion is satisfied) because of the convergence behavior, then these blocks are dealt with separately.

Before starting the numerical tests some remarks have to be made: first of all the complexity of the reduction of a symmetric matrix into a similar semiseparable one costs $9n^2 + \mathcal{O}(n)$ flops more than the reduction of a matrix to tridiagonal form. An implicit QR-step applied to a symmetric tridiagonal matrix costs $31n$ flops while it costs $\approx 10n$ flops more for a symmetric semiseparable matrix. However, this increased complexity is compensated when comparing the number of iteration steps the traditional algorithm needs with the number of steps the semiseparable algorithm needs. Figures about these results can be found in the following tests.

---

[19]The algorithm does not need to be implemented recursively. It is better to maintain a list of indices marking the beginning and end of every diagonal block for which one needs to compute the eigenvalues. The eigenvalues of these blocks are then computed one after another by means of the QR-algorithm.

Experiment 8.9 (The block experiment for symmetric matrices). *This experiment is taken from [129, p. 153]. Suppose we have a symmetric matrix $A^{(0)}$ of dimension n and we construct the following matrices $T(m, \delta)$ using $A^{(0)}$, where m denotes the number of blocks, and δ are the small subdiagonal elements, between the blocks. For example:*

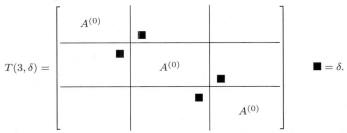

We get the following results: For $n = 10$ we compare the maximum number of iterations for any eigenvalue to converge for the semiseparable and the tridiagonal approach. This is done for a varying size of δ. $A^{(0)}$ has eigenvalues $1 : 10$. Both of the algorithms use the same normal deflation criterion.

The difficulty in this experiment are the small off-diagonal elements. The reduction to the tridiagonal form will leave these elements quite small. In this way our resulting tridiagonal matrix is almost a block diagonal. The algorithm we perform now will not deflate the blocks. In this way we can see the difficulties the implicit QR-algorithm encounters, trying to solve this problem. The semiseparable approach will perform better, because the reduction to semiseparable form, will already rearrange these small elements, because of the QL-steps performed while reducing these matrices. In this way the overall number of steps before convergence can occur is less than the corresponding number of steps using the tridiagonal approach. For

$m = 10$:

δ	10^{-13}	10^{-12}	10^{-11}	10^{-10}	10^{-9}	10^{-8}	10^{-7}
semiseparable QR	*3*	*3*	*2*	*2*	*2*	*3*	*3*
tridiagonal QR	*4*	*4*	*4*	*4*	*4*	*4*	*3*

For $m = 25$:

δ	10^{-15}	10^{-14}	10^{-13}	10^{-12}	10^{-11}
semiseparable QR	*4*	*4*	*4*	*3*	*3*
tridiagonal QR	*4*	*5*	*5*	*4*	*4*
δ	10^{-10}	10^{-9}	10^{-8}	10^{-7}	
semiseparable QR	*3*	*2*	*2*	*2*	
tridiagonal QR	*3*	*4*	*2*	*3*	

For $m = 40$

δ	10^{-19}	10^{-18}	10^{-17}	10^{-16}	10^{-15}	10^{-14}	10^{-13}
semiseparable QR	4	5	4	4	4	4	4
tridiagonal QR	4	5	4	5	5	4	4
δ	10^{-12}	10^{-11}	10^{-10}	10^{-9}	10^{-8}	10^{-7}	
semiseparable QR	3	1	1	1	3	3	
tridiagonal QR	4	5	4	4	4	5	

It can be clearly seen that the tridiagonal approach has more difficulties in finding particular eigenvalues. Figure 8.1 gives a comparison in the complete number of QR-steps for the last experiment ($m = 40$).

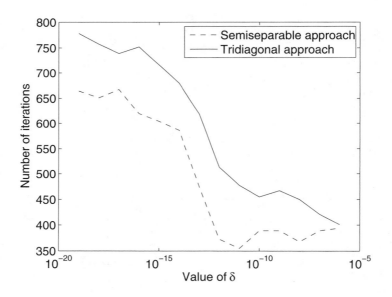

Figure 8.1. *Total number of steps compared to several values of δ.*

The figure shows that the semiseparable approach needs less iterations than the traditional approach. The matrices involved are of size 400. It can be seen that for δ in the neighborhood of 10^{-10} the number of QR-steps with the semiseparable approach are even less than 400. This can also be seen in the table for $m = 40$.

Experiment 8.10 (Deflation). *In Figure 8.2, we compared the accuracy of the eigenvalues, depending on the deflation criterion. A sequence of matrices of varying sizes was generated, with equally spaced eigenvalues in the interval $[0, 1]$. The* *eigenvalues for these matrices were calculated by using the aggressive and the normal deflation criterion. For both tests the absolute error of the residuals was computed and plotted in the next figure. One can clearly see that applying the aggressive deflation criterion is almost as good as the normal deflation criterion. This is due to the convergence behavior of the reduction to semiseparable form. The reduction creates some kind of gradedness in the matrix.*

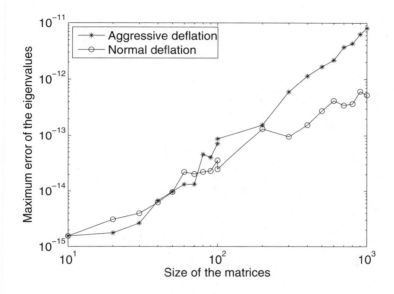

Figure 8.2. *Comparing different deflation criteria.*

8.4.2 Experiments for the singular value decomposition

In this section we perform tests on some arbitrary matrices to compute their singular values. First the matrices are reduced to upper triangular semiseparable form, and then the implicit QR-algorithm to compute the singular values is applied on these matrices.

Numerical tests are performed comparing the behavior of the traditional approach via bidiagonal and the proposed semiseparable approach for computing the singular values of several matrices. Special attention is paid to the accuracy of both algorithms for different sets of the singular values and to the number of QR-steps needed to compute all the singular values of the matrices.

Experiment 8.11. *Figure 8.3 shows a comparison of the number of implicit QR-iterations performed (the line connects the average number of iterations for each experiment). It shows that the semiseparable approach needs less steps in order to find the singular values but do not forget that a semiseparable implicit QR-step costs* *a little more than a corresponding step on the bidiagonal.*

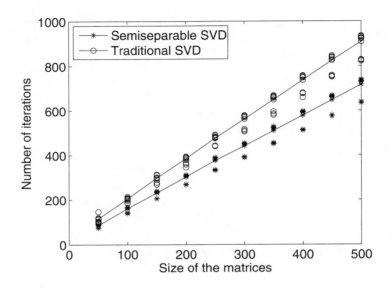

Figure 8.3. *Number of implicit QR-steps.*

Experiment 8.12. *Figure 8.4 shows comparisons in accuracy of the two approaches. (A line is drawn connecting the average error for different experiments. The error is the maximum relative error between the computed and the exact singular values.) In the left figure of Figure 8.4 the singular values were chosen equally spaced in $[0.1]$, and in the right figure, the singular values ranged from 1 to n. Both figures show that the two approaches are equally accurate.*

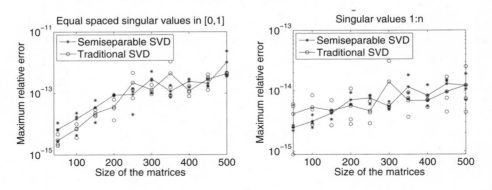

Figure 8.4. *Comparison in accuracy.*

8.5 Conclusions

In this chapter we first discussed another interpretation of the chasing technique based on the Givens-vector representation. This new technique was discussed for both the semiseparable and the semiseparable plus diagonal case.

The mathematical details behind the implementations of the different implicit QR-algorithms and the reduction to unreduced form were provided. Moreover a cutting criterion was also provided for the division in blocks and a mathematical investigation deciding whether a block is of rank 1. Briefly the computation of eigenvectors was mentioned. Finally some numerical tests were performed using the implicit QR-algorithm for semiseparable matrices. We tested the algorithm on an "almost" block matrix and compared the results with the algorithm based on tridiagonal matrices. We also tested the algorithm on a matrix with the eigenvalues in stair-form and we showed that the diagonal elements already approximated quite well the eigenvalues after the reduction to semiseparable form. We received accurate results. Some experiments were performed comparing the number of steps and the accuracy of the tridiagonal and the semiseparable approach. Also experiments concerning the computation of the singular values were presented.

In the upcoming part we will continue exploiting the results of the first and second part of this book. A Lanczos-like process for making the matrix will be given, followed by a rank-revealing method based on the reduction to semiseparable form. Also a divide-and-conquer method for computing the eigenvalues of semiseparable plus diagonal matrices will be presented. The last chapter of this part discusses the multishift algorithm, an implicit QH-algorithm, an implicit algorithm on unitary plus rank one matrices as well as some interesting viewpoints on some of the discussed QR-algorithms for semiseparable matrices.

Chapter 9

More on QR-related algorithms

In the previous chapters, the basic traditional QR-algorithms for structured rank matrices were designed. The singular value decomposition and single shift QR-methods for semiseparable, semiseparable plus diagonal and Hessenberg-like matrices were explored.

In this chapter we will deal with some interesting topics that are more general than the ones discussed in the previous chapters of this part. It covers generalizations, other viewpoints and QR-algorithms for related structured rank matrices. This chapter covers the multishift technique. It will be shown that also different QR-algorithms related to different types of QR-factorizations are available. The literature for computing eigenvalues of companion and related matrices will be discussed. Finally another iteration for computing the eigenvalues of structured rank matrices will be presented.

The first section contains some important remarks related to the use of complex Givens transformations. Moreover also some important remarks related to 2×2 unitary matrices, Givens transformations and the shift through operation will be made.

Section 9.2 discusses some variants of the QR-algorithm. The QR-factorization of structured rank matrices can be computed in different manners: the $\vee$- and the $\wedge$-pattern. Since the QR-algorithm is based on the QR-factorization there are also different variants for performing the QR-algorithm.

Throughout the book, we did not yet discuss the class of quasiseparable matrices, even though, it is studied quite a lot. Section 9.3 discusses an effective implicit QR-algorithm for the class of symmetric quasiseparable matrices of quasiseparability rank 1. The idea is simple and rather straightforward. A quasiseparable matrix can be considered as a block diagonal matrix, for which all blocks are either of semiseparable plus diagonal or of tridiagonal form. Hence combining both QR-methods in a good way solves the problem.

All QR-algorithms discussed before were restricted to the single shift case. Unfortunately, in the case of a real Hessenberg-like matrix having complex eigenvalues, this means that one has to switch to complex arithmetic. This increases the computational cost. In Section 9.4 we will adapt the single shifted QR-algorithm for semiseparable and/or Hessenberg-like matrices towards the multishift setting. This makes it possible to use, e.g., the double shift version, thereby restricting the computations to real arithmetic in case of a real Hessenberg-like matrix. Subsection 9.4.1 introduces some basic ideas about multishift algorithms. It is shown that a single column of the involved unitary transformation determines the complete similarity transformation. Subsection 9.4.2 shows how to compute and how to manipulate this vector in the Hessenberg-like case. We will see that more disturbing Givens transformations will be necessary now, instead of one in the single shifted version. Hence the chasing technique needs to be adapted. The new chasing approach is discussed in Subsection 9.4.3.

Section 9.5 discusses a new kind of iteration for computing the eigenvalues of structured rank matrices. The iteration is called a QH-iteration, since in every step of the algorithm an orthogonal Hessenberg-like factorization is computed. The Q-factor of this factorization is then used for determining the similarity transformation that is performed on the original structured rank matrix. It will be shown that this method uses only half the number of Givens transformations in the standard QR-method and moreover it converges faster.

Exploiting the rank structure for developing effective QR-methods is essential in several topics. In Section 9.6 some results will be presented on specific structured rank matrices. Firstly the relation between roots of polynomials and structured rank matrices will be explored. Companion, comrade, confederate and other types of specific matrices will be defined. It will be shown that the eigenvalues of these matrices correspond with the roots of polynomials expressed in a certain basis. Further a QR-step for these types of matrices will be developed. Since unitary matrices are involved in these computations, the QR-algorithm for unitary Hessenberg matrices is also presented.

It is impossible to cover all topics related to eigenvalue and singular value computations in a single book. Hence we will briefly discuss some remaining topics in Section 9.7. The relation between rational Krylov matrices and semiseparable plus diagonal matrices will be discussed. It is shown how Sturm sequences and the bisection method can be used for computing eigenvalues of some specific structured rank matrices. Finally some other references are discussed.

✎ *Even though this chapter is the last chapter of this part it might contain the most interesting results of this part. This chapter covers extensions of the previously discussed methods, new techniques and applications of the gained knowledge to compute for example the eigenvalues of companion matrices. One can, however, also skip this chapter and continue with the next part.*

Since we have not yet discussed complex arithmetic it is a good idea to read the notes related to Givens transformations in Section 9.1.

Section 9.2 is an important section, especially the first and last subsection discussing the different QR-factorizations and the different related QR-algorithms.

Readers interested in quasiseparable matrices might want to check out Sec-

tion 9.3. The results are a simple combination of standard QR-algorithms for tridiagonal and semiseparable matrices. To understand this section one needs to read Subsection 9.2.2.

Section 9.4 discusses the extension of the single shift case to the multishift case. This might be interesting for readers wanting to compute eigenvalues of real Hessenberg-like matrices. By using a double shifted QR-method they can restrict computations to the real field.

The section discussing the QH-method is a very important one. It presents the fastest and most accurate QR-like algorithm for computing eigenvalues of some structured rank matrices.

The results presented in Section 9.6 are also very interesting. Efficient algorithms for computing the eigenvalues of unitary, companion and fellow matrices are presented, thereby exploiting the rank structure of the matrices involved.

The references discussed in Section 9.7 denote interesting results related to structured rank matrices. Moreover these results show current research projects as well as interesting relations that were not discussed in the book. The reader can however easily skip this section.

9.1 Complex arithmetic and Givens transformations

Throughout the book, we focused mostly on real number arithmetic, even though most results are also valid for complex numbers. Since, however, even real nonsymmetric matrices can have complex eigenvalues and this problem will be addressed in this chapter, we will switch here to complex arithmetic.

Before discussing the different topics, we will briefly comment on some issues related to complex arithmetic.

A complex Givens transformation (often called Givens rotation) is defined as follows:

$$G = \left[\begin{array}{cc} c & -\bar{s} \\ s & \bar{c} \end{array} \right],$$

in which c and s satisfy $|c|^2 + |s|^2 = 1$. It is easy to check that this transformation is unitary, i.e., $G^H G = G G^H = I$.

Unfortunately a Givens transformation is not uniquely determined. It is mostly chosen to annihilate a specific element in a vector having two nonzero elements. Suppose our Givens transformation is chosen such that the following equality is satisfied:

$$\left[\begin{array}{cc} c & -\bar{s} \\ s & \bar{c} \end{array} \right] \left[\begin{array}{c} x \\ y \end{array} \right] = \left[\begin{array}{c} r \\ 0 \end{array} \right].$$

This Givens transformation, annihilating the element y, is not uniquely determined. In fact replacing c, s and r by ωc, ωs and ωr, with $|\omega| = 1$, will also satisfy the above equation. More information on this subject can be found in [20]. There are of course several possibilities for choosing the ω. Many of them are discussed in [20].

Complex arithmetic introduces also new terms, such as unitary, hermitian conjugate and so forth.

Note 9.1. *Throughout the complete book we used mostly Givens transformations for building for example the Givens-vector representation. In fact one can easily relax this constraint and use also 2×2 unitary transformations. There is no loss in generality since the shift through lemma and the fusion remain valid. A unitary matrix has more parameters, but it might sometimes be more convenient to work with a unitary matrix instead of a Givens transformation.*

Notes and references

A detailed analysis on how to compute Givens transformations in a reliable way is discussed extensively in the following article:

☞ D. Bindel, J. W. Demmel, W. Kahan, and O. A. Marques. On computing Givens rotations reliably and efficiently. *ACM Transactions on Mathematical Software*, 28(2):206–238, June 2002.

9.2   Variations of the QR-algorithm

All QR-algorithms discussed before were based on the QR-factorization presented in Subsection 6.1.1. This QR-factorization of a (rank 1) semiseparable (Hessenberg-like) matrix plus a diagonal consists of $2n - 2$ Givens transformations [156]. A first sequence of Givens transformations from bottom to top transforms the semiseparable (Hessenberg-like) plus diagonal matrix into a Hessenberg matrix, whereas the second sequence of transformations will bring the Hessenberg matrix to upper triangular form.

Correspondingly, all (implicit) QR-algorithms connected to this type of QR-factorization can also be decomposed into two steps. A first step performs a similarity transformation involving $n - 1$ Givens transformations and corresponds to performing a step of the QR-method without shift. Secondly a disturbance is introduced and $n - 2$ Givens transformations are needed for restoring the structure.

In this section we will first briefly restate the existence of other types of QR-factorizations and correspondingly other types of QR-algorithms. An extensive study on the other factorizations, thereby using so-called rank-expanding Givens transformations can be found in the first volume [169].

9.2.1   The QR-factorization and its variants

For simplicity we will discuss Hessenberg-like plus diagonal matrices. Semiseparable plus diagonal matrices can be considered in the same way. Since the QR-algorithms are closely related to the QR-factorization, we will depict here the two main QR-factorizations graphically. The standard factorization contains some repetition of previously discussed results.

The traditional factorization: ∧-pattern

For this type of QR-factorization firstly an ascending sequence of Givens transformations, followed by a descending sequence of Givens transformations, is applied onto the Hessenberg-like plus diagonal matrix $Z + D$. The first sequence of Givens transformations acting on $Z + D$, denoted by $\hat{Q}_1^H$, consists of $n-1$ Givens transformations in which each Givens transformation acts on two successive rows of the matrix Z thereby exploiting the rank structure in the lower triangular part to annihilate all elements below the diagonal (these unitary transformations coincide with the ones from the top to bottom representation). We obtain

$$\hat{Q}_1^H Z = R \quad \text{and} \quad \hat{Q}_1^H(Z + D) = H,$$

in which H is a Hessenberg matrix. This is followed by a second sequence of $n-1$ Givens transformations from top to bottom for annihilating the subdiagonal elements of the matrix H. This gives

$$\hat{Q}_2^H H = \hat{Q}_2^H \hat{Q}_1^H(Z + D) = \hat{Q}^H(Z + D) = \hat{R},$$

in which $\hat{R}$ is the resulting upper triangular matrix.

This matrix product $\hat{Q}^H(Z + D)$ is graphically represented as follows:

$$\tag{9.1}$$

The name $\wedge$-pattern originates from the pattern of the Givens transformations for annihilating the lower triangular part in the matrix. To indicate that the Givens transformations are coming from the $\wedge$-pattern we denote them with a $\hat{\ }$.

We recall that the right part consisting of $\times$ and $\boxtimes$ represents the matrix $Z + D$. The elements $\boxtimes$ denote the part of the matrix satisfying the rank structure. The other elements $\times$ denote arbitrary elements. In this scheme the elements on the diagonal are not includable in the rank structure because they are perturbed by the diagonal D. The left part, consisting of the brackets with arrows, denote the Givens transformations. (See Chapter 4 for a more detailed explanation of these schemes.)

In the scheme above, the Givens transformations from columns 1 up to 4 represent the Givens transformations in the matrix $\hat{Q}_1^H$. The ones in the columns 5 up to 8 denote these of the matrix $\hat{Q}_2^H$.

Applying the Givens transformations in positions 1 up to 4 removes the strictly lower triangular part of the matrix. As a result we obtain the following scheme. This represents exactly the same matrix as in the previous scheme but equals now $\hat{Q}_2^H H$.

$$\tag{9.2}$$

Applying the remaining four Givens transformations in Scheme 9.2 onto the Hessenberg matrix on the right will remove the remaining subdiagonal elements. Hence we obtain the upper triangular matrix $\hat{R}$.

Due to some specific properties of Givens transformations, we can obtain other patterns.

The ∨-pattern

The transition from the ∨ to the ∧-pattern was discussed extensively in Volume I. As we will need this operation quite often in the upcoming sections we will briefly illustrate how it is done.

We start with Scheme 9.1 by applying a fusion between the Givens transformations in positions 4 and 5. Since we will only change the order of the Givens transformations we will not depict the semiseparable plus diagonal matrix $Z + D$ on the right of the following schemes.

After having applied the shift through operation we obtain the next scheme. Next we apply the shift through lemma at positions $6, 5$ and 4 (this is already depicted).

Rearranging slightly the position of the Givens transformations, we can again re-apply the shift through lemma.

Continuing to apply the shift through lemma gives us another pattern.

Continuing this procedure by applying the shift through lemma one more time, presents a ∨-pattern that we can use again to obtain a QR-factorization of the involved Hessenberg-like plus diagonal matrix $Z + D$.

$$(9.3)$$

This pattern can also be decomposed into two parts. First a descending sequence of Givens transformations $\check{Q}_1^H$ (position 1 up to 3) is applied followed by an ascending sequence of Givens transformations $\check{Q}_2^H$ (position 4 up to 7). The order of the Givens transformations has changed, but we compute the same factorization (more information can be found in [169]):

$$\check{Q}_2^H \check{Q}_1^H (Z + D) = \hat{R}.$$

Note 9.2. *To distinguish between the $\vee$ and the $\wedge$-pattern we put a $\vee$ on top of the unitary transformations in case of the $\vee$-pattern.*

Some important remarks related to both the $\vee$ and $\wedge$-patterns have to be made.

- *We have the following equality*

$$\check{Q}_1 \check{Q}_2 = \hat{Q}_1 \hat{Q}_2,$$

since $\hat{R}$ was not affected, we obtain an identical QR-factorization.

- *But we also have generically:*

$$\check{Q}_2 \neq \hat{Q}_2$$
$$\check{Q}_1 \neq \hat{Q}_1,$$

which means that the factorization of the unitary matrix in the QR-factorization is different in both patterns.

The first three Givens transformations are in fact rank expanding Givens transformations. They lift up the rank structure. Hence after having applied these first Givens transformations we obtain the following scheme.

$$(9.4)$$

The scheme clearly illustrates that the strictly lower triangular rank structure has lifted up and that the diagonal is includable into the lower triangular rank structure.

The remaining four Givens transformations from bottom to top will remove the rank structure in the lower triangular part such that we obtain the upper triangular matrix $\hat{R}$.

Writing the above scheme in mathematical formulas, we obtain ($\check{Z}$ denotes another Hessenberg-like matrix):

$$\check{Q}_2^H \check{Q}_1^H (Z + D) = \check{Q}_2^H \check{Z},$$
$$\check{Q}_1^H (Z + D) = \check{Z},$$
$$(Z + D) = \check{Q}_1 \check{Z}.$$

The final equation denotes a structured rank factorization of the matrix $Z+D$, since the matrix $\check{Z}$ is of Hessenberg-like form and $\check{Q}_1$ is a unitary transformation. This unitary-Hessenberg-like (QH) factorization will form the basis of the eigenvalue computations proposed in Section 9.5 of this chapter.

Definition 9.3. *A factorization of the form*

$$A = \check{Q}\check{Z},$$

with $\check{Q}$ unitary and $\check{Z}$ a Hessenberg-like matrix is called a unitary-Hessenberg-like factorization, briefly a QH-factorization.

Note 9.4. *This factorization is a straightforward extension of the QR-factorization as the QR-factorization can be considered as a QH-factorization in which the matrix $\check{Z}$ is of semiseparability rank 0, i.e., that this matrix has the strictly lower triangular part equal to zero.*

Before discussing the different QR-algorithms related to the $\vee$ and the $\wedge$-pattern for computing the QR-factorization, we will discuss a flexibility in the QR-algorithm.

9.2.2   Flexibility in the QR-algorithm

Throughout the complete book we used the $\wedge$-pattern for constructing the QR-factorization.

The scheme used represented the matrix product $\hat{Q}^H(Z + D)$ graphically as follows.

$$(9.5)$$

In the discussed QR-methods we always firstly performed the ascending sequence of Givens transformations, i.e., the Givens transformations in position 1 to 4. These Givens transformations correspond to the the orthogonal similarity transformation $\hat{Q}_1^H(Z + D)\hat{Q}_1$.

After this sequence we perform the initial disturbing Givens transformation, which is located here in the scheme in position 5. One switches to the implicit approach and the chasing procedure starts.

The scheme above has some flexibility, with regard to the Givens transformations in positions 4 and 5. For example the scheme below represents exactly the same factorization.

$$
\begin{array}{c}
\text{❶} \\
\text{❷} \\
\text{❸} \\
\text{❹} \\
\text{❺}
\end{array}
\left|
\begin{array}{ccccc}
\times & \times & \times & \times & \times \\
\boxtimes & \times & \times & \times & \times \\
\boxtimes & \boxtimes & \times & \times & \times \\
\boxtimes & \boxtimes & \boxtimes & \times & \times \\
\boxtimes & \boxtimes & \boxtimes & \boxtimes & \times
\end{array}
\right.
\qquad (9.6)
$$

$$
\begin{array}{c|ccccccc}
 & 7 & 6 & 5 & 4 & 3 & 2 & 1
\end{array}
$$

Instead of following the Scheme 9.5 we can now also follow Scheme 9.6 for performing a step of the QR-method. Let us see how this will work. We will start by applying the Givens transformations from position 1 up to position 3. One can easily verify that this will transform the Hessenberg-like plus diagonal matrix $Z + D$ into a Hessenberg matrix, similarly as one would obtain by using Scheme 9.5.

Let us investigate in more detail what the effect of performing this QR-step will be. Assume $Z + D$ to be of the following form:

$$
\begin{bmatrix}
\boxtimes & \times & \times & \times & \times \\
\boxtimes & \boxtimes & \times & \times & \times \\
\boxtimes & \boxtimes & \boxtimes & \times & \times \\
\boxtimes & \boxtimes & \boxtimes & \boxtimes & \times \\
\boxtimes & \boxtimes & \boxtimes & \boxtimes & \boxtimes
\end{bmatrix}
+
\begin{bmatrix}
d_1 & & & & \\
 & d_2 & & & \\
 & & d_3 & & \\
 & & & d_4 & \\
 & & & & d_5
\end{bmatrix}.
$$

The resulting matrix $Q_1^H (Z + D) Q_1$, in which Q_1 is a combination of the first three Givens transformations from Scheme 9.6, is of the following form:

$$
\begin{bmatrix}
\boxtimes & \times & \times & \times & \times \\
\boxtimes & \boxtimes & \times & \times & \times \\
 & \boxtimes & \boxtimes & \times & \times \\
 & \boxtimes & \boxtimes & \boxtimes & \times \\
 & \boxtimes & \boxtimes & \boxtimes & \boxtimes
\end{bmatrix}
+
\begin{bmatrix}
0 & & & & \\
 & d_1 & & & \\
 & & d_2 & & \\
 & & & d_3 & \\
 & & & & d_4
\end{bmatrix}.
$$

As with the semiseparable plus diagonal case the diagonal shifts down (see Section 8.1.3). Because instead of four only three Givens transformations are performed we obtain only two nonzero elements in the first column. In a certain sense, the complete Hessenberg-like plus diagonal matrix has shifted down one position.

To start the chasing procedure one first needs to rewrite the matrix slightly such that the first two diagonal elements equal d_1. We obtain the following form for the matrix:

$$
\begin{bmatrix}
\boxplus & \times & \times & \times & \times \\
\boxtimes & \boxtimes & \times & \times & \times \\
 & \boxtimes & \boxtimes & \times & \times \\
 & \boxtimes & \boxtimes & \boxtimes & \times \\
 & \boxtimes & \boxtimes & \boxtimes & \boxtimes
\end{bmatrix}
+
\begin{bmatrix}
d_1 & & & & \\
 & d_1 & & & \\
 & & d_2 & & \\
 & & & d_3 & \\
 & & & & d_4
\end{bmatrix}
\qquad (9.7)
$$

One can now apply the Givens transformation from position 4 onto this matrix. The result is a disturbed Hessenberg-like plus diagonal matrix, the disturbance can be chased by using the standard techniques from Chapter 7.

To conclude this part we will illustrate the flow of the chasing in case one is working with the Givens-vector representation. In Chapter 8, we discussed the chasing for semiseparable matrices represented with the right left representation, and for semiseparable plus diagonal matrices represented with the top bottom representation. In the following we assume the top bottom representation.

The first term in Figure (9.7) can be represented by a Givens-vector representation as follows (the reader can verify that the result of applying the Givens transformations is the matrix above).

$$
\begin{array}{c}
\begin{array}{l}
❶ \\
❷ \\
❸ \\
❹ \\
❺
\end{array}
\left|
\begin{array}{ccccccc}
 & & \times & \cdot & \cdot & \cdot & \cdot \\
 & & & \times & \cdot & \cdot & \cdot \\
 & & & & \times & \cdot & \cdot \\
 & & & & & \times & \cdot \\
 & & & & & & \times
\end{array}
\right. \\
 3\ 2\ 1
\end{array}
$$

First we apply a similarity transformation which is the disturbing Givens transformation shown in position 4 in Scheme 9.6. Applying this similarity transformation will not affect the diagonal, hence we omit it temporarily.

Apply now the initial disturbing Givens transformation. The transformation performed on the right will create a bulge in position $(2,1)$, marked with $\otimes$. The transformation performed on the left is depicted in position 4.

$$
\begin{array}{c}
\begin{array}{l}
❶ \\
❷ \\
❸ \\
❹ \\
❺
\end{array}
\left|
\begin{array}{ccccccc}
 & & & \times & \cdot & \cdot & \cdot & \cdot \\
 & & & \otimes & \times & \cdot & \cdot & \cdot \\
 & & & & & \times & \cdot & \cdot \\
 & & & & & & \times & \cdot \\
 & & & & & & & \times
\end{array}
\right. \\
 4\ 3\ 2\ 1
\end{array}
$$

To continue, we will annihilate the bulge, marked with $\otimes$ resulting in the Givens transformation in position 1 in Scheme 9.8. In this scheme also the fusion for the Givens transformations above at positions 3 and 4 is depicted.

$$
\begin{array}{c}
\begin{array}{l}
❶ \\
❷ \\
❸ \\
❹ \\
❺
\end{array}
\left|
\begin{array}{ccccccc}
 & & & \times & \cdot & \cdot & \cdot & \cdot \\
 & & & & \times & \cdot & \cdot & \cdot \\
 & & & & & \times & \cdot & \cdot \\
 & & & & & & \times & \cdot \\
 & & & & & & & \times
\end{array}
\right. \\
 4\ 3\ 2\ 1
\end{array}
\tag{9.8}
$$

After applying the depicted fusion, we obtain the following well-known scheme, which is exactly the same as the one in the semiseparable plus diagonal case, hence one can apply the standard chasing techniques.

$$
\begin{array}{c}
\textbf{①} \\
\textbf{②} \\
\textbf{③} \\
\textbf{④} \\
\textbf{⑤}
\end{array}
\left|
\begin{array}{cccccc}
 & & & \times & \cdot & \cdot & \cdot & \cdot \\
 & & & & \times & \cdot & \cdot & \cdot \\
 & & & & & \times & \cdot & \cdot \\
 & & & & & & \times & \cdot \\
 & & & & & & & \times
\end{array}
\right.
$$

$$\overline{4\ \ 3\ \ 2\ \ 1}$$

We can conclude that this approach results in the same chasing technique. Only the initialization procedure, before the chasing is started is slightly different.

This kind of chasing will be used when developing the implicit QR-algorithm for quasiseparable matrices discussed in Section 9.3. Next we will discuss the QR-algorithms related to the different patterns for computing the QR-factorization.

9.2.3 The QR-algorithm and its variants

As there are different manners of computing the QR-factorization, the connected QR-algorithms will also be slightly different, yielding the same final result. In fact one obtains exactly the same result, but the way of computing the matrices after one step of the QR-method can differ. In this section we will briefly discuss the QR-algorithms connected to both the $\wedge$ and the $\vee$ pattern for computing the QR-factorization. We remark once more that the final outcome of both transformations will be equal. The order in which the Givens transformations are performed will, however, change as well as the Givens transformations themselves.

The QR-algorithm connected to the $\wedge$-pattern

We consider the following iteration step on a Hessenberg-like minus shift matrix (we will comment on the Hessenberg-like plus diagonal case afterwards):

$$
\begin{aligned}
Z - \mu I &= \hat{Q}_1 \hat{Q}_2 \hat{R}, \\
\hat{Z} &= \hat{R} \hat{Q}_1 \hat{Q}_2 + \mu I = \hat{Q}_2^H \hat{Q}_1^H Z \hat{Q}_1 \hat{Q}_2,
\end{aligned}
$$

in which $\hat{Z}$ denotes the new iterate.

The single shift QR-algorithm based on the $\wedge$-pattern was firstly discussed in an implicit form in [164].

Let us discuss the global flow of the iteration related to the $\wedge$-pattern. The iteration can be decomposed into two steps, each step corresponding to performing a sequence of $n-1$ Givens transformations. The first sequence is ascending denoted by $\backslash$ in the $\wedge$ and annihilates the low rank part in the Hessenberg-like matrix. The second sequence corresponds to the descending Givens transformations, denoted by $/$ in the $\wedge$-pattern, which removes the subdiagonal elements.

Since the new iterate is defined as $\hat{Q}_2^H \hat{Q}_1^H Z \hat{Q}_1 \hat{Q}_2 = \hat{Q}_2^H (\hat{Q}_1^H Z \hat{Q}_1) \hat{Q}_2$, two similarity transformations need to be applied onto the matrix Z. One is determined by $\hat{Q}_1$ and the other by $\hat{Q}_2$.

- The first similarity transformation (related to $\hat{Q}_1$) computes the following (see Subsection 9.2.1):

$$\tilde{Z} = \hat{Q}_1^H Z \hat{Q}_1 = \left(\hat{Q}_1^H Z \right) \hat{Q}_1 = R\hat{Q}_1.$$

  This corresponds to performing a step of the QR-method without shift onto the matrix Z. As a result we obtain another Hessenberg-like matrix $\tilde{Z}$.

- The second similarity transformation (related to $\hat{Q}_2$) can be performed in an implicit way as follows. Determine the first Givens transformation $\tilde{G}$ of $\hat{Q}_2$, for annihilating the element in position $(2,1)$ of the Hessenberg matrix $\hat{Q}_1^H(Z - \mu I) = H$. Applying this Givens transformation $\tilde{G}$ as a similarity transformation onto the Hessenberg-like matrix $\tilde{Z}$ disturbs the specific rank structure of this Hessenberg-like matrix.

  The implicit part of the method consists of finding the remaining $n-2$ Givens transformations and applying them onto $\tilde{G}^H \tilde{Z} \tilde{G}$, such that the resulting matrix is back of Hessenberg-like form. Based on the implicit Q-theorem for Hessenberg-like matrices (see Chapter 6, Section 6.2) one knows that the result of this approach is again a Hessenberg-like matrix, which is essentially the same as the one resulting from an explicit step of the QR-method.

Note 9.5. *The first similarity transformation based on $\hat{Q}_1$ is independent from the chosen shift μ. The second similarity transformation is dependent of the shift μ.*

The QR-method for Hessenberg-like plus diagonal matrices $Z + D$ is identical. First a number of Givens transformations are performed, corresponding to a step of QR-without shift onto Z, followed by a similarity transformation determined by $\hat{Q}_2$. For restoring the structure in the Hessenberg-like plus diagonal case, one needs to consider the structure of the diagonal, as the diagonal is preserved under a step of the QR-method.

The QR-algorithm connected to the ∨-pattern

We consider the following iteration step:

$$Z - \mu I = \check{Q}_1 \check{Q}_2 \hat{R},$$
$$\hat{Z} = \hat{R}\check{Q}_1 \check{Q}_2 + \mu I = \check{Q}_2^H \check{Q}_1^H Z \check{Q}_1 \check{Q}_2,$$

in which $\hat{Z}$ denotes the new iterate.

The QR-algorithm based on the ∨-pattern has not yet been discussed in this book. The idea is however a straightforward generalization of the QR-algorithm based on the ∧-pattern. Because in some sense we have switched the order of both sequences of $n-1$ Givens transformations, we can also switch the interpretation of this algorithm.

Two similarity transformations have to be performed: $\check{Q}_2^H (\check{Q}_1^H Z \check{Q}_1) \check{Q}_2$. Now, $\check{Q}_1$ is a descending sequence of Givens transformations for expanding the rank structure and $\check{Q}_2$ is an ascending sequence of Givens transformations for removing the newly created rank structure of the intermediate Hessenberg-like matrix.

- The first step can be performed in an implicit manner, similar to the second sequence in the ∧-case. An initial disturbing Givens transformation is applied, followed by $n - 2$ structure restoring Givens transformations[20]. As a result we obtain the Hessenberg-like matrix[21]

$$\tilde{Z} = \check{Q}_1^H Z \check{Q}_1.$$

- One can prove that the second step (corresponding to the Givens transformations from bottom to top) can again be seen as performing a step of the QR-method without shift onto the newly created Hessenberg-like matrix $\tilde{Z}$. After performing the similarity transformation corresponding to $\check{Q}_2$, we obtain the result of performing one step of the QR-method without shift applied onto the Hessenberg-like matrix $\tilde{Z}$.

Note 9.6. *In the similarity transformation related to the ∨-pattern the first step is dependent on the shift μ, whereas the second step is independent from μ. See also Remark 9.5 for the iteration related to the ∧-pattern.*

Note 9.7. *The remark above makes it clear that this algorithm (as well as the algorithm related to the ∧-pattern) has a kind of contradicting convergence behavior in it. When we look at the bottom-right of the matrix, we have*

- *The first step is determined by the shift and hence creates convergence to the eigenvalue(s) closest to the shift.*

- *The second step corresponds to a QR-step without shift and hence converges to the smallest eigenvalue(s) in modulus.*

Both convergence behaviors do not necessarily cooperate. In some sense the second step can slightly destroy the good improvements made by the first step.

One can opt to remove the second similarity transformation. Unfortunately we will not have a QR-factorization and the corresponding QR-method anymore. This approach will lead to the QH-method, which will be discussed in Section 9.5.

Notes and references

Explicit QR-algorithms for structured rank matrices, based on the ∧-pattern are legion: for general rank structured matrices and unitary matrices [55, 49] and for quasiseparable matrices in [70]. More general forms, for computing the eigenvalues of polynomials can be found in [21, 22, 23].

In the beginning of this section, the interchanging from the ∨ to the ∧-pattern was briefly illustrated. This interchanging is a powerful tool resulting in different annihilation patterns for computing for example the QR-factorization of structured rank matrices. These results were extensively discussed in the first volume of the book [169]. Exploiting these different QR-factorizations can lead to a parallel algorithm for computing the QR-factorization of for example semiseparable plus diagonal matrices.

---

[20]The chasing can be performed similarly as the chasing step in case of the ∧-pattern.

[21]In Section 9.5 we will prove that the matrix $\tilde{Z}$ is indeed of Hessenberg-like form.

☞ R. Vandebril, M. Van Barel, and N. Mastronardi. A parallel QR-factorization/solver of structured rank matrices. *Electronic Transactions on Numerical Analysis*, 30:144–167, 2008.

9.3 The QR-method for quasiseparable matrices

Up till now, we have not discussed quasiseparable matrices. In this section we will briefly define quasiseparable matrices, and we will see how to develop an (implicit) QR-method for these matrices. The algorithm will be based on a combination of results for semiseparable plus diagonal and tridiagonal matrices.

More information on quasiseparable matrices can be found in Volume I of the book and in the references discussed at the end of this section.

In this section we will firstly discuss some properties of quasiseparable matrices followed by the general idea on how to perform an explicit QR-step, by considering all diagonal blocks separately. The different diagonal blocks are either of semiseparable plus diagonal or of tridiagonal form. This will be exploited in the algorithm. The final subsection discusses the implicit version of the method by simply combining both QR-methods.

9.3.1 Definition and properties

For simplicity we will only discuss symmetric quasiseparable matrices. The unsymmetric case proceeds similarly.

Definition 9.8. *A matrix A is called a symmetric quasiseparable matrix if all the subblocks, taken out of the strictly lower triangular part of the matrix (respectively, the strictly upper triangular part), are of rank ≤ 1, and the matrix A is symmetric.*

This means that a symmetric matrix $A \in \mathbb{R}^{n \times n}$ is symmetric quasiseparable, if the following relation is satisfied (for $i = 1, \ldots, n-1$):

$$\text{rank}\left(A(i+1:n, 1:i)\right) \leq 1 \text{ for } i = 1, \ldots, n-1.$$

This means that all subblocks taken out of the part marked with ⊠ in the following 5×5 matrix have at most rank 1. Hence the diagonal is not always includable in the strictly lower triangular rank structure.

$$A = \begin{bmatrix} \times & \times & \times & \times & \times \\ \boxtimes & \times & \times & \times & \times \\ \boxtimes & \boxtimes & \times & \times & \times \\ \boxtimes & \boxtimes & \boxtimes & \times & \times \\ \boxtimes & \boxtimes & \boxtimes & \boxtimes & \times \end{bmatrix}.$$

Let us consider some examples of symmetric quasiseparable matrices.

Example 9.9 The following matrices are all of quasiseparable form.

- A symmetric semiseparable matrix.

- A symmetric tridiagonal matrix.

- A symmetric semiseparable plus diagonal matrix.

■

More information on the specific structure of quasiseparable matrices can be found in the first volume of the book. Important for the development of the QR-method is the block diagonal structure of a quasiseparable matrix.

Consider a symmetric quasiseparable matrix A. It is fairly easy to prove that such a matrix is always of the following block diagonal form (see Figure 9.1), in which all blocks are of semiseparable plus diagonal or of tridiagonal form (see Proposition 1.19 in Volume I). The blocks overlap sharing a single diagonal element.

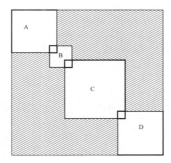

Figure 9.1. *Block form of a quasiseparable matrix.*

Note 9.10. *In some sense one can consider a quasiseparable matrix as a semiseparable plus diagonal matrix, in which some of the diagonal values can also be infinity.*

One can derive from Theorem 1.20 in Volume I, that a tridiagonal matrix can be considered as a semiseparable plus diagonal matrix (see Example 9.11), in which all diagonal elements, except the first and the last one, are equal to infinity.

Since the quasiseparable matrix has blocks corresponding to semiseparable plus diagonal and to tridiagonal matrices, we can consider it as a global semiseparable plus diagonal matrix, having some of the diagonal elements equal to infinity.

This interpretation will be the basis for the construction of the quasiseparable QR-algorithm, based on the QR-method for semiseparable plus diagonal matrices.

The following example is studied more extensively in Volume I showing the relation between tridiagonal and semiseparable plus diagonal matrices. In the first volume the relations between the classes of tridiagonal, semiseparable, semiseparable plus diagonal and quasiseparable matrices are studied extensively.

Example 9.11 Consider the following tridiagonal matrix, which is the limit of a sequence of semiseparable plus diagonal matrices. Note that, to obtain this tridiagonal matrix, the diagonal also needs to change.

$$\lim_{\epsilon \to \infty} \left(\begin{bmatrix} a & b & \frac{1}{\epsilon} \\ b & \epsilon bd & d \\ \frac{1}{\epsilon} & d & e \end{bmatrix} + \begin{bmatrix} 0 & 0 & 0 \\ 0 & c - \epsilon bd & 0 \\ 0 & 0 & 0 \end{bmatrix} \right) = \begin{bmatrix} a & b & \\ b & c & d \\ & d & e \end{bmatrix} \qquad (9.9)$$

The matrices on the left are of semiseparable plus diagonal form, whereas the limit is a tridiagonal matrix, which does not belong to the class of semiseparable plus diagonal matrices.   ■

To develop the results for quasiseparable matrices there are several possibilities.

- A first possibility is to consider the quasiseparable matrices as a class existing on its own having a specific representation. Based on this representation one can compute the QR-factorization and prove preservation of structure and one can design a QR-method. This is done for example in [70].

- Secondly one can use the block decomposition as presented before and combine the existing results for tridiagonal and semiseparable plus diagonal matrices.

- In a third approach one considers the quasiseparable matrix as a very specific kind of semiseparable plus diagonal matrix, admitting values $\pm\infty$ on the diagonal.

In this section we choose to combine the second and third item to obtain quickly an easy understandable QR-algorithm for quasiseparable matrices.

9.3.2   The QR-factorization and the QR-algorithm

Before deriving the QR-method for quasiseparable matrices, we will see how to change the QR-method for semiseparable plus diagonal matrices, such as to make it applicable to tridiagonal matrices. Moreover we will see that the QR-method for tridiagonal matrices is a sort of simplification of the QR-method for semiseparable plus diagonal matrices.

Suppose we have a tridiagonal matrix $T = \lim_{\epsilon}(S_\epsilon + D_\epsilon)$. Since we admit this decomposition for the tridiagonal matrix, we can easily prove some statements related to tridiagonal matrices.

- We know that the structure of a semiseparable plus diagonal matrix is preserved under the QR-iteration, thereby even preserving the diagonal. Hence, also the tridiagonal structure is preserved under the QR-iteration[22].

---

[22]To prove this and the following statements in a more rigid way one needs to use theorems presented in Volume I on the closure of the classes of semiseparable plus diagonal matrices. We will not go into details but leave this to the reader.

- Let us investigate how to compute the QR-factorization of the matrix $S_\epsilon + D_\epsilon$. The first sequence of Givens transformations is an ascending sequence. The structure of these Givens is however trivial. The Givens transformation acting on rows 2 and 3 in Equation (9.9) chosen to annihilate the low rank part of the third row is of the following form

$$G_\epsilon = \frac{1}{\sqrt{1 + (b\epsilon)^2}} \begin{bmatrix} b\epsilon & 1 \\ -1 & b\epsilon \end{bmatrix}$$

Taking the limit now, gives us the following Givens transformation:

$$G = \lim_{\epsilon \to \infty} G_\epsilon = \begin{bmatrix} 1 & 0 \\ 0 & 1 \end{bmatrix}.$$

Hence the first sequence of Givens transformations vanishes, leaving only the second sequence of descending Givens transformations. This sequence is identical to the one from the traditional QR-method for tridiagonal matrices.

Hence translating the QR-method for semiseparable plus diagonal matrices towards the tridiagonal case, simply results in the standard QR-method for tridiagonal matrices.

The following formula $T = \lim_\epsilon (S_\epsilon + D_\epsilon)$ illustrates what we mean when saying that we can consider a tridiagonal matrix as a semiseparable plus diagonal matrix having diagonal entries equal to $\pm\infty$.

Let us come back now to the case of a quasiseparable matrix. We have just discussed that in fact the tridiagonal QR-method is sort of a simplification of the QR-method for semiseparable plus diagonal matrices. For quasiseparable matrices we can do exactly the same. Only we do not have all diagonal elements equal to infinity, but only the ones corresponding to the tridiagonal blocks on the diagonal. Suppose $A = \lim_\epsilon (S_\epsilon + D_\epsilon)$.

- Since the structure of a semiseparable plus diagonal matrix is preserved under the QR-iteration, the structure of a quasiseparable matrix is preserved under a QR-iteration.

  Preservation of structure means that not only the next iterate is quasiseparable but also that its block structure is preserved. This means that the blocks on the diagonal of semiseparable plus diagonal form, remain semiseparable plus diagonal form (preserving even the diagonal), and the tridiagonal blocks remain tridiagonal.

- The QR-method for quasiseparable matrices is a simplification of the QR-method for semiseparable plus diagonal matrices. In fact it is a combination of the QR-method for semiseparable plus diagonal matrices and the QR-method for tridiagonal matrices.

  The generalization is almost trivial. First the ascending sequence of Givens transformations is performed. This sequence will remove the low rank part in the diagonal blocks that are of semiseparable plus diagonal form. The Givens transformations acting on the blocks that are of tridiagonal form will be equal to the identity. This results in the following equation. Suppose A is the quasiseparable matrix, Q_1^H is the ascending sequence of Givens transformations and μ is a suitably chosen shift:

$$Q_1^H (A - \mu I) = H.$$

As a result we obtain the matrix H, which is a Hessenberg matrix.

The second sequence of Givens transformations is chosen to remove the diagonal.

Combining techniques for semiseparable plus diagonal and tridiagonal matrices leads to an explicit QR-algorithm.

For quasiseparable matrices one can also perform the second sequence in an implicit manner, to restore the structure of the quasiseparable matrix. With restoring the structure we mean restoring the semiseparable plus diagonal structure of the corresponding blocks and restoring the tridiagonal form in the corresponding blocks.

Note 9.12. *In the previous deduction we assumed our quasiseparable matrix to be of the block diagonal form, where each block is of semiseparable plus diagonal or of tridiagonal structure. Unfortunately this structure is quite often not known. Hence one needs to determine first the diagonals of the semiseparable plus diagonal matrices involved in the block splitting of the quasiseparable matrix.*

9.3.3 The implicit method

The previous section was not so detailed on the implementation of the quasiseparable QR-method, therefore, we will derive it here in more detail.

We assume to be working with a very specific quasiseparable matrix, but there is no loss of generality, as it covers the generic case. Assume our matrix A to be of the following form. (For simplicity we assume the shift matrix to be included in the matrix A.):

$$A = \begin{bmatrix} \boxed{1} & \boxtimes & \boxtimes & & & & & & \\ \boxtimes & \boxed{2} & \boxtimes & & & & & & \\ \boxtimes & \boxtimes & \boxed{3} & \times & & & & & \\ & & \times & \times & \times & & & & \\ & & & \times & \times & \times & & & \\ & & & & \times & \boxed{4} & \boxtimes & \boxtimes & \boxtimes \\ & & & & & \boxtimes & \boxed{5} & \boxtimes & \boxtimes \\ & & & & & \boxtimes & \boxtimes & \boxed{6} & \boxtimes \\ & & & & & \boxtimes & \boxtimes & \boxtimes & \boxed{7} \end{bmatrix}.$$

We have introduced here a new kind of notation to reduce the space consumption. The box with a number in it depicts the following

$$\boxed{1} = \boxtimes + d_1$$
$$\boxed{2} = \boxtimes + d_2$$
$$\vdots$$

Instead of putting the diagonal from the semiseparable plus diagonal blocks in a separate matrix they are included here in the graphical figure. The quasiseparable matrix A above, can be decomposed into three main diagonal blocks. The blocks $A(1:3, 1:3)$ and $A(6:9, 6:9)$ are of semiseparable plus diagonal form, with diagonal elements $d_1, \ldots, d_3$ for the first and $d_4, \ldots, d_7$ for the second block. The block $A(3:6, 3:6)$ is of tridiagonal form. It can be seen that the diagonal blocks share the upper left and the lower right element with another block.

Let us apply now the first sequence of Givens transformations onto the matrix A. This sequence of Givens transformations will remove the low rank part of the semiseparable plus diagonal blocks and will not act on the tridiagonal blocks. For removing the low rank part, we will use the form explained in Subsection 9.2.2. This form will only perform two instead of three Givens transformations for removing the low rank part in the block $A(6:9, 6:9)$ and only one instead of two for removing the low rank part in the block $A(1:3, 1:3)$. After having performed this transformation, we obtain the following matrix:

$$\begin{bmatrix}
\times & \times & & & & & & & \\
\times & \boxed{1} & \boxtimes & \boxtimes & & & & & \\
& \boxtimes & \boxed{2} & \boxtimes & & & & & \\
& \boxtimes & \boxtimes & \boxed{3} & \times & & & & \\
& & & \times & \times & \times & & & \\
& & & & \times & \times & \times & & \\
& & & & & \times & \boxed{4} & \boxtimes & \boxtimes \\
& & & & & & \boxtimes & \boxed{5} & \boxtimes \\
& & & & & & \boxtimes & \boxtimes & \boxed{6}
\end{bmatrix}.$$

The complete structure of the matrix has shifted down one position, thereby introducing at the top a new 2×2 tridiagonal (or semiseparable plus diagonal) block.

Note 9.13. *Some remarks have to be made:*

- *We assumed to be working with the QR-form explained in Subsection 9.2.2, for the semiseparable plus diagonal blocks. One can, however, also perform three (two) instead of only two (one) Givens transformations. This will not make the method more complicated.*

- *Related to the previous item is the computation of the disturbing Givens transformation. If one chooses to perform three instead of two Givens transformations, the disturbing initial Givens transformation will change also!*

- *A final remark is related to the shift. Performing the first sequence of Givens transformations is independent of the chosen shift. The influence of the shift* *is incorporated into the disturbing Givens transformation. We do not pay a lot of attention to computing this Givens transformation as the reader should be able to compute it based on the knowledge gained in previous chapters. The disturbing Givens transformation is chosen to annihilate the first subdiagonal element in the Hessenberg matrix $H = Q_1^H A$. (Remember that we assumed the shift μ to be included in the matrix A.)*

Let us now start the chasing procedure. We will consider each block separately. First apply the disturbing Givens transformation at the top followed by a single structure restoring Givens transformation. We obtain again a semiseparable plus diagonal matrix in the upper left 3×3 corner. How to perform the disturbing Givens transformation, and how to restore the structure is explained in Subsection 9.2.2. As a result we obtain:

$$
\begin{bmatrix}
\boxed{1} & \boxtimes & \boxtimes & \boxtimes & & & & & \\
\boxtimes & \boxed{2} & \boxtimes & \boxtimes & & & & & \\
\boxtimes & \boxtimes & \boxed{3} & \times & & & & & \\
\boxtimes & \boxtimes & \times & \times & \times & & & & \\
 & & & \times & \times & \times & & & \\
 & & & & \times & \times & \times & & \\
 & & & & & \times & \boxed{4} & \boxtimes & \boxtimes \\
 & & & & & & \boxtimes & \boxed{5} & \boxtimes \\
 & & & & & & \boxtimes & \boxtimes & \boxed{6}
\end{bmatrix} . \tag{9.10}
$$

The next structure restoring similarity Givens transformation will act on rows and columns 3 and 4. We know that after performing this transformation, the upper left 3×3 block needs to be of semiseparable plus diagonal form, but the elements in row and column 4 need to belong to a tridiagonal block. Hence the Givens transformation is chosen to annihilate the undesired elements in the fourth row/column. Fortunately these elements still belong to the semiseparable plus diagonal structure, as depicted in Figure (9.10) and hence we can remove all the undesired elements by performing a single Givens similarity transformation.

Note 9.14. *When the figures depicted here are not clear at first sight, the reader can design the graphical schemes based on the Givens-vector representation of the chasing, similar to Subsection 9.2.2. Based on these figures one can easily deduce from the above structure in which we obtain a 3×3 semiseparable plus diagonal block, for which the rank structure slightly extends to the fourth row and column.*

Removing the remaining $\boxtimes$ elements in the fourth row and column by a similarity Givens transformation gives us the following matrix:

$$
\begin{bmatrix}
\boxed{1} & \boxtimes & \boxtimes & & & & & & \\
\boxtimes & \boxed{2} & \boxtimes & & & & & & \\
\boxtimes & \boxtimes & \boxed{3} & \times & \otimes & & & & \\
 & & \times & \times & \times & & & & \\
 & & \otimes & \times & \times & \times & & & \\
 & & & & \times & \times & \times & & \\
 & & & & & \times & \boxed{4} & \boxtimes & \boxtimes \\
 & & & & & & \boxtimes & \boxed{5} & \boxtimes \\
 & & & & & & \boxtimes & \boxtimes & \boxed{6}
\end{bmatrix}.
$$

We see that we have now created a bulge in the tridiagonal block. One can now switch to the chasing procedure for tridiagonal matrices, in which one shifts down the bulge one position after each similarity transformation.

After having completely restored the tridiagonal part within the index range $(3 : 6, 3 : 6)$ we obtain:

$$
\begin{bmatrix}
\boxed{1} & \boxtimes & \boxtimes & & & & & & \\
\boxtimes & \boxed{2} & \boxtimes & & & & & & \\
\boxtimes & \boxtimes & \boxed{3} & \times & & & & & \\
 & & \times & \times & \times & & & & \\
 & & \times & \times & \times & \otimes & & & \\
 & & & \times & \times & \times & & & \\
 & & & & \otimes & \times & \boxed{4} & \boxtimes & \boxtimes \\
 & & & & & & \boxtimes & \boxed{5} & \boxtimes \\
 & & & & & & \boxtimes & \boxtimes & \boxed{6}
\end{bmatrix}.
$$

The structure is completely restored similar to the semiseparable plus diagonal case. The upper left part in the matrix separated by the horizontal and vertical line is already of the correct structure. Unfortunately a part of the tridiagonal structure extends beyond its boundaries. Here there is a bulge remaining in positions $(7, 5)$ and $(5, 7)$. Applying a similarity Givens transformation will result in a disturbing Givens transformation acting on the semiseparable plus diagonal matrix in the lower right block. This initializes another chasing procedure for restoring the semiseparable plus diagonal part, resulting in a matrix having the same structure as the original quasiseparable matrix where even the diagonals of the semiseparable plus diagonal parts are preserved:

$$
\begin{bmatrix}
\boxed{1} & \boxtimes & \boxtimes & & & & & & & \\
\boxtimes & \boxed{2} & \boxtimes & & & & & & & \\
\boxtimes & \boxtimes & \boxed{3} & \times & & & & & & \\
 & & \times & \times & \times & & & & & \\
 & & \times & \times & \times & & & & & \\
 & & & \times & \boxed{4} & \boxtimes & \boxtimes & \boxtimes & \\
 & & & & \boxtimes & \boxed{5} & \boxtimes & \boxtimes & \\
 & & & & \boxtimes & \boxtimes & \boxed{6} & \boxtimes & \\
 & & & & \boxtimes & \boxtimes & \boxtimes & \boxed{7}
\end{bmatrix}.
$$

The result of this chasing procedure is essentially equivalent to a matrix coming from a step of the QR-method.

Note 9.15. *To explain this method we assumed to have a semiseparable plus diagonal block, followed by a tridiagonal block, followed again by a semiseparable plus diagonal block. In fact one can also have two semiseparable blocks following each other such as depicted here:*

$$
A = \begin{bmatrix}
\boxed{1} & \boxtimes & \boxtimes & & & \\
\boxtimes & \boxed{2} & \boxtimes & & & \\
\boxtimes & \boxtimes & \boxed{3} & \boxtimes & \boxtimes & \boxtimes \\
 & & \boxtimes & \boxed{4} & \boxtimes & \boxtimes \\
 & & \boxtimes & \boxtimes & \boxed{5} & \boxtimes \\
 & & \boxtimes & \boxtimes & \boxtimes & \boxed{6}
\end{bmatrix}.
$$

The chasing method for this kind of structure is similar to the one discussed above.

In this section we did not discuss essential topics such as an implicit Q-theorem for quasiseparable matrices, unreducedness of a quasiseparable matrix. These issues are of a theoretical nature and were already discussed extensively for semiseparable, semiseparable plus diagonal and tridiagonal matrices. Combining all properties of these matrices provides the desired theoretical background. We have provided a working algorithm.

When implementing this method, one firstly needs to know the block decomposition of the involved quasiseparable matrix. Moreover, an implementation based on the Givens-vector representation, thereby using the graphical schemes is the most simple to implement since it will easily reveal possible breakdowns in case of undesired zeros and badly defined chasing Givens transformations.

Notes and references

Quasiseparable matrices have not been discussed in this volume. The class of matrices as well as solvers for this class of matrices were presented extensively in Volume I [169].

These matrices were firstly defined in the articles [67, 146] (see Section 1.3, Chapter 1).

The QR-factorization for quasiseparable matrices is similar to the one for semiseparable plus diagonal matrices (see also [156, 59]). The method proposed by Dewilde–van der Veen was adapted by Eidleman and Gohberg.

☞ P. Dewilde and A.-J. van der Veen. Inner-outer factorization and the inversion of locally finite systems of equations. *Linear Algebra and its Applications*, 313:53–100, February 2000.

☞ Y. Eidelman and I. C. Gohberg. A modification of the Dewilde-van der Veen method for inversion of finite structured matrices. *Linear Algebra and its Applications*, 343–344:419–450, April 2002.

Implicit QR-algorithms for quasiseparable matrices have not yet been discussed in the literature. An explicit method for quasiseparable matrices was already presented in the following article by Eidelman, Gohberg and Olshevsky.

☞ Y. Eidelman, I. C. Gohberg, and V. Olshevsky. The QR iteration method for Hermitian quasiseparable matrices of an arbitrary order. *Linear Algebra and its Applications*, 404:305–324, July 2005.

9.4 The multishift QR-algorithm

In this book we already discussed several useful algorithms for computing the eigenvalues of arbitrary matrices and of semiseparable matrices. We discussed efficient transitions via orthogonal similarity transformations to semiseparable and related forms. Moreover QR-algorithms for all resulting matrices were also presented. Unfortunately only the single shift case was discussed and developed.

In general a real Hessenberg-like matrix not only has real eigenvalues but also complex conjugate ones. Hence one needs to switch to complex arithmetic for computing these eigenvalues of Hessenberg-like matrices via the single shift approach.

The multishift QR-step for Hessenberg-like matrices, which will be proposed, will solve this problem. The setting discussed here is very general and covers k shifts. As a special case it will be shown how to design a double shift version for Hessenberg-like matrices such that one can stick to real arithmetic.

The section is organized as follows. Subsection 9.4.1 discusses the problem setting. Details are presented on the approach we will follow and important results related to Hessenberg-like matrices are given. Subsection 9.4.2 and Subsection 9.4.3 discuss the implicit multishift approach for Hessenberg-like matrices. Subsection 9.4.2 focuses on the introduction of the disturbing Givens transformations, whereas Subsection 9.4.3 is dedicated to the structure restoring chasing technique. Subsection 9.4.4 presents some details concerning the implementation of the double shift version for the real Hessenberg case.

9.4.1 The multishift setting

Explicitly, the multishift QR-step on an arbitrary matrix A is of the following form. Suppose we want to apply k shifts at the same time, denote them with μ_1 up to μ_k. Assume we computed the following QR-factorization:

$$QR = (A - \mu_1 I)(A - \mu_2 I)\dots(A - \mu_k I).$$

The multishift QR-step consists now of applying the following similarity transformation onto the matrix A:

$$\hat{A} = Q^H A Q.$$

An implicit multishift QR-step consists of performing an initial disturbing unitary similarity transformation onto the matrix A, followed by structure restoring similarity transformations, not involving the first column, assuming thereby the matrix A to be of structured form.

Let us explain in more detail the implicit version of the multishift QR-step if the matrix $A = H$ is a Hessenberg matrix. More precisely, assume

$$\mathbf{v} = (H - \mu_1 I)(H - \mu_2 I)\dots(H - \mu_k I)\mathbf{e}_1, \tag{9.11}$$

in which $\mathbf{e}_1 = [1, 0, \ldots, 0, 0]^T$ denotes the first basis vector of $\mathbb{C}^n$. Suppose we design a specific unitary matrix $\tilde{Q}$ such that $\tilde{Q}^H \mathbf{v} = \pm \|\mathbf{v}\| \mathbf{e}_1$. Apply this transformation onto the matrix H, resulting in

$$\tilde{H} = \tilde{Q}^H H \tilde{Q}.$$

Because the matrix H is of Hessenberg form and the matrix $\tilde{Q}$ was constructed in a specific manner, this results in the introduction of a bulge, whose size depends on k, below the subdiagonal.

The implicit approach is then completed by transforming the matrix H back to its original form, e.g., Hessenberg form, by successively applying unitary similarity transformations. The unitary transformation matrix $\hat{Q}$ needs to be chosen such that $\hat{Q}\mathbf{e}_1 = \mathbf{e}_1$ and

$$\hat{H} = \hat{Q}^H \tilde{Q}^H H \tilde{Q} \hat{Q},$$

with $\hat{H}$ again a Hessenberg matrix. The condition $\hat{Q}\mathbf{e}_1 = \mathbf{e}_1$ is necessary to prove that this implicit QR-step is essentially the same as an explicit QR-step. One can prove this via the implicit Q-theorem (see Subsection 6.2.3) or via other approaches (see [181]).

Due to the preservation of the Hessenberg-like structure under a QR-step, we know that the structure is also preserved under a multishift QR-step (see, e.g., [94]). To design an implicit multishift QR-algorithm for Hessenberg-like matrices, we use the idea of introducing the distortion by reducing the first column $\mathbf{v}$ (as in Equation (9.11), but for the Hessenberg-like case) to a multiple of $\mathbf{e}_1$.

In the Hessenberg case we will see that the vector $\mathbf{v}$ has only $k+1$, in case of k shifts, elements different from zero, located all in the top positions of the vector. To annihilate these elements, and transforming the vector $\mathbf{v}$ to $\pm\beta\mathbf{e}_1$, one can easily perform a single Householder transformation or k Givens transformations. Unfortunately, in the case of a Hessenberg-like matrix Z we will see that the vector $\mathbf{v}$ is generally full. How to efficiently choose transformations for reducing $\mathbf{v}$ to $\pm\beta\mathbf{e}_1$, such that the corresponding similarity transformations can be efficiently performed onto the Hessenberg-like matrix Z, is the subject of the next subsection.

9.4.2   An efficient transformation from v to $\pm\beta\mathbf{e}_1$

The reduction of an arbitrary vector $\mathbf{v}$ to $\pm\beta\mathbf{e}_1$, a multiple of the first basis vector, can easily be accomplished by performing a Householder transformation or $n-1$ Givens transformations. Unfortunately applying the Householder or the Givens transformations onto the Hessenberg-like matrix Z creates an overhead on distortion in the matrix Z, which cannot be removed efficiently. It seems that a more appropriate technique is needed to reduce the complexity of restoring the structure of the disturbed matrix Z.

Reconsidering the single shift case, we know that the Givens transformations needed for bringing the vector $\mathbf{v}$ back to $\pm\beta\mathbf{e}_1$ are closely related to the Givens transformations used in the Givens-vector representation. Hence, we will search how to link the transformation of $\mathbf{v}$ to $\pm\beta\mathbf{e}_1$ with the Givens-vector representation of the matrix Z.

Assume we have the following matrix $\tilde{Z}$, whose first column we would like to reduce efficiently to $\pm\beta\mathbf{e}_1$. Denote $Z_0 = Z$ and the μ_i are suitably chosen shifts:

$$\tilde{Z} = (Z_0 - \mu_1 I)(Z_0 - \mu_2 I)\ldots(Z_0 - \mu_k I).$$

The matrix Z_0 is represented with the Givens-vector representation, and hence we have $Z_0 = Q_1 R_1$, in which the matrix Q_1 contains the Givens transformations from the representation, and the matrix R_1 is an upper triangular matrix. This leads to

$$\tilde{Z} = Q_1\left(R_1 - \mu_1 Q_1^H\right)Q_1\left(R_1 - \mu_2 Q_1^H\right)\ldots Q_1\left(R_1 - \mu_k Q_1^H\right),$$

in which $Q_1^H \mu_i I$ is a Hessenberg matrix, for $i = 1, \ldots, k$.

We will now transform the first column of the matrix $\tilde{Z}$ to $\pm\beta\mathbf{e}_1$ but in such a way that we can perform these transformations efficiently as similarity transformations onto the matrix Z_0.

Perform the transformation Q_1^H onto the matrix $\tilde{Z}$

$$\begin{aligned}
Q_1^H\tilde{Z} &= \left(R_1 - \mu_1 Q_1^H\right)Q_1\left(R_1 - \mu_2 Q_1^H\right)\ldots Q_1\left(R_1 - \mu_k Q_1^H\right)\\
&= (R_1 Q_1 - \mu_1 I)(R_1 Q_1 - \mu_2 I)\ldots\\
&\qquad\ldots(R_1 Q_1 - \mu_{k-1} I)\left(R_1 - \mu_k Q_1^H\right).
\end{aligned}$$

The matrix product $R_1 Q_1$ produces a new Hessenberg-like matrix Z_1. (Check the correspondence with the right to left representation of Hessenberg-like matrices.) The RQ-factorization of the matrix considered can easily be transformed to the QR-factorization. Due to the connection with the Givens-vector representation this can be done efficiently. Moreover, we will see further on in the text that we do need this transition anyway. Denote this as $Z_1 = Q_2 R_2$. The matrix Z_1 is in fact the result of performing a step of the QR-algorithm without shift onto the matrix Z_0. This is equivalent as in the single shift approach. We have the equality $Z_1 = Q_1^H Z_0 Q_1$.

We obtain the following relations:

$$\begin{aligned}
Q_1^H\tilde{Z} &= Q_2\left(R_2 - \mu_1 Q_2^H\right)Q_2\left(R_2 - \mu_2 Q_2^H\right)\ldots\\
&\qquad\ldots Q_2\left(R_2 - \mu_{k-1} Q_2^H\right)\left(R_1 - \mu_k Q_1^H\right)\\
Q_2^H Q_1^H\tilde{Z} &= (R_2 Q_2 - \mu_1 I)(R_2 Q_2 - \mu_2 I)\ldots\\
&\qquad\ldots(R_2 Q_2 - \mu_{k-2} I)\left(R_2 - \mu_{k-1} Q_2^H\right)\left(R_1 - \mu_k Q_1^H\right).
\end{aligned}$$

It is clear that the procedure can easily be continued by defining

$$Z_2 = Q_2^H Z_1 Q_2 = R_2 Q_2,$$

which is the result of applying another step of the QR-method without shift onto the matrix Z_1.

As a result we obtain that, if for every $i = 1, \ldots, k$

$$\begin{aligned}
Z_{i-1} &= Q_i R_i,\\
Z_i &= Q_i^H Z_{i-1} Q_i = R_i Q_i,
\end{aligned}$$

the following relation holds

$$Q_k^H \ldots Q_1^H \tilde{Z} = \left(R_k - \mu_1 Q_k^H\right)\left(R_{k-1} - \mu_2 Q_{k-1}^H\right)\ldots$$
$$\ldots \left(R_3 - \mu_{k-2}Q_3^H\right)\left(R_2 - \mu_{k-1}Q_2^H\right)\left(R_1 - \mu_k Q_1^H\right). \quad (9.12)$$

The first column of $Q_k^H \ldots Q_1^H \tilde{Z}$ has only the first $k+1$ elements different from zero as it is the product of k Hessenberg matrices. These elements can be computed easily as the matrices Q_k are a sequence of Givens transformations from bottom to top. For computing the first column of the matrix product $Q_k^H \ldots Q_1^H \tilde{Z}$, we efficiently compute the multiplication of the right-hand side of Equation (9.12) with $\mathbf{e}_1$.

Moreover combining all these transformations and perform them as a similarity transformation onto the matrix Z_0 gives us

$$Q_k^H \ldots Q_2^H Q_1^H Z_0 Q_1 Q_2 \ldots Q_k = Q_k^H \ldots Q_2^H Z_1 Q_2 \ldots Q_k$$
$$= Q_k^H \ldots Q_3^H Z_2 Q_3 \ldots Q_k$$
$$= Z_k.$$

This corresponds to performing k steps of the QR-method without shift onto the matrix Z_0, which can be performed efficiently in $\mathcal{O}(kn^2)$. (Especially in the semi-separable case one can perform these steps in linear time $\mathcal{O}(kn)$.)

To complete the reduction of $\mathbf{v}$ to $\pm\beta\mathbf{e}_1$, we have to annihilate k of the $k+1$ nonzero elements of the vector $Q_k^H \ldots Q_1^H \tilde{Z}\mathbf{e}_1$. This can be done by performing k Givens transformations $\hat{G}_k \ldots \hat{G}_1$ annihilating first the nonzero element in position $k+1$, followed by annihilating the element in position k and so forth. This results in the matrix

$$\hat{G}_k^H \ldots \hat{G}_1^H Z_k \hat{G}_1 \ldots \hat{G}_k,$$

which we have to bring back via similarity transformations to Hessenberg-like form without touching the first column and row.

Let us summarize this step. In order to start the implicit procedure, in the case of k shifts, we have to perform k steps of QR-without shift onto the matrix Z_0. Moreover we have to use these unitary transformations to compute the first column $Q_k^H \ldots Q_1^H \tilde{Z}\mathbf{e}_1$ in an efficient way. The result of these k QR-steps without shift onto Z_0 gives us again a Hessenberg-like matrix Z_k. The vector $Q_k^H \ldots Q_1^H \tilde{Z}\mathbf{e}_1$ is still not a multiple of $\mathbf{e}_1$ and hence an extra k Givens transformations are needed to transform this vector to $\pm\beta\mathbf{e}_1$[23]. These final k Givens transformations disturb the Hessenberg-like structure of the matrix Z_k. The result of a QR-step is again a Hessenberg-like matrix, hence we will develop in the next section a chasing technique for restoring the structure.

9.4.3   The chasing method

The chasing method consists of two parts. First we need to perform some transformations to prepare the matrix for the chasing procedure. Let us call this first part

---

[23]The performance of k steps of the QR-algorithm without shift do not dramatically increase the complexity of the multishift with regard to single shift strategy. Also in the single shift strategy a QR-step without shift is needed.

the initialization procedure. The second part is the effective chasing part. For simplicity we will demonstrate these techniques on a 7×7 matrix for $k = 3$ multishifts. The technique can be extended easily to larger matrices and more shifts.

Initialization

The matrix we are working with, Z_3 is of Hessenberg-like form. Denote the initial disturbing Givens transformations, three in this case, with $\hat{G}_1, \hat{G}_2$ and $\hat{G}_3$.

The chasing procedure needs to be performed onto the following matrix

$$\hat{G}_3^H \hat{G}_2^H \hat{G}_1^H Z_3 \hat{G}_1 \hat{G}_2 \hat{G}_3.$$

Recall that the transformation $\hat{G}_1^H$ acts on rows 3 and 4, $\hat{G}_2^H$ on rows 2 and 3 and $\hat{G}_3^H$ on rows 1 and 2. We will switch now to the graphical representation of the matrix Z_3. For theoretical considerations we consider the graphical schemes as QR-factorizations of the matrix $Z_3 = Q_3 R_3$. In the following scheme, we have to apply the transformations on the left to the matrix Q_3 and the transformations on the right are applied onto the upper triangular matrix R_3. We obtain the following scheme, in which the upper structured rank part, denoted by the elements $\boxtimes$ is created by performing the three Givens transformations on the right. More precisely in the scheme we see on the left $\hat{G}_3^H \hat{G}_2^H \hat{G}_1^H Q_3$, where the three disturbing Givens transformations are found in position $7, 8$ and 9 and the matrix Q_3 consisting of 6 Givens transformations is found in the positions 1 up to 6. The matrix on the right of the scheme is the matrix $R_3 \hat{G}_1 \hat{G}_2 \hat{G}_3$, the result of applying the three Givens transformations onto the upper triangular matrix. This new upper left part, denoted by $\boxtimes$ is in fact a small Hessenberg-like matrix.

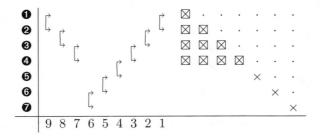

The initial step consists of removing this newly created small structured rank part denoted by $\boxtimes$. One can easily remove this part be performing three Givens transformations, just like when reducing a Hessenberg-like matrix to upper triangular form. In fact this is a swapping procedure (see Chapter 1, Section 1.3) in which the representation from right to left acting on the columns is transformed into a representation from top to bottom acting on the rows. This leads to the following scheme in which the first three Givens transformations (in positions $1, 2$ and 3) are chosen to annihilate the newly created part denoted by $\boxtimes$. At the same time we compressed the notation in the left part.

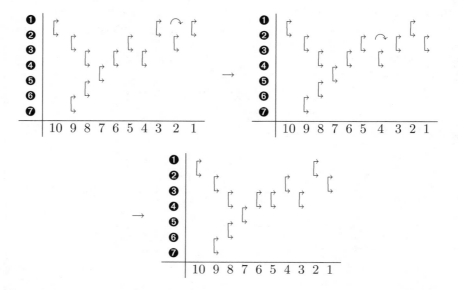

(9.13)

For the moment we do not need the upper triangular part on the right and we will therefor not depict it in the following schemes. We will now try to rearrange the Givens transformations shown in Scheme 9.13, in order to remove the Givens transformation in position 1. This is done by applying few times the shift through lemma. Interchanging the Givens transformations in positions 3 and 4 from Scheme 9.13, leads to the first scheme below, in which we already depicted the first application of the shift through lemma. The second scheme shows the result of applying the shift through lemma and indicates the next application. The bottom scheme is the result of applying the shift through lemma as indicated in the second scheme.

We have now finished applying the shift through lemma. Generally we need to apply it $k-1$ times. To complete the procedure, a fusion of the Givens transformations in positions 5 and 6 needs to be performed. In the left scheme you see on which Givens operations the fusion will act. The right scheme shows the result, and in the bottom scheme we have rewritten the scheme by interchanging the order of transformations 3 and 2.

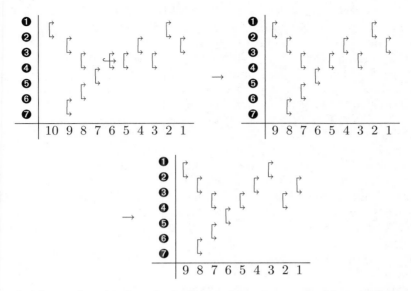

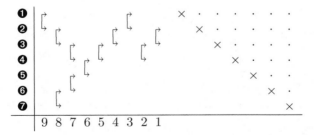

We can clearly see that this form is similar to the one we started from in Scheme 9.13, except for the fact that the Givens transformation in the first position of Scheme 9.13 is now removed. This is the structure on which we will perform the chasing method. This initialization procedure only has to be performed once before starting the chasing.

Restoring the structure

The scheme at the end of the previous section can be decomposed into four main parts. The complete scheme, including the matrix on which the transformations act is of the following form.

Let us discuss these four different parts and indicate the restoring procedure. A first part is the matrix on the right on which the Givens transformations proposed on the left act. The remaining three parts denote specific parts in the Givens transformations on the left. A first part contains the Givens transformations in positions 1 and 2. Generally this will contain in case of k shifts $k-1$ Givens transformations. Secondly we have the Givens transformations from the representation. This is the

sequence from top to bottom, ranging from position 3 until 8. Finally we have three
more Givens transformations in positions 7, 8 and 9 shown on the top left part of
the scheme. Generally there are k Givens transformations in this part.

The aim of the procedure is to remove the Givens transformations on the left
upper part and the ones in positions 1 and 2, to obtain only a sequence from top
to bottom which will denote the new representation of the Hessenberg-like matrix.
The right-hand side should remain an upper triangular matrix. This should be
accomplished by performing unitary similarity transformations onto the matrix.
For all transformations performed on the right-hand side, we cannot change the
first column: $\hat{Q}\mathbf{e}_1$ needs to be equal to $\mathbf{e}_1$.

The idea is to reshuffle the Givens transformations based on the shift through
operation and then apply them onto the upper triangular matrix, followed by a
structure restoring similarity transformation. This structure restoring similarity
transformation will result in a scheme similar to the one above. Only the undesired
Givens transformations in positions 1, 2, 7, 8 and 9 will have shifted down one
position. This procedure can hence be continued until these Givens transformations
slide off the scheme.

We start by rearranging the Givens transformations. First we change the ∨-
pattern on top ranging from positions 3 to 9 (see Scheme 9.14) to a ∧-pattern. This
transition is only possible if we shift the Givens transformations on the bottom, in
position 7 and 8 backwards. The new ∧-pattern is of course still located between
position 3 and 9 as shown in the right of Scheme 9.14.

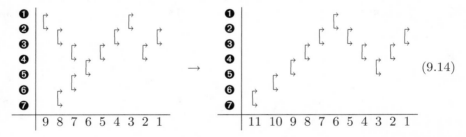

$$(9.14)$$

Now, see the right of Scheme 9.14, we have again a ∨-pattern ranging from position
1 to 5. We transform it to a ∧-pattern shown in the next scheme.

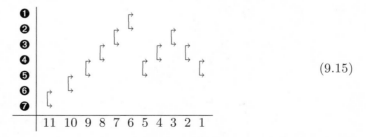

$$(9.15)$$

In fact we dragged the top three transformations in the positions 7,8 and 9
from the left scheme in Scheme 9.14 completely to the right located now in positions
1,2 and 3 in Scheme 9.15. Remark that instead of acting on the first four rows, they

act now on the rows 2, 3, 4 and 5. Moreover the Givens transformations located in the left scheme of Scheme 9.14 in the positions 1 and 2 can be found now in the positions 4 and 5 in Scheme 9.15, also shifted down one row.

To remove now the Givens transformations in the first three positions, we need to apply a similarity transformation. To achieve this goal, we need the upper triangular matrix R_3. Applying the first three Givens transformations onto the matrix R_3 results in the following scheme. The elements $\otimes$ are filled in by applying these Givens transformations onto the upper triangular matrix.

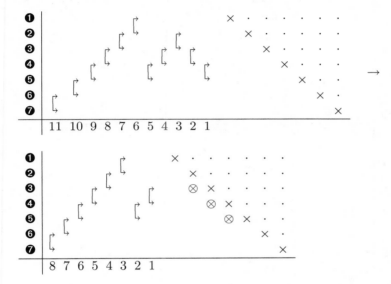

One can consider the bottom scheme above as $\hat{Q}_3\hat{R}_3$, in which $\hat{Q}_3$ denotes the combination of all Givens transformations and the matrix $\hat{R}_3$ denotes the upper triangular matrix disturbed in a few subdiagonal elements. To remove the bulges, denoted by $\otimes$ in the matrix $\hat{R}_3$, we apply three Givens transformations on the right acting on its columns $2, 3, 4$ and 5. As only similarity transformations are allowed for restoring the structure also three Givens transformations have to be applied onto the matrix $\hat{Q}_3$. This completes one step of the chasing procedure and results in the scheme below. The similarity transformation removed the bulges but introduced three new Givens transformations shown in positions $8, 9$ and 10.

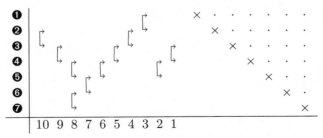

As a result we obtain that the unwanted Givens transformations have shifted down one row, and this completes the first similarity transformation in the chasing procedure.

Let us continue now in an identical way to determine the second similarity transformation. Rearranging the Givens transformations, by twice applying the transformation from a $\vee$ to a $\wedge$-pattern gives the top scheme and then applying the three right Givens transformations to the matrix gives us the bottom scheme.

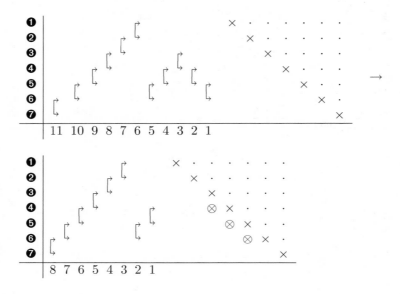

Again we have to apply a similarity transformation acting on the columns of the disturbed upper triangular matrix to remove the bulges, denoted by $\otimes$.

This results in the upper scheme in Scheme 9.16. Rewriting all the Givens transformations and applying them onto the upper triangular matrix reveals another similarity transformation that needs to be performed.

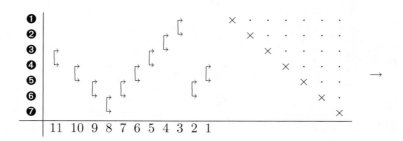

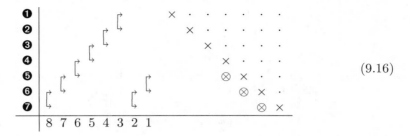

$$(9.16)$$

This chasing procedure needs to be performed as many times until the unwanted Givens transformations reach the bottom rows. We see that the Givens transformations in the positions 1 and 2 have now arrived at the bottom row. From now on the chasing procedure is finishing, and gradually all the undesired Givens transformations will vanish.

Scheme 9.17 denotes the result of applying the similarity transformation proposed by the scheme above to remove the bulges. We see in this scheme that the removing of the Givens transformations has started. The fusion of the Givens transformations in positions 9 and 8 is depicted.

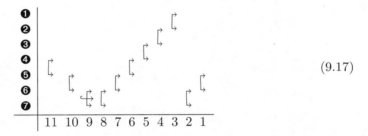

$$(9.17)$$

We do not show the upper triangular matrix as it does not change in the following operations. After the fusion we perform the change from the ∨ to the ∧-pattern (from the left of Scheme 9.18 to the right of Scheme 9.18). Again another fusion is indicated.

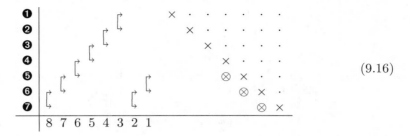

$$(9.18)$$

The fusion of the Givens transformations in positions 2 and 3 creates again the possibility of changing the ∨ to the ∧-pattern. This is depicted in the following scheme.

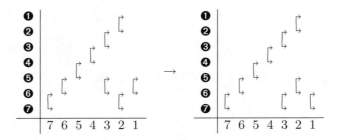

Applying the Givens transformations in positions 1 and 2 creates again bulges. This is shown in Scheme 9.19.

$$(9.19)$$

It is clear from Scheme 9.19 that the next similarity transformation, to remove the bulges, involves only two Givens transformations instead of three as in all previous chasings. Performing the similarity transformation and continuing this procedure will remove all unwanted Givens transformations. Let us depict the main steps in the schemes. Applying the similarity transformation creates the scheme below on the left. The fusion is depicted. Applying the fusion and rewriting the V-pattern leads to the rightmost scheme, in which again a fusion is depicted.

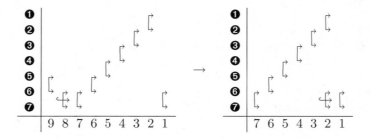

Applying the fusion and performing this transformation onto the upper triangular matrix results in one bulge in the bottom row shown in Scheme 9.20.

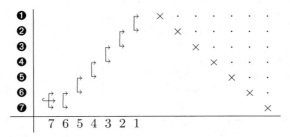

$$(9.20)$$

A final similarity transformation consisting of one Givens transformation needs to be performed. The similarity transformation, acting on the last two rows and the last two columns, leads to the following scheme, in which one final fusion needs to be performed.

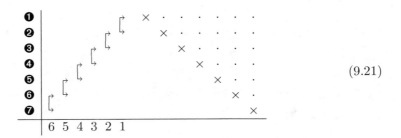

After the fusion, we obtain the new Givens-vector representation of the Hessenberg-like matrix shown in Scheme 9.21. We are now ready to apply a new step of the multishift QR-method.

$$(9.21)$$

Note 9.16. *There are other variations in the chasing procedure. Instead of applying the initialization procedure, thereby performing the initial transformations on the columns, and transforming them to Givens transformations acting on the rows, one can just leave them on the right, acting on the columns.*

The changed chasing procedure needs one swapping of the $\vee$-pattern to the $\wedge$-pattern and then immediately applying these transformations onto the rows. The created bulges then need to be removed by Givens transformations acting on the columns. As there are still existing Givens transformations acting on the columns *we have to perform another pattern change (but then on the right side of the matrix, instead of the left side), before one knows which similarity transformations need to be performed. Applying the similarity transformation introduces again some Givens transformations on the left. The reader can try to develop these schemes himself. It gets more complicated to depict the schemes because Givens transformations acting on the columns as well as on the rows are present now.*

The chasing procedure discussed above only involved Givens transformations performed on the columns not changing the first column. This means that combining all similarity transformations into the matrix $\hat{Q}$ satisfies $\hat{Q}\mathbf{e}_1 = \mathbf{e}_1$. Based on the implicit Q-theorem for Hessenberg-like matrices, we know that this procedure corresponds with a step of the multishift QR-algorithm on the matrix Z_0.

9.4.4   The real Hessenberg-like case

Especially important for real Hessenberg-like matrices is the fact that they only have real eigenvalues and complex conjugate eigenvalues. Based on a double shift strategy, we can hence restrict the computations of these eigenvalues to real operations (see [94]).

Assume we are working with the double shift strategy and the eigenvalues returned from the lower right 2×2 block B are complex conjugate eigenvalues μ_1 and μ_2. Considering Equation (9.12), we obtain the following relation:

$$Q_2^H Q_1^H \tilde{Z} = \left(R_2 - \mu_1 Q_2^H \right) \left(R_1 - \mu_2 Q_1^H \right), \tag{9.22}$$

in which $Z_0 = Q_1 R_1$, and $Z_1 = R_1 Q_1 = Q_2 R_2$. Rounding errors will however prevent the first column of the above equation to be real. The complex part of the first column, will however be small, with regard to the machine precision. Rearranging the terms in the above formula gives us the following:

$$\begin{aligned}
Q_2^H Q_1^H \tilde{Z} &= \left(R_2 - \mu_1 Q_2^H \right) \left(R_1 - \mu_2 Q_1^H \right), \\
&= R_2 R_1 + \mu_1 \mu_2 Q_2^H Q_1^H - \mu_1 Q_2^H R_1 - \mu_2 R_2 Q_1^H, \\
&= R_2 R_1 + \det\left(B \right) Q_2^H Q_1^H - \left(\mu_1 + \mu_2 \right) Q_2^H R_1, \\
&= R_2 R_1 + \det\left(B \right) Q_2^H Q_1^H - \operatorname{trace}\left(B \right) Q_2^H R_1.
\end{aligned}$$

Using the first column of the equation above ensures that all computations remain in the real field.

Notes and references

The results on this section are based on the ones of the following technical report. The report contains more details related to the implementation and many numerical experiments.

☞ R. Vandebril, M. Van Barel, and N. Mastronardi. A multiple shift QR-step for structured rank matrices. Technical Report TW489, Department of Computer Science, Katholieke Universiteit Leuven, Celestijnenlaan 200A, 3000 Leuven (Heverlee), Belgium, April 2007.

9.5 A QH-algorithm

In this section we will investigate a new type of iteration. We will see if it preserves the matrix structure, how its convergence properties are and how to implement it in an implicit way.

Let us comment a little on the origin of this iteration. Reconsider the $\vee$-pattern for computing the QR-factorization of a Hessenberg-like minus a shift matrix $Z - \mu I = \check{Q}_2 \check{Q}_1 R$, in which μ is a suitably chosen shift. It was mentioned (see Section 9.2) that the QR-algorithm related to this factorization could be decomposed into two steps. The complete algorithm performs a similarity transformation of the form $\check{Q}_2^H \check{Q}_1^H Z \check{Q}_1 \check{Q}_2$ this corresponds to a step of the QR-algorithm with shift. The first similarity transformation $\tilde{Z} = \check{Q}_1^H Z \check{Q}_1$ performs a chasing step, taking into consideration the shift. The second transformation $\check{Q}_2^H \tilde{Z} \check{Q}_2$ performs however a step of the QR-method without shift onto $\tilde{Z}$.

The above algorithm has, however, some kind of contradiction in it. Let us explain this in more detail. The convergence behavior of the QR without shift and the one of the QR with shift do not always cooperate. In some senses the second step can destroy the good improvements made by the first step.

Therefore, one can opt to simply omit the second similarity transform. This will give us the new method. Unfortunately one does not perform now a step of the QR-method anymore, but of the QH-method.

Based on the comments above we would like to use only the factor $\check{Q}_1$ for performing an orthogonal similarity transformation onto the matrix Z. As $\check{Q}_1$ is closely related to the QH-factorization, a naive approach would be the following:

$$Z - \mu I = \check{Q} \check{Z},$$

which is a QH-factorization of the matrix $Z - \mu I$. Define the new iteration as

$$\hat{Z} = Q^H Z \check{Q}.$$

Unfortunately this will create some problems as we will see further on in the text.

9.5.1 More on the QH-factorization

The QH-factorization will be the basic step in the new QH-method. Unfortunately this QH-factorization as proposed before, is not essentially unique. Hence we need more constraints to be posed on this factorization, similarly as we had to do for computing the QR-factorization of Hessenberg matrices (see Section 5.2) and Hessenberg-like matrices (see Subsection 6.1.1).

Since we will only focus onto the class of Hessenberg-like matrices (minus a shift), we will discuss only the essentials related to this class.

Example 9.17 Suppose we have the following matrix:

$$Z = \begin{bmatrix} 1 & 0 & 0 \\ 0 & 1 & 0 \\ 0 & 0 & 1 \end{bmatrix}.$$

This matrix is obviously already of Hessenberg-like form. Hence the factorization $Z = IZ$ is a QH-factorization. But in fact one can apply an arbitrary 2×2 Givens transformation acting on the last two rows, without disturbing the structure. This means that we have an infinite number of QH-factorizations for this matrix. ∎

One can also clearly see in Schemes 9.3 and 9.4 that the first three Givens transformations already applied to the matrix create the desired structure. This means that in general one needs $n - 2$ Givens transformations to obtain a matrix of the following form (e.g., a 4×4 matrix):

$$
Z = \begin{bmatrix}
\boxtimes & \times & \times & \times \\
\boxtimes & \boxtimes & \times & \times \\
\boxtimes & \boxtimes & \boxtimes & \times \\
\boxtimes & \boxtimes & \boxtimes & \boxtimes
\end{bmatrix}.
$$

This matrix is clearly of Hessenberg-like form and an arbitrary Givens transformation, acting on the last two rows can never destroy this low rank structure.

The freedom for computing this factorization has its direct impact onto the QH-method, as we cannot guarantee the preservation of the structure anymore. Later on we will show that we can guarantee this, after having defined our QH-factorization in a more rigid way.

Example 9.18 Suppose we have the following 3×3 matrix Z and a QH-factorization of this matrix. The given matrix Z is clearly a Hessenberg-like matrix, which has its structure preserved under the standard QR-algorithm. Let us construct a QH-factorization of this matrix:

$$
Z = \begin{bmatrix}
0 & 1 & 0 \\
1 & 0 & 0 \\
0 & 0 & 1
\end{bmatrix}
=
\begin{bmatrix}
0 & -1 & \\
1 & 0 & \\
 & & 1
\end{bmatrix}
\begin{bmatrix}
1 & & \\
 & 0 & -1 \\
 & 1 & 0
\end{bmatrix}
\begin{bmatrix}
1 & 0 & 0 \\
0 & 0 & 1 \\
0 & 1 & 0
\end{bmatrix}
$$
$$
= (\check{G}_1 \check{G}_2)\check{Z} = \check{Q}\check{Z},
$$

in which $\check{G}_1 \check{G}_2 = \check{Q}$, with G_1 and G_2 two Givens transformations and $\check{Z}$ a Hessenberg-like matrix. Performing the similarity transformation with the unitary matrix Q, we obtain:

$$
\hat{Z} = \check{Q}^H Z \check{Q} = \begin{bmatrix}
0 & 0 & 1 \\
0 & 1 & 0 \\
1 & 0 & 0
\end{bmatrix}.
$$

The new iterate $\hat{Z}$ after a step of the QH-method with this factorization is clearly not of Hessenberg-like form anymore. ∎

Hence, it is clear, that we have to pose some extra constraints onto the QH-factorization. Let us consider the following constructive procedure. Suppose we want to compute the QH-factorization of the matrix $Z - \mu I$, in which Z is a Hessenberg-like matrix and μ a suitably chosen shift, different from 0.

The computation of the QH-factorization consists of two steps. First we compute the RQ-factorization of the Hessenberg-like matrix Z as follows:

$$Z = RQ,$$

in which Q consists of a sequence of Givens transformations. In fact one poses the same constraints on the QR-factorization as posed on the QR-factorization of a Hessenberg-like matrix (see Subsection 6.1.1).

We obtain now:

$$Z - \mu I = RQ - \mu Q^H Q = (R - \mu Q^H)Q$$
$$= \check{Q}\check{R}Q,$$

where $\check{Q}\check{R} = R - \mu Q^H$, which is the QR-factorization of the left factor $H = R - \mu Q^H$ in the product. Since the matrix Q consists of a single sequence of Givens transformations going from left to right, it is a lower Hessenberg matrix. The summation $R - \mu Q^H$ is hence also of (upper) Hessenberg form, and one can easily compute its QR-factorization subjected to the constraints from Section 5.2.

We have computed now a QH-factorization of the original matrix $Z - \mu I$:

$$Z - \mu I = \check{Q}\check{R}Q = \check{Q}\check{Z},$$

with $\check{Z}$ a Hessenberg-like matrix. The matrix $\check{Z} = \check{R}Q$ has exactly the same Q-factor in its representation from right to left as the original matrix $Z = RQ$, only the upper triangular matrices $\check{R}$ and R differ. In some sense one can pose this as an extra constraint onto the computed factorizations.

Note 9.19. *Few remarks have to be made.*

- *When considering an irreducible Hessenberg-like matrix Z, one can easily prove uniqueness of the above factorization. Since the matrix Z is irreducible it has an essentially unique RQ-factorization, in which all Givens transformations differ from I. This implies that the corresponding Hessenberg matrix H is irreducible guaranteeing an essentially unique QR-factorization of H. Hence we obtain an essentially unique QH-factorization of the matrix Z.*

- *We posed the constraint that μ needed to be different from zero. In fact one can without loss of generality also consider $\mu = 0$. This will however always lead to a trivial factorization of the following form:*

$$Z = IZ = \check{Q}\check{Z},$$

  with $\check{Q} = I$ and $\check{Z} = Z$.

- *Reconsidering now both examples above, we see that they don't match our constructive procedure.*

The new iteration for computing the eigenvalues of a Hessenberg-like matrix is of the following form.

Definition 9.20. *Assume a Hessenberg-like matrix Z is given and we have shift μ (with RQ an RQ-factorization of Z):*

$$
\begin{aligned}
Z - \mu I &= RQ - \mu Q^H Q \\
&= (R - \mu I Q^H)Q \\
&= \check{Q}\check{R}Q \\
&= \check{Q}\check{Z},
\end{aligned}
$$

which gives us a specific QH-factorization of the matrix $Z - \mu I$.
The new iterate is defined as follows

$$
\begin{aligned}
\hat{Z} &= \check{Z}\check{Q} + \mu I \\
&= \check{Q}^H Z \check{Q},
\end{aligned}
$$

9.5.2   Convergence and preservation of the structure

There are several ways to prove the convergence and preservation of the structure. The report [167] discusses both a direct proof based on the formulas above and it discusses the results via transforming the problem to the standard QR-method. Based on the results presented in [168] we can easily prove the desired results.

A QR-iteration driven by a rational function

Let us interpret the QH-iteration in terms of a QR-iteration driven by a rational function. The analysis presented here is similar to the one in [179, 182, 177] and is a special case of the results presented in [168].

As discussed in the previous section, the global iteration is of the following form:

$$
\begin{aligned}
Z &= RQ, \\
Z - \mu I = (R - \mu Q^H)Q &= \check{Q}\check{R}Q, \\
\hat{Z} &= \check{Q}^H Z \check{Q},
\end{aligned}
$$

where $\hat{Z}$ defines the new iterate in the method.

One can rewrite the above formulas and obtain that the matrix $\check{Q}$ is the Q-factor in the QR-factorization of the matrix product $(Z - \mu I)Z^{-1}$:

$$
\begin{aligned}
(Z - \mu I)Z^{-1} &= \left(\check{Q}\check{R}Q\right)\left(Q^H R^{-1}\right) \\
&= \check{Q}\check{R}R^{-1}.
\end{aligned}
$$

This formula illustrates that we have computed the unitary factor of a special function in Z. Using the knowledge that we perform a similarity transformation defined by a rational function in the matrix Z one can use most of the results known for multishift QR-methods. Let us see what we get.

Preservation of the structure

When proving preservation of structure in the multishift case, one simply decouples the different factors and shows that the multishift step corresponds to several times performing a single shifted step. We will do the same here.

We know that
$$(Z - \mu I)Z^{-1} = \check{Q}\check{R}R^{-1}.$$

We will prove that $\hat{Z} = \check{Q}^H Z \check{Q}$ has also the Hessenberg-like structure.

The equation above can be decomposed into two steps.

- **Step 1.**

$$Z = RQ,$$
$$\tilde{Z} = QZQ^H.$$

This corresponds to performing a step of the RQ-method onto the matrix Z. This readily implies that the matrix $\tilde{Z}$ inherits the structure of the original matrix Z. Hence $\tilde{Z}$ is also of Hessenberg-like form.

- **Step 2.**

$$(\tilde{Z} - \mu I) = \tilde{Q}\tilde{R},$$
$$\hat{\tilde{Z}} = \tilde{Q}^H \tilde{Z}\tilde{Q}.$$

Since $\tilde{Z}$ is of Hessenberg-like form and this structure is preserved under the QR-iteration, we know that also the matrix $\hat{\tilde{Z}}$ will be of Hessenberg-like form.

The only thing that remains to be proven: Is $\hat{\tilde{Z}}$ essentially equivalent to $\hat{Z}$. The following relations show that the matrices $\check{Q}$ and $Q^H\tilde{Q}$ are essentially the same and hence they determine essentially the same similarity transformation.

$$\begin{aligned}
\check{Q}\check{R}R^{-1} = (Z - \mu I)Z^{-1} &= (Z - \mu I)Q^H R^H \\
&= Q^H Q(Z - \mu I)Q^H R^H \\
&= Q^H(\tilde{Z} - \mu I)R^H \\
&= Q^H \tilde{Q}\tilde{R}R^H.
\end{aligned}$$

As a result the structure of the matrix Z is preserved.

Convergence

We have the following QR-factorization for the rational function in Z.

$$(Z - \mu I)Z^{-1} = \check{Q}\check{R}R^{-1}.$$

This means that the convergence properties of the iteration performed on the matrix Z are defined by the subspace convergence properties, defined by the rational function $p(\lambda) = (\lambda - \mu)\lambda^{-1}$.

These convergence properties, and results for a general rational iteration of the form $p(\lambda) = (\lambda - \mu)(\lambda - \kappa)^{-1}$ were discussed in [168].

Without loss of generality we can adapt Theorem 3.11 to make it suitable for rational functions. We obtain the following.

Theorem 9.21 (Theorem 5.1 from [182]). *Given a semisimple matrix $A \in \mathbb{C}^{n \times n}$ with $\lambda_1, \lambda_2, \ldots, \lambda_n$ the eigenvalues and the associated linearly independent eigenvectors $\mathbf{v}_1, \mathbf{v}_2, \ldots, \mathbf{v}_n$. Let $V = [\mathbf{v}_1, \mathbf{v}_2, \ldots, \mathbf{v}_n]$ and let κ_V be the condition number of V, with regard to the spectral[24] norm. Let k be an integer $1 \leq k \leq n - 1$, and define the invariant subspaces $\mathcal{U} = \langle \mathbf{v}_{k+1}, \ldots, \mathbf{v}_n \rangle$ and $\mathcal{T} = \langle \mathbf{v}_1, \ldots, \mathbf{v}_k \rangle$. Denote with $(p_i)_i$ a sequence of rational functions and let $\hat{p}_i = p_i \ldots p_2 p_1$. Suppose that the*

$$p_i(\lambda_j) \neq 0 \quad j = 1, \ldots, k$$
$$p_i(\lambda_j) \neq \pm\infty \quad j = k+1, \ldots, n$$

for all i, and let

$$\hat{r}_i = \frac{\max_{k+1 \leq j \leq n} |\hat{p}_i(\lambda_j)|}{\min_{1 \leq j \leq k} |\hat{p}_i(\lambda_j)|}.$$

Let $\mathcal{S}$ be a k-dimensional subspace of $\mathbb{C}^n$, satisfying

$$\mathcal{S} \cap \mathcal{U} = \{0\}.$$

Let $\mathcal{S}_i = \hat{p}_i(A)\mathcal{S}_0, i = 1, 2, \ldots$, with $\mathcal{S}_0 = \mathcal{S}$. Then there exists a constant C (depending on $\mathcal{S}$) such that for all i,

$$d(\mathcal{S}_i, \mathcal{T}) \leq C \, \kappa_V \, \hat{r}_i.$$

In particular $\mathcal{S}_i \to \mathcal{T}$ if $\hat{r}_i \to 0$. More precisely we have that

$$C = \frac{d(V^{-1}\mathcal{S}, V^{-1}\mathcal{T})}{\sqrt{1 - d(V^{-1}\mathcal{S}, V^{-1}\mathcal{T})}}$$

Using Lemma 3.12, we can relate the convergence of the subspaces towards the vanishing of subblocks in a matrix.

Let us compare the convergence behavior of this new iteration with regard to the standard QR-iteration with shift μ. We consider only one iteration, i.e., r denotes the contraction rate from step i in the iteration process. For the standard QR-algorithm we obtain the following contraction ratio:

$$r^{(QR)} = \frac{\max_{k+1 \leq j \leq n} |\lambda_j - \mu|}{\min_{1 \leq j \leq k} |\lambda_j - \mu|}. \tag{9.23}$$

We introduce the following constants:

$$\omega = \min_{k+1 \leq j \leq n} \{|\lambda_j|\},$$
$$\Omega = \max_{1 \leq j \leq k} \{|\lambda_j|\}.$$

---

[24]The spectral norm is naturally induced by the $\|.\|_2$ norm on vectors.

Calculating now an upper bound for the convergence towards the eigenvalue closest to the shift μ, for the QH-method gives us:

$$r^{(QH)} = \max_{k+1\leq j\leq n} \left| \frac{\lambda_j - \mu}{\lambda_j} \right| \max_{1\leq j\leq k} \left| \frac{\lambda_j}{\lambda_j - \mu} \right| \leq \frac{\Omega}{\omega} \frac{\max_{k+1\leq j\leq n} |\lambda_j - \mu|}{\min_{1\leq j\leq k} |\lambda_j - \mu|} = \frac{\Omega}{\omega} r^{(QR)}$$

This indicates that convergence of the new iteration is comparable (up to a constant) with the convergence of the standard QR-method. This constant will only create a small, neglectable delay in the convergence. This means that if the traditional QR-method converges to an eigenvalue in the lower right corner, the QH-method will also converge. Hence, to obtain convergence to a specific eigenvalue λ_j, we choose μ close to this eigenvalue. The convergence results prove that this eigenvalue will then be revealed in both the QR- and the QH-method in the lower right corner.

Moreover we also have extra convergence, which is not present in the standard QR-case, and which is created by the factor λ^{-1} in the rational functions.

Define the following constants:

$$\Delta = \max_{k+1\leq j\leq n} \{|\lambda_j - \mu|\},$$
$$\delta = \min_{1\leq j\leq k} \{|\lambda_j - \mu|\}.$$

Similarly as above, we can define the following contraction ratio, for all k.

$$r = \frac{\Delta}{\delta} \frac{\max_{1\leq j\leq k} |\lambda_j|}{\min_{k+1\leq j\leq n} |\lambda_j|}.$$

Assume now (without loss of generality), all eigenvalues to be ordered, i.e., $|\lambda_1| \leq |\lambda_2| \leq \ldots \leq |\lambda_n|$. This means that our convergence rate can be simplified as follows:

$$r = \frac{\Delta}{\delta} \frac{|\lambda_k|}{|\lambda_{k+1}|}.$$

This means that we get a contraction for all k determined by the ratio λ_k/λ_{k+1}. This is a basic nonshifted subspace iteration taking place for all k at the same time. Remark that this convergence takes place in addition to the convergence imposed by the shift μ, which can force for example extra convergence towards the bottom right element.

More information on this specific type of subspace iteration can be found in the report [168].

Interesting is the relation between the reduction to Hessenberg-like form and the QH-iteration.

Note 9.22. *The unitary similarity reduction of an arbitrary matrix to Hessenberg-like form had an extra convergence property with regard to the traditional reduction to tridiagonal form. In every step of the reduction process also a kind of nested nonshifted subspace iteration took place. This nested nonshifted subspace iteration can also be found here in the new QH-iteration. The standard convergence results of the QR-iteration are present, plus an extra subspace iteration convergence.*

Let us see now how to implement this algorithm in an implicit manner.

9.5.3   An implicit QH-iteration

Even though the presented theoretical results might be complicated to prove, the actual implementation is quite simple, more simple than the implementation of the QR-method.

In this section, we will derive the implicit chasing technique developed for Hessenberg-like matrices.

An implicit algorithm

In this section we will design an implicit way for performing an iteration of the QH-method onto a Hessenberg-like matrix.

Based on the results above we can compute the following factorization:

$$Z - \mu I = \check{Q}\check{Z},$$

the matrix $\check{Q}$ is then used for performing a unitary similarity transformation onto the matrix Z:

$$\hat{Z} = \check{Q}^H Z \check{Q}.$$

The idea of the implicit method is to compute the unitary similarity transformation $\check{Q}^H Z \check{Q}$, based on only the first column of $\check{Q}$ and on the fact that the matrix $\hat{Z}$ satisfies some structural constraints. This idea is completely similar to the implicit QR-step for tridiagonal matrices [129, 94] (and also semiseparable matrices [164]).

Because the matrix $\check{Q} = \check{G}_1 \check{G}_2 \ldots \check{G}_{n-1}$ consists of a descending sequence of $n-1$ Givens transformations only the first Givens transformation $\check{G}_1$ is necessary for determining the first column of the matrix $\check{Q}$. Having determined this Givens transformation, it will be applied onto the matrix Z disturbing thereby the Hessenberg-like structure. The remaining $n-2$ Givens transformations are constructed in such a manner to restore the structure of the Hessenberg-like matrix.

After having performed these transformations, we know, based on the implicit Q-theorems for Hessenberg-like matrices (see Theorem 6.28), that we have performed a step of the QH-method in an implicit manner.

Computing the initial disturbing Givens transformation

Similarly as in the traditional QR-method for Hessenberg-like matrices, we assume our matrix to be irreducible (see Subsection 6.2.1).

For the actual implementation, we assume the Hessenberg-like matrix Z to be represented with the Givens-vector representation. This can be seen as the QR-factorization of this matrix $Z = QR$. The matrix $Q = G_{n-1} G_{n-2} \ldots G_1$ can be factored as a sequence of Givens transformations, where each Givens transformation G_i acts on two successive rows i and $i+1$. Graphically this representation $Z = QR$

is depicted as follows:

$$
\begin{array}{c|ccccc}
\mathbf{1} & & & \times & \times & \times & \times & \times \\
\mathbf{2} & & & & \times & \times & \times & \times \\
\mathbf{3} & & & & & \times & \times & \times \\
\mathbf{4} & & & & & & \times & \times \\
\mathbf{5} & & & & & & & \times \\
\hline
& 4 & 3 & 2 & 1 &
\end{array}
\tag{9.24}
$$

The Givens transformations in positions 1 to 4 make up the matrix Q, and the upper triangular matrix R is shown on the right.

We have the following equations

$$
Z - \mu I = QR - \mu I = Q\left(R - \mu Q^H\right) = Q\left(R - H\right),
$$

in which H is a Hessenberg matrix. For computing the QH-factorization, we want to apply now a sequence of descending Givens transformations onto this matrix $Z - \mu I$ such that we obtain a Hessenberg-like matrix $\check{Z}$. We remark that the construction of the QH-factorization here corresponds to the construction used before, preserving the dependencies in the low rank part. Here however the procedure is optimized, for working with the Givens-vector representation.

Using the graphical representation we can represent $Q\left(R - H\right)$ as follows (the Givens transformations making up the matrix Q are shown on the left, whereas the Hessenberg matrix $R - H$ is shown on the right).

$$
\begin{array}{c|ccccc}
\mathbf{1} & & & \times & \times & \times & \times & \times \\
\mathbf{2} & & & \otimes & \times & \times & \times & \times \\
\mathbf{3} & & & & \times & \times & \times & \times \\
\mathbf{4} & & & & & \times & \times & \times \\
\mathbf{5} & & & & & & \times & \times \\
\hline
& 4 & 3 & 2 & 1 &
\end{array}
$$

The element marked with $\otimes$ should be annihilated because we want to obtain a Givens-vector representation of a new Hessenberg-like matrix, namely $\check{Z}$, as in Scheme 9.24. Removing this element by placing a new Givens transformation in position one, and applying the indicated fusion, gives us the following result.

$$
\begin{array}{c|ccccc}
\mathbf{1} & & & \times & \times & \times & \times & \times \\
\mathbf{2} & & & 0 & \times & \times & \times & \times \\
\mathbf{3} & & & & \times & \times & \times & \times \\
\mathbf{4} & & & & & \times & \times & \times \\
\mathbf{5} & & & & & & \times & \times \\
\hline
& 4 & 3 & 2 & 1 &
\end{array}
\;\rightarrow\;
\begin{array}{c|ccccc}
\mathbf{1} & & & \times & \times & \times & \times & \times \\
\mathbf{2} & & & 0 & \times & \times & \times & \times \\
\mathbf{3} & & & \otimes & \times & \times & \times \\
\mathbf{4} & & & & \times & \times & \times \\
\mathbf{5} & & & & & \times & \times \\
\hline
& 4 & 3 & 2 & 1 &
\end{array}
$$

Annihilating the element marked in position $(3, 2)$, by a Givens transformation and performing the shift-through operation at the indicated position we obtain the following scheme.

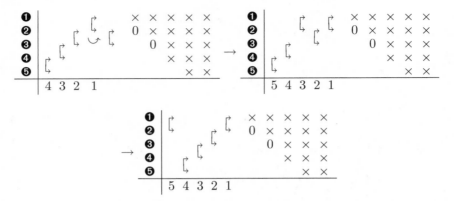

We remark that the last scheme still represents the original matrix $Z - \mu I$. Due to the rewriting of the matrix we can however clearly see that performing the hermitian conjugate of the Givens transformation in position 5 to the left of the matrix $Z - \mu I$, will give already a Hessenberg-like structure in the upper left corner of this matrix. This is because this upper left part is already represented in the Givens-vector representation.

When continuing this process, another Givens transformation annihilating the element in position $(4, 3)$ can be dragged through the representation, and so forth. As a result we obtain the QH-factorization of the matrix $Z - \mu I$, constructed simply by applying few times the shift through operation. The reader can verify that the constructed QH-factorization coincides with the desired one from the previous subsections. Moreover, as a result of these computations we have immediately the matrix $\check{Z}$ in its Givens-vector representation.

For our purposes, however, only the transpose of the initial Givens transformation, working on rows 1 and 2, and presented in position 5 on the rightmost scheme above, is needed. Having calculated this Givens transformation, we can apply it as a similarity transformation onto the matrix Z and then to complete the implicit chasing procedure, restore the structure of this matrix, never interfering anymore with the first column and row. We illustrate how to restore the structure of this matrix based on an initial disturbing Givens transformation.

Restoring the structure

We have a Hessenberg-like matrix Z. We know that after a step of the QH-method our resulting matrix will be a Hessenberg-like matrix $\hat{Z}$.

After having computed the initial disturbing Givens transformation, we will apply this transformation onto the matrix Z. Let us apply the transformation onto the matrix $Z = Z_0$, which is represented by its Givens-vector representation.

In the following schemes we show the effect of performing the similarity transformation $\check{G}_1$ acting on the matrix Z_0. For simplicity reasons we assume our matrix to be of size 5×5. Let us name $Z_1 = \check{G}_1^H Z_0 \check{G}_1$.

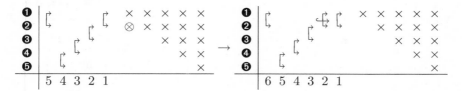

The transformation $\check{G}_1$ applied to the right creates the bulge, marked with $\otimes$, whereas the Givens transformation $\check{G}_1^H$ applied to the left can be found in position 5. The bulge marked with $\otimes$ can be annihilated by a Givens transformation as depicted above.

In the following scheme, we have combined the Givens transformations in positions 1 and 2, by a fusion. We have moved the transformation from position 6 to position 3 and we depicted where to apply the shift-through lemma. The right scheme shows the result after having applied the shift-through lemma and after having created the bulge, marked with $\otimes$.

We see clearly that we get the traditional chasing pattern, which is also present in the standard QR-method. See for example Subsection 8.1.2 in which the chasing is presented for the right left representation and Subsection 8.1.3 in which the chasing is presented for semiseparable plus diagonal matrices.

Applying similar techniques as in these sections provides us the chasing step in the QH-method. The only thing which is significantly different is the computation of the initial disturbing Givens transformation.

To conclude this section we will show another interesting relation between the QH-iteration and the QR-iteration.

9.5.4   The QR-iteration is a disguised QH-iteration

Even though it was not discussed, one can also apply similar techniques onto Hessenberg-like plus diagonal matrices. The convergence analysis gets a little more complicated, but essentially the same results remain true. Based on the fact that it is valid for Hessenberg-like plus diagonal matrices we can state the following.

In the previous part of the section we constructed a QH-factorization to make the QH-method suitable for working with Hessenberg-like (plus diagonal) matrices.

Let us see now how we can easily construct a QH-factorization of a Hessenberg matrix.

Let us compute now the QH-factorization of a Hessenberg matrix, based on a sequence of descending Givens transformations. Remark that a Hessenberg matrix has already the strictly lower triangular part of semiseparability rank 1. Hence the descending sequence of Givens transformations will be constructed in such a way to expand the strictly lower triangular rank structure, to include the diagonal into it. Let us first consider the structure of the involved Givens transformations.

Corollary 9.23. *Suppose the row* $[e, f]$ *and the following* 2×2 *matrix are given*

$$A = \begin{bmatrix} a & b \\ c & d \end{bmatrix}$$

Then there exists a Givens transformation

$$G = \frac{1}{\sqrt{1 + |t|^2}} \begin{bmatrix} \bar{t} & -1 \\ 1 & t \end{bmatrix} \tag{9.25}$$

such that the second row of the matrix $G^H A$, *and the row* $[e, f]$ *are linearly dependent. The value* t *in the Givens transformation* G *as in* (9.25), *is defined as*

$$\bar{t} = \frac{af - be}{cf - de},$$

under the assumption that $cf - de \neq 0$, *otherwise one could have taken* $G = I_2$.

Proof. The proof involves straightforward computations. (See also Corollary 9.43 in Volume I.)   □

Hence, we want to apply a sequence of successive Givens transformations onto the Hessenberg matrix H, to obtain the QH-factorization. Denote the diagonal elements of the Hessenberg matrix as $[a_1, \ldots, a_n]$ and the subdiagonal elements as $[b_1, \ldots, b_{n-1}]$. The first Givens transformation acts on rows 1 and 2 and only the first two columns are important, we have the following matrix A (as in the corollary):

$$A = \begin{bmatrix} a_1 & h_{1,2} \\ b_1 & a_2 \end{bmatrix}, \tag{9.26}$$

and we want to make the last row dependent of $[0, b_2]$. A Givens transformation with t defined as $t = \frac{a_1 b_2}{b_1 b_2} = \frac{a_1}{b_1}$, is found (assuming b_1 and b_2 to be different from zero).

Computing the product $G^H A$ gives us the following equalities:

$$G^H A = \frac{1}{\sqrt{1 + t^2}} \begin{bmatrix} \bar{t} & 1 \\ -1 & t \end{bmatrix} \begin{bmatrix} a_1 & b_1 \\ b_1 & a_2 \end{bmatrix} = \begin{bmatrix} \times & \times \\ 0 & \times \end{bmatrix}.$$

One can continue this process, and as a result we obtain the following equations:

$$H = \check{Q}\check{Z} = QR.$$

The Hessenberg-like matrix $\check{Z}$ will become an upper triangular matrix. Hence in this case the new QH-factorization coincides with the traditional QR-factorization and therefore also the QR-algorithm for Hessenberg (as well as tridiagonal) matrices fits perfectly into the framework.

9.5.5   Numerical experiments

In this section we will illustrate the speed and accuracy of the proposed method in case of symmetric semiseparable matrices.

Experiment 9.24. *In the following experiment we constructed arbitrary symmetric semiseparable matrices, and computed their eigenvalues via the traditional QR-method for semiseparable matrices (the implementation from [166] was used). These eigenvalues were compared with the algorithm described in this section. The eigenvalues computed by the* MATLAB *routine* EIG *were used to compare both solutions with. The following relative error norm was used: denote the vectors containing the eigenvalues as Λ, Λ_{QH} and Λ_{QR} for, respectively,* EIG, *the QH and the QR-method. The plotted error value equals*

$$\frac{\|\Lambda - \Lambda_{QH}\|}{\|\Lambda\|} \quad and \quad \frac{\|\Lambda - \Lambda_{QR}\|}{\|\Lambda\|},$$

for both methods. Five experiments were performed, and the line denotes the average accuracy of all five experiments combined. The x-axis denotes the problem sizes, ranging from 100 to 700, via steps of size 50. The cut-off criterion was chosen equal to 10^{-8}. In the Figures 9.2 and 9.3, the $\circ$'s denote the results of individual experiments of the QR-iteration, whereas the $\star$'s denote these of the QH-iteration.

It can be seen in Figure 9.2 that the QH-iteration is a little more accurate than the QR-method.

Figure 9.3 shows the average number of iterations per eigenvalue and the cputimings in seconds for both methods, we see that the new method needs in average much less iterations.

Notes and references

The results presented in this section are a condensed form of the ones presented in the technical report.

☞ R. Vandebril, M. Van Barel, and N. Mastronardi. A new iteration for computing the eigenvalues of semiseparable (plus diagonal) matrices. Technical Report TW507, Department of Computer Science, Katholieke Universiteit Leuven, Celestijnenlaan 200A, 3000 Leuven (Heverlee), Belgium, October 2007.

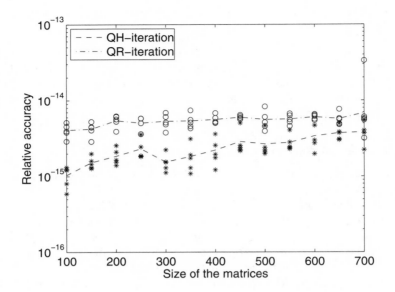

Figure 9.2. Accuracy comparison

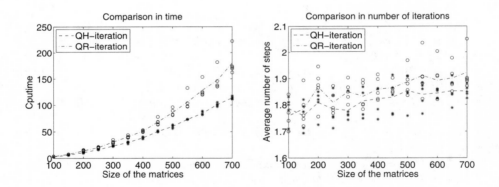

Figure 9.3. Timings and iterations

This report contains a more detailed analysis of the method presented here. Various interpretations are given, including different proofs for the preservation of the structure and the convergence of the proposed method. The presented results are also not restricted to the class of Hessenberg-like matrices, but results related to Hessenberg-like plus diagonal matrices are also proposed.

9.6 Computing zeros of polynomials

In this section we will discuss the computation of roots of polynomials. This problem is closely related to the computation of eigenvalues of specific structured rank matrices.

The section is subdivided into different parts. In the first subsection we will link the computation of roots of polynomials to eigenproblems for structured rank matrices. Since some of these eigenproblems involve unitary matrices, Subsection 9.6.2 discusses the Givens-weight representation for unitary matrices and shows how to interpret the QR-algorithm of unitary Hessenberg matrices in this case. Subsection 9.6.3 discusses how to compute the eigenvalues of Hessenberg matrices, which are of unitary plus low rank form. Finally some references related to this subject will be discussed.

9.6.1 Connection to eigenproblems

In this section we will provide a general theory on the connection between zeros of polynomials and eigenvalues of structured matrices. These results will lead directly to the definition of companion and fellow matrices. More information related to this subject can be found in [13].

We will start by investigating the relation for polynomials represented in the classical basis.

The classical basis

Suppose we are working with a polynomial $p(z)$ either in $\mathbb{R}_n[z]$ or $\mathbb{C}_n[z]$. The subscript n in $\mathbb{R}_n[z]$ and $\mathbb{C}_n[z]$ denotes that the considered polynomials have degree less than or equal to n.

For simplicity we will start by working in the classical basis. Assume our basis polynomials $\varphi_0(z), \varphi_1(z), \dots$ are chosen equal to $1, z, z^2, \dots$ which is the classical basis.

An arbitrary polynomial $p(z)$ can be written as follows:

$$p(z) = p_0 + p_1 z + p_2 z^2 + \dots + p_n z^n$$
$$= p_0 \varphi_0(z) + p_1 \varphi_1(z) + p_2 \varphi_2(z) + \dots + p_n \varphi_n(z).$$

Without loss of generality we can assume p_n to be different from zero. Clearly the zeros of $p(z)$ equal the zeros of $p(z)/p_n$, therefore we can assume in the remainder the polynomial $p(z)$ to be monical.

The following relations are obtained for the classical basis:

$$z \left[\varphi_0(z), \varphi_1(z), \ldots, \varphi_{n-1}(z)\right]$$

$$= \left[\varphi_0(z), \varphi_1(z), \ldots, \varphi_{n-1}(z), \varphi_n(z)\right] \begin{bmatrix} 0 & & & & & \\ 1 & 0 & & & & \\ 0 & 1 & 0 & & & \\ & \ddots & \ddots & \ddots & & \\ & & & 0 & 1 & 0 \\ 0 & \cdots & & & 0 & 1 \end{bmatrix}$$

$$= \left[\varphi_0(z), \varphi_1(z), \ldots, \varphi_{n-1}(z), \varphi_n(z)\right] \left[\frac{\hat{J}}{\mathbf{e}_n^T} \right]$$

$$= \left[\varphi_0(z), \varphi_1(z), \ldots, \varphi_{n-1}(z), \varphi_n(z)\right] J. \tag{9.27}$$

The matrix J is a rectangular matrix having $n+1$ rows and n columns and a square leading principal submatrix $\hat{J}$ of size $n \times n$.

Rewriting the polynomial $p(z)$ in matrix formulation we obtain the following (assume $p_n = 1$):

$$p(z) = \left[\varphi_0(z), \varphi_1(z), \ldots, \varphi_n(z)\right] \begin{bmatrix} p_0 \\ p_1 \\ \vdots \\ p_{n-1} \\ \hline 1 \end{bmatrix} = \left[\frac{\hat{\mathbf{p}}}{1} \right]$$

$$= \boldsymbol{\varphi}_{0:n}(z)^T \mathbf{p}.$$

We will use the following notation $\mathbf{p} = \left[\hat{\mathbf{p}}^T, 1\right]^T = [p_0, p_1, \ldots, p_{n-1}, p_n]^T$, for any kind of polynomial $p(z)$ and $\boldsymbol{\varphi}_{i:j}(z)^T = [\varphi_i(z), \varphi_{i+1}(z), \ldots, \varphi_n(z)]$. The formula above can easily be expanded, using Equation (9.27), to obtain a relation in which $[\varphi_0(z), \ldots, \varphi_{n-1}(z)] = \boldsymbol{\varphi}_{0:n}(z)^T$ and $p(z)$ are shown on the left-hand side:

$$\left[\boldsymbol{\varphi}_{0:n-1}(z)^T, p(z)\right] = \boldsymbol{\varphi}_{0:n}(z)^T \begin{bmatrix} 1 & & & & p_0 \\ & 1 & & & p_1 \\ & & \ddots & & \vdots \\ & & & 1 & p_{n-1} \\ & & & & 1 \end{bmatrix}$$

$$= \boldsymbol{\varphi}_{0:n}(z)^T \left[\begin{array}{c|c} I & \hat{\mathbf{p}} \\ \hline 0 & 1 \end{array} \right]$$

$$= \boldsymbol{\varphi}_{0:n}(z)^T R_{\mathbf{p}}. \tag{9.28}$$

The matrix $R_{\mathbf{p}}$ is an invertible upper triangular matrix, associated to the polynomial $p(z)$. It is fairly easy to determine the inverse of this matrix, which we will naturally denote as $R_{-\mathbf{p}}$:

$$R_{\mathbf{p}}^{-1} = \begin{bmatrix} 1 & & & & -p_0 \\ & 1 & & & -p_1 \\ & & \ddots & & \vdots \\ & & & 1 & -p_{n-1} \\ & & & & 1 \end{bmatrix} = R_{-\mathbf{p}}. \qquad (9.29)$$

Equation (9.27), can be rewritten in the following way:

$$\boldsymbol{\varphi}_{0:n}(z)^T = \left[\boldsymbol{\varphi}_{0:n-1}(z)^T, p(z)\right] R_{-\mathbf{p}}.$$

Substituting this in Equation (9.27) gives us to the following important relations, which will lead to the companion matrix:

$$\begin{aligned} z\boldsymbol{\varphi}_{0:n-1}(z)^T &= \left[\boldsymbol{\varphi}_{0:n-1}(z)^T, p(z)\right] R_{-\mathbf{p}} J \\ &= \left[\boldsymbol{\varphi}_{0:n-1}(z)^T, p(z)\right] \left[\begin{array}{c|c} I & -\hat{\mathbf{p}} \\ \hline 0 & 1 \end{array}\right] \left[\begin{array}{c} \hat{J} \\ \hline \mathbf{e}_n^T \end{array}\right] \\ &= \left[\boldsymbol{\varphi}_{0:n-1}(z)^T, p(z)\right] \left[\begin{array}{c} \hat{J} - \hat{\mathbf{p}}\mathbf{e}_n^T \\ \hline \mathbf{e}_n^T \end{array}\right] \\ &= \left[\boldsymbol{\varphi}_{0:n-1}(z)^T, p(z)\right] \left[\begin{array}{c} C_p \\ \hline \mathbf{e}_n^T \end{array}\right], \end{aligned} \qquad (9.30)$$

in which $C_p = PR_{-\mathbf{p}}J = \hat{J} - \mathbf{p}\mathbf{e}_n^T$ is the companion matrix associated to the polynomial p. The matrix P is the projection matrix $P = [I, 0]$ having $n + 1$ columns and n rows.

More precisely one can define a companion matrix formally as follows.

Definition 9.25. *Given a polynomial $p(z) = p_0 + p_1 z + p_2 z^2 + \ldots + p_n z^n$, with $(p_n = 1)$ the associated companion matrix is an $n \times n$ matrix of the following form:*

$$C_p = \begin{bmatrix} 0 & & & -p_0 \\ 1 & \ddots & & -p_1 \\ & \ddots & & \vdots \\ & & 1 & -p_{n-1} \end{bmatrix}. \qquad (9.31)$$

Note 9.26. *One can fairly easy see that the eigenvalues of the companion matrix coincide with the zeros of the associated polynomial $p(z)$, because $p(z) = \det(C_p - zI)$. Nevertheless, the construction provided here is much more general as will be shown later on.*

Theorem 9.27. *Suppose a monic polynomial $p(z)$ is given, with associated companion matrix C_p. Then we have that λ is a zero of $p(z)$ if and only if λ is an eigenvalue of the companion matrix C_p.*

Proof.

Consider a zero λ of the polynomial $p(z)$. Evaluating the Expression (9.30) for z equal to λ, gives us the following simplification:

$$\lambda \varphi_{0:n-1}(\lambda)^T = \left[\varphi_{0:n-1}(\lambda)^T, p(\lambda) \right] \left[\begin{array}{c} C_p \\ \mathbf{e}_n^T \end{array} \right],$$

$$= \left[\varphi_{0:n-1}(\lambda)^T, 0 \right] \left[\begin{array}{c} C_p \\ \mathbf{e}_n^T \end{array} \right],$$

$$= \varphi_{0:n-1}(\lambda)^T C_p.$$

The last equation clearly shows that λ is also an eigenvalue of the companion matrix C_p with $\varphi_{0:n-1}(z)^T$ as the corresponding left eigenvector. Hence, we have proved in a more general fashion that the zeros of the polynomial $p(z)$ are the eigenvalues of the companion matrix C_p. If all zeros of the polynomial are simple, one has n zeros of $p(z)$ and n eigenvalues of C_p and hence the statement is proved.

Suppose $p(z)$ to have zeros of higher multiplicity. In this case one has to use derivatives and so forth and one will obtain Jordan chains. Nevertheless the results remain valid. We leave the proof of this more general statement to the reader.

$\square$

The above technique can be generalized to other bases. This is the subject of the upcoming text.

Generalization to other bases of polynomials

The theoretical results deduced in the previous subsection admit extension of the results to other types of bases. The previous section focused on companion matrices. In this section we will introduce the so-called congenial matrices. The class of congenial matrices covers the confederate, comrade and colleague matrices. All these matrices are associated with a specific type of basis.

Assume we have now the following arbitrary basis $\varphi_0(z), \varphi_1(z), \ldots, \varphi_k(z)$, for which $\deg(\varphi_k(z)) \leq k$. For this basis we get the following equality

$$z \left[\varphi_0(z), \varphi_1(z), \ldots, \varphi_{n-1}(z) \right] = \left[\varphi_0(z), \varphi_1(z), \ldots, \varphi_{n-1}(z), \varphi_n(z) \right] H,$$

in which H depicts a rectangular Hessenberg matrix. The role of the matrix J in the previous subsection is now replaced by the matrix H.

Similarly as before one can deduce the relations, obtaining now $C_p = PR_{-\mathbf{p}}H$. In general, this matrix is now called the congenial matrix.

Definition 9.28. *Given a polynomial $p(z) = p_0\varphi_0(z) + p_1\varphi_1(z) + \ldots + p_n\varphi_n(z)$ ($p_n = 1$), with the following relation between the polynomials of the basis $\varphi_k(z)$:*

$$z\varphi_{0:n-1}(z) = \varphi_{0:n}(z)H, \tag{9.32}$$

the associated congenial matrix, with $R_{-\mathbf{p}}$ as in Equation (9.29) is the following matrix

$$C_p = PR_{-\mathbf{p}}H, \qquad (9.33)$$

with $P = [I, 0]$ the projection matrix, having n rows and $n+1$ columns.

Similarly as Theorem 9.27 one can prove the following statement.

Theorem 9.29. *Suppose a polynomial $p(z)$ is given, with associated congenial matrix C_p. Then we have that λ is a zero of $p(z)$ if and only if λ is an eigenvalue of the congenial matrix C_p.*

Depending on the case of orthogonal polynomials considered, the matrix H and the associated congenial matrices $R_{-\mathbf{p}}H$ will change. Let us consider some important classes of bases and the associated matrices.

Orthogonal basis on the real line

Suppose the polynomials $\varphi_k(z)$ to be orthogonal on the real line. In this case the matrix H becomes a real tridiagonal matrix T. Let us consider this in more detail.

Suppose we have the following three terms recurrence relation between the polynomials of the basis:

$$\varphi_0(z) = 1,$$
$$\varphi_1(z) = a_1 z + b_1,$$
$$\varphi_i(z) = (a_i z + b_i)\,\varphi_{i-1}(z) - c_i \varphi_{i-2}(z), \qquad \text{for } i \geq 2.$$

Using the above recurrence relations, we obtain the following real tridiagonal matrix $T = H$:

$$T = \begin{bmatrix} -\frac{b_1}{a_1} & \frac{c_2}{a_2} & & & & & \\ \frac{1}{a_1} & -\frac{b_2}{a_2} & \frac{c_3}{a_3} & & & & \\ & \ddots & \ddots & \ddots & & & \\ & & \frac{1}{a_{i-1}} & -\frac{b_i}{a_i} & \frac{c_{i+1}}{a_{i+1}} & & \\ & & & \ddots & \ddots & \ddots & \\ & & & & \ddots & \ddots & \\ & & & & & & \ddots \end{bmatrix}.$$

One can construct a diagonal matrix D, rescaling the columns of the matrix T, such that $\tilde{T} = TD$ and $P\tilde{T} = \hat{T}$ (with $P = [I, 0]$) is a real symmetric tridiagonal matrix. Assume our matrix $\tilde{T}$ to have diagonal elements $\tilde{a}_i$ and subdiagonal elements $\tilde{b}_i$. The matrix $D = \text{diag}([d_1, \ldots, d_n])$.

Let us investigate the structure of the associated congenial matrix C_p, using all previous relations:

$$C_p = PR_{-\mathbf{p}}T$$
$$= PR_{-\mathbf{p}}\tilde{T}D^{-1}$$

$$= P \begin{bmatrix} 1 & & & -p_0 \\ & 1 & & -p_1 \\ & & \ddots & \vdots \\ & & 1 & -p_{n-1} \\ & & & 1 \end{bmatrix} \left[\begin{array}{ccccc} \tilde{a}_1 & \tilde{b}_1 & & & \\ \tilde{b}_1 & \tilde{a}_2 & \tilde{b}_2 & & \\ & \tilde{b}_2 & \ddots & \ddots & \\ & & \ddots & & \tilde{b}_{n-1} \\ & & & \tilde{b}_{n-1} & \tilde{a}_n \\ \hline 0 & \cdots & & 0 & \tilde{b}_n \end{array} \right] D^{-1}$$

$$= P \left[\begin{array}{c|c} I & -\hat{\mathbf{p}} \\ \hline 0 & 1 \end{array} \right] \left[\begin{array}{c} \hat{T} \\ \hline \tilde{b}_n \mathbf{e}_n^T \end{array} \right] D^{-1}$$

$$= P \left[\begin{array}{c} \hat{T} - \tilde{b}_n \hat{\mathbf{p}} \mathbf{e}_n^T \\ \hline \tilde{b}_n \mathbf{e}_n^T \end{array} \right] D^{-1}$$

$$= \begin{bmatrix} \tilde{a}_1 & \tilde{b}_1 & & & & -p_0\tilde{b}_n \\ \tilde{b}_1 & \tilde{a}_2 & \tilde{b}_2 & & & -p_1\tilde{b}_n \\ & \tilde{b}_2 & \ddots & \ddots & & \vdots \\ & & & \tilde{b}_{n-2} & & \\ & & & \tilde{a}_{n-1} & \tilde{b}_{n-1} - p_{n-2}\tilde{b}_n \\ & & & \tilde{b}_{n-1} & \tilde{a}_n - p_{n-1}\tilde{b}_n \end{bmatrix} D^{-1}.$$

This matrix is also called the comrade matrix. The comrade matrix is, up to a scaling, a symmetric tridiagonal matrix perturbed by a rank 1 matrix.

It is clear that the comrade matrix is a generalization of the companion matrix. Assuming $a_i = 1$ and $b_i = c_i = 0$ for all i, we obtain the companion matrix. Considering $a_i = 1, b_i = 0$, and $c_i = 1$ for all i, than we obtain the colleague matrix.

Note 9.30. *The name congenial matrix, covers the complete class of comrade, companion and confederate matrices. A confederate matrix is identical to a congenial matrix, with the only restriction that the considered polynomials are real.*

Orthogonal basis on the unit circle

Suppose the polynomials $\varphi_k(z)$ to be orthogonal on the unit circle. These polynomials are called Szegő polynomials (see [144]). Based on the coupled two-term recurrence relation for these orthogonal polynomials, it can be shown that the matrix H (see Equation (9.32))

$$H = \begin{bmatrix} \tilde{H} \\ \alpha\mathbf{e}_n \end{bmatrix},$$

will have orthonormal columns. This means that we obtain the following relations:

$$\left[\tilde{H}^H, \bar{\alpha} \mathbf{e}_n^H \right] \left[\begin{array}{c} \tilde{H} \\ \alpha \mathbf{e}_n \end{array} \right] = I.$$

Rewriting the equation above leads to

$$\tilde{H}^H \tilde{H} = I - \left[\begin{array}{cccc} 0 & \dots & 0 & 0 \\ \vdots & \ddots & \vdots & \vdots \\ 0 & \dots & 0 & 0 \\ 0 & \dots & 0 & |\alpha|^2 \end{array} \right]$$

with $|\alpha| \leq 1$. Hence one can multiply the last column by a factor such that $\tilde{H}$ becomes a unitary matrix. This means that the matrix $\tilde{H}$ is an almost unitary matrix.

Let us take a closer look now at the matrix $C_p = P R_{-\mathbf{p}} H$ associated to this problem. In this context, one names the matrix a fellow matrix.

$$\begin{aligned} C_p &= P R_{-\mathbf{p}} H \\ &= P \left[\begin{array}{c|c} I & -\hat{\mathbf{p}} \\ \hline 0 & 1 \end{array} \right] \left[\begin{array}{c} \tilde{H} \\ \hline \alpha \mathbf{e}_n^T \end{array} \right] \\ &= P \left[\begin{array}{c} \tilde{H} - \alpha \hat{\mathbf{p}} \mathbf{e}_n^T \\ \hline \alpha \mathbf{e}_n^T \end{array} \right] \\ &= \tilde{H} - \alpha \hat{\mathbf{p}} \mathbf{e}_n^T. \end{aligned}$$

Important is the fact that only the last column of the Hessenberg matrix $\tilde{H}$ does not exactly satisfy the unitary structure. Adding the nonunitary part into the right term we can rewrite C_p as follows:

$$C_p = \hat{H} + \mathbf{u} \mathbf{v}^T,$$

with $\mathbf{v} = \mathbf{e}_n^T$ and $\mathbf{u}$ chosen such that $\hat{H}$ becomes unitary.

In this section we will present results on the solution of eigenproblems related to the matrices presented above. Since unitary matrices will play an important role in these results we will first present some results related to unitary matrices.

9.6.2 Unitary Hessenberg matrices

We will design here an implicit QR-algorithm for unitary Hessenberg matrices. Before doing so, we will provide the Schur parameterization of a unitary Hessenberg matrix.

Representation

Assume we have a unitary matrix U, which is of Hessenberg form. Such a matrix can easily be represented by a sequence of Givens transformations. This is the so-called Schur parameterization.

Suppose the unitary matrix U to be of the following form:

$$U = \begin{bmatrix} \times & \times & \times & \times & \times & \times \\ \times & \times & \times & \times & \times & \times \\ & \times & \times & \times & \times & \times \\ & & \times & \times & \times & \times \\ & & & \times & \times & \times \\ & & & & \times & \times \end{bmatrix}.$$

We will exploit now the fact that the product of two unitary matrices is again a unitary matrix, for creating the factorization of the matrix into Givens transformations.

In the following scheme we use a similar notation as throughout previous sections. In the first step we annihilate the subdiagonal element marked with $\otimes$:

$$\begin{bmatrix} \times & \times & \times & \times & \times & \times \\ \otimes & \times & \times & \times & \times & \times \\ & \times & \times & \times & \times & \times \\ & & \times & \times & \times & \times \\ & & & \times & \times & \times \\ & & & & \times & \times \end{bmatrix} \rightarrow$$

The fact that zeroing out the first subdiagonal element creates many more zeros is proved by the statement above: the product of two unitary matrices remains unitary.

The next element is already marked for deletion. Again a Givens transformation is performed, and also a third one is depicted in the following scheme. Since the Givens transformations can be scaled by a factor ω, with $|\omega| = 1$ (see Section 9.1), one can choose them in such a way to obtain ones on the diagonal.

The process can be continued fairly easy and two more Givens transformations give us the following scheme. The right scheme is just a compact representation.

$$\tag{9.34}$$

The remaining value α has norm $|\alpha| = 1$. Since we have factored the matrix as a product of Givens rotations each having determinant equal to 1, it is logical that there is a value α remaining in the lower right corner. In fact the factorization above is the Givens-vector representation of the unitary matrix U. In the remainder of this section we will always neglect this parameter α until it appears. In most cases we will not see the influence of this parameter as the leading principal submatrix on the left of order $(n-1) \times (n-1)$ equals the identity matrix.

Note 9.31. *In Section 9.1 it was mentioned that we could also replace the role of 2×2 Givens rotations by 2×2 unitary matrices. Suppose we do so, then we can remove also the factor α.*

We can construct a 2×2 unitary diagonal matrix such as to make also the last element α equal to zero. Applying then a fusion between the first and second unitary transformation gives us the following scheme.

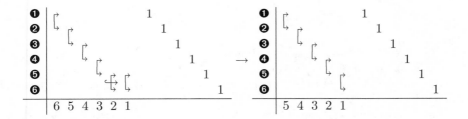

We have constructed now a sequence of $n-1$ descending Givens transformations annihilating all subdiagonal elements of the unitary Hessenberg matrix. Since this final scheme is identical to the unitary Hessenberg matrix, we have provided now a factorization of the unitary Hessenberg matrix. In fact we have the following scheme representing our matrix.

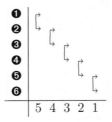

This is a representation of the unitary matrix, based on 2×2 unitary matrices.

Hence we can represent a unitary Hessenberg matrix by a descending sequence of Givens transformations and a parameter α or by a descending sequence of $n-1$ unitary transformations.

Since the difference between these two representations lies in the parameter α, the following implicit QR-algorithm is almost identical for both representations.

Implicit QR-algorithm for unitary Hessenberg matrices

We have now a suitable representation for the unitary Hessenberg matrix. It is known that the unitary as well as the Hessenberg structure is maintained under a step of the QR-algorithm. Based on this information, we will design an implicit method, acting on the representation as presented in Scheme 9.34.

Suppose we want to perform a step of the QR-method with shift μ onto the matrix H:

$$H - \mu I = QR$$
$$\hat{H} = RQ + \mu I.$$

Determine the Givens transformation such that the first column of $H - \mu I$ is transformed into a multiple of $\mathbf{e}_1$. Suppose G_1 is the Givens transformation such that the following equation is satisfied:

$$G_1^H (H - \mu I)\mathbf{e}_1 = \beta \mathbf{e}_1,$$

with $\beta = \pm \|(H - \mu I)\mathbf{e}_1\|_2$. Since H is a unitary Hessenberg matrix, we know, that based on the implicit Q-theorem this first Givens transformation determines uniquely the QR-step. Assume in the remainder that the Hessenberg matrix is unreduced.

Let us construct the chasing procedure, based on the graphical schemes. Applying the first disturbing Givens transformation onto the matrix $H = H_0$, we obtain $H_1 = G_1^H H_0 G_1$. Remark that in previously discussed chasing algorithms the operation G_1 performed on the right of the matrix H_0 appeared in the matrix on the right in several schemes. Here this transformation, will simply appear in position 0. We put this transformation in position 0, to stress that this Givens transformation is coming from the similarity transformation. We obtain the following scheme (we have compressed the complete right-hand side, only α is still depicted)[25].

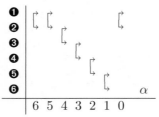

The Givens transformation in position 0, corresponds to G_1. The transformation in position 6 corresponds to G_1^H, the Givens transformations in position 1 up to 5 depict H_0. Let us rewrite the scheme above and apply a fusion between the Givens transformations in positions 6 and 5. Interchange also the positions of the Givens

---

[25]In case one uses the unitary representation, one can consider $\alpha = 1$ and leave it out from the schemes.

transformations, such that we can apply the shift through lemma.

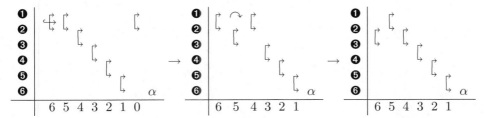

Since we want to obtain again a scheme similar to the one in Scheme 9.34 we have to remove the Givens transformation in position 6. Hence we determine the Givens transformation G_2 such that applying the transformation $G_2^H H_1$, will remove this transformation. Applying the similarity transformation $G_2^H H_1 G_2$ results therefore in the following scheme.

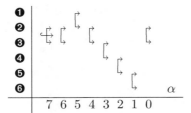

The transformation G_2^H is shown in position 7, the transformation G_2 is shown in position 0 and the matrix H_1 is represented by the transformations in position 1 to 6. The fusion is already depicted. Applying the fusion will remove both Givens transformations in position 6 and 7. Reshuffling the matrix and applying the shift through lemma gives us the following schemes.

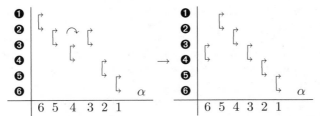

Clearly the bulge has shifted down one position. In this way one can easily continue until the bulge is completely removed. Suppose we have the matrix H_4, which will be of the following form.

The Givens transformation G_5^H is determined such as to annihilate the Givens transformation in position 6. Applying the similarity transformation $G_5^H H_4 G_5$ gives.

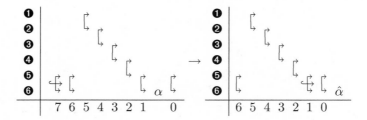

Before being able to apply a fusion on the right, one should slightly interchange the order of the Givens transformation and the parameter α. This results in another parameter and a different Givens transformation. This final step is slightly different from the previous ones. Applying the fusion to the Givens in positions 7 and 6 will remove both Givens. Applying the fusion to the Givens in positions 1 and 0 leads to the following scheme.

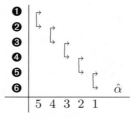

This scheme represents another unitary Hessenberg matrix, which is the result of applying one step of the shifted QR-method.

9.6.3   Unitary plus rank 1 matrices

In this section we will develop a QR-method for computing eigenvalues of Hessenberg matrices, which can be written as unitary plus rank one matrices. This is an important class of matrices as the companion and the fellow matrix are both of this form. Moreover the corresponding QR-algorithms are based on techniques for working with structured rank matrices, discussed in this book. Hence the problem of computing the zeros of polynomials can be translated to the eigenproblem of a unitary plus rank one matrix. Let us divide this subsection into small parts discussing all ingredients for developing an implicit QR-method.

Structure under a QR-step

Let us denote the Hessenberg matrix we are working with as follows

$$H = U + \mathbf{u}\mathbf{v}^H, \tag{9.35}$$

with H a Hessenberg matrix, U a unitary matrix and two vectors $\mathbf{u}$ and $\mathbf{v}$.

Suppose we have a shift μ and we would like to perform a step of the QR-iteration onto the matrix H. We have the following formulas

$$H - \mu I = QR$$
$$\hat{H} = RQ + \mu I = Q^H H Q.$$

Applying now the similarity transformation onto the terms of Equation (9.35) we obtain the following relations:

$$\hat{H} = Q^H H Q = Q^H U Q + Q^H \mathbf{u}\mathbf{v}^H Q$$
$$= \hat{U} + \hat{\mathbf{u}}\hat{\mathbf{v}}^H.$$

Hence the unitary plus rank 1 structure of the Hessenberg matrix is preserved under a step of the QR-iteration. This is essential for developing an efficient implicit QR-method exploiting the matrix structure.

A representation for the unitary matrix

Since the unitary matrix in the splitting of the Hessenberg matrix is a dense matrix, we want to represent this matrix as efficient as possible. Even though the matrix is full, it has a structured rank part.

Consider

$$U = H - \mathbf{u}\mathbf{v}^H,$$

since the matrix H is Hessenberg and has zeros below the subdiagonal the matrix U needs to be of rank 1 below the subdiagonal. The matrix U is therefore of the following form and the elements $\boxtimes$ make up the structured rank part (rank 1) in this 6×6 example:

$$U = \begin{bmatrix} \times & \times & \times & \times & \times & \times \\ \times & \times & \times & \times & \times & \times \\ \boxtimes & \times & \times & \times & \times & \times \\ \boxtimes & \boxtimes & \times & \times & \times & \times \\ \boxtimes & \boxtimes & \boxtimes & \times & \times & \times \\ \boxtimes & \boxtimes & \boxtimes & \boxtimes & \times & \times \end{bmatrix}.$$

We will construct the representation of this matrix in a graphical form, similarly as we did for the unitary Hessenberg case. In the first scheme we already annihilated some elements in the last row of the matrix U, by a single Givens transformation. Three elements are annihilated by one Givens transformation, since they make up a low rank part in the matrix U.

The elements to be annihilated by the second Givens transformation are marked with $\otimes$.

```
❶           ×  ×  ×  ×  ×  ×
❷           ×  ×  ×  ×  ×  ×
❸           ⊠  ×  ×  ×  ×  ×
❹           ⊗  ⊠  ×  ×  ×  ×
❺                 ×  ×  ×  ×
❻                 ×  ×  ×
            2  1
```

After the Givens transformation acting on row three and four, we have completely removed the low rank part. The matrix remaining on the right is now a generalized Hessenberg matrix, having two subdiagonals.

```
❶           ×  ×  ×  ×  ×  ×
❷           ×  ×  ×  ×  ×  ×
❸           ⊗  ×  ×  ×  ×  ×
❹              ⊗  ×  ×  ×  ×
❺                 ⊗  ×  ×  ×
❻                    ⊗  ×  ×
            3  2  1
```

Peeling of the second subdiagonal, removing successively all elements marked with $\otimes$, will give us a descending sequence of Givens transformations.

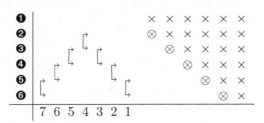

In this final step we will remove the remaining subdiagonal from the Hessenberg matrix on the right. This will be done by a descending sequence of five Givens transformations. Moreover, because the matrix on the right is a unitary matrix, we can choose the Givens transformations such that the matrix becomes almost the identity matrix. Again there is a value α remaining in the bottom right corner with $|\alpha| = 1$. We obtain the following scheme.

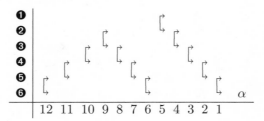

Denoting the sequence from 12 to 10 by V, the sequence from 9 to 6 by W and the sequence from 5 to 1 by X, we have factored the unitary matrix U as the product of four unitary matrices $U = VWXD$, in which D is the diagonal matrix $D = \mathrm{diag}([1,\ldots,1,\alpha])$. We use four different symbols for the unitary matrices, because the use of too many sub- and superscripts would make the forthcoming implicit method difficult to read.

The above unitary matrix VWX can be written in a compressed way to obtain the following.

$$
\begin{array}{l}
❶ \\
❷ \\
❸ \\
❹ \\
❺ \\
❻ \\
\hline
\quad 9\ 8\ 7\ 6\ 5\ 4\ 3\ 2\ 1
\end{array}
\tag{9.36}
$$

This will be the representation of the unitary matrix in the sum $H = U + \mathbf{u}\mathbf{v}^H$. Due to the relation between the matrices $\mathbf{u}\mathbf{v}^H$ and U, we obtain the following representation for the vector $\mathbf{u}$.

$$
\begin{array}{l}
❶ \qquad \times \\
❷ \qquad \times \\
❸ \qquad \times \\
❹ \qquad 0 \\
❺ \qquad 0 \\
❻ \qquad 0 \\
\hline
\quad 9\ 8\ 7
\end{array}
\tag{9.37}
$$

Important to remark is that the Givens transformations in position 9, 8 and 7 in Scheme 9.37 can be chosen exactly the same as the ones in the corresponding position of Scheme 9.36.

This means that we get the following equalities:

$$
\begin{aligned}
H &= U + \mathbf{u}\mathbf{v}^H \\
&= VWXD + \mathbf{u}\mathbf{v}^H \\
&= V\left(WXD + V^H\mathbf{u}\mathbf{v}^H\right) \\
&= V\left(WXD + \hat{\mathbf{u}}\mathbf{v}^H\right),
\end{aligned}
$$

where the vector $\hat{\mathbf{u}}$ has only the first three elements different from zero.

This representation, depicted here is the one we will use when developing an implicit QR-method for unitary plus low rank matrices.

There is a lot of flexibility in the choice of representation. Moreover the chosen representation determines quite often the used implicit approach. Different choices and alternative algorithms will be discussed later on.

The implicit method

There exist several variants for performing an implicit QR-method onto the unitary plus low rank matrices. Unfortunately all these algorithms are not capable of preserving both the unitary, the low rank and the Hessenberg structure. Numerical roundoff creates loss of the low rank structure, loss of unitarity or loss of the Hessenberg structure and hence compression is needed. This compression can create undesired results.

In the following method, we exploit all involved structures and the chasing method is constructed such that no compression to preserve the structure is needed.

Since an implicit QR-step performed on a Hessenberg matrix is only determined by the first Givens transformation, we have to determine it explicitly. Determine G_1 such that the following equation is satisfied:

$$G_1^H (H - \mu I)\mathbf{e}_1 = \pm \|H - \mu I\|\mathbf{e}_1.$$

We will now perform the similarity transformation $G_1^H H G_1$, exploiting the factorization of the matrix H. Important in the following algorithm is that we would like to keep in each step of the algorithm a factorization similar to the initial one:

$$H = V \left(W X D + \mathbf{u}\mathbf{v}^H \right).$$

Of course after each step the structure will not be exactly as desired and a sequence of Givens transformations needs to be constructed to restore the structure. This will be the chasing.

Let us perform the first similarity transformation onto the matrix $H = H_0$ with

$$H_0 = V_0 \left(W_0 X_0 D_0 + \mathbf{u}_0 \mathbf{v}_0^H \right).$$

We obtain (with $D_0 = D_1$)

$$\begin{aligned}
H_1 &= G_1^H V_0 \left(W_0 X_0 D_1 + \mathbf{u}_0 \mathbf{v}_0^H \right) G_1 \\
&= G_1^H V_0 \left(W_0 X_0 G_1 D_1 + \mathbf{u}_0 \mathbf{v}_0^H G_1 \right) \\
&= G_1^H V_0 \left(W_0 X_0 G_1 D_1 + \mathbf{u}_0 \mathbf{v}_1^H \right),
\end{aligned}$$

with $\mathbf{v}_1^H = \mathbf{v}_0^H G_1$. In the remainder of the derivations, all intermediate variables will be depicted with $\tilde{\ }$ or $\hat{\ }$. The final values for representing H_1 will be $V_1, W_1, X_1, \mathbf{u}_1$ and $\mathbf{v}_1$.

Since the Givens transformation G_1^H acts on the first two rows, and V_0 acts on the rows 3 up to 6, they commute and we obtain the following:

$$\begin{aligned}
H_1 &= V_0 \left(G_1^H W_0 X_0 G_1 D_1 + G_1^H \mathbf{u}_0 \mathbf{v}_1^H \right) \\
&= V_0 \left(G_1^H W_0 X_0 G_1 D_1 + \tilde{\mathbf{u}}_1 \mathbf{v}_1^H \right),
\end{aligned}$$

with $\tilde{\mathbf{u}}_1 = G_1^H \mathbf{u}_0$, having only the first three elements different from zero.

It seems that the matrix H_1 is already in the good form, except for the unitary matrix $G_1^H W_0 X_0 G_1$. Let us see how to change this matrix. The Givens transformation G_1^H is shown in position 10, the transformations in position 9 up to 6 represent

W_0, the ones in position 5 up to 1 represent X_0 and position 0 is reserved for G_1. The undesired Givens transformations are marked with a $-$.

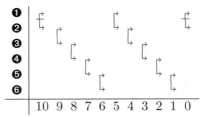

We will start by removing the Givens transformation in position 10. Reordering permits the application of the shift through operation.

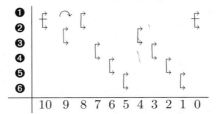

This leads to the following scheme. We have marked another Givens transformation now with $-$.

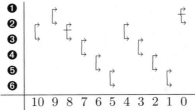

Rewriting gives us the possibility to apply again the shift through operation.

This results after rewriting in the following scheme.

The undesired Givens transformation moves down, creating a new sequence, starting in positions 10 and 9 and removing the sequence in positions 3 up to 1. Continuing this procedure, and applying two more times the shift through operation gives us the following scheme.

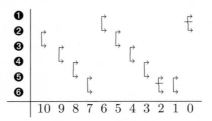

One can easily fuse the undesired Givens transformation with the Givens transformation in position 1. Hence we have removed already 1 undesired Givens transformation. We obtain the following scheme.

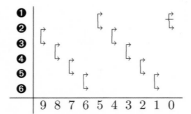

This corresponds to the following relation:

$$H_1 = V_0 \left(\tilde{W}_1 \tilde{X}_1 G_1 D_1 + \tilde{\mathbf{u}}_1 \mathbf{v}_1^H \right),$$

where the unitary matrices $\tilde{W}_1$ and $\tilde{X}_1$ are two descending sequences of transformations.

The only remaining undesired Givens transformation is G_1 in the middle of the formula. The idea is now to drag this Givens transformation completely through the other matrices such that it appears before the matrix V_0 and moreover acts on row 2 and row 3. It has to act on row 2 and row 3, so that choosing the next Givens transformation G_2, determining the following similarity transformation, can be chosen such as to annihilate this transformation.

Let us continue and remove the second undesired Givens transformation, the one in position 0. One can fairly easy apply two times the shift through operation to obtain the following scheme.

(9.38)

Now there is an unwanted transformation in position 10. Denote this transformation with $\tilde{G}_1$. In formulas we obtain now

$$H_1 = V_0 \left(\tilde{G}_1 \hat{W}_1 X_1 D_1 + \tilde{\mathbf{u}}_1 \mathbf{v}_1^H \right),$$

in which the sequences of Givens transformations $\hat{W}_1$ and X_1 have changed again.

This gives us the following relations:

$$H_1 = V_0 \tilde{G}_1 \left(\hat{W}_1 X_1 D_1 + \tilde{G}_1^H \tilde{\mathbf{u}}_1 \mathbf{v}_1^H \right),$$
$$= V_0 \tilde{G}_1 \left(\hat{W}_1 X_1 D_1 + \hat{\mathbf{u}}_1 \mathbf{v}_1^H \right),$$

Two terms are of importance here: $V_0 \tilde{G}_1$ and $\tilde{G}_1^H \tilde{\mathbf{u}}_1 = \hat{\mathbf{u}}_1$. The vector $\hat{\mathbf{u}}_1$ will loose one of his zeros. The vector will have now four nonzero elements.

The matrix product $V_0 \tilde{G}_1$ does not essentially change the Givens pattern of the matrix V_0. A single fusion of two Givens transformations removes the undesired transformation $\tilde{G}_1$, without destroying the structure of the matrix V_0. In the scheme below, the undesired Givens transformation can be found in position 0, whereas the Givens transformations in positions 3 up to 1 make up the matrix V_0.

The resulting matrix $\tilde{V}_1 = V_0 \tilde{G}_1$ has the same Givens pattern than the matrix V_0. We obtain the following relations:

$$H_1 = \tilde{V}_1 \left(\hat{W}_1 X_1 D_1 + \hat{\mathbf{u}}_1 \mathbf{v}_1^H \right). \tag{9.39}$$

We are however not yet satisfied. We want to obtain a similar factorization as of the original matrix H. Hence, the four nonzero elements in the vector $\hat{\mathbf{u}}_1$ need to be transformed into three nonzero elements.

To do so, we need to rewrite Equation (9.39). Denote the Givens transformation in Scheme 9.38 in position 9, with G_1^l. We denote this with $\cdot^l$ to clearly indicate that one extra Givens will move to the left, outside the brackets. Until we move the Givens back to the right we will indicate this on the affected elements. We can rewrite the formulas above as follows:

$$H_1 = \tilde{V}_1 G_1^l {G_1^l}^H \left(\hat{W}_1 X_1 D_1 + \hat{\mathbf{u}}_1 \mathbf{v}_1^H \right)$$
$$= \tilde{V}_1 G_1^l \left({G_1^l}^H \hat{W}_1 X_1 D_1 + {G_1^l}^H \hat{\mathbf{u}}_1 \mathbf{v}_1^H \right)$$
$$= \tilde{V}_1^l \left(\hat{W}_1^l X_1 D_1 + \hat{\mathbf{u}}_1^l \mathbf{v}_1^H \right)$$

in which $G_1^{l\,H}\hat{\mathbf{u}}_1 = \hat{\mathbf{u}}_1^l$ still has four elements different from zero, the matrix $\tilde{V}_1^l$ is now a sequence having one more Givens transformation than $\tilde{V}_1$, and $\hat{G}_1\hat{W}_1 = \hat{W}_1^l$ has lost a single Givens transformation.

Graphically $\tilde{V}_1^l\hat{W}_1^l = \tilde{V}_1\hat{W}_1$ is depicted as follows.

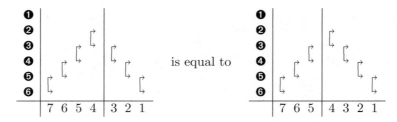

The matrix $\tilde{V}_1$ consists of the Givens in position 7 to 5, whereas the new $\tilde{V}_1^l$ consists of the Givens transformations in position 7 up to 4. Similarly we have that the matrix $\hat{W}_1$ consists of the transformations in positions 4 up to 1 and $\hat{W}_1^l$ consists of the transformations in positions 3 up to 1.

Construct now a Givens transformation acting on row 3 and 4 of the vector $\hat{\mathbf{u}}_1^l$ and annihilating the element in position 4. Let us denote this Givens transformation by $\hat{G}_1$:

$$\hat{G}_1^H\hat{\mathbf{u}}_1^l = \mathbf{u}_1^l,$$

in which $\mathbf{u}_1^l$ is a vector having only the first three elements different from zero. Let us plug this in into the formulas.

$$
\begin{aligned}
H_1 &= \tilde{V}_1^l\left(\hat{W}_1^lX_1D_1 + \hat{G}_1\hat{G}_1^H\hat{\mathbf{u}}_1^l\mathbf{v}_1^H\right) \\
&= \tilde{V}_1^l\left(\hat{G}_1\hat{G}_1^H\hat{W}_1^lX_1D_1 + \hat{G}_1\mathbf{u}_1^l\mathbf{v}_1^H\right) \\
&= \tilde{V}_1^l\hat{G}_1\left(\left(\hat{G}_1^H\hat{W}_1^l\right)X_1D_1 + \mathbf{u}_1^l\mathbf{v}_1^H\right)
\end{aligned}
$$

Let us take a closer look now at $\hat{G}_1^H\hat{W}_1^l$. Graphically this is depicted in the scheme below. The Givens transformations from 1 up to 3 make up the matrix $\hat{W}_1^l$ and the Givens $\hat{G}_1^H$ is shown in position 4. Applying one fusion, removes the Givens transformation in position 4.

This results in a sequence of Givens transformations denoted as $\hat{G}_1^H\hat{W}_1 = W_1^l$.

The matrix product $\tilde{V}_1^l \hat{G}_1$ is of the following form. Applying the shift through lemma gives us the following scheme.

$$(9.40)$$

This scheme can be written as $G_2 V_1^l$, in which V_1^l incorporates the Givens from position 1 to 4 and G_2 is shown in the fifth position.

Plugging all of this into the equations above gives us:

$$H_1 = G_2 V_1^l \left(W_1^l X_1 D_1 + \mathbf{u}_1^l \mathbf{v}_1^H \right).$$

Rewriting now the formula above by bringing the Givens transformation in position 1 of the matrix V_1^l inside the brackets does not change the formulas significantly. We obtain:

$$H_1 = G_2 V_1 \left(W_1 X_1 D_1 + \mathbf{u}_1 \mathbf{v}_1^H \right).$$

The reader can verify that the number of nonzero elements in the vector $\mathbf{u}_1$ is still three, as desired. The product depicted here is almost of the desired structure. Only the Givens transformation G_2 is undesired. We will remove this transformation, by applying another similarity transformation with G_2.

We have now performed the initial step. We will continue the algorithm, in an implicit way. Only one step is depicted as the other steps are similar. Since we want our Hessenberg matrix H to become again of unitary plus low rank form, we want to remove the disturbance G_2. Performing the similarity transformation with this Givens transformation results in the following (with $D_2 = D_1$):

$$\begin{aligned}
H_2 &= G_2^H H_1 G_2 \\
&= G_2^H G_2 V_1 \left(W_1 X_1 D_1 + \mathbf{u}_1 \mathbf{v}_1^H \right) G_2 \\
&= V_1 \left(W_1 X_1 D_1 G_2 + \mathbf{u}_1 \mathbf{v}_1^H G_2 \right) \\
&= V_1 \left(W_1 X_1 G_2 D_2 + \mathbf{u}_1 \mathbf{v}_2 \right),
\end{aligned}$$

where $\mathbf{v}_2 = \mathbf{v}_1 G_2$.

Similarly as in the initial step we can drag G_2 through W_1 and X_1. We obtain $W_1 X_1 G_2 = \tilde{G}_2 W_2 X_2$. This gives us

$$\begin{aligned}
H_2 &= V_1 \left(\tilde{G}_2 W_2 X_2 D_2 + \mathbf{u}_1 \mathbf{v}_2 \right), \\
&= V_1 \left(\tilde{G}_2 W_2 X_2 D_2 + \tilde{G}_2 \tilde{G}_2^H \mathbf{u}_1 \mathbf{v}_2 \right), \\
&= V_1 \tilde{G}_2 \left(W_2 X_2 D_2 + \tilde{G}_2^H \mathbf{u}_1 \mathbf{v}_2 \right).
\end{aligned}$$

Since the Givens transformation $\tilde{G}_2^H$ acts on row 4 and row 5, $\tilde{G}_2^H \mathbf{u}_1 = \mathbf{u}_1$. Applying a final shift through operation for $V_1 \tilde{G}_2$ we obtain $G_3 V_2 = V_1 \tilde{G}_2$ giving us:

$$H_2 = G_3 V_2 \left(W_2 X_2 D_2 + \mathbf{u}_1 \mathbf{v}_2 \right)$$
$$= G_3 V_2 \left(W_2 X_2 D_2 + \mathbf{u}_2 \mathbf{v}_2 \right).$$

Clearly we have performed now a step of the chasing method since the first Givens transformation G_3 has shifted down one position. We remark that in the remaining chasing procedure the vector $\mathbf{u}_2$ will not change anymore.

One can easily continue this chasing procedure. Only the last similarity transformation G_{n-1} can unfortunately not be determined based on the procedure above. To construct this transformation, we have to compute explicitly the $(n-1)$-th column of the matrix H_{n-2}. Based on the representation above, this can however be done in an efficient way. Only when performing this final similarity transformation, the matrix $D_1 = D_{n-2}$ will change.

Note 9.32. *Some remarks have to be made.*

- *The procedure above only shows how to perform the initial step and the chasing step. Important issues such as the deflation criterion and how to perform deflation were not discussed.*

- *When considering real polynomials it might be convenient to stick to real computations. In order to do so a double shift technique is desired.*

9.6.4   Other methods

Let us discuss some other variants for computing the eigenvalues of unitary plus low rank matrices. Since these papers use other terminology, some extra definitions are needed. These definitions can be found in more elaborate form in Volume I.

Definitions

Most of the structured rank matrices discussed in this book were of easy form, having semiseparability rank 1. In case of the unitary plus low rank problem also higher order structured rank matrices appear.

Definition 9.33.   *An $n \times n$ matrix A is called a $\{p, q\}$-quasiseparable matrix, with $p \geq 0$ and $q \geq 0$, if the following two properties are satisfied (for $i = 1, \ldots, n-1$):*

$$\operatorname{rank} \left(A(i+1 : n, 1 : i) \right) \leq p,$$
$$\operatorname{rank} \left(A(1 : i, i+1 : n) \right) \leq q.$$

In the upcoming text also the sub(super)diagonal rank is mentioned.

Definition 9.34. *An $n \times n$ matrix A has subdiagonal rank p if*

$$\max\big\{\, \mathrm{rank}\,(A(i+1:n, 1:i)) \;\mid\; \forall i \in \{1, \ldots, n-1\}\big\} = p.$$

The matrix A has a superdiagonal rank q if

$$\max\big\{\, \mathrm{rank}\,(A(1:i, i+1:n)) \;\mid\; \forall i \in \{1, \ldots, n-1\}\big\} = q.$$

Hence, a $\{p, q\}$-quasiseparable matrix A has subdiagonal rank less or equal to p and a superdiagonal rank less or equal to q.

Different approaches

Recently several algorithms were described to solve the unitary plus low rank eigenvalue problem. These algorithms are very similar as the algorithm developed in Subsection 9.6.3.

☞ D. Bindel, S. Chandrasekaran, J. W. Demmel, D. Garmire, and M. Gu. A fast and stable nonsymmetric eigensolver for certain structured matrices. Technical report, Department of Computer Science, University of California, Berkeley, California, USA, May 2005.

☞ D. A. Bini, F. Daddi, and L. Gemignani. On the shifted QR iteration applied to companion matrices. *Electronic Transactions on Numerical Analysis*, 18:137–152, 2004.

Independently of each other Bindel et al. and Bini et al. proved that the QR-iterates H_k have $\{1, 3\}$-quasiseparable structure when $H_0 = U + \mathbf{u}\mathbf{v}^H$ is a Hessenberg matrix that is unitary plus rank 1. Hence, each Hessenberg matrix H_k can be represented using $\mathcal{O}(n)$ parameters. Based on this fact, several algorithms where developed that performed QR-iteration steps on the Hessenberg matrices H_k. These algorithms differ in the way the Hessenberg matrix is represented, in the way the Hessenberg as well as the unitary plus rank 1 structure is tried to be maintained and in the explicit or implicit way of performing each QR-iteration step.

In the article by Bini et al., the following relationship between C_p of (9.31) and its inverse C_p^{-1} is used (when p_0 is different from zero):

$$C_p = C_p^{-H} + UV^H,$$

with $U_k, V_k \in \mathbb{C}^{n \times 2}$. This relationship remains valid for the QR-iterates H_k

$$H_k = H_k^{-H} + U_k V_k^H,$$

with $U_k, V_k \in \mathbb{C}^{n \times 2}$. Because the upper triangular part of H_k^{-H} is the lower triangular part of the inverse of an (unreduced) Hessenberg matrix, this lower triangular part is the lower triangular part of a rank 1 matrix $\mathbf{x}_k \mathbf{y}_k^H$. In the article the authors give five different methods to compute these vectors $\mathbf{x}_k$ and $\mathbf{y}_k$. The QR-iteration step with shift is implemented in an explicit way where the QR-iterates

H_k are represented by their subdiagonal elements, the vectors $\mathbf{x}_k$ and $\mathbf{y}_k$ and the $n \times 2$ matrices U_k and V_k. It turns out that for a Hessenberg matrix H_k, which is $\{1, 3\}$-quasiseparable, the Q_k and R_k factor of the QR-factorization of $H_k - \alpha_k I$ are $\{1, 1\}$ and $\{0, 4\}$-quasiseparable, respectively. The experiments described in the article show that the implementation of the algorithm has numerical difficulties (overflow/underflow, no convergence).

In the article of Bindel et al. a larger class of matrices is considered, more precisely, the symmetric, skew symmetric or orthogonal matrices plus a low rank modification. The authors call a matrix A rank-symmetric if $\operatorname{rank}(A_{21}) = \operatorname{rank}(A_{12})$ for any 2×2 block partitioning of the square matrix A

$$A = \begin{bmatrix} A_{11} & A_{12} \\ A_{21} & A_{22} \end{bmatrix},$$

with A_{11} and A_{22} square. Consider the matrix $A + L$ with $\operatorname{rank}(L) = k$, and then it is shown that the ranks of the subdiagonal and superdiagonal blocks of $A + L$ can differ by maximum $2k$, i.e.,

$$\| \operatorname{rank}(A_{12} + L_{12}) - \operatorname{rank}(A_{21} + L_{21}) \| \leq 2k.$$

The symmetric, skew symmetric or orthogonal plus low rank structure is maintained when performing a QR-iteration step (with shift). Let H be the Hessenberg matrix that is orthogonally similar to the matrix $A + L$, i.e.,

$$\begin{aligned} H &= Q^T (A + L) Q \\ &= Q^T A Q + Q^T L Q \\ &= \hat{A} + \hat{L}. \end{aligned}$$

The structure (symmetric, skew symmetric, or orthogonal) of $\hat{A}$ is the same as that of A. Also the rank of $\hat{L}$ is still the same as the rank of L. However, because we know now that H has Hessenberg structure, the subdiagonal rank is 1 and, hence, the superdiagonal rank is not greater than $1 + 2 \operatorname{rank}(L)$. The Hessenberg matrix is block partitioned and the strictly upper block triangular part of it is represented by a (block)-quasiseparable representation. Each step of the QR-iteration involves applying a sequence of Givens transformations to the left of H to obtain the R-factor. Applying the similarity transformation based on this Givens sequence to the Hessenberg matrix using the quasiseparable representation involves the merging and the splitting of the blocks. To split the blocks a rank-revealing decomposition has to be computed. A similar compression technique is needed in several other algorithms.

☞ D. A. Bini, Y. Eidelman, L. Gemignani, and I. C. Gohberg. Fast QR eigenvalue algorithms for Hessenberg matrices which are rank-one perturbations of unitary matrices. *SIAM Journal on Matrix Analysis and Applications*, 29(2):566–585, 2007.

In this article Bini et al. developed an alternative algorithm to solve the Hessenberg, unitary plus rank 1 case $H_k = U_k + \mathbf{u}_k \mathbf{v}_k^H$. The authors showed that the matrix

U_k can be written as a product of three sequences of 2×2 unitary transformations as represented in (9.40). Hence, each of the matrices U_k has $\{2,2\}$-quasiseparable structure. The representation that is the input for the $(k+1)$-th iteration step of the QR-algorithm contains the diagonal entries and a quasiseparable representation of the strictly upper triangular part of the unitary matrix U_k, the subdiagonal and diagonal entries of the Hessenberg matrix H_k and the two vectors $\mathbf{u}_k$ and $\mathbf{v}_k$. First the sequence of Givens transformations $G_1, G_2, \ldots, G_{n-1}$ is computed such that $R_k = G_{n-1}^H \cdots G_1^H (H_k - \mu_k I)$ is upper triangular (with μ_k a suitably chosen shift). The next iterate H_{k+1} could be computed as the product of quasiseparable matrices R_k and $G_1 \cdots G_{n-1}$ and then adding $\mu_k I$ to this product. However, if this is done, the resulting matrix would have a $\{1,4\}$-quasiseparable structure while we know it should have a $\{1,3\}$-quasiseparable structure. Hence, a compression strategy is needed. This is done as follows. From the $\{1,4\}$-quasiseparable representation, one can determine a $\{2,3\}$-quasiseparable representation for U_{k+1}. However, we need a $\{2,2\}$-quasiseparable representation. To obtain this representation, the unitary matrix is written as a product of three sequences of 2×2 unitary transformations as indicated above. This guarantees that the corresponding unitary matrix has the $\{2,2\}$-quasiseparable structure again. However, instead of the almost identity matrix I that is the R-factor of the QR-factorization of the matrix $U_{k+1} = \tilde{U}_{k+1} \hat{R}_{k+1}$, we will obtain $\hat{R}_{k+1} = I + \tilde{R}_{k+1}$, in which $\tilde{R}_{k+1}$ is an upper triangular matrix. To avoid amplification of errors, we continue working with the Hessenberg matrix

$$H_{k+1} \tilde{R}_{k+1}^{-1} = \tilde{U}_{k+1} + \mathbf{u}_{k+1} \mathbf{v}_{k+1} \tilde{R}_{k+1}^{-1}.$$

☞ S. Chandrasekaran, M. Gu, J. Xia, and J. Zhu. A fast QR algorithm for companion matrices. *Operator Theory: Advances and Applications*, 179:111–143, 2007.

As in the article of Bindel et al., a bulge chasing procedure is designed. In contrast to the previous articles this is done for a single as well as for a double shift. The representation that is used for the Hessenberg matrix H_k in each step is its QR-factorization, i.e., $H_k = Q_k R_k$. The unitary matrix Q_k consists of one sequence of Givens transformations while the upper triangular matrix R_k has a $\{0,2\}$-quasiseparable structure. Each step in the bulge chasing consists of chasing the bulge in the lower triangular part of R_k and shifting the Givens transformation (two of them in case of a double shift) to remove this bulge, through the sequence of Givens transformations representing Q_k. However, during this bulge chasing the superdiagonal rank of R_k is increased by two in the single shift case and by four in the double shift case. Hence, compression is needed to recover a compact representation for R_{k+1} again. Using the fact that R_{k+1} is the sum of a unitary and a rank 1 matrix, we know that R_{k+1} can be represented by computing the sum of this unitary and rank 1 matrix. The corresponding similarity transformation is performed on the rank 1 part. A $\{2,2\}$-quasiseparable unitary approximation is then computed for the difference between R_{k+1} and this rank 1 part. The final $\{0,3\}$-quasiseparable representation for R_{k+1} is then computed as the sum of this $\{2,2\}$-quasiseparable matrix and rank 1 part. Note that instead of the minimal superdiagonal rank 2 during this algorithm a rank of 3 is used.

☞ D. A. Bini, L. Gemignani, and V. Y. Pan. Fast and stable QR eigenvalue
   algorithms for generalized companion matrices and secular equations. *Numerische Mathematik*, 100(3):373–408, 2005.

All the previous articles handled the case $H = U + \mathbf{u}\mathbf{v}^H$, i.e., when the Hessenberg matrix H is the sum of a unitary U and rank 1 matrix $\mathbf{u}\mathbf{v}^H$. Only the article of Bindel et al. considered the more general class of symmetric, skew symmetric, or unitary plus rank 1 case. In the article by Bini et al. of 2005, the matrices $A = (a_{i,j})_{i,j}$ considered have the following form:

$$
\begin{aligned}
a_{ii} &= d_i + z_i \overline{w}_i & &\text{(9.41)}\\
a_{ij} &= u_i t_{ij}^\times \overline{v}_j & &\text{when } i > j\\
a_{ij} &= \overline{u}_j \overline{t_{ji}^\times} v_i + z_i \overline{w}_j - \overline{z}_j w_i & &\text{when } i < j,
\end{aligned}
$$

with $t_{ij}^\times = t_{i-1}t_{i-2}\cdots t_{j+1}$ and given vectors $\mathbf{u}, \mathbf{v}, \mathbf{t}, \mathbf{z}, \mathbf{w}, \mathbf{d}$[26]. This set is called the set of generalized companion matrices. It includes the arrowhead matrices, comrade matrices (symmetric tridiagonal plus rank 1), diagonal plus rank 1 matrices .... Note that each matrix of this set is $\{1,3\}$-quasiseparable. The authors prove that this set is closed under the application of each step of the QR-algorithm. Each step of the QR-algorithm is performed in an explicit way, i.e., the QR-factorization of $A_k - \mu_k I$ is computed and then the two factors multiplied in reverse order are added to $\mu_k I$. Computing the QR-factorization of A_k consists of applying two sequences of Givens transformations. The first one transforms A_k into an upper Hessenberg matrix while the second one transforms this Hessenberg matrix into an upper triangular one. This upper triangular matrix is a $\{0,4\}$-quasiseparable matrix but is not represented as such but as a sum of four matrix terms in which two of those are products of two matrices. This representation is used to compute the representation of A_{k+1} in the form (9.41).

Notes and references

The different classes of matrices to represent polynomials as presented in Subsection 9.6.1 were named according to the following book.

☞ S. Barnett. *Polynomials and Linear Control Systems*. Monographs and
   Textbooks in Pure and Applied Mathematics. Marcel Dekker, Inc., New
   York, USA, 1983.

Some other references related to unitary Hessenberg matrices are the following ones.

☞ L. Gemignani. A unitary Hessenberg QR-based algorithm via semiseparable matrices. *Journal of Computational and Applied Mathematics*, 184:505–517, 2005.

Unitary Hessenberg matrices have the upper triangular part in the matrix of semiseparable (quasiseparable) form. In this article by Gemignani the quasiseparable structure of the unitary Hessenberg matrix is exploited to develop a QR-algorithm. Moreover the presented algorithm is also valid for unitary Hessenberg plus rank 1 matrices. The presented

---

[26]This representation is called the quasiseparable representation. It is discussed in Volume I.

method transforms the matrix into a hermitian semiseparable plus diagonal matrix via the Möbius transform, then a QR-method is applied for computing its eigenvalues. This article contains also lots of references related to the QR-algorithm for unitary Hessenberg matrices.

Unitary Hessenberg matrices are matrices having a very specific structure. First of all they are of Hessenberg form, and second they have the upper triangular structure of semiseparable form.

More information on eigenvalue problems related to unitary matrices can be found in the following articles.

☞ G. S. Ammar and W. B. Gragg. Schur flows for orthogonal Hessenberg matrices. In A. M. Bloch, editor, *Hamiltonian and Gradient Flows, Algorithms and Control*, volume 3, pages 27–34. American Mathematical Society, Providence, Rhode Island, 1994.

☞ G. S. Ammar, W. B. Gragg, and L. Reichel. On the eigenproblem for orthogonal matrices. In *Proceedings of the 25th IEEE Conference on Decision & Control*, pages 1963–1966. IEEE, New York, USA, 1986.

☞ G. S. Ammar, L. Reichel, and D. C. Sorensen. An implementation of a divide and conquer algorithm for the unitary eigenproblem. *ACM Transactions on Mathematical Software*, 18(3):292–307, September 1992.

☞ R. J. A. David and D. S. Watkins. Efficient implementation of the multishift QR algorithm for the unitary eigenvalue problem. *SIAM Journal on Matrix Analysis and Applications*, 28(3):623–633, 2006.

☞ W. B. Gragg. The QR algorithm for unitary Hessenberg matrices. *Journal of Computational and Applied Mathematics*, 16:1–8, 1986.

☞ W. B. Gragg and L. Reichel. A divide and conquer algorithm for the unitary eigenproblem. In M. T. Heath, editor, *Hypercube Multiprocessors 1987*, pages 639–647. SIAM, Philadelphia, Pennsylvania, USA, 1987.

☞ M. Stewart. Stability properties of several variants of the unitary Hessenberg QR-algorithm in structured matrices in mathematics. In V. Olshevsky, editor, *Structured Matrices in Mathematics, Computer Science and Engineering, II*, volume 281 of *Contemporary Mathematics*, pages 57–72. American Mathematical Society, Providence, Rhode Island, USA, 2001.

☞ T. L. Wang, Z. J. and W. B. Gragg. Convergence of the shifted QR algorithm, for unitary Hessenberg matrices. *Mathematics of Computation*, 71(240):1473–1496, 2002.

☞ T. L. Wang, Z. J. and W. B. Gragg. Convergence of the unitary QR algorithm with unimodular Wilkinson shift. *Mathematics of Computation*, 72(241):375–385, 2003.

For more general unitary structured rank matrices, the reader is referred to [57, 55]. In the first report [57], the authors developed an efficient representation for these matrices, based on the Givens-weight idea. Based on this representation an implicit QR-algorithm for these matrices was developed in [55].

Some other related references.

☞ G. S. Ammar, D. Calvetti, W. B. Gragg, and L. Reichel. Polynomial zerofinders based on Szegő polynomials. *Journal of Computational and Applied Mathematics*, 127:1–16, 2001.

☞ G. S. Ammar, D. Calvetti, and L. Reichel. Continuation methods for
the computation of zeros of Szegő polynomials. *Linear Algebra and its
Applications*, 249:125–155, 1996.

☞ G. S. Ammar, W. B. Gragg, and C. He. An efficient QR algorithm for a
Hessenberg submatrix of a unitary matrix. In W. Dayawansa, A. Lindquist,
and Y. Zhou, editors, *New Directions and Applications in Control Theory*,
volume 321 of *Lecture Notes in Control and Information Sciences*, pages
1–14. Springer-Verlag, Berlin, Germany, 2005.

We conclude by mentioning that there are a lot of other references devoted to ap-
proximating the zeros of a polynomial. Because these methods are not directly related to
the algorithms described above, we will not consider these references here.

9.7   References to related subjects

In the previous chapters references were discussed directly related to the implicit
QR-algorithms or related to the reduction algorithms. There exist however more
references related to structured rank matrices and eigenvalue/singular value problems. These references do not necessarily fit directly into the notes and references
discussed before.

Few references are investigated in more detail and discussed in separate subsections.

9.7.1   Rational Krylov methods

Let us first discuss the article by Fasino.

☞ D. Fasino. Rational Krylov matrices and QR-steps on hermitian diagonal-
plus-semiseparable matrices. *Numerical Linear Algebra with Applications*,
12(8):743–754, October 2005.

There is a well-known relation between the unitary matrix Q in the tridiagonalization procedure, and the Q-factor in the QR-factorization of a suitably chosen
Krylov matrix. This relation is further investigated and results are obtained concerning the reduction to semiseparable plus diagonal form and the QR-factorization
of a rational Krylov matrix. Let us discuss these results more in detail and start
with the traditional, sparse matrix case.

Assume a symmetric $n \times n$ matrix A is given. Suppose an arbitrary vector $\mathbf{v}$
is chosen and we have the following Krylov matrix:

$$K = K(A, \mathbf{v}) = \left[\mathbf{v}, A\mathbf{v}, A^2\mathbf{v}, \dots, A^{n-1}\mathbf{v} \right].$$

In case the matrix K is nonsingular, we have that for $K = QR$, the QR-factorization
of the matrix K:

$$T = Q^H A Q,$$

is an irreducible tridiagonal matrix T. We remark that this statement is also valid
in the other direction. If A is reduced via orthogonal similarity transformations to

an irreducible tridiagonal matrix $Q^H A Q = T$, then we have that

$$Q^H K(A, Q\mathbf{e}_1) = R,$$

is a nonsingular upper triangular matrix R.

This is closely related to the Lanczos tridiagonalization procedure. These theorems state in some sense also that a reduction to an irreducible tridiagonal matrix T is uniquely determined by the first column. This is also the contents of the implicit Q-theorem for tridiagonal matrices. Note that in some sense one can also interpret the QR-algorithm, which transforms a tridiagonal matrix to another tridiagonal matrix by orthogonal similarity transformations, can be interpreted in this context.

Suppose $A = V \Delta V^H$ to be the spectral factorization of the matrix A, having eigenvalues $\lambda_1, \ldots, \lambda_n$. Denote $\mathbf{w} = V^H \mathbf{v}$, we also obtain the following relations:

$$\begin{aligned}
K(A, \mathbf{v}) &= \left[\mathbf{v}, A\mathbf{v}, A^2\mathbf{v}, \ldots, A^{n-1}\mathbf{v} \right]. \\
&= V \left[\mathbf{w}, \Delta\mathbf{w}, \Delta^2\mathbf{w}, \ldots, \Delta^{n-1}\mathbf{w} \right] \\
&= V \operatorname{diag}([w_1, w_2, \ldots, w_n]) \begin{bmatrix} 1 & \lambda_1 & \ldots & \lambda_1^{n-1} \\ \vdots & \vdots & & \vdots \\ 1 & \lambda_n & \ldots & \lambda_n^{n-1} \end{bmatrix}.
\end{aligned}$$

The latter matrix $F = [\lambda_i^{(j-1)}]_{i,j}$ is a Vandermonde matrix. Let us remark that the matrix K is nonsingular when all eigenvalues λ_i are pairwise distinct and all the elements of the vector $\mathbf{w}$ are different from zero.

Let us show now the related results for generator representable diagonal plus semiseparable matrices. These results are studied in the article [72] and all proofs are provided.

Let us assume that in the remainder of the discussion a symmetric matrix A is considered, having eigenvalues $\lambda_1, \ldots, \lambda_n$. Moreover, take a fixed diagonal $D = \operatorname{diag}([d_1, \ldots, d_n])$. Assume that for every element d_i the following condition is satisfied

$$\det(A - d_i I) \neq 0,$$

this means in fact that the matrix $A - d_i I$ is nonsingular for every d_i, i.e., d_i is not an eigenvalue of A. We need to define some rational functions:

$$\begin{aligned}
\phi_1(\lambda) &= (\lambda - d_1)^{-1} \\
\phi_2(\lambda) &= (\lambda - d_1)^{-1} (\lambda - d_2)^{-1} \\
&\vdots \\
\phi_k(\lambda) &= \prod_{i=1}^{k} (\lambda - d_i)^{-1}
\end{aligned}$$

We will denote the rational Krylov matrix $K_R(A, \mathbf{v})$ as follows:

$$K_R = K_R(A, \mathbf{v}) = [\phi_1(A)\mathbf{v}, \phi_2(A)\mathbf{v}, \ldots, \phi_n(A)\mathbf{v}].$$

Considering the following generalized Vandermonde matrix we can obtain a similar factorization to characterize the matrix $K_R(A, \mathbf{v})$:

$$F = [\phi_j(\lambda_i)]_{i,j}$$
$$= \begin{bmatrix} \phi_1(\lambda_1) & \phi_2(\lambda_1) & \ldots & \phi_n(\lambda_1) \\ \phi_1(\lambda_2) & \phi_2(\lambda_2) & \ldots & \phi_n(\lambda_2) \\ \vdots & \vdots & & \vdots \\ \phi_1(\lambda_n) & \phi_2(\lambda_n) & \ldots & \phi_n(\lambda_n) \end{bmatrix}.$$

Suppose $A = V\Delta V^H$ is the spectral factorization of the matrix A, with eigenvalues $\lambda_1, \ldots, \lambda_n$. Denoting $\mathbf{w} = V^H\mathbf{v}$, then we obtain the following factorization:

$$K_R(A, \mathbf{v}) = [\phi_1(A)\mathbf{v}, \phi_2(A)\mathbf{v}, \phi_3(A)\mathbf{v}, \ldots, \phi_n(A)\mathbf{v}].$$
$$= V \operatorname{diag}([w_1, w_2, \ldots, w_n])F.$$

This means that the matrix $K_R(A, \mathbf{v})$ is nonsingular if all eigenvalues λ_i are pairwise distinct and moreover all entries of the vector $\mathbf{w}$ are nonzero.

The most important results is formulated in [72, Theorem 1]. Suppose the matrix $K_R(A, \mathbf{v})$ is nonsingular, and Q is the orthogonal matrix, coming from the QR-factorization of the matrix $K_R(A, \mathbf{v})$. Then we have that the matrix $Q^H A Q$ is a symmetric generator representable semiseparable plus diagonal matrix $S + D$, for which the diagonal matrix $D = \operatorname{diag}([d_1, d_2, \ldots, d_n])$. Also the converse of this statement holds.

This article [72] further exploits this theorem to prove an implicit Q-theorem for semiseparable plus diagonal matrices. Furthermore results concerning orthogonal similarity transformations to semiseparable plus diagonal form and on QR-steps performed on these matrices are included.

9.7.2   Sturm sequences

In this subsection we will describe in an easy manner the bisection method and the combination with Sturm sequences to compute specific eigenvalues.

First we will deduce most of these results for tridiagonal matrices, and afterwards we will see how to apply them towards semiseparable (plus diagonal) matrices.

Bisection method

The bisection method is an easy and almost straightforward method for computing roots of polynomials (and other functions). Suppose a polynomial $p(\lambda)$ is given with simple roots, and we would like to compute the single root λ in the interval $[a, b]$.

Considering $c = (a + b)/2$ and comparing the sign of the function values $p(c)$ and $p(a)$, we can simply determine whether the root is located in $[a, c]$ or in $[c, b]$. This halves the interval and one can repeat the procedure.

We obtain the following algorithm.

Algorithm 9.35 (Bisection method).
Input: a, b, *the polynomial* $p(\lambda)$ *and a threshold* ϵ.
Output: $\lambda = c$ *up to a predefined precision.*

While $|a - b| > \epsilon(|a| + |b|)$ *do*

1. $c = (a + b)/2;$

2. *If* $p(a)p(c) < 0$

$\qquad\qquad b = c;$

   else

$\qquad\qquad a = c;$

  endif;

endwhile;

Based on this bisection method one can determine specific eigenvalues' of a tridiagonal matrix in linear time, using a recurrence relation for its characteristic polynomial. Essential in this analysis is the fact that the considered tridiagonal matrix is symmetric and irreducible, as this guarantees that all considered roots are simple. For a proof of this statement we refer to [94, 129].

Recurrence relation for the characteristic polynomial of a tridiagonal matrix

Consider the following irreducible symmetric tridiagonal matrix T_n of the following form:

$$
T_n = \begin{bmatrix}
a_1 & b_1 & & & & \\
b_1 & a_2 & b_2 & & & \\
 & b_2 & a_3 & & & \\
 & & & \ddots & \ddots & b_{n-1} \\
 & & & & b_{n-1} & a_n
\end{bmatrix}.
$$

Let us use the following notation:

$$
p_n(\lambda) = \det(T_n - \lambda I).
$$

It is easy to verify that for a tridiagonal matrix T_n, the following relation holds for $p_0(\lambda) = 1$, $p_1(\lambda) = a_1 - \lambda$ and $i = 2, \ldots, n$:

$$
p_i(\lambda) = (a_i - \lambda)p_{i-1}(\lambda) - b_{i-1}^2 p_{i-2}(\lambda).
$$

Based on this recurrence relation, one can evaluate the polynomial $p_n(\lambda)$ in linear time $\mathcal{O}(n)$. Hence, one can use the bisection method to compute approximate roots of $p_n(\lambda)$, which are eigenvalues of the matrix T_n. The method converges linearly since the error is approximately halved by each iteration.

Recurrence relation for the characteristic polynomial of a symmetric semiseparable matrix

Similarly as in the tridiagonal case we can develop a recurrence relation between the characteristic polynomials of leading principal submatrices of a semiseparable matrix. In order to have a well separated spectrum of eigenvalues, the symmetric semiseparable matrix needs to be of irreducible form.

Suppose we have an irreducible symmetric semiseparable matrix[27] of the following form:

$$S_n = \begin{bmatrix} u_1v_1 & u_2v_1 & u_3v_1 & \cdots & u_nv_1 \\ u_2v_1 & u_2v_2 & u_3v_2 & \cdots & u_nv_2 \\ u_3v_1 & u_3v_2 & u_3v_3 & & \vdots \\ \vdots & \vdots & & \ddots & \\ u_nv_1 & u_nv_2 & u_nv_3 & \cdots & u_nv_n \end{bmatrix}.$$

Based on some simple techniques for computing determinants we can derive the following recurrence relations. We use a similar notation as above, i.e., $p_n(\lambda) = \det(S_n - \lambda I)$ and $p_0(\lambda) = 1$.

We have $p_1(\lambda) = u_1v_1 - \lambda$. Let us compute now $p_2(\lambda)$:

$$p_2(\lambda) = \det(S_2 - \lambda I)$$
$$= \det \begin{bmatrix} u_1v_1 - \lambda & u_2v_1 \\ u_2v_1 & u_2v_2 - \lambda \end{bmatrix}$$
$$= (u_2v_2 - \lambda)p_1(\lambda) - (u_2v_1)^2 p_0(\lambda).$$

For computing $p_3(\lambda)$, we rewrite the matrix determinant slightly (by substituting row 3 by a linear combination of rows 3 and 2. A similar operation is used for the columns 3 and 2):

$$p_3(\lambda) = \det(S_3 - \lambda I)$$
$$= \det \begin{bmatrix} u_1v_1 - \lambda & u_2v_1 & u_3v_1 \\ u_2v_1 & u_2v_2 - \lambda & u_3v_2 \\ u_3v_1 & u_3v_2 & u_3v_3 - \lambda \end{bmatrix}$$
$$= \det \begin{bmatrix} u_1v_1 - \lambda & u_2v_1 & u_3v_1 \\ u_2v_1 & u_2v_2 - \lambda & u_3v_2 \\ 0 & \frac{u_3}{u_2}\lambda & u_3v_3 - \lambda - \frac{u_3^2}{u_2}v_2 \end{bmatrix}$$
$$= \det \begin{bmatrix} u_1v_1 - \lambda & u_2v_1 & 0 \\ u_2v_1 & u_2v_2 - \lambda & \frac{u_3}{u_2}\lambda \\ 0 & \frac{u_3}{u_2}\lambda & u_3v_3 - \lambda - \frac{u_3^2}{u_2}v_2 - \frac{u_3^2}{u_2^2}\lambda \end{bmatrix}$$
$$= \left(u_3v_3 - \lambda - \frac{u_3^2}{u_2}v_2 - \frac{u_3^2}{u_2^2}\lambda\right) p_2(\lambda) - \left(\frac{u_3}{u_2}\lambda\right)^2 p_1(\lambda).$$

---

[27]For simplicity we assume to be working with a semiseparable matrix. The quasiseparable or semiseparable plus diagonal case is similar.

Generally we can write down the following recurrence relation, since only the last row and column change in the formulas above:

$$p_i(\lambda) = \left(u_i v_i - \lambda - \frac{u_i^2}{u_{i-1}} v_{i-1} - \frac{u_i^2}{u_{i-1}^2} \lambda \right) p_{i-1}(\lambda) - \left(\frac{u_i}{u_{i-1}} \lambda \right)^2 p_{i-2}(\lambda).$$

The above formula gives a three terms recurrence relation, which can also be evaluated in $\mathcal{O}(n)$. As a consequence one can also use the bisection method for irreducible symmetric semiseparable matrices.

Sturm sequence methods

Suppose we have a sequence of characteristic polynomials of a symmetric matrix $A = A_n$ (A_i is the principal leading submatrix of A), with strictly separating eigenvalues, i.e., that the eigenvalues of $p_i(\lambda)$ strictly separate those of the polynomial $p_{i+1}(\lambda)$. The matrix A_n can, for example, be irreducible symmetric tridiagonal or irreducible symmetric semiseparable. Consider the following sequence:

$$\{p_0(\lambda), p_1(\lambda), \dots, p_n(\lambda)\},$$

with $p_0(\lambda) = 1$. Consider an arbitrary real number c and the sequence:

$$\{p_0(c), p_1(c), \dots, p_n(c)\}.$$

If $s(c)$ denotes the number of sign changes in the sequence above, then we have that $s(c)$ is the number of eigenvalues of the matrix A_n, smaller than c. (The proof can be found in [185].)

Based on this knowledge, one can deduce the following algorithm for computing the k-th largest eigenvalue.

Algorithm 9.36 (Sturm sequences).
Input: a, b, the polynomials $p_i(\lambda)$ and a threshold ϵ.
Output: The k-th largest eigenvalue $\lambda_k = c$ up to a predefined precision.

While $|a - b| > \epsilon(|a| + |b|)$ do

    1. $c = (a + b)/2$;

    2. If $s(c) > n - k$

        $b = c$;

    else

        $a = c$;

    endif;

endwhile;

We remark that the initial input values of a, b are important. One needs to start with a not too large interval. In case of a tridiagonal matrix, one can choose a and b based on the Gershgorin theorem (see [94].) Similarly one can construct bounds for semiseparable (plus diagonal) and quasiseparable matrices.

The results presented here are a condensed form of the ones discussed by Eidelman, Gohberg and Olshevsky in the following article:

> ☞ Y. Eidelman, I. C. Gohberg, and V. Olshevsky. Eigenstructure of order-one-quasiseparable matrices. Three-term and two-term recurrence relations. *Linear Algebra and its Applications*, 405:1–40, 2005.

The general class of quasiseparable matrices is considered, recurrence relations for the characteristic polynomials are provided as well as statements guaranteeing the simplicity of the eigenvalues. Also formulas on how to compute the corresponding eigenvectors are presented.

9.7.3   Other references

Let us discuss some other references related to the subject but not mentioned before.

> ☞ R. Bevilacqua, E. Bozzo, and G. M. Del Corso. Transformations to rank structures by unitary similarity. *Linear Algebra and its Applications*, 402:126–134, 2005.

The authors Bevilacqua, Bozzo and Del Corso prove in this article, via the Krylov methods, the existence of unitary similarity transformations of square matrices to structured rank matrices. The matrices are characterized by low rank blocks and can hence be of band, semiseparable or of semiseparable plus band form.

> ☞ B. Plestenjak, M. Van Barel, and E. Van Camp. A cholesky LR algorithm for the positive definite symmetric diagonal-plus-semiseparable eigenproblem. *Linear Algebra and its Applications*, 428:586–599, 2008.

This article by Plestenjak, Van Camp and Van Barel discusses another technique for computing the eigenvalues of positive definite semiseparable plus diagonal matrices. The approach is based on the LR-factorization of the involved matrix. This factorization consists of an upper triangular factor R and a lower triangular factor L. Positive definiteness of the involved matrix is necessary to guarantee existence of this factorization. In fact this factorization fits also into the general framework of GR-algorithms as provided by Watkins (see [179]). To make sure that the shifted version of the semiseparable plus diagonal matrix remains positive definite a special kind of shift has to be taken, e.g., Laguerre's shift.

> ☞ V. Y. Pan. A reduction of the matrix eigenproblem to polynomial rootfinding via similarity transforms into arrow-head matrices. Technical Report TR-2004009, Department of Computer Science, City University of New York, New York, USA, July 2004.

In this article Pan discusses a method for transforming matrices via similarity transforms into diagonal plus rank 1 matrices. First a unitary similarity transformation is performed on an arbitrary square matrix for reducing it to triangular plus rank one form. Next, nonunitary similarity transforms are needed for bringing this matrix to diagonal plus rank 1 form.

The authors Delvaux and Van Barel generalized many of the concepts known for semiseparable (plus diagonal) and quasiseparable matrices to higher order variants. Let us summarize some results related to spectral methods for these higher order structured rank matrices. The PhD text of Delvaux contains all these results. Also a generalization of the Givens-weight representation towards the unitary weight representation is presented.

> ☞ S. Delvaux. *Rank Structured Matrices.* PhD thesis, Department of Computer Science, Katholieke Universiteit Leuven, Celestijnenlaan 200A, 3000 Leuven (Heverlee), Belgium, June 2007.

To represent general structured rank matrices the authors adapted the Givens-vector representation [165] to the Givens-weight representation [53], discussed in Chapter 1, Section 1.3. In the article [56], discussed in Chapter 6, Section 6.1, the authors discuss an effective way of computing the QR-factorization of these matrices. The preservation of the structure under the QR-method is discussed in [52, 50], see also Chapter 6, Section 6.1. An explicit version of the QR-algorithm for structured rank matrices of higher order was presented in the technical report [49], see Chapter 6, Section 6.3. As it is not always advantageous, in terms of complexity, to perform the QR-method onto higher order structured rank matrices the authors also developed a technique to reduce an arbitrary structured rank matrix as efficiently as possible to Hessenberg form [54], see Chapter 2, Section 2.6.

9.8   Conclusions

In this chapter we discussed several techniques closely related to QR-algorithms for structured rank matrices. It was shown that different types of QR-factorizations exist, resulting also in variants of implementing the QR-algorithms. The QR-algorithms for companion and quasiseparable matrices were discussed, as well as the multishift QR-algorithm. Finally also a new type of QH-iteration for structured rank matrices and some other references were presented.

In this chapter several new techniques were presented. But, not all details for all classes of structured rank matrices and for all algorithms were given. For example: The multishift QR-algorithm was only discussed for Hessenberg-like matrices, what about quasiseparable and the Hessenberg-like plus diagonal cases. Is there a multishift version of the QH-method? What are the convergence properties of the QH-method related to Hessenberg-like plus diagonal matrices? Some of these problems can be solved rather easily, whereas others will require a lot of effort and research.

In the next part a divide-and-conquer technique and a Lanczos semiseparabilization are presented. Furthermore details on how to use the reduction to semiseparable form for rank-revealing purposes are also given.

Part III

Some generalizations and miscellaneous topics

In the first two parts of this book the standard algorithms for computing the eigendecomposition of matrices were discussed. Reduction algorithms to structured rank matrices were discussed, as well as the accompanying QR-methods. In this part we will discuss some topics that are slightly more general than the ones from the first two parts. We will discuss an iterative method for reducing the matrices to structured rank form, a rank-revealing method will be presented and finally we will discuss a divide-and-conquer method for computing the eigendecomposition of quasiseparable matrices.

Chapter 10 discusses divide-and-conquer methods for computing the eigendecomposition. Important for the development of these methods are the interlacing properties of eigenvalues for arrowhead and diagonal plus rank one matrices. This property is first studied in Section 10.1. Section 10.2 proposes two methods for computing the eigendecomposition of tridiagonal matrices. One method is based on the use of arrowhead matrices whereas the other method makes use of diagonal plus rank one matrices. Section 10.3 discusses four different divide-and-conquer methods for computing eigenvalues and eigenvectors of quasiseparable matrices. Finally also numerical experiments are provided.

Chapter 11 discusses two topics. Firstly, in Section 11.1, an iterative method for reducing matrices to semiseparable form is presented. The classical Lanczos method for tridiagonal matrices is discussed first, followed by the algorithm for transforming matrices to semiseparable form. Section 11.2 discusses some details to exploit the convergence properties of the reduction algorithms. It is shown how to adapt the reduction method to be able to use it as a rank-revealing factorization.

✎ *This part contains extra material that is not essential anymore for further understanding of the book. Nevertheless, Chapter 10, discussing the divide-and-conquer methods is very interesting as it provides a fast and accurate method for computing the eigendecomposition of quasiseparable matrices. In Chapter 11, Section 11.1.2 contains the iterative Lanczos-like version for reducing a matrix to semiseparable form.*

Chapter 10

Divide-and-conquer algorithms for the eigendecomposition

This chapter will be devoted to the description of divide-and-conquer techniques to compute the eigendecomposition of some special symmetric structured rank matrices. More precisely we will describe divide-and-conquer methods for both tridiagonal and quasiseparable matrices. The divide-and-conquer techniques for structured rank matrices as they will presented here originate from [40, 28].

The main idea of the algorithm is to divide the original problem into two similar independent subproblems (divide step). Once the latter subproblems have been solved, they are glued together to solve the original one (conquer step). This procedure can be recursively applied on the two independent subproblems until their size is sufficiently small to be handled by standard techniques, like the QR or the QL-methods, previously described.

A special role in the conquer step of the divide-and-conquer algorithms is played either by symmetric arrowhead matrices or diagonal plus rank one matrices. Section 10.1 emphasizes the properties of the latter matrices. Attention is paid to the interlacing property of the eigenvalues and to the computation of the eigenvectors of these matrices. It will be shown that the eigendecomposition of both these matrices can be computed fast and accurately.

In Section 10.2 the main ideas of the divide-and-conquer algorithms for computing the eigendecomposition of symmetric tridiagonal matrices are described. Two types of decompositions in smaller subproblems are presented. The first method uses the arrowhead matrix properties in the conquer step. The second method exploits the diagonal plus rank 1 properties.

The main ideas of the divide-and-conquer algorithms for quasiseparable matrices developed in [40, 117] are described in Section 10.3. Several types of divide-and-conquer methods are discussed. A direct splitting method is developed in which the original matrix is immediately split up, a method in which a rank 1 matrix is subtracted before splitting the problem and two methods involving initial orthogonal similarity transformations.

In the last section of the chapter some numerical experiments concerning timings and accuracy are presented.

✎ *This chapter is an important one. The eigenvalue decomposition as presented in this chapter has proven to be very fast and very accurate. Section 10.1 discusses some general properties of arrowhead and rank 1 perturbations of diagonal matrices. These properties are not essential for a thorough understanding of the divide-and-conquer methods. Section 10.2 discusses the traditional tridiagonal case, one can also skip this part. The important results of this chapter are contained in Section 10.3. This section contains four different types of divide-and-conquer methods and contains the basic ideas of these methods. It is worth reading all four different methods as they differ significantly. Section 10.4 discusses the complexities as well some numerical experiments related to the different methods.*

10.1   Arrowhead and diagonal plus rank 1 matrices

The kernel of the divide-and-conquer algorithms to compute the spectral decomposition of symmetric matrices proposed in the literature is either the computation of the spectral decomposition of symmetric arrowhead matrices or the computation of the spectral decomposition of symmetric diagonal plus rank 1 matrices, that are both special cases of symmetric quasiseparable matrices[1]. In the next subsections we describe the spectral properties of these matrices.

10.1.1   Symmetric arrowhead matrices

In this subsection we describe the main properties of symmetric arrowhead matrices and how the spectral decomposition of such matrices can be retrieved in a stable way. Symmetric arrowhead matrices have entries different from zero on the main diagonal, in the last row and in the last column,

$$
A = \begin{bmatrix} \alpha_1 & & & \beta_1 \\ & \ddots & & \vdots \\ & & \alpha_{n-1} & \beta_{n-1} \\ \beta_1 & \cdots & \beta_{n-1} & \gamma \end{bmatrix}, \tag{10.1}
$$

by using pivoting, one can assume $\alpha_1 \geq \alpha_2 \geq \ldots \geq \alpha_{n-1}$. Obviously, arrowhead matrices are special quasiseparable matrices.

Let λ_i, $i = 1, \ldots, n$ be the eigenvalues of A. By the Cauchy interlacing property [94, 129], the eigenvalues of $A(1:n-1,1:n-1) = D$, i.e., $\{\alpha_i\}_{i=1}^{n-1}$, interlace those of A, this means that

$$
\lambda_1 \geq \alpha_1 \geq \lambda_2 \geq \alpha_2 \geq \ldots \geq \lambda_{n-1} \geq \alpha_{n-1} \geq \lambda_n. \tag{10.2}
$$

In the following cases some eigenvalues of A are explicitly known.

1. If $\beta_j = 0$, then $\lambda_j = \alpha_j$, with corresponding eigenvector $\mathbf{e}_j$, were $\mathbf{e}_j$ is the j-th vector of the canonical basis of $\mathbb{R}^n$.

---

[1]In fact they have even more structure. They are of semiseparable plus diagonal form.

2. If $\alpha_j = \alpha_{j+1}$, for some j, then $\lambda_j = \alpha_j$. In this case, a 2×2 orthogonal similarity transformation is applied to the arrowhead matrix in order to make $\beta_j = 0$.

The other eigenvalues can be computed deleting the j-th row and the j-th column from A. This process is called deflation [64].

After the deflation, the problem is to compute the spectral decomposition of an arrowhead matrix A, named irreducible now, with $\alpha_1 > \alpha_2 > \ldots > \alpha_{n-1}$, and $|\beta_j| > 0$, for $j = 1, \ldots, n-1$. The eigenvalues of symmetric arrowhead matrices are the zeros of a secular equation. In fact the following lemma holds.

Lemma 10.1. *Let*

$$A = \begin{bmatrix} D & \mathbf{b} \\ \mathbf{b}^T & \gamma \end{bmatrix}$$

be a symmetric arrowhead matrix of order n, with $D = \mathrm{diag}([\alpha_1, \ldots, \alpha_{n-1}])$, $\alpha_1 > \alpha_2 > \cdots > \alpha_{n-1}$, $\mathbf{b} = [\beta_1, \ldots, \beta_{n-1}]^T$. Then the eigenvalues λ_i, $i = 1, \ldots, n$ of A interlace the diagonal entries of D, i.e.,

$$\lambda_1 > \alpha_1 > \lambda_2 > \alpha_2 > \cdots > \lambda_{n-1} > \alpha_{n-1} > \lambda_n. \tag{10.3}$$

Moreover, λ_i are the zeros of the secular equation

$$f(\lambda) = \gamma - \lambda - \sum_{i=1}^{n-1} \frac{\beta_i^2}{\alpha_i - \lambda}. \tag{10.4}$$

The eigenvector of A corresponding to λ_i is given by

$$\mathbf{v}_i = \left[\frac{\beta_1}{\alpha_1 - \lambda_i}, \frac{\beta_2}{\alpha_2 - \lambda_i}, \cdots, \frac{\beta_{n-1}}{\alpha_{n-1} - \lambda_i}, -1 \right]^T \Big/ \sqrt{1 + \sum_{j=1}^{n-1} \frac{\beta_j^2}{(\alpha_j - \lambda_i)^2}} \tag{10.5}$$

Proof. Since $\beta_j \neq 0$, $j = 1, \ldots, n-1$, the eigenvalues of A cannot be equal to α_j, $j = 1, \ldots, n-1$. Therefore, taking the Cauchy interlacing property (see Equation (10.2)) into account, Equation (10.3) holds.
To prove (10.4), let us define $\hat{\mathbf{b}} = (D - \lambda I_{n-1})^{-1}\mathbf{b}$. Since[2]

$$A - \lambda I_n = \left[\begin{array}{c|c} D - \lambda I_{n-1} & \mathbf{b} \\ \hline \mathbf{b}^T & \gamma - \lambda \end{array} \right]$$

$$= \left[\begin{array}{c|c} I & \\ \hline \hat{\mathbf{b}}^T & 1 \end{array} \right] \left[\begin{array}{c|c} D - \lambda I_{n-1} & \\ \hline & \gamma - \lambda - \mathbf{b}^T\hat{\mathbf{b}} \end{array} \right] \left[\begin{array}{c|c} I & \hat{\mathbf{b}} \\ \hline & 1 \end{array} \right],$$

---

[2]The identity matrix of size n is denoted by I_n.

with $\lambda \neq \alpha_i$, $i = 1, \ldots, n-1$, the characteristic polynomial of A is given by

$$
\begin{aligned}
\det(A - \lambda I_n) &= \det \left[\begin{array}{c|c} D - \lambda I_{n-1} & \\ \hline & \gamma - \lambda - \mathbf{b}(D - \lambda I_{n-1})^{-1}\mathbf{b} \end{array} \right] \\
&= \prod_{i=1}^{n-1} (\alpha_i - \lambda) \left(\gamma - \lambda - \sum_{i=1}^{n-1} \frac{\beta_i^2}{\alpha_i - \lambda} \right).
\end{aligned}
$$

Since α_i are not eigenvalues of A, an eigenvalue of A is a zero of (10.4). Equation (10.5) can be checked easily. $\qquad\square$

A straightforward way to compute the zeros of (10.4) is the bisection method. Some other efficient and reliable algorithms to compute the zeros of the secular Equation (10.4), based on Newton, Halley and rational interpolation are available in the literature [44, 64, 113, 27, 28, 63].

10.1.2   Computing the eigenvectors

Computing the approximation $\hat{\lambda}_i$ of λ_i in some way, the corresponding eigenvector can be computed in a simple manner via Equation (10.5). Suppose we have computed

$$
\hat{\mathbf{v}}_i = \left[\frac{\beta_1}{\alpha_1 - \hat{\lambda}_i}, \frac{\beta_2}{\alpha_2 - \hat{\lambda}_i}, \cdots, \frac{\beta_{n-1}}{\alpha_{n-1} - \hat{\lambda}_i}, -1 \right]^T \bigg/ \sqrt{1 + \sum_{j=1}^{n-1} \frac{\beta_j^2}{(\alpha_j - \hat{\lambda}_i)^2}}.
$$

Even though $\hat{\lambda}_i$ is close to λ_i, the approximate ratios $\beta_j/(\alpha_j - \hat{\lambda}_i)$ could be very different from the exact ratios $\beta_j/(\alpha_j - \lambda_i)$, and the resulting eigenvector matrix may not be numerically orthogonal. Each difference, each ratio, each product, and each sum in (10.5) can be computed to high relative accuracy, if λ_i is given exactly and the corresponding approximation of $\mathbf{v}_i$ can be computed to componentwise high relative accuracy. However, in practice we can only hope to compute an approximation $\hat{\lambda}_i$ to λ_i. To overcome this problem, we consider the following Lemma [25].

Lemma 10.2. *Given two sets of real numbers $\{\hat{\lambda}_i\}_{i=1}^n$ and $\{\alpha_i\}_{i=1}^{n-1}$, satisfying the interlacing property*

$$
\hat{\lambda}_1 > \alpha_1 > \hat{\lambda}_2 > \alpha_2 > \cdots > \hat{\lambda}_{n-1} > \alpha_{n-1} > \hat{\lambda}_n,
$$

there exists a symmetric arrowhead matrix

$$
\hat{A} = \left[\begin{array}{cc} D & \hat{\mathbf{b}} \\ \hat{\mathbf{b}}^T & \hat{\gamma} \end{array} \right]
$$

whose eigenvalues are $\{\hat{\lambda}_i\}_{i=1}^n$. The vector $\hat{\mathbf{b}} = [\hat{\beta}_1, \ldots, \hat{\beta}_n]^T$ and the scalar $\hat{\gamma}$ are *given by*

$$|\hat{\beta}_i| = \sqrt{(\alpha_i - \hat{\lambda}_1)(\hat{\lambda}_n - \alpha_i) \prod_{j=1}^{i-1} \frac{(\hat{\lambda}_j - \alpha_i)}{(\alpha_j - \alpha_i)} \prod_{j=i+1}^{n-1} \frac{(\hat{\lambda}_j - \alpha_i)}{(\alpha_{j+1} - \alpha_i)}} \qquad (10.6)$$

$$\hat{\gamma} = \hat{\lambda}_n + \sum_{i=1}^{n-1} \left(\hat{\lambda}_j - \alpha_j\right),$$

where the sign of $\hat{\beta}_i$ can be chosen arbitrarily.

After having computed the approximate eigenvalues $\{\hat{\lambda}_i\}_{i=1}^n$ of A, we can construct a new symmetric arrowhead matrix $\hat{A}$ whose exact eigenvalues are $\{\hat{\lambda}_i\}_{i=1}^n$, by means of Lemma 10.2. Then the corresponding eigenvectors can be computed by Equation (10.5). We point out that each difference, each product, and each ratio in (10.6) can be computed to high relative accuracy, and the sign of $\hat{\beta}_i$ can be chosen as the sign of β_i. Thus $\hat{\beta}_i$ can be computed to componentwise high relative accuracy. Replacing the exact eigenvalues $\{\hat{\lambda}_i\}_{i=1}^n$ and the computed $\hat{b}$ into (10.5), each eigenvector of $\hat{A}$ can be computed to componentwise high relative accuracy resulting in an eigenvector matrix that is numerically orthogonal. Then the spectral decomposition of $\hat{A}$ replaces the spectral decomposition of A. Since

$$A = \begin{bmatrix} D & \mathbf{b} \\ \mathbf{b} & \gamma \end{bmatrix} = \hat{A} + \begin{bmatrix} 0 & \mathbf{b} - \hat{\mathbf{b}} \\ \mathbf{b}^T - \hat{\mathbf{b}}^T & \gamma - \hat{\gamma} \end{bmatrix},$$

it turns out

$$\|\hat{A} - A\|_2 \le |\hat{\gamma} - \gamma| + \|\hat{\mathbf{b}} - \mathbf{b}\|_2.$$

Hence, such substitution is stable as long as $\hat{\gamma}$ and $\hat{b}$ are close to γ and $\mathbf{b}$, respectively. The problem is to compute the eigenvalues of A accurately.

10.1.3   Rank 1 modification of a diagonal matrix

The class of matrices described in this subsection arises in some divide-and-conquer algorithms to compute the spectral decomposition of symmetric matrices. In fact, some divide-and-conquer algorithms require, as a conquer step, the computation of the spectral decomposition of symmetric diagonal plus rank 1 matrices, i.e., another special kind of quasiseparable matrices.

Let $A = D + \mathbf{z}\mathbf{z}^T$, with $D = \text{diag}([d_1, d_2, \ldots, d_n])$ and $\mathbf{z} = [z_1, z_2, \ldots, z_n]^T$. Without loss of generality, we assume that $d_1 \le d_2 \le \cdots \le d_n$. In the following cases some eigenvalues of A are explicitly known.

1. If $z_j = 0$, for some j, then $\lambda_j = d_j$, with corresponding eigenvector $\mathbf{e}_j$, were $\mathbf{e}_j$ is the j-th vector of the canonical basis of $\mathbb{R}^n$.

2. If $d_j = d_{j+1}$, for some j, then $\lambda_j = d_j$. In this case, a 2×2 orthogonal similarity transformation is applied to the diagonal plus rank 1 matrix in order to make $z_j = 0$.

The other eigenvalues can be computed deleting the j-th row and the j-th column from A. This process is called deflation. If $d_i \neq d_j$, $i \neq j$ and $z_j \neq 0$, $j = 1, \ldots, n$, the diagonal plus rank 1 matrix is said to be irreducible. After the deflation, the problem is to compute the spectral decomposition of an irreducible symmetric diagonal plus rank 1 matrix.

The following lemma holds [101].

Lemma 10.3. *Given a diagonal matrix $D = \mathrm{diag}([d_1, d_2, \ldots, d_n])$ and a vector $\mathbf{z} = [z_1, z_2, \ldots, z_n]^T$. Assume that*

$$d_1 < d_2 < \cdots < d_n$$

and $z_j \neq 0$, $j = 1, \ldots, n$. Then the eigenvalues $\{\lambda_i\}_{i=1}^n$ of $D + \mathbf{z}\mathbf{z}^T$ satisfy the interlacing property

$$d_1 < \lambda_1 < d_2 < \lambda_2 < \cdots < \lambda_{n-1} < d_n < \lambda_n \tag{10.7}$$

and are the roots of the secular equation

$$f(\lambda) = 1 + \sum_{j=1}^n \frac{z_j^2}{d_j - \lambda}.$$

For each eigenvalue λ_i, the corresponding eigenvector is given by

$$\mathbf{q}_i = \left[\frac{z_1}{d_1 - \lambda_i}, \ldots, \frac{\cdot z_n}{d_n - \lambda_i} \right]^T \bigg/ \sqrt{\sum_{j=1}^n \frac{z_j^2}{(d_j - \lambda)^2}}. \tag{10.8}$$

The eigenvalues λ_i, $i = 1, \ldots, n$, are efficiently computed by algorithms based on rational interpolation [101, 27, 28]. For each eigenvalue λ_i, the latter algorithms find a numerical approximation $\tilde{\lambda}_i$. The corresponding approximation of the eigenvector computed by Equation (10.8) is

$$\tilde{\mathbf{q}}_i = \left[\frac{z_1}{d_1 - \tilde{\lambda}_i}, \ldots, \frac{z_n}{d_n - \tilde{\lambda}_i} \right]^T \bigg/ \sqrt{\sum_{j=1}^n \frac{z_j^2}{(d_j - \tilde{\lambda}_i)^2}},$$

where the exact λ_i is replaced by the computed approximation $\tilde{\lambda}_i$. Similar to what happens to arrowhead matrices, although $\tilde{\lambda}_i$ is very close to λ_i, the approximate ratio $z_j/(d_j - \tilde{\lambda}_i)$ can be very different from the exact ratio $z_j/(d_j - \lambda_i)$, resulting in a computed eigenvector very different from the true eigenvector. When all the eigenvectors are computed, the orthogonality among the columns of the resulting eigenvector matrix is lost. To retrieve the orthogonality among the computed eigenvectors similar techniques considered for arrowhead matrices, can be used also in this framework.

Each difference, each ratio, each product, and each sum in (10.8) can be computed to high relative accuracy, if λ_i is given exactly. Therefore, the corresponding approximation of $\mathbf{q}_i$ is computed to componentwise high relative accuracy. However, in practice we can only hope to compute an approximation $\tilde{\lambda}_i$ to λ_i. Nevertheless, suppose that we could find an approximation $\tilde{\mathbf{z}}_j$ to $\mathbf{z}_j$, $j = 1, \ldots, n$, such that $\{\tilde{\lambda}_i\}_{i=1}^n$ are the exact eigenvalues of the new rank 1 modification matrix $\tilde{A} = D + \tilde{\mathbf{z}}\tilde{\mathbf{z}}^T$. Since

$$
\begin{aligned}
A &= D + \mathbf{z}\mathbf{z}^T \\
&= D + \tilde{\mathbf{z}}\tilde{\mathbf{z}}^T + \mathbf{z}\mathbf{z}^T - \tilde{\mathbf{z}}\tilde{\mathbf{z}}^T \\
&= \tilde{A} + (\mathbf{z} - \tilde{\mathbf{z}})\mathbf{z}^T + \mathbf{z}(\mathbf{z} - \tilde{\mathbf{z}})^T - (\mathbf{z} - \tilde{\mathbf{z}})(\mathbf{z} - \tilde{\mathbf{z}})^T.
\end{aligned}
$$

The matrix $\tilde{A}$ will be close to A as long as $\tilde{\mathbf{z}}$ is close to $\mathbf{z}$. Moreover,

$$
\tilde{\mathbf{q}}_i = \left[\frac{\tilde{z}_1}{d_1 - \tilde{\lambda}_i}, \ldots, \frac{\tilde{z}_n}{d_n - \tilde{\lambda}_i}\right] \bigg/ \sqrt{\sum_{j=1}^n \frac{\tilde{z}_j^2}{(d_j - \tilde{\lambda}_i)^2}}, \tag{10.9}
$$

gives the exact eigenvector corresponding to the eigenvalue $\tilde{\lambda}_i$ of $\tilde{A}$. As already observed, $\tilde{\mathbf{q}}_i$ can be computed to componentwise high relative accuracy. Thus, when all the eigenvectors of $\tilde{A}$ are computed, the resulting eigenvector matrix will be numerically orthogonal.

We show how the vector $\tilde{\mathbf{z}}$ can be computed. We observe that, since d_j, $j = 1, \ldots,$ are not eigenvalues of $\tilde{A}$,

$$
\begin{aligned}
\det(\tilde{A} - \lambda I) &= \det(D - \lambda I + \tilde{\mathbf{z}}\tilde{\mathbf{z}}^T) \\
&= \det(D - \lambda I) \det(I + (D - \lambda I)^{-1}\tilde{\mathbf{z}}\tilde{\mathbf{z}}^T) \\
&= \prod_{j=1}^n (d_i - \lambda) \left(1 + \sum_{j=1}^n \frac{\tilde{z}_j^2}{(d_j - \lambda)}\right).
\end{aligned}
$$

On the other hand,

$$
\det(\tilde{A} - \lambda I) = \prod_{j=1}^n (\tilde{\lambda}_j - \lambda).
$$

Combining the following two representations of the determinant of the characteristic polynomial of $\tilde{A}$, and setting $\lambda = d_i$, we get

$$
\tilde{z}_i = \frac{\prod_{j=1}^n (\hat{\lambda}_j - d_i)}{\prod_{j=1, j\neq i}^n (d_j - d_i)}. \tag{10.10}
$$

Therefore, assuming that the interlacing property for $\tilde{\lambda}_i$ $i = 1, \ldots, n$, is preserved,

$$
d_1 < \tilde{\lambda}_1 < d_2 < \tilde{\lambda}_2 < \cdots < d_n < \tilde{\lambda}_n,
$$

the right-hand side of (10.10) is positive and

$$
\tilde{z}_i = \sqrt{(\tilde{\lambda}_n - d_i) \prod_{j=1}^{i-1} \frac{(\tilde{\lambda}_j - d_i)}{(d_j - d_i)} \prod_{j=i+1}^n \frac{(\tilde{\lambda}_j - d_i)}{(d_{j+1} - d_i)}}. \tag{10.11}
$$

On the other hand, if each $\tilde{z}_i$ is given by (10.11), then the eigenvalues of $D + \tilde{\mathbf{z}}\tilde{\mathbf{z}}^T$ are $\tilde{\lambda}_i$, $i = 1, \ldots, n$. Each difference, each ratio, each product, and each sum in (10.11) can be computed to high relative accuracy. Therefore, $\tilde{\mathbf{z}}_j$ is computed to componentwise high relative accuracy. Replacing the computed $\tilde{\mathbf{z}}_j$ in (10.9), $\tilde{\mathbf{q}}_j$ can be computed to high relative accuracy. Consequently, when all the eigenvectors are computed, the resulting eigenvector matrix $\tilde{Q} = [\tilde{\mathbf{q}}_1, \tilde{\mathbf{q}}_2, \ldots, \tilde{\mathbf{q}}_n]$ turns out to be numerically orthogonal.

Notes and references

A first algorithm to compute the eigenvalues of symmetric arrowhead matrices, based on the bisection algorithm to compute the zeros of the involved secular functions, was proposed by O'Leary and Stewart.

☞ D. P. O'Leary and G. W. Stewart. Computing the eigenvalues and eigenvectors of symmetric arrowhead matrices. *Journal of Computational Physics*, 90(2):497–505, October 1990.

More efficient algorithms based on rational interpolation for computing the zeros of secular functions were proposed in the following articles.

☞ J. J. M. Cuppen. A divide and conquer method for the symmetric tridiagonal eigenproblem. *Numerische Mathematik*, 36:177–195, 1981.

☞ J. J. Dongarra and D. C. Sorensen. A fully parallel algorithm for the symmetric eigenvalue problem. *SIAM Journal on Scientific and Statistical Computation*, 3(2):139–154, March 1987.

☞ R.-C. Li. Solving secular equations stably and efficiently. LAPACK Working Note 89, April 1993.

A reliable algorithm based on the Newton method was proposed by Diele, Mastronardi, Van Barel and Van Camp.

☞ F. Diele, N. Mastronardi, M. Van Barel, and E. Van Camp. On computing the spectral decomposition of symmetric arrowhead matrices. *Lecture Notes in Computer Science*, 3044:932–941, 2004.

Other efficient algorithms to compute the zeros of the secular equations were proposed by Melman.

☞ A. Melman. Numerical solution of a secular equation. *Numerische Mathematik*, 69:483–493, 1995.

☞ A. Melman. A numerical comparison of methods for solving secular equations. *Journal of Computational and Applied Mathematics*, 86:237–249, 1997.

☞ A. Melman. A unifying convergence analysis of second-order methods for secular equations. *Mathematics of Computation*, 66(217):333–344, January 1997.

☞ A. Melman. Analysis of third-order methods for secular equations. *Mathematics of Computation*, 67(221):271–286, January 1998.

The recapture of the orthogonality of the computed eigenvector matrix of diagonal plus rank 1 matrices was studied independently by Borges and Gragg and by Gu and Eisenstat.

☞ C. F. Borges and W. B. Gragg. Divide and conquer for generalized real symmetric definite tridiagonal eigenproblems. In E.-X. Jiang, editor, *Proceedings of '92 Shanghai International Numerical Algebra and its Applications Conference*, pages 70–76. China Science and Technology Press, Beijing, China, October 1992.

☞ C. F. Borges and W. B. Gragg. A parallel divide and conquer algorithm for the generalized real symmetric definite tridiagonal eigenproblem. In L. Reichel, A. Ruttan, and R. S. Varga, editors, *Numerical Linear Algebra and Scientific Computing*, pages 11–29. de Gruyter, Berlin, Germany, 1993.

☞ M. Gu and S. C. Eisenstat. A stable and efficient algorithm, for the rank-one modification of the symmetric eigenproblem. *SIAM Journal on Matrix Analysis and Applications*, 15(4):1266–1276, October 1994.

☞ M. Gu and S. C. Eisenstat. A divide-and-conquer algorithm for the symmetric tridiagonal eigenproblem. *SIAM Journal on Matrix Analysis and Applications*, 16(1):172–191, January 1995.

10.2 Divide-and-conquer algorithms for tridiagonal matrices

In this section we show how the computation of the spectral decomposition of symmetric tridiagonal matrices is reduced either to the computation of the spectral decomposition of symmetric arrowhead matrices or to the computation of the spectral decomposition of symmetric diagonal plus rank 1 matrices, in the divide-and-conquer techniques.

10.2.1 Transformation into a similar arrowhead matrix

In the first approach, the symmetric tridiagonal matrix T of order n is partitioned in the following way (without loss of generality we can assume the matrix T to be irreducible):

$$
T = \left[\begin{array}{ccc|c|ccc}
a_1 & b_1 & & & & & \\
b_1 & \ddots & \ddots & & & & \\
& \ddots & a_k & b_k & & & \\
\hline
& & b_k & a_{k+1} & a_{k+1} & & \\
\hline
& & & b_{k+1} & \ddots & \ddots & \\
& & & & \ddots & a_{n-1} & b_{n-1} \\
& & & & & b_{n-1} & a_n
\end{array}\right] \tag{10.12}
$$

$$
= \left[
\begin{array}{c|c|c}
T_1 & b_k \mathbf{e}_k^{(k)} & \\
\hline
b_k \mathbf{e}_k^{(k)T} & a_{k+1} & b_{k+1} \mathbf{e}_1^{(n-k-1)T} \\
\hline
& b_{k+1} \mathbf{e}_1^{(n-k-1)} & T_2
\end{array}
\right], \tag{10.13}
$$

with $\mathbf{e}_1^{(n-k-1)}$ and $\mathbf{e}_k^{(k)}$ the first and the k-th columns of the identity matrices of order $n - k - 1$ and k, respectively, $1 \le k \le n$. Let

$$
T_1 = Q_1 \Delta_1 Q_1^T, \quad T_2 = Q_2 \Delta_2 Q_2^T
$$

be the spectral decompositions of T_1 and T_2, respectively, with Q_1 and Q_2 orthogonal matrices and

$$
\Delta_1 = \mathrm{diag}([\lambda_1^{(1)}, \lambda_2^{(1)}, \ldots, \lambda_k^{(1)}]), \quad \Delta_2 = \mathrm{diag}([\lambda_1^{(2)}, \lambda_2^{(2)}, \ldots, \lambda_{n-k-1}^{(2)}]).
$$

The matrix T can be written in the following way:

$$
T = \left[
\begin{array}{c|c|c}
Q_1 \Delta_1 Q_1^T & b_k \mathbf{e}_k^{(k)} & \\
\hline
b_k \mathbf{e}_k^{(k)T} & a_{k+1} & b_{k+1} \mathbf{e}_1^{(n-k-1)T} \\
\hline
& b_{k+1} \mathbf{e}_1^{(n-k-1)} & Q_2 \Delta_2 Q_2^T
\end{array}
\right]
$$

$$
= \left[
\begin{array}{c|c|c}
Q_1 & & \\
\hline
& 1 & \\
\hline
& & Q_2
\end{array}
\right]
\left[
\begin{array}{c|c|c}
\Delta_1 & & b_k \mathbf{v}_1 \\
\hline
& \Delta_1 & b_k \mathbf{v}_2 \\
\hline
b_{k+1} b_k \mathbf{v}_1^T & b_k \mathbf{v}_2^T & a_k
\end{array}
\right]
\left[
\begin{array}{c|c|c}
Q_1 & & \\
\hline
& 1 & \\
\hline
& & Q_2
\end{array}
\right]^T,
$$

with $\mathbf{v}_1$ and $\mathbf{v}_2$ the last row of Q_1 and the first row of Q_2, respectively. Therefore, knowing the spectral decomposition of T_1 and T_2, the problem of computing the spectral decomposition of the tridiagonal matrix T is reduced to computing the spectral decomposition of an arrowhead matrix. The same procedure can be applied recursively to T_1 and T_2, until the considered blocks are sufficiently small and their spectral decompositions can be easily retrieved.

10.2.2　Transformation into a diagonal plus rank 1

Another approach to compute the eigendecomposition of a symmetric tridiagonal matrix via divide-and-conquer methods is described in the sequel, whose conquer step is the computation of the spectral decomposition of a diagonal plus rank 1 matrix.

Let T be a symmetric tridiagonal matrix of order n

$$T = \left[\begin{array}{cccc|cccc} a_1 & b_1 & & & & & & \\ b_1 & \ddots & \ddots & & & & & \\ & \ddots & a_k & b_k & & & & \\ \hline & & b_k & a_{k+1} & \ddots & & \\ & & & \ddots & \ddots & b_{n-1} \\ & & & & b_{n-1} & a_n \end{array}\right] \tag{10.14}$$

$$= \left[\begin{array}{c|c} T_1 & b_k \mathbf{e}_k^{(k)} \mathbf{e}_1^{(n-k)^T} \\ \hline \mathbf{e}_1^{(n-k)} \mathbf{e}_k^{(k)^T} & T_2 \end{array}\right]$$

and with $\mathbf{e}_1^{(n-k)}$ and $\mathbf{e}_k^{(k)}$ the first and the k-th columns of the identity matrices of order $n-k$ and k, respectively, $1 \le k \le n-1$. Let

$$\hat{T}_1 = T_1 - b_k \mathbf{e}_k^{(k)} \mathbf{e}_k^{(k)^T}, \quad \hat{T}_2 = T_2 - b_k \mathbf{e}_1^{(n-k)} \mathbf{e}_1^{(n-k)^T}.$$

Then the tridiagonal matrix T can be written as

$$T = \left[\begin{array}{c|c} \hat{T}_1 & \\ \hline & \hat{T}_2 \end{array}\right] + b_k \left[\begin{array}{c} \mathbf{e}_k^{(k)} \\ \mathbf{e}_1^{(n-k)} \end{array}\right] \left[\begin{array}{cc} \mathbf{e}_k^{(k)^T} & \mathbf{e}_1^{(n-k)^T} \end{array}\right].$$

Denote by

$$\hat{T}_1 = \hat{Q}_1 \hat{\Delta}_1 \hat{Q}_1^T, \quad \hat{T}_2 = \hat{Q}_2 \hat{\Delta}_2 \hat{Q}_2^T,$$

the spectral decompositions of $\hat{T}_1$ and $\hat{T}_2$, respectively, with $\hat{Q}_1$ and $\hat{Q}_2$ orthogonal matrices and

$$\hat{\Delta}_1 = \mathrm{diag}([\hat{\lambda}_1^{(1)}, \hat{\lambda}_2^{(1)}, \ldots, \hat{\lambda}_k^{(1)}]), \quad \hat{\Delta}_2 = \mathrm{diag}([\hat{\lambda}_1^{(2)}, \hat{\lambda}_2^{(2)}, \ldots, \hat{\lambda}_{n-k}^{(2)}]),$$

the matrix T can be written in the following way:

$$T = \left[\begin{array}{c|c} \hat{Q}_1 \hat{\Delta}_1 \hat{Q}_1^T & \\ \hline & \hat{Q}_2 \hat{\Delta}_2 \hat{Q}_2^T \end{array}\right] + b_k \left[\begin{array}{c} \mathbf{e}_k^{(k)} \\ \mathbf{e}_1^{(n-k)} \end{array}\right] \left[\begin{array}{cc} \mathbf{e}_k^{(k)^T} & \mathbf{e}_1^{(n-k)^T} \end{array}\right]$$

$$= \left[\begin{array}{c|c} \hat{Q}_1 & \\ \hline & \hat{Q}_2 \end{array}\right] \left(\left[\begin{array}{c|c} \hat{\Delta}_1 & \\ \hline & \hat{\Delta}_2 \end{array}\right] + b_k \mathbf{z}\mathbf{z}^T\right) \left[\begin{array}{c|c} \hat{Q}_1^T & \\ \hline & \hat{Q}_2^T \end{array}\right],$$

with

$$\mathbf{z} = \left[\begin{array}{c} z_1^{(1)} \\ \vdots \\ z_k^{(1)} \\ \hline z_1^{(2)} \\ \vdots \\ z_{n-k}^{(2)} \end{array}\right] = \left[\begin{array}{c} Q_1^T \mathbf{e}_k^{(k)} \\ \hline Q_2^T \mathbf{e}_1^{(n-k)} \end{array}\right].$$

Therefore, known the spectral decomposition of $\hat{T}_1$ and $\hat{T}_2$, the spectral decomposition of T is reduced to compute the spectral decomposition of a diagonal plus rank 1 matrix. The same procedure can be applied recursively to $\hat{T}_1$ and $\hat{T}_2$ to compute their spectral decomposition, until the size of the involved matrices is small enough to be handled by standard methods, like the QR or the QL-method.

Notes and references

A first divide-and-conquer algorithm to compute the eigenvalues and eigenvectors of symmetric tridiagonal matrices was proposed by Cuppen [44] (see notes and references of the previous section).

Other divide-and-conquer algorithms for computing the spectral decomposition of symmetric tridiagonal matrices can be found in the following articles (see also [28, 102], discussed in the previous section):

☞ D. A. Bini and V. Y. Pan. Practical improvement of the divide-and-conquer eigenvalue algorithms. *Computing*, 48(1):109–123, 1992.

☞ C. F. Borges, R. Frezza, and W. B. Gragg. Some inverse eigenproblems for Jacobi and arrow matrices. *Numerical Linear Algebra with Applications*, 2(3):195–203, 1995.

Divide-and-conquer algorithms are very suitable for implementation on parallel computers. Nevertheless, the performance of divide-and-conquer algorithms for the symmetric tridiagonal eigenproblem are comparable to those of classical algorithms for such problems, like the QR and the QL-methods, even on serial computers (see [64] in the previous section).

10.3   Divide-and-conquer methods for quasiseparable matrices

In this section we consider some different divide-and-conquer approaches to compute the eigendecomposition of symmetric generator representable quasiseparable matrices.

Consider a symmetric $n \times n$ generator representable quasiseparable matrix of the form

$$
A = \begin{bmatrix}
d_1 & v_1 u_2 & v_1 u_3 & \cdots & v_1 u_{n-1} & v_1 u_n \\
u_2 v_1 & d_2 & v_2 u_3 & v_2 u_4 & \cdots & v_2 u_n \\
u_3 v_1 & u_3 v_2 & \ddots & \cdots & \cdots & \cdots \\
\vdots & u_4 v_2 & \vdots & \ddots & \cdots & \cdots \\
u_{n-1} v_1 & \vdots & \vdots & \vdots & d_{n-1} & v_{n-1} u_n \\
u_n v_1 & u_n v_2 & \vdots & \vdots & u_n v_{n-1} & d_n
\end{bmatrix}. \tag{10.15}
$$

Let $D = \mathrm{diag}([d_1, \ldots, d_n])$, $\mathbf{u} = [u_1, u_2, \ldots, u_{n-1}, u_n]^T$, $\mathbf{v} = [v_1, v_2, v_3, \ldots, v_n]^T$. A can be written as the sum of three matrices:

$$
A = D + \mathrm{tril}(\mathbf{u}\mathbf{v}^T, -1) + \mathrm{triu}(\mathbf{v}\mathbf{u}^T, 1).
$$

In the remainder of this section, we will assume that $v_1 \neq 0$. This is not a restriction because if $v_1 = 0$, then d_1 is an eigenvalue and the matrix can be reduced into a matrix with the first component of $\mathbf{v}$ different from 0.

Note 10.4. *Even though the mathematical descriptions depicted in this chapter are restricted to the class of symmetric generator representable quasiseparable matrices, the deductions stay also valid in case one is working with a general symmetric quasiseparable matrix. It is easy to check that these results remain valid for the more general class of symmetric quasiseparable matrices, including thereby semiseparable plus diagonal matrices and so forth. Moreover an implementation of the divide-and-conquer method based on the Givens-vector representation for semiseparable plus diagonal matrices is available. We chose however to present the algorithms for the generator quasiseparable case, instead of the general quasiseparable case, as this was in our opinion the clearest to present. The general case uses exactly the same ideas but would be unnecessarily complicated to present.*

We will now describe four different types of divide-and-conquer algorithms.

10.3.1 A first divide-and-conquer algorithm

Divide-and-conquer algorithms have been used to compute the eigendecomposition of symmetric tridiagonal matrices. The first divide-and-conquer approach for computing the spectral decomposition of quasiseparable matrices was proposed in [40]. We shortly describe this algorithm in this subsection.

Let $k = \lfloor \frac{n}{2} \rfloor$. Define the following vectors

$$\mathbf{u}_1 = \mathbf{u}(1:k), \; \mathbf{u}_2 = \mathbf{u}(k+1:n), \; \mathbf{v}_1 = \mathbf{v}(1:k), \; \mathbf{v}_2 = \mathbf{v}(k+1:n),$$

$$A_1 = \operatorname{diag}([d_1,\ldots,d_k]) + \operatorname{tril}(\mathbf{u}_1\mathbf{v}_1^T, -1) + \operatorname{triu}(\mathbf{v}_1\mathbf{u}_1^T, 1),$$

and

$$A_2 = \operatorname{diag}([d_{k+1},\ldots,d_n]) + \operatorname{tril}(\mathbf{u}_2\mathbf{v}_2^T, -1) + \operatorname{triu}(\mathbf{v}_2\mathbf{u}_2^T, 1).$$

The Matrix (10.15) can be rewritten as

$$A = \begin{bmatrix} A_1 & \mathbf{v}_1\mathbf{u}_2^T \\ \mathbf{u}_2\mathbf{v}_1^T & A_2 \end{bmatrix} \tag{10.16}$$

$$= \begin{bmatrix} A_1 - \rho\mathbf{v}_1\mathbf{v}_1^T & \\ & A_2 - \rho\mathbf{u}_2\mathbf{u}_2^T \end{bmatrix} + \rho\mathbf{w}\mathbf{w}^T \tag{10.17}$$

where

$$\rho = \pm 1, \quad \mathbf{w} = \begin{bmatrix} \rho\mathbf{v}_1 \\ \mathbf{u}_2 \end{bmatrix}.$$

We observe that the matrices

$$A_1 - \rho\mathbf{v}_1\mathbf{v}_1^T \quad \text{and} \quad A_2 - \rho\mathbf{u}_2\mathbf{u}_2^T,$$

are in quasiseparable form, because

$$\begin{aligned}
A_1 - \rho \mathbf{v}_1 \mathbf{v}_1^T &= \mathrm{diag}([d_1 - \rho v_1^2, \ldots, d_k - \rho v_k^2]) \\
&= + \mathrm{triu}(\mathbf{v}_1 (\mathbf{u}_1 - \rho \mathbf{v}_1)^T, 1) + \mathrm{tril}((\mathbf{u}_1 - \rho \mathbf{v}_1)\mathbf{v}_1^T, -1), \\
A_2 - \rho \mathbf{u}_2 \mathbf{u}_2^T &= \mathrm{diag}([d_{k+1} - \rho u_{k+1}^2, \ldots, d_n - \rho u_n^2]) \\
&= + \mathrm{triu}((\mathbf{v}_2 - \rho \mathbf{u}_2)\mathbf{v}_2^T, 1) + \mathrm{tril}(\mathbf{u}_2 (\mathbf{v}_2 - \rho \mathbf{u}_2)^T, -1).
\end{aligned}$$

Hence, the initial quasiseparable matrix A has been transformed in (10.17) as the symmetric rank 1 modification of a 2×2 block diagonal matrix, whose blocks have a quasiseparable structure. Therefore, to compute the spectral decomposition of A, a diagonal plus rank 1 eigenproblem needs to be solved.

This procedure can be recursively applied on the blocks until the size of the subblock is sufficiently small to compute their spectral decomposition via standard procedures, i.e., for instance, QR or QL-methods.

10.3.2   A straightforward divide-and-conquer algorithm

Now we derive a straightforward divide-and-conquer algorithm in order to compute the eigenvalues and eigenvectors of a symmetric quasiseparable matrix.

The matrix A is divided in the following way,

$$A = \left[\begin{array}{c|c} A_{11} & \\ \hline & A_{22} \end{array} \right] + \left[\begin{array}{c|c} & A_{12} \\ \hline A_{12}^T & \end{array} \right] = A_1 + A_2,$$

with A_{11} an $s \times s$ matrix and A_{22} an $(n-s) \times (n-s)$ matrix. The matrices A_{11} and A_{22} are still symmetric quasiseparable matrices and A_{12} is a rank 1 matrix. Hence A_2 is a symmetric rank two matrix. The best way to split the problem is to divide it into two parts of about the same size, so $s = \lfloor n/2 \rfloor$, where $\lfloor x \rfloor$ denotes the largest natural number not exceeding x.

This is the case considered in the remainder of this chapter.

The next lemma shows that A_2 can be written as the sum of two rank 1 matrices.

Lemma 10.5. *Let $\mathbf{a} \in \mathbb{R}^{n \times 1}, \mathbf{b} \in \mathbb{R}^{m \times 1}$ be two nonzero vectors. Define $\hat{\mathbf{a}} = \mathbf{a}/\|\mathbf{a}\|_2$ and $\hat{\mathbf{b}} = \mathbf{b}/\|\mathbf{b}\|_2$. Hence*

$$\tilde{A} = \left[\begin{array}{c|c} & \mathbf{a}\mathbf{b}^T \\ \hline \mathbf{b}\mathbf{a}^T & \end{array} \right]$$

has rank 2 and

$$\tilde{A} = \rho \mathbf{q}_1 \mathbf{q}_1^T - \rho \mathbf{q}_2 \mathbf{q}_2^T \tag{10.18}$$

where $\rho = \|\mathbf{a}\|_2 \|\mathbf{b}\|_2$, $\mathbf{q}_1 = 1/\sqrt{2} \left[\hat{\mathbf{a}}^T, \hat{\mathbf{b}}^T \right]^T$ and $\mathbf{q}_2 = 1/\sqrt{2} \left[\hat{\mathbf{a}}^T, -\hat{\mathbf{b}}^T \right]^T$.

Proof. Substitute the values of $\rho, \mathbf{q}_1$ and $\mathbf{q}_2$ in (10.18). □

Consequently, a symmetric generator representable quasiseparable matrix as defined in (10.15) can be written as follows:

$$A = \left[\begin{array}{c|c} A_{11} & \\ \hline & A_{22} \end{array}\right] + \rho \mathbf{q}_1 \mathbf{q}_1{}^T - \rho \mathbf{q}_2 \mathbf{q}_2{}^T,$$

with $\rho = \|\mathbf{v}(1:s)\| \, \|\mathbf{u}(s+1:n)\|$, $\mathbf{q}_1 = \frac{1}{\sqrt{2}} \left[\hat{\mathbf{v}}^T, \hat{\mathbf{u}}^T\right]^T$, $\mathbf{q}_2 = \frac{1}{\sqrt{2}} \left[\hat{\mathbf{v}}^T, -\hat{\mathbf{u}}^T\right]^T$, where

$$\hat{\mathbf{v}} = \frac{\mathbf{v}(1:s)}{\|\mathbf{v}(1:s)\|},$$

$$\hat{\mathbf{u}} = \frac{\mathbf{u}(s+1:n)}{\|\mathbf{u}(s+1:n)\|}.$$

So, let

$$A_{11} = Q_1 \Delta_1 Q_1^T \quad \text{and} \quad A_{22} = Q_2 \Delta_2 Q_2^T$$

be the eigendecompositions of A_{11} and A_{22}, respectively. Then

$$A = \left[\begin{array}{c|c} Q_1 & \\ \hline & Q_2 \end{array}\right] \left(\left[\begin{array}{c|c} \Delta_1 & \\ \hline & \Delta_2 \end{array}\right] + \rho \tilde{\mathbf{q}}_1 \tilde{\mathbf{q}}_1^T - \rho \tilde{\mathbf{q}}_2 \tilde{\mathbf{q}}_2^T\right) \left[\begin{array}{c|c} Q_1 & \\ \hline & Q_2 \end{array}\right]^T$$

with

$$\tilde{\mathbf{q}}_1 = \left[\begin{array}{c|c} Q_1 & \\ \hline & Q_2 \end{array}\right]^T \mathbf{q}_1 = \frac{1}{\sqrt{2}} \left[\begin{array}{c} Q_1^T \hat{\mathbf{v}} \\ Q_2^T \hat{\mathbf{u}} \end{array}\right],$$

$$\tilde{\mathbf{q}}_2 = \left[\begin{array}{c|c} Q_1 & \\ \hline & Q_2 \end{array}\right]^T \mathbf{q}_2 = \frac{1}{\sqrt{2}} \left[\begin{array}{c} Q_1^T \hat{\mathbf{v}} \\ -Q_2^T \hat{\mathbf{u}} \end{array}\right].$$

Hence the problem of computing the spectral decomposition of a quasiseparable matrix is reduced to the problem of computing the spectral decomposition of a diagonal matrix plus two rank 1 modifications. The spectral decomposition of the latter matrix can be computed as follows. Let

$$\tilde{Q} \tilde{\Delta} \tilde{Q}^T = \left[\begin{array}{c|c} \Delta_1 & \\ \hline & \Delta_2 \end{array}\right] + \rho \tilde{\mathbf{q}}_1 \tilde{\mathbf{q}}_1^T,$$

then

$$\left[\begin{array}{c|c} \Delta_1 & \\ \hline & \Delta_2 \end{array}\right] + \rho \tilde{\mathbf{q}}_1 \tilde{\mathbf{q}}_1^T - \rho \tilde{\mathbf{q}}_2 \tilde{\mathbf{q}}_2^T = \tilde{Q} \left(\tilde{\Delta} - \rho \hat{\mathbf{q}}_2 \hat{\mathbf{q}}_2^T\right) \tilde{Q}^T,$$

with $\hat{\mathbf{q}}_2 = \tilde{Q}^T \tilde{\mathbf{q}}_2$. Hence, when the eigendecomposition of the latter problem is computed:

$$\left(\tilde{\Delta} - \rho \hat{\mathbf{q}}_2 \hat{\mathbf{q}}_2^T\right) = \hat{Q} \hat{\Delta} \hat{Q}^T,$$

the spectral decomposition of the original matrix A is known,

$$
A = \left[\begin{array}{c|c} Q_1 & \\ \hline & Q_2 \end{array} \right] \left(\left[\begin{array}{c|c} \Delta_1 & \\ \hline & \Delta_2 \end{array} \right] + \rho \tilde{\mathbf{q}}_1 \tilde{\mathbf{q}}_1^T - \rho \tilde{\mathbf{q}}_2 \tilde{\mathbf{q}}_2^T \right) \left[\begin{array}{c|c} Q_1 & \\ \hline & Q_2 \end{array} \right]^T
$$

$$
= \left[\begin{array}{c|c} Q_1 & \\ \hline & Q_2 \end{array} \right] \tilde{Q} \left(\tilde{\Delta} - \rho \hat{\mathbf{q}}_2 \hat{\mathbf{q}}_2^T \right) \tilde{Q}^T \left[\begin{array}{c|c} Q_1 & \\ \hline & Q_2 \end{array} \right]^T
$$

$$
= \left[\begin{array}{c|c} Q_1 & \\ \hline & Q_2 \end{array} \right] \tilde{Q} \hat{Q} \hat{\Delta} \hat{Q}^T \tilde{Q}^T \left[\begin{array}{c|c} Q_1 & \\ \hline & Q_2 \end{array} \right]^T
$$

$$
= \left[\begin{array}{c|c} Q_1 & \\ \hline & Q_2 \end{array} \right] \tilde{Q} \hat{Q} \hat{\Delta} \left(\left[\begin{array}{c|c} Q_1 & \\ \hline & Q_2 \end{array} \right] \tilde{Q} \hat{Q} \right)^T .
$$

So, this divide-and-conquer algorithm reduces the problem of computing the spectral decomposition of a quasiseparable matrix into computing the spectral decomposition of two matrices having the diagonal plus a rank 1 modification structure.

Note that in the implementation of this algorithm, the normalization used in Lemma 10.5 is crucial for the numerical stability.

The divide-and-conquer algorithm just described requires to solve two rank 1 modifications at any conquer step. The divide-and-conquer algorithms introduced in the next subsections require either to compute the spectral decomposition of an arrowhead matrix or to solve a rank 1 modification at each conquer step. The price to pay for the advantage of only one modification in the conquer step, is that more computations are required in the divide step.

10.3.3   A one-way divide-and-conquer algorithm

In this subsection we derive a one-way divide-and-conquer algorithm that reduces the eigendecomposition of a symmetric quasiseparable matrix to the spectral decomposition of an arrowhead one.

Starting from the original quasiseparable matrix, to derive the divide step, we consider a simple algorithm that, at the k-th step annihilates the elements of the k-th row (and k-th column) from the $(k+2)$nd column (row) up to the last column (row) of the involved matrix. Hence this algorithm divides the original quasiseparable structure into a kind of block quasiseparable structure, a top left and a bottom right block. At each step the dimension of the top left block grows with one, while the dimension of the bottom right block decreases with one. Moreover, the blocks made by the first k rows and $n - k + 1$ columns and the first k columns and the last $n - k + 1$ columns, respectively, are zero.

Algorithm 10.6.
Input: *Let A be the quasiseparable matrix defined in* (10.15).
Output: *A changed matrix, ready for applying a divide-and-conquer method.*

Define $\tilde{v}_1 = v_1$ and $A^{(1)} = A$.
For $k = 1, \ldots, n - 2$, do:

1. *Compute*

$$G_k = \left[\begin{array}{cc} c_k & -s_k \\ s_k & c_k \end{array}\right],$$

with $c_k = \frac{v_{k+1}}{\sqrt{\tilde{v}_k^2 + v_{k+1}^2}}, s_k = \frac{\tilde{v}_k}{\sqrt{\tilde{v}_k^2 + v_{k+1}^2}},$
which is the Givens rotation such that

$$\left[\begin{array}{c} 0 \\ \tilde{v}_{k+1} \end{array}\right] = \left[\begin{array}{c} 0 \\ \sqrt{\tilde{v}_k^2 + v_{k+1}^2} \end{array}\right] = \left[\begin{array}{cc} c_k & -s_k \\ s_k & c_k \end{array}\right]\left[\begin{array}{c} \tilde{v}_k \\ v_{k+1} \end{array}\right].$$

2. *Perform* $A^{(k+1)} = \tilde{G}_k A^{(k)} \tilde{G}_k^T$, *with* $\tilde{G}_k = \mathrm{diag}([I_{k-1}, G_k, I_{n-k-1}]).$

endfor;

It is easy to prove by induction that each step of the latter algorithm leads to a division of the original symmetric generator representable quasiseparable matrix A into two submatrices with similar structure in the upper left corner and in the lower right corner, respectively. More precisely, after k steps of the algorithm, the matrices $A^{(k+1)}(1:k+1, 1:k+1)$ and $A^{(k+1)}(k+3:n, k+3:n)$ are generator representable quasiseparable:

$$A^{(k+1)} = \left[\begin{array}{c|c|c} \hat{A}^{(k+1)} & \mathbf{w}^{(k+1)} & \\ \hline \mathbf{w}^{(k+1)T} & \alpha^{(k+1)} & \mathbf{z}^{(k+1)T} \\ \hline & \mathbf{z}^{(k+1)} & \tilde{A}^{(k+1)} \end{array}\right],$$

with $\hat{A}^{(k+1)}$ of dimension $k \times k$ and $\tilde{A}^{(k+1)}$ of dimension $(n-k-1) \times (n-k-1)$, where $\alpha^{(k+1)} = a_{k+1,k+1}^{(k+1)}$ and the matrices

$$\hat{A}^{(k+1)} = A^{(k+1)}(1:k+1, 1:k+1), \quad \tilde{A}^{(k+1)} = A^{(k+1)}(k+3:n, k+3:n) \quad (10.19)$$

are generator representable quasiseparable, too.

The best way to split the original problem is into two parts of about the same dimension. Hence, take $k = \lfloor (n-1)/2 \rfloor$ steps of Algorithm 10.6. It can be easily checked that the matrices $\hat{A}^{(k+1)}$ and $\tilde{A}^{(k+1)}$ have the quasiseparable structure. The preservation of the structure allows to construct a divide-and-conquer algorithm because computing the eigendecompositions of $\hat{A}^{(k+1)}$ and $\tilde{A}^{(k+1)}$ are similar problems as the original one, but of dimension about half the dimension of the starting matrix A.

Knowing the eigendecomposition of the two subproblems $\hat{A}^{(k+1)}$ and $\tilde{A}^{(k+1)}$, we solve the spectral decomposition of A by transforming the remaining problem into the eigenproblem of an arrowhead matrix.

Let

$$\hat{A}^{(k+1)} = \hat{Q}\hat{\Delta}\hat{Q}^T, \quad \tilde{A}^{(k+1)} = \tilde{Q}\tilde{\Delta}\tilde{Q}^T,$$

be the eigendecompositions of $\hat{A}$ and $\tilde{A}$, respectively. Define

$$G^{(k)} = \tilde{G}_k \cdots \tilde{G}_2 \tilde{G}_1,$$

where $\tilde{G}_i, i = 1, \ldots, k$ are the matrices as defined in Algorithm 10.6.

$$
\begin{aligned}
A &= G^{(k)^T} A^{(k+1)} G^{(k)} \\
&= G^{(k)^T}
\left[
\begin{array}{c|c|c}
\hat{A} & \mathbf{w} & \\
\hline
\mathbf{w}^T & \alpha & \mathbf{z}^T \\
\hline
 & \mathbf{z} & \tilde{A}
\end{array}
\right]
G^{(k)} \\
&= G^{(k)^T}
\left[
\begin{array}{c|c|c}
\hat{Q} & & \\
\hline
 & 1 & \\
\hline
 & & \tilde{Q}
\end{array}
\right]
\left[
\begin{array}{c|c|c}
\hat{\Delta} & \hat{\mathbf{w}} & \\
\hline
\hat{\mathbf{w}}^T & \alpha & \tilde{\mathbf{z}}^T \\
\hline
 & \tilde{\mathbf{z}} & \tilde{\Delta}
\end{array}
\right]
\left[
\begin{array}{c|c|c}
\hat{Q} & & \\
\hline
 & 1 & \\
\hline
 & & \tilde{Q}
\end{array}
\right]^T
G^{(k)},
\end{aligned}
$$

where

$$\hat{\mathbf{w}} = \hat{Q}^T \mathbf{w}, \quad \tilde{\mathbf{z}} = \tilde{Q}^T \mathbf{z}.$$

Let P be the permutation matrix such that

$$
P^T
\left[
\begin{array}{c|c|c}
\hat{\Delta} & \hat{\mathbf{w}} & \\
\hline
\hat{\mathbf{w}}^T & \alpha & \tilde{\mathbf{z}}^T \\
\hline
 & \tilde{\mathbf{z}} & \tilde{\Delta}
\end{array}
\right]
P =
\left[
\begin{array}{cc|c}
\hat{\Delta} & & \hat{\mathbf{w}} \\
 & \tilde{\Delta} & \tilde{\mathbf{z}} \\
\hline
\hat{\mathbf{w}}^T & \tilde{\mathbf{z}}^T & \alpha
\end{array}
\right].
\tag{10.20}
$$

Hence the computation of the eigendecomposition of the symmetric quasiseparable matrix A is reduced to the computation of the eigendecomposition of the arrowhead matrix in (10.20).

10.3.4   A two-way divide-and-conquer algorithm

In this section we derive a two-way divide-and-conquer algorithm which reduces the eigendecomposition of a symmetric quasiseparable matrix to the spectral decomposition of a diagonal matrix plus a rank 1 modification.

In this case, the Givens rotations used in Algorithm 10.6, are simultaneously applied to the top left and the bottom right of the matrix in order to annihilate elements in the first rows and columns, respectively, in the last rows and columns. If the dimension n of the original matrix A is even, we apply the same number of steps for the two groups of Givens rotations. If n is odd, we apply one extra Givens rotation of the kind we used in Algorithm 10.6. The details are given in Algorithm 10.7, where $\lceil x \rceil$ denotes the smallest natural number m such that $m \geq x$.

Algorithm 10.7.
Input: *Let A be the quasiseparable matrix defined in (10.15).*
Output: *A changed matrix, ready for applying a divide-and-conquer method.*

Define $\tilde{v}_1 = v_1$, $A^{(1)} = A$, $n_1 = \lceil \frac{n}{2} \rceil$ and $n_2 = n_1 + 3$
For $k = 1, \ldots, n_1$ do:

Compute

$$G_k = \begin{bmatrix} c_k & -s_k \\ s_k & c_k \end{bmatrix},$$

with $c_k = \frac{v_{k+1}}{\sqrt{\tilde{v}_k^2 + v_{k+1}^2}}$, the Givens rotation such that

$$\begin{bmatrix} 0 \\ \tilde{v}_{k+1} \end{bmatrix} = \begin{bmatrix} 0 \\ \sqrt{\tilde{v}_k^2 + v_{k+1}^2} \end{bmatrix} = \begin{bmatrix} c_k & -s_k \\ s_k & c_k \end{bmatrix} \begin{bmatrix} \tilde{v}_k \\ v_{k+1} \end{bmatrix},$$

endfor;

For $l = n, n - 1, n - 2, \ldots, n_2$ do:

Compute

$$H_l = \begin{bmatrix} c_l & s_l \\ -s_l & c_l \end{bmatrix},$$

with $c_l = \frac{u_{l-1}}{\sqrt{\tilde{u}_l^2 + u_{l-1}^2}}, s_l = \frac{\tilde{u}_l}{\sqrt{\tilde{u}_l^2 + u_{l-1}^2}}$,
the Givens rotation such that

$$\begin{bmatrix} \tilde{u}_{l-1} \\ 0 \end{bmatrix} = \begin{bmatrix} \sqrt{\tilde{u}_l^2 + u_{l-1}^2} \\ 0 \end{bmatrix} = \begin{bmatrix} c_k & s_k \\ -s_k & c_k \end{bmatrix} \begin{bmatrix} u_{l-1} \\ \tilde{u}_l \end{bmatrix},$$

endfor;

For $k = \lceil \frac{n+1}{2} \rceil - 2$, do:

1. $\tilde{G}_k = \operatorname{diag}([I_{k-1}, G_k, I_{n-2k-2}, H_{n-k-1}, I_{k-1}]).$
2. $A^{(k+1)} = \tilde{G}_k A^{(k)} \tilde{G}_k^T.$

endfor;

If $2 \times \lfloor \frac{n}{2} \rfloor \neq n$,

For $k = \lceil \frac{n+1}{2} \rceil - 1$,

(a) $\tilde{G}_k = \operatorname{diag}([I_{k-1}, G_k, I_{n-k-1}]).$
(b) $A^{(k+1)} = \tilde{G}_k A^{(k)} \tilde{G}_k^T.$

endfor;

endif;

Algorithm 10.7 leads to a new way to divide a symmetric quasiseparable matrix into two submatrices similar to the original one and of about half the dimension and some additional structure.

It can be easily proven by induction that the matrix A is transformed by Algorithm 10.7, in the following form,

$$
\left[
\begin{array}{c|c}
\hat{A}^{(n_1+1)} & \alpha^{(n_1+1)} \mathbf{e}_{n_1+1}\mathbf{e}_1^T \\
\hline
\alpha^{(n_1+1)} \mathbf{e}_1\mathbf{e}_{n_1+1}^T & \tilde{A}^{(n_1+1)}
\end{array}
\right]
$$

where $\hat{A}^{(n_1+1)} \in \mathbb{R}^{(n_1+1)\times(n_1+1)}$ and $\tilde{A}^{(n_1+1)} \in \mathbb{R}^{(n-n_2+1)\times(n-n_2+1)}$ are quasiseparable matrices, $\mathbf{e}_{n_1+1}$ the (n_1+1)-th vector of the canonical basis of $\mathbb{R}^{n_1+1}$, $\mathbf{e}_1$ the first vector of the canonical basis of $\mathbb{R}^{n-n_2+2}$ and $\alpha^{(n_1+1)} = a_{n_1+1,n_1+2}^{(n_1+1)}$.

Define

$$
G^{(n_1)} = \tilde{G}_{n_1} \cdots \tilde{G}_2 \tilde{G}_1,
$$

where $\tilde{G}_i$ are defined in Algorithm 2. Then

$$
\begin{aligned}
A &= G^{(n_1)T} A^{(n_1+1)} G^{(n_1)} \\
&= G^{(n_1)T} \left[
\begin{array}{c|c}
\hat{A}^{(n_1+1)} & \alpha^{(n_1+1)} \mathbf{e}_{n_1+1}\mathbf{e}_1^T \\
\hline
\alpha^{(n_1+1)} \mathbf{e}_1\mathbf{e}_{n_1+1}^T & \tilde{A}^{(n_1+1)}
\end{array}
\right] G^{(n_1)} \\
&= G^{(n_1)T} \left(\left[
\begin{array}{c|c}
\hat{A}^{(n_1+1)} - \alpha^{(n_1+1)} \mathbf{e}_{n_1+1}\mathbf{e}_{n_1+1}^T & 0 \\
\hline
0 & \tilde{A}^{(n_1+1)} - \alpha^{(n_1+1)} \mathbf{e}_1\mathbf{e}_1^T
\end{array}
\right] \right. \\
&\left. + \alpha^{(n_1+1)} \left[
\begin{array}{c|c}
\mathbf{e}_{n_1+1}\mathbf{e}_{n_1+1}^T & \mathbf{e}_{n_1+1}\mathbf{e}_1^T \\
\hline
\mathbf{e}_1\mathbf{e}_{n_1+1}^T & \mathbf{e}_1\mathbf{e}_1^T
\end{array}
\right] \right) G^{(n_1)}.
\end{aligned}
$$

The subtraction of the element $\alpha^{(n_1+1)}$ from the last diagonal element of $\hat{A}^{(n_1+1)}$ and from the first diagonal element of $\tilde{A}^{(n_1+1)}$ is done in order to create a rank 1 modification. The quasiseparable structure is not affected by these subtractions.

Knowing the eigendecomposition of the symmetric quasiseparable matrices $\hat{A}^{(n_1+1)} - \alpha^{(n_1+1)}\mathbf{e}_{n_1+1}\mathbf{e}_{n_1+1}^T$ and $\tilde{A}^{(n_1+1)} - \alpha^{(n_1+1)}\mathbf{e}_1\mathbf{e}_1^T$, only the spectral decomposition of a diagonal matrix plus a rank 1 modification must be calculated in order to know the eigendecomposition of the original matrix A. During the remainder of this subsection we omit the superscript n_1+1.

Let

$$
\begin{aligned}
\hat{A} - \alpha\mathbf{e}_{n_1+1}\mathbf{e}_{n_1+1}^T &= \hat{Q}\hat{\Delta}\hat{Q}^T, \\
\tilde{A} - \alpha\mathbf{e}_1\mathbf{e}_1^T &= \tilde{Q}\tilde{\Delta}\tilde{Q}^T,
\end{aligned}
$$

be the eigendecompositions of the adapted $\hat{A}$ and $\tilde{A}$, respectively.

The matrix A can be transformed into:

$$
A = G^{(n_1)T} \left[
\begin{array}{c|c}
\hat{Q} & \\
\hline
& \tilde{Q}
\end{array}
\right]
\left(\left[
\begin{array}{c|c}
\hat{\Delta} & \\
\hline
& \tilde{\Delta}
\end{array}
\right] + \alpha\mathbf{y}\mathbf{y}^T \right)
\left[
\begin{array}{c|c}
\hat{Q} & \\
\hline
& \tilde{Q}
\end{array}
\right]^T G^{(n_1)}
$$

with

$$
\mathbf{y} = \left[
\begin{array}{c}
\hat{Q}^T\mathbf{e}_{n_1+1} \\
\hline
\tilde{Q}^T\mathbf{e}_1
\end{array}
\right].
\tag{10.21}
$$

Hence, the eigenproblem of A is reduced to compute the eigendecomposition of a rank 1 modification of a diagonal matrix $\left[\begin{array}{c|c} \hat{\Delta} & \\ \hline & \tilde{\Delta} \end{array}\right] + \alpha \mathbf{y}\mathbf{y}^T$. The way to solve this latter problem was mentioned in Subsection 10.3.2 on the straightforward divide-and-conquer algorithm.

Notes and references

The literature on divide-and-conquer methods to compute the eigendecomposition of quasiseparable matrices is very recent. The first divide-and-conquer algorithm for computing the eigendecomposition of symmetric quasiseparable matrices proposed in this section has been introduced by Chandrasekaran and Gu.

☞ S. Chandrasekaran and M. Gu. A divide-and-conquer algorithm for the eigendecomposition of symmetric block-diagonal plus semiseparable matrices. *Numerische Mathematik*, 96(4):723–731, February 2004.

The last three divide-and-conquer algorithms described in this section have been presented by Mastronardi, Van Barel and Van Camp. The results in this section are mostly based on their article (see also [155]).

☞ N. Mastronardi, M. Van Barel, and E. Van Camp. Divide-and-conquer algorithms for computing the eigendecomposition of symmetric diagonal-plus-semiseparable matrices. *Numerical Algorithms*, 39(4):379–398, 2005.

Moreover, in the article by Fasino and Gemignani, the authors present a procedure that transforms by a congruence transformation the matrix pencil $(A - \lambda I)$ into $(T - \lambda V)$ where T and V are both symmetric tridiagonal and V is positive definite is described. The authors suggest solving the latter generalized eigenvalue problem by a divide-and-conquer technique proposed in [28].

☞ D. Fasino and L. Gemignani. Direct and inverse eigenvalue problems, for diagonal-plus-semiseparable matrices. *Numerical Algorithms*, 34:313–324, 2003.

10.4 Computational complexity and numerical experiments

In this section we will present firstly a comparison in computational complexity of the different methods for computing the eigendecomposition of structured rank matrices. Secondly we will propose a numerical experiment related to these methods.

We compare the computational complexity of the four proposed divide-and-conquer algorithms described in this chapter. Let us assume that we apply the algorithms on a symmetric $n \times n$ quasiseparable matrix.

For the one-way divide-and-conquer algorithm we have to solve the eigenproblems of arrowhead matrices. Hence two vectors and a scalar must be constructed (referred to as step 1) and the zeros of a secular equation must be solved (referred to as step 2). For the latter, we use an algorithm implemented by Gragg, based on [26, 28]. This algorithm has a computational complexity (denoted by $a(n)$) of order n^2. In the last step (referred to as step 3), the eigenvectors need to be updated.

The first divide-and-conquer algorithm proposed (referred to as first divide-and-conquer algorithm), the straightforward algorithm and the two-way divide-and-conquer algorithm require to solve the eigenproblem of a diagonal matrix plus a rank 1 modification. Also here two vectors and a scalar need to be constructed (step 1) and the zeros of a secular equation must be computed (step 2). Last, also here the eigenvectors need to be updated (step 3).

As shown in [26], the matrix of eigenvectors computed by the algorithm of Gragg, and hence also by the slightly adapted algorithm, can be reduced into the product of a diagonal matrix times a Cauchy matrix times another diagonal matrix. This special structure reduces the computational complexity of the matrix of eigenvectors times a vector (denoted by $b(n)$) to order $nlog^2(n)$ (see [91, 92]).

Using these notations, the Table 10.1 shows the computational complexity of the four divide-and-conquer algorithms at one level (so one divide step and one conquer step). Table 10.1 indicates that the computational complexity of the algorithm proposed in [40] is less than that of the one-way divide-and-conquer algorithm, which itself is less than the computational complexity of the two-way divide-and-conquer algorithm. However, the difference in computational complexity between these three algorithms only occurs at the second highest order term and hence their computational complexity is very comparable. The straightforward divide-and-conquer algorithm we proposed, clearly has the largest computational complexity because even the highest order term is larger than for the other algorithms.

	First	Straightforward	One-way	Two-way
Divide	$14n$	0	$28n$	$58n$
Conquer				
– Step 1:	$\frac{n^2}{2} + 2n$	$n^2 + 4n + b(n)$	n^2	$\frac{n}{2}$
– Step 2:	$a(n)$	$2a(n)$	$a(n)$	$a(n)$
– Step 3:	$2nb(\frac{n}{2})$	$nb(n) + 2nb(\frac{n}{2})$	$2nb(\frac{n}{2})$ $+\frac{3}{2}n^2$	$2nb(\frac{n}{2}) + 3n^2$

Table 10.1. *Comparison in complexity.*

Experiment 10.8 (Computational complexity). *Figure 10.1 shows the cputime (in seconds) computed in* MATLAB *for problems of dimension $n = 2^j, j = 2, \ldots 9$, for the three divide-and-conquer algorithms we present. Also this figure shows that the one-way divide-and-conquer algorithm slightly needs less cputime than the two-way algorithm for increasing dimension. The straightforward algorithm is much slower.*

In the following two experiments we investigated the accuracy of both eigenvalues and eigenvectors. For the three divide-and-conquer algorithms, we built symmetric generator representable quasiseparable matrices of dimension $n = 2^j$,

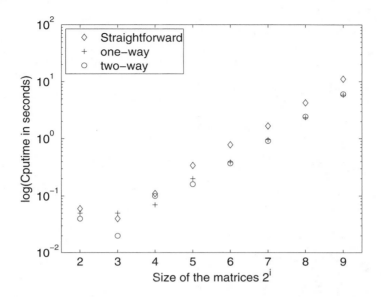

Figure 10.1. *Cputime of the straightforward, the one-way and the two-way divide-and-conquer algorithm.*

$j = 2, \ldots, 9$ and for each dimension matrices that have condition number equal to $10^3, 10^6, 10^9$ and 10^{12}. For each of these 32 classes of test matrices we took 100 samples.

The test matrices were built as follows: starting from a diagonal matrix $\Delta = [\alpha, \alpha^2, \ldots, \alpha^n]$ with α the $(n-1)$-th root of the requested condition number, we applied $(n-1)$ random Givens rotations $G_i \neq I$ to the left, such that G_i works on the i-th and $(i+1)$-th row, and G_i^T to the right of Δ. Hence Δ was transformed into a matrix $A = G\Delta G^T$. This matrix A is a generator representable quasiseparable matrix because the i-th Givens rotation G_i makes the elements of row i and $i+1$ proportional. The transpose G_i^T does the same with column i and $i+1$, so we created a semiseparable structure except on the diagonal.

Experiment 10.9. *First we applied the straightforward divide-and-conquer algorithm on the quasiseparable matrix A in order to calculate the eigenvalues $\lambda_i^{(1)}$ and the eigenvectors $U_i^{(1)}$ of A. Then the accuracy of the eigenvalues was tested by calculating $r_i = \frac{|\lambda_i - \lambda_i^{(1)}|}{\max\{|\Lambda|\}}$, $i = 1, \ldots, n$, where λ_i are the diagonal elements of the diagonal matrix Δ and $\max\{|\Lambda|\}$ is the maximum of the absolute values of the diagonal elements of Δ. The maximum r of all these fractions r_i was divided by the machine precision $\epsilon \approx 2.22 \cdot 10^{-16}$ and the dimension of the matrix and $\frac{r}{n\epsilon}$ was defined as the relative error of the straightforward algorithm for the considered example. Remark that this error contains the error caused by the latter divide-and-conquer algorithm but also the errors caused by the application of the Givens rotations G_i.*

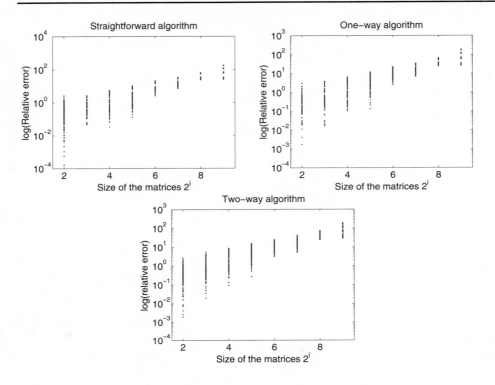

Figure 10.2. *Plots of the relative errors of the eigenvalues computed by the straightforward, the one-way and the two-way divide-and-conquer algorithm.*

The same was done for the one-way and the two-way algorithm. The tests didn't indicate any dependency of accuracy on the condition number, so only a separation according to the dimension is performed.

As shown in Figure 10.2, the straightforward divide-and-conquer algorithm, the one-way and the two-way algorithm compute the eigenvalues of the test matrices in an accurate way. The relative error also increases with the dimension of the problem.

Experiment 10.10. *Next the accuracy of the eigenvectors of our examples was tested by looking at the relative residual norm*

$$\frac{\|AU^{(1)} - U^{(1)}\mathrm{diag}(\Delta^{(1)})\|_\infty}{N\epsilon\|A\|_\infty}$$

where A is the test matrix of dimension n, $U^{(1)}$ the matrix of the computed eigenvectors and $\Delta^{(1)}$ the eigenvalues computed with the straightforward divide-and-conquer algorithm. The same was done for the one-way and the two-way divide-and-conquer algorithm. As shown in Figure 10.3, the relative residual norms for the three algorithms are small.

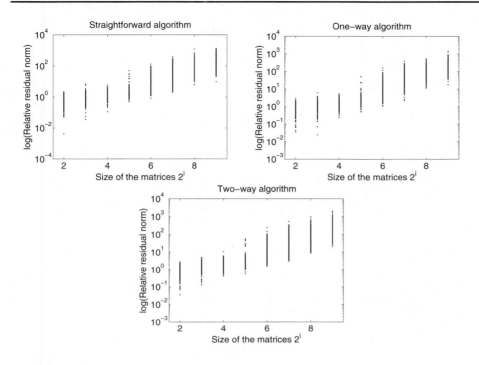

Figure 10.3. *Plots of the relative residual norms computed by the straight-forward, the one-way and the two-way divide-and-conquer algorithm.*

Also here the error caused by the divide-and-conquer algorithm as well as the errors caused by the application of the Givens rotations G_i are included in the relative residual norms of Figure 10.3.

10.5   Conclusions

In this chapter we discussed divide-and-conquer algorithms for computing the eigendecomposition of structured rank matrices. We have described four divide-and-conquer algorithms that all reduce the eigendecomposition of the original quasiseparable matrix to two problems of the same structure and approximately half the size and some extra structure. For the straightforward algorithm this extra structure is a diagonal matrix plus two rank 1 matrices, for the two-way algorithm only one rank 1 adaptation is needed and for the one-way algorithm the extra structure consists of an arrowhead matrix.

Chapter 11

A Lanczos-type algorithm and rank revealing

In this chapter we will present two main topics. First we will present a Lanczos version for constructing a semiseparable matrix similar to a given symmetric matrix, and secondly we will show how one can exploit the rank-revealing properties of the transformation to semiseparable form.

Tridiagonalization of a symmetric matrix can be done in two different ways. One can perform similarity Householder transformations in order to create zeros in the matrix, or one can apply the Lanczos tridiagonalization process. The Lanczos tridiagonalization process is based on matrix vector products to compute the elements on the diagonal and subdiagonal of the tridiagonal matrix. The fact that the matrix vector product is the most heavy computational step in the Lanczos tridiagonalization process, makes this technique suitable when this vector product can be computed in an efficient manner, e.g., in the case of structured matrices such as Hankel, Toeplitz, band matrices and so forth. In the first part of the book we already designed a way to obtain a semiseparable matrix, based on orthogonal similarity transformations using thereby Householder and Givens transformations. In the first section of this chapter we will show how to obtain a Lanczos semiseparabilization process.

In the second section we will propose a rank-revealing algorithm based on the reduction to semiseparable form, which, of course, exploits the specific convergence behavior of this reduction. Some numerical experiments are provided.

✎ *This chapter contains two main ideas, namely the Lanczos semiseparabilization algorithm and how to exploit the rank-revealing properties of the reduction algorithm.*

Both sections are in some sense extra topics, which are only relevant for the interested reader.

11.1 Lanczos semiseparabilization

An algorithm that transforms symmetric matrices into similar semiseparable ones has been proposed in Chapter 2. The latter algorithm works without taking into account the structure of the original matrix. In this section we propose a Lanczos-like algorithm to transform a symmetric matrix into a similar semiseparable one relying on the product of the original matrix times a vector at each step. Therefore an efficient algorithm is obtained if the original matrix is sparse or structured.

The matrices handled in this section are symmetric. However, there is no loss of generality, because similar techniques can be applied to real unsymmetric matrices, as well. Moreover, the extension of this algorithm to reduce rectangular matrices into upper triangular semiseparable ones in order to compute the singular value decomposition is quite straightforward, just as the adaptation to obtain a semiseparable plus diagonal matrix, with a free chosen diagonal.

The section is organized as follows. In Subsection 11.1.1 the basic steps of the classical Lanczos algorithm are described followed by the Lanczos algorithm for reducing symmetric matrices into semiseparable ones in Subsection 11.1.2.

11.1.1 Lanczos reduction to tridiagonal form

The Lanczos algorithm is a simple and effective method for finding extreme eigenvalues and corresponding eigenvectors of symmetric matrices since the part already reduced has the Lanczos-Ritz values as eigenvalues. Because it only accesses the matrix through matrix vector multiplications, it is commonly used when matrices are sparse or structured. The algorithm can be summarized as follows.

Algorithm 11.1 (Lanczos reduction to tridiagonal form).
Input: *A symmetric matrix $A \in \mathbb{R}^{n \times n}$ and $\mathbf{r}_0 \in \mathbb{R}^n$ as a starting vector.*
Output: *The diagonal entries a_i and the subdiagonal entries b_i of the tridiagonal matrix, similar to A. The orthogonal similarity transformation is determined by the matrix $Q = [\mathbf{q}_1, \mathbf{q}_2, \ldots]$.*

Initialize $b_0 = \|\mathbf{r}_0\|_2$ and $\mathbf{q}_0 = 0$.
For $i = 1, 2, \ldots$

1. $\mathbf{q}_i = \mathbf{r}_{i-1}/\|\mathbf{r}_{i-1}\|_2$

2. $p = A\mathbf{q}_i$

3. $a_i = \mathbf{q}_i^T p$

4. $\mathbf{r}_i = p - a_i \mathbf{q}_i - b_{i-1}\mathbf{q}_{i-1}$

5. $b_i = \|\mathbf{r}_i\|_i$

endfor;

The values a_i and b_i define the diagonal and subdiagonal of the tridiagonal matrix, and $[\mathbf{q}_1, \mathbf{q}_2, \ldots,]$ is the corresponding orthogonal matrix determining the similarity transformation to the tridiagonal form.

Let $Q_k = [\mathbf{q}_1, \mathbf{q}_2, \ldots, \mathbf{q}_k]$ and

$$
T_k = \begin{bmatrix}
a_1 & b_1 & & & & \\
b_1 & a_2 & b_2 & & & \\
& b_2 & \ddots & \ddots & & \\
& & \ddots & a_{n-1} & b_{k-1} \\
& & & b_{k-1} & a_k
\end{bmatrix},
$$

and let $T_k = U_k \Theta_k U_k^T$ its spectral decomposition, with $\Theta_k = \text{diag}(\theta_1, \theta_2, \ldots, \theta_k)$ and $U_k \in \mathbb{R}^{k \times k}$ orthogonal. Then

$$
A(Q_k U_k) = (Q_k U_k)\Theta_k + b_k \mathbf{q}_{k+1} \mathbf{e}_k^T U_k,
$$

where $\mathbf{e}_k$ is the last vector of the canonical basis of $\mathbb{R}^k$. It turns out (see, e.g., [94, p. 475]) that

$$
\min_{\mu \in \lambda(A)} |\theta_i - \mu| \leq |b_k||u_{k,i}|, \quad i = 1, \ldots, k.
$$

Hence b_k and the last components of the eigenvectors of T_k associated to θ_i give an indication on the convergence of the latter eigenvalues to the eigenvalues of A.

To prevent the loss of orthogonality in the computation of the Krylov basis Q_k, a lot of techniques have been developed (see, e.g., [94] and the references therein, [127, 130, 137, 139]).

Note 11.2. *It seems that the algorithm as presented above always runs to completion. Unfortunately this is not always the case. When computing the vector $\mathbf{r}_i$, it can occur that this vector equals zero. One calls this a breakdown of the algorithm. This means that we have found an invariant subspace consisting of the vectors $\mathbf{q}_1, \ldots, \mathbf{q}_{i-1}$. To finalize the procedure for tridiagonalizing the matrix, one has to restart the algorithm, but now choosing a new initial vector $\mathbf{r}_0$, which is orthogonal to the vectors $\mathbf{q}_1, \ldots, \mathbf{q}_{i-1}$. Due to the orthogonality, the algorithm will now generate another set of vectors. The breakdown corresponds in fact also to a value of $b_i = 0$. This means that our tridiagonal matrix will become of block diagonal form, in which all blocks are irreducible tridiagonal matrices.*

11.1.2 Lanczos reduction to semiseparable form

The Lanczos-like algorithm for semiseparable matrices is based on the classical Lanczos algorithm for reducing symmetric matrices into similar symmetric tridiagonal ones, as described in the previous subsection. We have to adapt the reduction to semiseparable form slightly in order to use it for the Lanczos-like version of the algorithm.

Another version of the reduction to semiseparable form

The standard algorithm for transforming a symmetric matrix into a similar semiseparable one was described in Chapter 2 of this book. This algorithm started by

annihilating elements in the last column and last row of the matrix. The standard
Lanczos procedure (see also the previous section) starts by annihilating elements in
the first row and in the first column, therefore, we will also reconsider our algorithms
in this section as starting in the top left corner.

In order to be able to derive a Lanczos-like algorithm for obtaining a semiseparable matrix, we have to change our standard algorithm slightly. In the original
algorithm as discussed in Chapter 2, the intermediate matrices are of the following
form (assume the reduction started at the top left position):

$$
\begin{bmatrix}
\boxtimes & \boxtimes & \boxtimes & \boxtimes & \boxtimes \\
\boxtimes & \boxtimes & \boxtimes & \boxtimes & \boxtimes \\
\boxtimes & \boxtimes & \times & \times & \times \\
\boxtimes & \boxtimes & \times & \times & \times \\
\boxtimes & \boxtimes & \times & \times & \times
\end{bmatrix}.
\tag{11.1}
$$

The bottom right 3×3 block is still of unreduced form, whereas the first two rows
and the first two columns satisfy already the semiseparable structure. In fact we
have the upper left 2×2 block already of semiseparable form.

A slightly changed version of the algorithm, consists of having intermediate
matrices of the following form:

$$
\begin{bmatrix}
\boxtimes & \boxtimes & 0 & 0 & 0 \\
\boxtimes & \boxtimes & \times & \times & \times \\
0 & \times & \times & \times & \times \\
0 & \times & \times & \times & \times \\
0 & \times & \times & \times & \times
\end{bmatrix}.
\tag{11.2}
$$

Similar as above, we have the lower 3×3 matrix unreduced and the upper left
2×2 block of semiseparable form. The only difference is the fact that in this
intermediate step, the first two columns and the first two rows are not includable
in the semiseparable structure.

Let us briefly present the second algorithm in more detail and let us point out
the difference with the original algorithm. Assume we are working on a symmetric
matrix $A_0^{(1)} = A$, of size 5×5. We start by annihilating elements in the first column
and row, by using a Householder transformation matrix $H_1^{(1)}$ (the elements marked
with $\otimes$ will be annihilated):

$$
\begin{bmatrix}
\times & \times & \otimes & \otimes & \otimes \\
\times & \times & \times & \times & \times \\
\otimes & \times & \times & \times & \times \\
\otimes & \times & \times & \times & \times \\
\otimes & \times & \times & \times & \times
\end{bmatrix}
\xrightarrow{H_1^{(1)^T} A_0^{(1)} H_1^{(1)}}
\begin{bmatrix}
\boxtimes & \boxtimes & & & \\
\boxtimes & \boxtimes & \times & \times & \times \\
 & \times & \times & \times & \times \\
 & \times & \times & \times & \times \\
 & \times & \times & \times & \times
\end{bmatrix}
$$

$$
\Updownarrow
$$

$$
A_0^{(1)} \xrightarrow{\; H_1^{(1)^T} A_0^{(1)} H_1^{(1)} \;} A_1^{(1)}.
$$

The standard algorithm (see Theorem 2.6) would now perform already a single Givens transformation in order to obtain the matrix as depicted in Figure (11.1). In this case however, we will not yet start the chasing procedure as our matrix is of the form shown in Figure (11.2). Hence step 1 of the slightly changed algorithm is finished and we have $A^{(2)} = A_1^{(1)}$.

We continue now by immediately performing another Householder transformation $H_1^{(2)}$ such as to annihilate elements in column 2 and row 2.[3]

$$
\begin{bmatrix}
\boxtimes & \boxtimes & & & \\
\boxtimes & \boxtimes & \times & \otimes & \otimes \\
& \times & \times & \times & \times \\
& \otimes & \times & \times & \times \\
& \otimes & \times & \times & \times
\end{bmatrix}
\xrightarrow{H_1^{(2)^T} A_0^{(2)} H_1^{(2)}}
\begin{bmatrix}
\boxtimes & \boxtimes & & & \\
\boxtimes & \boxtimes & \times & & \\
& & \times & \times & \times & \times \\
& & & \times & \times & \times \\
& & & \times & \times & \times
\end{bmatrix}
$$

$$\Updownarrow$$

$$A_0^{(2)} \xrightarrow{H_1^{(2)^T} A_0^{(2)} H_1^{(2)}} A_1^{(2)}$$

In the standard algorithm one would start the chasing procedure acting on rows (columns) 2 and 3. Here, however, we will only perform a single Givens transformation on rows (columns) 1 and 2. Similarly as in the standard case one can compute this Givens transformation acting on rows (columns) 1 and 2, such that we obtain the following result:

$$
\begin{bmatrix}
\boxtimes & \boxtimes & & & \\
\boxtimes & \boxtimes & \times & & \\
& \times & \times & \times & \times \\
& & \times & \times & \times \\
& & \times & \times & \times
\end{bmatrix}
\xrightarrow{G_2^{(2)^T} A_1^{(2)} G_2^{(2)}}
\begin{bmatrix}
\boxtimes & \boxtimes & \boxtimes & & \\
\boxtimes & \boxtimes & \boxtimes & & \\
\boxtimes & \boxtimes & \boxtimes & \times & \times \\
& & & \times & \times & \times \\
& & & \times & \times & \times
\end{bmatrix}
$$

$$\Updownarrow$$

$$A_1^{(2)} \xrightarrow{G_2^{(2)^T} A_1^{(2)} G_2^{(2)}} A_2^{(2)}.$$

This finishes the second step.

Defining now $A_0^{(3)} = A_2^{(2)}$, we will perform another step. First a Householder transformation matrix is constructed such to annihilate elements in the third row and column.

---

[3]Even though the procedure has slightly changed, the Householder transformations chosen in Theorem 2.6 and the ones here are essentially the same.

$$
\begin{bmatrix}
\boxtimes & \boxtimes & \boxtimes & & \\
\boxtimes & \boxtimes & \boxtimes & & \\
\boxtimes & \boxtimes & \boxtimes & \times & \otimes \\
& & \times & \times & \times \\
& \otimes & \times & \times
\end{bmatrix}
\xrightarrow{H_1^{(3)^T} A_0^{(3)} H_1^{(3)}}
\begin{bmatrix}
\boxtimes & \boxtimes & \boxtimes & & \\
\boxtimes & \boxtimes & \boxtimes & & \\
\boxtimes & \boxtimes & \boxtimes & \times & \\
& & \times & \times & \times \\
& & & \times & \times
\end{bmatrix}
$$

$$
\Updownarrow
$$

$$
A_1^{(3)} \xrightarrow{H_1^{(3)^T} A_0^{(2)} H_1^{(3)}} A_1^{(3)}
$$

Now we have to update the already existing semiseparable part. To do so we first perform a Givens transformation acting on rows (columns) 2 and 3.

$$
\begin{bmatrix}
\boxtimes & \boxtimes & \boxtimes & & \\
\boxtimes & \boxtimes & \boxtimes & & \\
\boxtimes & \boxtimes & \boxtimes & \times & \\
& & \times & \times & \times \\
& & & \times & \times
\end{bmatrix}
\xrightarrow{G_2^{(3)^T} A_1^{(3)} G_2^{(3)}}
\begin{bmatrix}
\boxtimes & \boxtimes & & & \\
\boxtimes & \boxtimes & \boxtimes & \boxtimes & \\
& \boxtimes & \boxtimes & \boxtimes & \\
& \boxtimes & \boxtimes & \boxtimes & \times \\
& & & \times & \times
\end{bmatrix}
$$

$$
\Updownarrow
$$

$$
A_1^{(3)} \xrightarrow{G_2^{(3)^T} A_1^{(2)} G_2^{(3)}} A_2^{(3)}.
$$

Finally a Givens transformation acting on rows (columns) 1 and 2 is necessary to obtain the desired structure.

$$
\begin{bmatrix}
\boxtimes & \boxtimes & & & \\
\boxtimes & \boxtimes & \boxtimes & \boxtimes & \\
& \boxtimes & \boxtimes & \boxtimes & \\
& \boxtimes & \boxtimes & \boxtimes & \times \\
& & & \times & \times
\end{bmatrix}
\xrightarrow{G_3^{(3)^T} A_2^{(3)} G_3^{(3)}}
\begin{bmatrix}
\boxtimes & \boxtimes & \boxtimes & \boxtimes & \\
\boxtimes & \boxtimes & \boxtimes & \boxtimes & \\
\boxtimes & \boxtimes & \boxtimes & \boxtimes & \\
\boxtimes & \boxtimes & \boxtimes & \boxtimes & \times \\
& & & \times & \times
\end{bmatrix}
$$

$$
\Updownarrow
$$

$$
A_2^{(3)} \xrightarrow{G_3^{(3)^T} A_2^{(2)} G_3^{(3)}} A_3^{(3)}.
$$

To obtain a complete semiseparable matrix, a final step of the chasing should be performed.

Hence, it is clear that the flow of this algorithm is quite comparable to the flow of the original reduction. In every step first zeros are created by a Householder or Givens transformation, and secondly the existing semiseparable structure is enlarged by a sequence of Givens transformations.

Note 11.3. *We remark that the slightly changed version for reducing the matrix to semiseparable form can be adapted to the nonsymmetric case and to the reduction to upper triangular semiseparable form in a straightforward way.*

Also important is the fact that the specific convergence behavior of the first algorithm is also present in this case. The theoretical results predicting the convergence behavior change slightly as the chasing is performed now on a matrix of smaller size. Nevertheless the results remain valid.

An algorithmic description of the reduction

We will present now an algorithmic description of the new reduction method. For a change the algorithm is depicted with the generator representation instead of the Givens-vector representation. For the Lanczos case there is no loss of generality as a breakdown, i.e. $b_i = 0$ corresponds to the deflation of a block. Since we know that a semiseparable matrix is block diagonal with all diagonal blocks of generator representable form, we can simply assume that in this Lanczos algorithm all blocks are of generator representable form. Unfortunately as mentioned before, the generator representation can give rise to numerical problems. But, to get another kind of implementation we will describe it here with the generator representation.

Given $A \in \mathbb{R}^{n \times n}$, the slightly changed algorithm to reduce a symmetric matrix into a symmetric semiseparable one can be summarized as follows.

Algorithm 11.4 (Alternative reduction to semiseparable form).
Input: *A symmetric matrix $A \in \mathbb{R}^{n \times n}$.*
Output: *An orthogonally similar semiseparable matrix A_{n-1}.*

Let $A^{(0)} = A$.
For $i = 1, \ldots, n-1$,

1. *Compute the Householder matrix $H^{(i-1)^T}$ in order to annihilate the entries of $A^{(i-1)}$ in the i-th column below the $(i+1)$-th row.*

2. *Compute $H^{(i-1)^T} A^{(i-1)} H^{(i-1)}$.*

3. *Compute the orthogonal chasing matrix $U^{(i-1)^T}$, working on rows $1, \ldots, i$ such that the leading principal submatrix of order $i+1$ of*

$$U^{(i-1)^T} H^{(i-1)^T} A^{(i-1)} H^{(i-1)} U^{(i)}$$

is symmetric semiseparable.

4. *$A^{(i)} = U^{(i-1)^T} H^{(i-1)^T} A^{(i-1)} H^{(i-1)} U^{(i-1)}$*

endfor;

Note 11.5. *The leading principal submatrices of order i of $H^{(i-1)^T} A^{(i-1)} H^{(i-1)}$ already have the symmetric semiseparable structure. Therefore step 3 in the latter algorithm can be interpreted as an updating step, i.e., increasing by one the order of the leading principal semiseparable submatrix.*

The matrix $U^{(i)}$ in the above algorithm is a combination of all Givens transformations needed to perform the chasing, and thereby enlarging the structure of the semiseparable part in the matrix.

Let $\tilde{S} \in \mathbb{R}^{k \times k}$ be a symmetric semiseparable matrix. We describe now in a mathematical manner how the augmented matrix

$$
S = S_0 = \left[\begin{array}{c|c} & 0 \\ \tilde{S} & \vdots \\ & 0 \\ & b_k \\ \hline 0 \ \cdots \ 0 \ \ b_k & a_{k+1} \end{array} \right], \tag{11.3}
$$

with $b_k \neq 0$, can be updated, i.e., reduced in symmetric semiseparable form by orthogonal similarity transformations working on the first k rows (and columns).

For simplicity, let us suppose $k = 4$,

$$
S_0 = \left[\begin{array}{cccc|c} v_1 u_1 & v_1 u_2 & v_1 u_3 & v_1 u_4 & 0 \\ v_1 u_2 & v_2 u_2 & v_2 u_3 & v_2 u_4 & 0 \\ v_1 u_3 & v_2 u_3 & v_3 u_3 & v_3 u_4 & 0 \\ v_1 u_4 & v_2 u_4 & v_3 u_4 & v_4 u_4 & b_4 \\ \hline 0 & 0 & 0 & b_4 & a_5 \end{array} \right].
$$

Define $\delta_4 = b_4$. Let

$$
G_1^T = \left[\begin{array}{cc} c_1 & s_1 \\ -s_1 & c_1 \end{array} \right] \quad \text{such that} \quad G_1^T \left[\begin{array}{c} u_3 \\ u_4 \end{array} \right] = \left[\begin{array}{c} \hat{u}_3 \\ 0 \end{array} \right]
$$

and let

$$
\hat{G}_1^T = \left[\begin{array}{ccc} I_2 & & \\ & G_1^T & \\ & & I_1 \end{array} \right],
$$

where I_k is the identity matrix of order k. Then,

$$
\hat{G}_1^T S_0 = \left[\begin{array}{ccccc} v_1 u_1 & v_1 u_2 & v_1 u_3 & v_1 u_4 & 0 \\ v_1 u_2 & v_2 u_2 & v_2 u_3 & v_2 u_4 & 0 \\ v_1 \hat{u}_3 & v_2 \hat{u}_3 & v_3 \hat{u}_3 & \rho_3 & s_1 \delta_4 \\ 0 & 0 & 0 & \delta_3 & c_1 \delta_4 \\ 0 & 0 & 0 & \delta_4 & a_5 \end{array} \right],
$$

with

$$
\left[\begin{array}{c} \rho_3 \\ \delta_3 \end{array} \right] = \left[\begin{array}{c} (c_1 v_3 + s_1 v_4) u_4 \\ (-s_1 v_3 + c_1 v_4) u_4 \end{array} \right].
$$

Therefore

$$
S_1 = \hat{G}_1^T S_0 \hat{G}_1 = \left[\begin{array}{ccccc} v_1 u_1 & v_1 u_2 & v_1 \hat{u}_3 & 0 & 0 \\ v_1 u_2 & v_2 u_2 & v_2 \hat{u}_3 & 0 & 0 \\ v_1 \hat{u}_3 & v_2 \hat{u}_3 & \eta_3 & s_1 \delta_3 & s_1 \delta_4 \\ 0 & 0 & s_1 \delta_3 & c_1 \delta_3 & c_1 \delta_4 \\ 0 & 0 & s_1 \delta_4 & c_1 \delta_4 & a_5 \end{array} \right],
$$

with $\eta_3 = v_3 \hat{u}_3 c_1 + \rho_3 s_1$. The subblock matrices $S_1(1:3, 1:3)$ and $S_1(3:5, 3:5)$ turn out to be of symmetric semiseparable form.

Let

$$G_2^T = \begin{bmatrix} c_2 & s_2 \\ -s_2 & c_2 \end{bmatrix} \text{ such that } G_2^T \begin{bmatrix} u_2 \\ \hat{u}_3 \end{bmatrix} = \begin{bmatrix} \hat{u}_2 \\ 0 \end{bmatrix},$$

and

$$\hat{G}_2^T = \begin{bmatrix} I_1 & & \\ & G_2^T & \\ & & I_2 \end{bmatrix}.$$

Multiplying S_1 to the left by $\hat{G}_2^T$ and to the right by $\hat{G}_2$, it turns out that

$$S_2 = \hat{G}_2^T \hat{G}_1^T S_0 \hat{G}_1 \hat{G}_2 = \begin{bmatrix} v_1 u_1 & v_1 \hat{u}_2 & 0 & 0 & 0 \\ v_1 \hat{u}_2 & \eta_2 & s_2 \delta_2 & s_2 s_1 \delta_3 & s_2 s_1 \delta_4 \\ 0 & s_2 \delta_2 & c_2 \delta_2 & c_2 s_1 \delta_3 & c_2 s_1 \delta_4 \\ 0 & s_2 s_1 \delta_3 & c_2 s_1 \delta_3 & c_1 \delta_3 & c_1 \delta_4 \\ 0 & s_2 s_1 \delta_4 & c_2 s_1 \delta_4 & c_1 \delta_4 & a_5 \end{bmatrix},$$

with $\eta_2 = v_2 \hat{u}_2 c_2 + \rho_2 s_2$ and $\begin{bmatrix} \rho_2 \\ \delta_2 \end{bmatrix} = \begin{bmatrix} c_2 v_2 \hat{u}_3 + s_2 \eta_3 \\ -s_2 v_2 \hat{u}_3 + c_2 \eta_3 \end{bmatrix}.$

Therefore the subblock matrices $S_2(1:2, 1:2)$ and $S_2(2:5, 2:5)$ are of symmetric semiseparable form. To end the updating, let us consider the Givens rotation

$$\hat{G}_3^T = \begin{bmatrix} G_1^T & \\ & I_3 \end{bmatrix},$$

with

$$G_3^T = \begin{bmatrix} c_3 & s_3 \\ -s_3 & c_3 \end{bmatrix} \text{ such that } G_3^T \begin{bmatrix} u_1 \\ \hat{u}_2 \end{bmatrix} = \begin{bmatrix} \hat{u}_1 \\ 0 \end{bmatrix}.$$

Then

$$S_3 = \hat{G}_3^T \hat{G}_2^T \hat{G}_1^T S_0 \hat{G}_1 \hat{G}_2 \hat{G}_3$$

$$= \begin{bmatrix} \eta_1 & s_3 \delta_1 & s_3 s_2 \delta_2 & s_3 s_2 s_1 \delta_3 & s_3 s_2 s_1 \delta_4 \\ s_3 \delta_1 & c_3 \delta_1 & c_3 s_2 \delta_2 & c_3 s_2 s_1 \delta_3 & c_3 s_2 s_1 \delta_4 \\ s_3 s_2 \delta_2 & c_3 s_2 \delta_2 & c_2 \delta_2 & c_2 s_1 \delta_3 & c_2 s_1 \delta_4 \\ s_3 s_2 s_1 \delta_3 & c_3 s_2 s_1 \delta_3 & c_2 s_1 \delta_3 & c_1 \delta_3 & c_1 \delta_4 \\ s_3 s_2 s_1 \delta_4 & c_3 s_2 s_1 \delta_4 & c_2 s_1 \delta_4 & c_1 \delta_4 & a_5 \end{bmatrix}$$

is a symmetric semiseparable matrix with

$$\eta_1 = v_1 \hat{u}_1 c_3 + \rho_1 s_3 \text{ and } \begin{bmatrix} \rho_1 \\ \delta_1 \end{bmatrix} = \begin{bmatrix} c_3 v_1 \hat{u}_2 + s_3 \eta_1 \\ -s_3 v_1 \hat{u}_2 + c_3 \eta_1 \end{bmatrix}.$$

Hence the generators of S_3 are

$$\mathbf{u} = \begin{bmatrix} \dfrac{\eta_1}{s_3 s_2 s_1}, & \dfrac{\delta_1}{s_2 s_1}, & \dfrac{\delta_2}{s_1}, & \delta_3, & \delta_4 \end{bmatrix}^T,$$

$$\mathbf{v} = \begin{bmatrix} s_3 s_2 s_1, & c_3 s_2 s_1, & c_2 s_1, & c_1, & \frac{a_5}{\delta_4} \end{bmatrix}^T.$$

The updating of a symmetric semiseparable matrix of order k has $\mathcal{O}(k)$ computational complexity and needs $\mathcal{O}(k)$ storage.

Having shown how the semiseparable structure in the Matrix (11.3) can be updated, it turns out quite straightforward how the Lanczos algorithm can be modified in order to compute semiseparable matrices.

A Lanczos reduction to semiseparable form

Combining the standard Lanczos algorithm, with the reduction presented above, we obtain a Lanczos algorithm for reducing matrices to semiseparable form. In the following algorithm the involved semiseparable matrix $S^{(i)}$ will grow in size in each step of the method.

Algorithm 11.6 (Lanczos reduction to semiseparable form).
Input: *A symmetric matrix $A \in \mathbb{R}^{n \times n}$ and $\mathbf{r}_0 \in \mathbb{R}^n$ as a starting vector.*
Output: *An orthogonally similar semiseparable matrix $S^{(n-1)}$.*

Let $b_0 = \|\mathbf{r}_0\|_2$, $\mathbf{q}_0 = 0$ and S_0 the empty matrix.
For $i = 1, 2, \ldots$

1. $\mathbf{q}_i = \mathbf{r}_{i-1} / \|\mathbf{r}_{i-1}\|_2$

2. $p = A\mathbf{q}_i$

3. $a_i = \mathbf{q}_i^T p$

4. *If $i = 1$*

$$S^{(i)} = [a_1]$$

   else

   Compute $S^{(i)}$, i.e., reduce into semiseparable form the augmented matrix
   $$\left[\begin{array}{c|c} S^{(i-1)} & b_{i-1}\mathbf{e}_{i-1} \\ \hline b_{i-1}\mathbf{e}_{i-1}^T & a_i \end{array} \right] \text{ with } \mathbf{e}_i = [\underbrace{0, \ldots, 0}_{i-1}, 1]^T,$$

   endif;

5. $\mathbf{r}_i = p - a_i\mathbf{q}_i - b_{i-1}\mathbf{q}_{i-1}$

6. $b_i = \|\mathbf{r}_i\|_i$

endfor;

Once the initial guess $\mathbf{r}_0$ is fixed the Krylov subspace generated by the Lanczos reduction to tridiagonal form is the same as the one generated by the Lanczos reduction to semiseparable form because the Givens rotations to perform the updating step 4 do not change this subspace. The Lanczos reduction to tridiagonal matrices halts before complete tridiagonalization if the initial guess $\mathbf{r}_0$ is contained in a proper invariant subspace. In this case, one of the b_j, $j \in \{1, \dots, n-1\}$ is equal to zero. However, the Lanczos reduction to tridiagonal form can be continued choosing a vector $\mathbf{r}_j$ orthogonal to the already computed Krylov basis yielding a block diagonal matrix with tridiagonal blocks [94]. The Lanczos reduction to semiseparable form with the same initial guess $\mathbf{r}_0$ has a similar behavior, breaking down at the same step j. Also in this case the algorithm can be run to completion choosing a vector $\mathbf{r}_j$ orthogonal to the already computed Krylov basis. The final reduced matrix is, in this case, a block diagonal matrix with generator representable semiseparable blocks.

The step 4 in the Lanczos reduction to semiseparable matrices corresponds to applying one iteration of the QR-method without shift to the matrix $S^{(i-1)}$. This is accomplished applying $i-2$ Givens rotations. Step 4 could be replaced by applying one step of the implicitly shifted QR-method, with the shift chosen in order to improve the convergence of the sequence of the generated semiseparable matrices towards a similar block diagonal one (see the reduction to semiseparable plus diagonal form in Chapter 2).

We observe that it is not necessary to compute the product of the Givens rotations at each step. The Givens coefficients are stored in a matrix and the product is computed only when the convergence of the sequence of the semiseparable matrix to a block diagonal form has occurred. As a consequence, the Krylov basis is then updated multiplying Q_k by the latter orthogonal matrix.

The Lanczos reduction as we proposed it here generates all possible orthogonally similar semiseparable matrices to a given symmetric matrix when varying the starting vector (or starting vectors when there is a breakdown in the Lanczos reduction, i.e., when one of the b_k is equal to zero). We will only give the proof for nonsingular matrices.

Theorem 11.7. *Let A be a given nonsingular symmetric matrix. Then each semiseparable matrix S that is orthogonally similar to A can be found by applying the Lanczos reduction as described in this section for a certain starting vector (or starting vectors when the Lanczos algorithm has found an invariant subspace).*

Proof. One can prove the above theorem directly by using the implicit Q-theorem for semiseparable matrices (see Subsection 6.2.3). □

In case of a singular matrix A, to generate all possible orthogonally similar matrices S, we have to use another degree of freedom in the algorithm. In this case, when performing the updating procedure, it can happen that the Givens transformation is not uniquely determined. When taking this degree of freedom into account, also all semiseparable matrices orthogonally similar to a symmetric singular matrix can be generated.

The reduction of a symmetric matrix into a similar semiseparable one proposed in this section has the same properties as the algorithm proposed in Chapter 2. Therefore, if gaps are present in the spectrum of the original matrix, they are revealed" after some steps of the algorithm (see next section) making the matrices numerically block diagonal with generator representable semiseparable blocks. This property makes the proposed algorithm suitable for computing the largest eigenvalues and the corresponding eigenvectors of sparse or structured matrices, if the large eigenvalues are quite well separated from the rest of the spectrum. Indeed, the most computationally expensive part, at each step of the proposed algorithm, is a matrix by vector product, which can be efficiently performed if the matrix is sparse or structured.

Notes and references

The results presented in this section are based on the following article.

☞ N. Mastronardi, M. Schuermans, M. Van Barel, R. Vandebril, and S. Van Huffel. A Lanczos-like reduction of symmetric structured matrices into semiseparable ones. *Calcolo*, 42(3–4):227–241, 2005.

The article contains also an extended section with numerical experiments showing the viability of the presented approach. Moreover the algorithm is applied to Hankel matrices, which allow an easy multiplication with a vector. The algorithm is not run to completion as one is only interested in the dominant eigenpairs, which are revealed by the subspace iteration properties of the reduction.

11.2 Rank-revealing properties of the orthogonal similarity reduction

The computation of the symmetric rank-revealing factorization of dense symmetric matrices is an important problem, e.g., in signal processing, where accurate computation of the numerical rank, as well as the numerical range space and the null space, is required [81, 104].

A family of algorithms for computing symmetric rank-revealing decompositions has been presented in [81, 104]. The choice of the algorithm depends on the definiteness of the matrix and the indefinite case seems to be the most difficult to handle.

In this section, we will exploit the convergence properties of the reduction to semiseparable form.

First, in Subsection 11.2.1 we will shortly introduce the definition and properties of a symmetric rank-revealing factorization. Secondly, the rank reduction decomposition based on semiseparable matrices is described in Subsection 11.2.2. Numerical tests can be found in Subsection 11.2.3.

11.2.1 Symmetric rank-revealing factorization

The symmetric rank-revealing decomposition of $A = W \Lambda W^T = \sum_{i=1}^n \lambda_i w_i w_i^T$, where λ_i and w_i, $i = 1, \ldots, n$ are the eigenvalues and the corresponding eigenvectors of A, respectively, with $|\lambda_1| = \sigma_1 \geq |\lambda_2| = \sigma_2 \geq \cdots |\lambda_n| = \sigma_n$, $\{\sigma_i\}_{i=1}^n$ the singular values of A, is defined in the following way[4] [104],

$$A = W_C C W_C^T,$$

where

$$C = \begin{bmatrix} C_{11} & C_{12} \\ C_{12}^T & C_{22} \end{bmatrix},$$

with $C_{11} \in \mathbb{R}^{k \times k}$, W_C an orthogonal matrix, the condition number of the matrix C_{11}: $\text{cond}(C_{11}) \approx \sigma_1 / \sigma_k$, $\|C_{12}\|_F^2 + \|C_{22}\|_F^2 \approx \sigma_{k+1}^2 + \cdots \sigma_n^2$, with $\|\cdot\|_F$ the Frobenius norm.

Of course, if the singular value decomposition of A is known, the symmetric rank-revealing decomposition is readily obtained. Unfortunately, the singular value decomposition is expensive to compute if A is quite large. Hence, it is suitable to derive cheaper methods for computing the symmetric rank-revealing decomposition.

We now show that the reduction of a symmetric matrix into semiseparable form, described in Section 2.2, can be used as a preprocessing step of a symmetric rank-revealing algorithm. In many cases this step is sufficient. If not, few steps of the QR-method without shift can be efficiently applied to the computed symmetric similar semiseparable matrix in order to reveal the numerical rank.

11.2.2 Rank-revealing via the semiseparable reduction

In order to clearly understand the rank-revealing algorithm, we will introduce some notation. Suppose we start by reducing the matrix A to semiseparable form, similarly as in the previous algorithms, we obtain a sequence of matrices $A^{(0)} = A, A^{(1)}, A^{(2)}, A^{(3)}, \ldots$. Remember that each matrix $A^{(i)}$ has a submatrix of size $i \times i$ of semiseparable form. This submatrix appears in the lower-right corner if the algorithms from Chapter 2 are used or in the upper-left corner if the exposition of the previous section is followed. Nevertheless let us denote the part of the matrix $A^{(i)}$ of semiseparable form by $S^{(i)}$. Hence $S^{(i)}$ is of semiseparable form and of size $i \times i$.

The algorithm consists of two important steps. First one starts by reducing the matrix A to semiseparable form. In every step of this reduction, the Frobenius norm of the semiseparable matrix $S^{(i)}$ is computed. As soon as the Frobenius norm of $S^{(i)}$ is close enough to the one of A, we stop the reduction algorithm, as this indicates that all dominant singular values of the matrix A are captured in the matrix $S^{(i)}$. Secondly one continues working on the semiseparable matrix $S^{(i)}$. Quite often in case the gap is large enough, the convergence behavior of the reduction to semiseparable form already reveals the numerical rank of the matrix $S^{(i)}$ and thereby also of the matrix A. In case the numerical rank is not yet accurate enough, one can continue by performing few steps of the QR-method without shift onto the matrix $S^{(i)}$. If this description is not yet clear enough, after the algorithm an example is presented, stating more clearly why step 2 in the algorithm is necessary.

---

[4]This definition is similar to the one used in [114], except that $\|\text{triu}(C_{22})\|_F^2$ instead of $\|C_{22}\|_F^2$ is used in [114].

The global rank-revealing algorithm based on the reduction into a semiseparable matrix, can now be formulated.

Note 11.8. *We remark that the method presented here is based on the full reduction to semiseparable form and not on the Lanczos variant. One can however easily adapt the algorithm below in order to use the Lanczos reduction to semiseparable form.*

Algorithm 11.9 (Rank-revealing factorization).
Input: *A symmetric matrix $A \in \mathbb{R}^{n \times n}$, τ_1 the threshold for the rank revealing, τ_2 the threshold used in the QR-algorithm applied onto the semiseparable matrix.*
Output: *A symmetric rank-revealing factorization of the original matrix A.*

1. *Compute the Frobenius norm of A: $\|A\|_F$.*

2. *Until $\|A\|_F - \|S^{(i)}\|_F < \tau_1$*

   - *Perform another step of the reduction to semiseparable form, and obtain $A^{(i)}$ and $S^{(i)}$.*

   enduntil;

3. *Apply few steps, lets say l steps of the QR-method without shift onto the matrix $S^{(i)}$ and obtain $\tilde{S}^{(i)}$.*

4. *For $k = i - 1, i - 2 \ldots, 1$*

   - *Compute $\|\tilde{S}^{(i)}(k + 1 : i, k + 1 : i)\|_F$ and $\tilde{S}^{(i)}(k, k)$.*

   endfor;

5. *Use the above information to deflate parts of the matrix $\tilde{S}^{(i)}$ corresponding to the singular values below the threshold τ_2.*

The second step of the algorithm is the most expensive one. Indeed, the computational complexity is $\mathcal{O}(in^2 + i^3)$. Moreover, it requires $\mathcal{O}(in)$ memory to store the involved Householder vectors and the Givens coefficients. Each iteration of the QR-method without shift in step 3 requires $\mathcal{O}(i)$ computational complexity and storage. The choice of l depends on the handled matrix. Often l is very low compared to the size of the matrix. Step 4 is also computed in $\mathcal{O}(i)$ flops. See Subsection 8.2.4 on how to compute the Frobenius norm effectively.

Example 11.10 Let us briefly illustrate with an example, why step 2 in the above method is necessary. Suppose we have three clusters in the spectrum of the matrix A. These clusters are depicted as Δ_1, Δ_2 and Δ_3. Assume that the cluster Δ_3 contains only the small eigenvalues. Hence, we would like our rank-revealing method to return to us both clusters Δ_1 and Δ_2. Assume now, that we start the reduction to semiseparable form as described in the previous section, and we omit step 2. We will stop the reduction procedure if a block diagonal matrix is formed in the reduction, assuming thereby that the revealed block contains the dominant eigenvalues. Unfortunately the reduction procedure first finds the cluster Δ_1 and part of the cluster Δ_3. After the eigenvalues of these clusters are approximated well enough, the subspace iteration can start working and reveal a block. Unfortunately the revealed block will not contain both clusters Δ_1 and Δ_2, but only an approximation to Δ_3.

If step 2 in the algorithm is checked, however, we would not have stopped the algorithm because the eigenvalues in Δ_2 are large and therefore the condition posed in step 2 will not be met.   ∎

It is well known that a single vector Lanczos algorithm can compute multiple eigenvalues of a symmetric matrix, but the multiple eigenvalues do not necessarily converge consecutively one after another. More precisely, if $\max\{|\Lambda|\}$, with Λ as the spectrum of A, is a multiple eigenvalue with maximum modulus of a symmetric matrix A, then usually a copy of $\max\{|\Lambda|\}$ will converge first, followed by several other smaller eigenvalues of A. Then another copy of $\max\{|\Lambda|\}$ will converge followed again by smaller eigenvalues of A, and so on. The consequence of this convergence behavior is that only fewer copies of the largest eigenvalues of A are also eigenvalues of $T^{(k)}$, where $T^{(k)}$ is the intermediate generated tridiagonal matrix. The semiseparable reduction behaves similarly to the Lanczos reduction to tridiagonal matrices. Therefore, in case of multiple eigenvalues of A, the symmetric semiseparable matrix computed by the reduction is block diagonal, with each block having $\max\{|\Lambda|\}$ as eigenvalue. A multivector semiseparable reduction can be considered in such cases [153].

Steps 4 and 5 of the latter algorithm are applied to order the diagonal entries of these blocks. If tiny eigenvalues are present in the blocks, they appear as the last entries of the blocks. In this case, we proceed removing the corresponding rows and columns from the matrix.

11.2.3   Numerical experiments

Some of the matrices used in [81, 104] are considered to test the proposed rank-revealing factorization.

For the matrices considered in the examples, the rank-revealing factorizations are computed correctly, and the numerical ranks are always equal to the desired ones. The results of the proposed rank-revealing factorization based on semiseparable matrices are compared to the results obtained by using the MATLAB functions hvsvsd (semipositive definite case) and hvsvid_L (indefinite case) of the MATLAB toolbox UTV [83, 82]. The number of steps of the QR-method without shift in the proposed algorithm is fixed to 10 for all the examples. As in [81, 104], the symmetric semidefinite test matrices are computed as $Q\Sigma Q^T$, with Q random orthogonal matrices and Σ diagonal matrices with the desired singular values. For the indefinite case, the diagonal matrices are replaced by the product of diagonal matrices of the desired singular values by diagonal random sign matrices. It can be noticed that the proposed algorithm is the fastest one. Moreover, it is as accurate as those available in the literature. The choice of threshold τ_2 is crucial for the accuracy of

Positive definite case

n	max $|\kappa_2(\hat{A}) - \kappa_2(H_{11})|$	max $|\kappa_2(\hat{A}) - \kappa_2(S_{11})|$	max $\|H_{12}\|_F$	max $\|S_{12}\|_F$
64	$6.88 \cdot 10^{-9}$	$1.00 \cdot 10^{-8}$	$8.33 \cdot 10^{-11}$	$2.14 \cdot 10^{-38}$
128	$4.49 \cdot 10^{-9}$	$1.17 \cdot 10^{-8}$	$1.22 \cdot 10^{-10}$	$1.66 \cdot 10^{-37}$
256	$8.26 \cdot 10^{-9}$	$3.48 \cdot 10^{-8}$	$1.24 \cdot 10^{-10}$	$3.26 \cdot 10^{-36}$

	average $\#flops^{(H)}$	average $\#flops^{(S)}$
64	$6.0390 \cdot 10^6$	$2.7410 \cdot 10^6$
128	$5.9928 \cdot 10^5$	$2.4800 \cdot 10^5$
256	$3.8894 \cdot 10^6$	$1.6242 \cdot 10^6$

Indefinite case

n	max $|\kappa_2(\hat{A}) - \kappa_2(H_{11})|$	max $|\kappa_2(\hat{A}) - \kappa_2(S_{11})|$	max $\|H_{12}\|_F$	max $\|S_{12}\|_F$
64	$4.86 \cdot 10^{-9}$	$1.07 \cdot 10^{-8}$	$5.48 \cdot 10^{-13}$	$3.18 \cdot 10^{-39}$
128	$2.99 \cdot 10^{-9}$	$2.34 \cdot 10^{-9}$	$2.66 \cdot 10^{-13}$	$8.15 \cdot 10^{-38}$
256	$3.52 \cdot 10^{-9}$	$3.74 \cdot 10^{-9}$	$4.53 \cdot 10^{-13}$	$2.82 \cdot 10^{-37}$

	average $\#flops^{(H)}$	average $\#flops^{(S)}$
64	$1.1771 \cdot 10^5$	$2.7478 \cdot 10^4$
128	$1.4552 \cdot 10^6$	$2.4793 \cdot 10^5$
256	$1.0641 \cdot 10^7$	$2.9243 \cdot 10^6$

Table 11.1. *Results obtained applying the rank-revealing algorithms* hvsvsd *(semipositive definite case),* hvsvid_L *(indefinite case) and the proposed one to 100 sycmetric random matrices with prescribed singular values, constructed as described in Experiment 11.11.*

the results. In all the examples, $\tau_2 = n\epsilon\|A\|_F$, where n is the order of the matrix, $\|\cdot\|_F$ is the Frobenius norm, and $\epsilon \sim 2.2204 \cdot 10^{-16}$ is the machine precision.

Experiment 11.11. *The matrices considered in this example are randomly generated test matrices of size* $n = 64, 128, 256$, *(100 matrices for each size), each with* $n - 4$ *singular values geometrically distributed between 1 and* 10^{-4}, *and the remaining four singular values given by* 10^{-7}, 10^{-8}, 10^{-9}, *and* 10^{-10}, *such that the numerical rank with respect to the threshold* $\tau_1 = 10^{-5}$ *is* $k = n - 4$.

Experiment 11.12. *The proposed rank-revealing factorization is now applied to* compute the numerical rank of four classes of matrices considered in [81]. The size of the considered matrices is $n = 20$ (100 matrices for each class), and the desired numerical rank is $k = 15$. The singular values are geometrically distributed between σ_1 and σ_k and also between σ_{k+1} and σ_n. Moreover $\sigma_1 = 1$ and $\sigma_n = 10^{-10}$.

The classes of considered matrices are constructed according to Table 11.2.

Cl.	$\sigma_1(A)$	$\sigma_n(A)$	$\sigma_k(S)$	$\sigma_k(S)/\sigma_{k+1}(A)$
1	1.0	$1.0 \cdot 10^{-10}$	$1.0 \cdot 10^{-3}$	2
2	1.0	$1.0 \cdot 10^{-10}$	$1.0 \cdot 10^{-3}$	10
3	1.0	$1.0 \cdot 10^{-10}$	$1.0 \cdot 10^{-6}$	2
4	1.0	$1.0 \cdot 10^{-10}$	$1.0 \cdot 10^{-6}$	10

Table 11.2. *Features of the singular values of the classes of matrices considered for Experiment 11.12. With $\sigma_i(A)$ the i-th singular value of A is meant.*

The results are depicted in Table 11.1. We assume that the reduction to semi-separable form generated a semiseparable part in the matrix in the upper left corner. The size of the matrices is given in the first column. In column 2, the maximum, over 100 matrices, of the difference between $\kappa_2(\hat{A})$, the condition number of the leading submatrices of order k of the symmetric generated matrices and $\kappa_2(H_{11})$, the condition number of the leading submatrices of order k of the rank-revealing matrices H computed by the functions `hvsvsd` or `hvsvid_L` of the UTV toolbox [83, 82] is reported and, in column 3, the maximum, over 100 matrices, of the difference between $\kappa_2(\hat{A})$ and $\kappa_2(S_{11})$, the condition number of the leading submatrices of order $n - k$ of the rank-revealing matrices S computed by the proposed algorithm.

In column 4 and 5, the maximum of the Frobenius norm of submatrices $H_{12} = H(1 : k, k + 1 : n)$ and $S_{12} = S(1 : k, k + 1 : n)$, respectively, are depicted. Finally, in column 6 and 7 the average number of flops, over the 100 matrices, for the algorithms of the UTV toolbox [83, 82] and the proposed algorithm, respectively, can be found.

The considered positive definite and indefinite symmetric matrices are constructed as in the previous example. The results are reported in Table 11.3. In the first column, the class of considered matrices is indicated. The legend of column 2 up to 7 is similar to the corresponding one of Table 11.1. For all these matrices, the threshold is $\tau_1 = (\sigma_k + \sigma_{k+1})/2$.

Notes and references

This section is based on the results presented in a more elaborate way in the following report.

☞ N. Mastronardi, M. Van Barel, and R. Vandebril. Computing the rank revealing factorization by the semiseparable reduction. Technical Report TW418, Department of Computer Science, Katholieke Universiteit Leuven, Celestijnenlaan 200A, 3000 Leuven (Heverlee), Belgium, May 2005.

Positive definite case

Cl.	max $\|\kappa_2(\hat{A}) - \kappa_2(H_{11})\|$	max $\|\kappa_2(\hat{A}) - \kappa_2(S_{11})\|$	max $\|H_{12}\|_F$	max $\|S_{12}\|_F$
1	$5.70 \cdot 10^{-11}$	$7.7875 \cdot 10^{-11}$	$2.25 \cdot 10^{-4}$	$1.08 \cdot 10^{-24}$
2	$6.65 \cdot 10^{-11}$	$1.4450 \cdot 10^{-10}$	$9.45 \cdot 10^{-6}$	$5.34 \cdot 10^{-23}$
3	$6.37 \cdot 10^{-5}$	$6.7613 \cdot 10^{-5}$	$2.46 \cdot 10^{-7}$	$2.34 \cdot 10^{-18}$
4	$5.97 \cdot 10^{-5}$	$5.5703 \cdot 10^{-5}$	$7.44 \cdot 10^{-9}$	$2.19 \cdot 10^{-15}$

Cl.	average $\#flops^{(H)}$	average $\#flops^{(S)}$
1	$6.6502 \cdot 10^4$	$2.7737 \cdot 10^4$
2	$6.6962 \cdot 10^4$	$2.7737 \cdot 10^4$
3	$6.6831 \cdot 10^4$	$2.7737 \cdot 10^4$
4	$6.6962 \cdot 10^4$	$2.7737 \cdot 10^4$

Indefinite case

Cl.	max $\|\kappa_2(\hat{A}) - \kappa_2(H_{11})\|$	max $\|\kappa_2(\hat{A}) - \kappa_2(S_{11})\|$	max $\|H_{12}\|_F$	max $\|S_{12}\|_F$
1	$9.37 \cdot 10^{-11}$	$5.85 \cdot 10^{-11}$	$7.19 \cdot 10^{-4}$	$1.08 \cdot 10^{-25}$
2	$5.34 \cdot 10^{-11}$	$5.60 \cdot 10^{-11}$	$2.28 \cdot 10^{-7}$	$3.19 \cdot 10^{-24}$
3	$1.01 \cdot 10^{-4}$	$4.51 \cdot 10^{-5}$	$6.27 \cdot 10^{-7}$	$6.41 \cdot 10^{-18}$
4	$4.90 \cdot 10^{-5}$	$4.66 \cdot 10^{-5}$	$3.14 \cdot 10^{-9}$	$2.56 \cdot 10^{-16}$

Cl.	average $\#flops^{(H)}$	average $\#flops^{(S)}$
1	$1.2417 \cdot 10^5$	$2.7737 \cdot 10^4$
2	$1.2443 \cdot 10^5$	$2.7737 \cdot 10^4$
3	$1.2708 \cdot 10^5$	$2.7737 \cdot 10^4$
4	$1.2758 \cdot 10^5$	$2.7737 \cdot 10^4$

Table 11.3. *Results obtained applying the rank-revealing algorithms* `hvsvsd` *(semipositive definite case),* `hvsvid_L` *(indefinite case) and the proposed one to 100 symmetric random matrices with prescribed singular values, constructed as described in Experiment 11.12.*

11.3 Conclusions

In this chapter we showed that it is possible to adapt the reduction to semiseparable form, such that it becomes suitable for applying a so-called Lanczos semiseparabilization process. Secondly an algorithm was presented for computing the rank-revealing factorization of matrices, based on the reduction to semiseparable form.

Part IV

Orthogonal (rational) functions (Inverse eigenvalue problems)

Eigenvalue problems and inverse eigenvalue problems for structured rank matrices are connected to several other domains of mathematics. In this part, we show the relationship between the inverse eigenvalue problem involving different structured rank matrices and orthonormal functions. The orthonormality is defined based on a discrete inner product, i.e., an inner product involving a finite number of nodes z_i.

This part is organized as follows. Chapter 12 handles the polynomial case. For orthonormal polynomials, the rank structure is upper Hessenberg. If the nodes z_i are all real, this becomes a symmetric tridiagonal matrix. When all nodes z_i are on the unit circle, the Hessenberg matrix is unitary. Hence, it can also be represented by $\mathcal{O}(n)$ parameters with n the size of the unitary Hessenberg matrix. For orthonormal polynomial vectors, discussed in Chapter 13, the rank structure turns out to be a generalization of an upper Hessenberg matrix. For rational functions, as studied in Chapter 14, the rank structure is Hessenberg-like. Also in these last two chapters, depending on the details of the discrete inner product, e.g., the nodes z_i are all real or all on the unit circle, the upper triangular part of of the structured rank matrix can also be rank structured. Besides giving algorithms to solve the different inverse eigenvalue problems, it is also explained how to solve the corresponding least squares approximation problems based on the knowledge of the recurrence coefficients for the different types of orthonormal functions.

✎ *Reading this part is not necessary to understand the rest of the book. However, it shows a nice application of structured rank matrices in inverse eigenvalue problems related to orthonormal polynomials, polynomial vectors and rational function. When one is interested in using orthonormal functions to solve least squares approximation problems, this part gives a concise introduction in solving this problem.*

Chapter 12

Orthogonal polynomials and discrete least squares

Let $\{z_k\}_{k=1}^m$ be a set of complex nodes with corresponding weights $\{w_k\}_{k=1}^m$. Note that up to now, we worked with real numbers. For the sake of generality, complex numbers are also allowed in the problem setting.

In this chapter, we shall solve the problem of finding the least squares polynomial approximant in the space with positive semidefinite inner product

$$\langle f, g \rangle = \sum_{k=1}^m \overline{f(z_k)} |w_k|^2 g(z_k), \tag{12.1}$$

with f and g functions belonging to some vector space. Note that this is a positive definite inner product for the space of vectors $[f(z_1), \ldots, f(z_m)]$ representing the function values at the given nodes for all possible functions f from the vector space considered. In the case that interests us here, this space will be the space of polynomials of degree less than m. This space will be denoted as $\mathbb{P}_{m-1}$.

The polynomial $p \in \mathbb{P}_n$ of degree $n \le m$ which minimizes

$$\|f - p\|, \quad \text{with} \quad \|v\| = \langle v, v \rangle^{1/2}$$

(note that this is a semi-norm) can be found as follows. Find a basis $\{\varphi_0, \ldots, \varphi_n\}$ for $\mathbb{P}_n$ which is orthonormal with respect to $\langle \cdot, \cdot \rangle$. The solution p is the generalized Fourier expansion of f with respect to this basis, truncated after the term of degree n. An algorithm that solves the problem will compute implicitly or explicitly the orthonormal basis and the Fourier coefficients. As we shall see in the following sections, the parameters of the recurrence relation for the sequence of these orthonormal polynomials can be stored in a structured rank matrix, more precisely, an upper Hessenberg matrix. This upper Hessenberg matrix can be found as the solution of an inverse eigenvalue problem, where the points z_i are the given eigenvalues and the weights w_i are the first components of the corresponding eigenvectors. An algorithm to solve this inverse eigenvalue problem is designed. We can reduce the complexity of such an algorithm by an order of magnitude when a "short recurrence" exists for the orthogonal polynomials. We shall consider the case where all the z_i are on the real line, in which case a three-term recurrence relation exists, and the case where all the z_i are on the complex unit circle, in which case a Szegő type recurrence relation exists.

The chapter is organized as follows. In Section 12.1, it is shown that the recurrence relation for a sequence of polynomials $p_0, p_1, \ldots, p_k, \ldots$ with $\deg p_k = k$ can be compactly represented by an upper Hessenberg matrix. When the polynomials p_k are orthonormal with respect to the discrete inner product as described in Section 12.2, this upper Hessenberg matrix is the solution of an inversion eigenvalue problem as formulated in Section 12.3. In Section 12.4, this result is used to define an extended Hessenberg matrix used in solving the discrete least squares polynomial approximation problem. This extended Hessenberg matrix is a rectangular matrix consisting of 2 initial columns followed by the upper Hessenberg matrix of the recurrence coefficients. An algorithm to solve the corresponding inverse eigenvalue problem is developed in Section 12.5. If the points of the inner product are lying on the real line or on the unit circle, the Hessenberg matrix reduces to a Jacobi matrix, i.e., a symmetric tridiagonal one, or a unitary Hessenberg matrix, respectively, as is indicated in Section 12.6.

✎ *Mastering the content of this chapter is important to understanding the following two chapters of the book. The essential ideas of this chapter are the following. A sequence of polynomials p_k having strict degree k satisfies a recurrence relation (see Equation (12.2)). The coefficients of this recurrence relation can be summarized into an upper Hessenberg matrix (see Equation (12.3)). When the polynomials are orthonormal with respect to a discrete inner product of the form (12.4), the upper Hessenberg matrix can be recovered by solving an inverse eigenvalue problem as defined in Definition 12.1. This problem can be solved an order of magnitude faster when the points of the inner product are lying all on the real line or all on the unit circle.*

12.1   Recurrence relation and Hessenberg matrix

Consider the sequence $p_0, p_1, p_2, \ldots$ of polynomials $p_k(z) \in \mathbb{P} = \mathbb{C}[z]$ where each polynomial p_k has strict degree k. In this section, we show that these polynomials satisfy a recurrence relation whose coefficients can be naturally stored in an upper Hessenberg matrix. When these polynomials are orthogonal with respect to the discrete inner product as defined in the next section, it will turn out that this Hessenberg matrix is the solution of an inverse eigenvalue problem.

Because the degree of each of the polynomials p_k is strictly equal to k, it is clear that this sequence of polynomials is linearly independent. Let us construct now a recurrence relation for this sequence. Consider the polynomial $zp_k(z)$ of degree $k + 1$. Because the polynomials $p_0, p_1, \ldots, p_{k+1}$ form a basis for the space of all polynomials having degree less than or equal to $k + 1$, we can write $zp_k(z)$ as a linear combination of these polynomials:

$$zp_k(z) = \sum_{j=0}^{k+1} p_j(z) h_{j,k}, \qquad k = 0, 1, 2, \ldots. \tag{12.2}$$

Note that $h_{k+1,k} \neq 0$. Hence, once we know p_0 and the recurrence coefficients $h_{j,k}$, we can compute the other polynomials as

$$h_{k+1,k}p_{k+1}(z) = zp_k(z) - \sum_{j=0}^{k} p_j(z)h_{j,k}, \qquad k = 0, 1, 2, \ldots.$$

When we denote the infinite vector of polynomials as

$$\mathbf{p}_{0:\infty} = [p_0, p_1, \ldots],$$

we can rewrite (12.2) in matrix notation as

$$z\mathbf{p}_{0:\infty} = \mathbf{p}_{0:\infty}H_{\infty \times \infty}, \tag{12.3}$$

with $H_{\infty \times \infty} = [h_{j,k}]$ an irreducible upper Hessenberg matrix.

12.2 Discrete inner product

In this section, a discrete inner product is defined and the corresponding sequence of orthogonal polynomials having increasing degree is considered. It is shown that in this case, the Hessenberg matrix containing the recurrence coefficients is unitarily similar to the diagonal matrix containing the nodes of the inner product.

Given the complex numbers z_i, called the points, and the nonzero complex numbers w_i, called the weights, $i = 1, 2, \ldots, m$, let us define the discrete inner product $\langle \cdot, \cdot \rangle$ as

$$\langle p, q \rangle = \sum_{i=1}^{m} |w_i|^2 \overline{p(z_i)}q(z_i), \tag{12.4}$$

with $\bar{z}$ denoting the complex conjugate of the complex number z. Assume now that the polynomials $\varphi_0, \varphi_1, \ldots$ are orthogonal with respect to each other, i.e.,

$$\langle \varphi_i, \varphi_j \rangle = 0, \qquad i, j = 0, 1, 2, \ldots \text{ and } i \neq j. \tag{12.5}$$

The $m \times \infty$ matrix whose j-th column for $j = 0, 1, \ldots$ consists of the function values of polynomial φ_j in each of the points z_i, is denoted by Φ. The diagonal matrix whose diagonal elements are the weights w_i is denoted by W. The orthogonality conditions (12.5) can be written in matrix notation as

$$(W\Phi)^H(W\Phi) = D, \qquad \text{with } D \text{ an (infinite) diagonal matrix.}$$

Because W has rank m and, hence, D has rank less than or equal to m, we can assume that only the first m polynomials have length 1 with respect to the norm corresponding to the inner product. In that case the other polynomials have length 0. Let us split up Φ as $\Phi = [\Phi_1 \quad \Phi_2]$ where Φ_1 consists of the first m columns of Φ. Hence, it follows that

$$\Phi_1^H W^H W \Phi_1 = I.$$

Therefore, the matrix $Q = W\Phi_1$ is unitary. Because

$$\Phi_2^H W^H Q = 0,$$

it follows that $\Phi_2 = 0$, i.e., each of the polynomials φ_j for $j = m, m+1, \ldots$ has to be zero in each of the points z_i. Hence,

$$\varphi_j(z) = (z - z_1)(z - z_2) \cdots (z - z_m)\psi(z),$$

with $\psi \in \mathbb{P}_{j-m}$ and of strict degree $j - m$. Taking the first m columns of (12.3) and evaluating in each of the points z_i, leads to

$$Z\Phi_1 = \Phi_1 H_{m \times m}, \tag{12.6}$$

with $H_{m \times m}$ the leading principal $m \times m$ submatrix of $H_{\infty \times \infty}$ and Z the diagonal matrix containing the nodes z_i on its diagonal. Multiplying (12.6) to the left by the diagonal matrix W consisting of the weights, gives us:

$$W(Z\Phi_1) = W(\Phi_1 H_{m \times m})$$

or because Z and W are diagonal matrices,

$$Z(W\Phi_1) = (W\Phi_1)H_{m \times m}.$$

Using the fact that $W\Phi_1 = Q$, it follows that

$$ZQ = QH,$$

with Q unitary and $H = H_{m \times m}$ an upper Hessenberg matrix. Summarizing, we get

$$Q\mathbf{e}_1 = \alpha \frac{\mathbf{w}}{\|\mathbf{w}\|^2}, \qquad \text{with } |\alpha| = 1$$
$$Q^H Z Q = H$$

where $\mathbf{w}$ denotes the vector consisting of the weights

$$\mathbf{w} = [w_1, w_2, \ldots, w_m]^T. \tag{12.7}$$

In words, the matrix $Q = W\Phi_1$ is unitary, its first column is a normalized vector consisting of the weights, the columns of Q^H are the eigenvectors of the upper Hessenberg matrix H whose eigenvalues are the points z_i.

12.3   Inverse eigenvalue problem

Suppose now that we start with the points z_i and the corresponding weights w_i for $i = 1, 2, \ldots, m$. How can we find the elements of the Hessenberg matrix H, i.e., the recurrence coefficients $h_{j,k}$ of (12.3) defining each of the polynomials $\varphi_k(z)$

for $k = 0, 1, \ldots, m - 1$, where $\varphi_k(z)$ has strict degree k. Because φ_0 is a constant polynomial, it is easy to determine a possible value for it:

$$\langle \varphi_0, \varphi_0 \rangle = \sum_{i=1}^{m} |w_i|^2 |\varphi_0|^2 = 1.$$

Hence,

$$\varphi_0 = \frac{\alpha}{\|\mathbf{w}\|^2}, \qquad \text{with } |\alpha| = 1.$$

Therefore, we know that the first column of Q should satisfy:

$$\begin{aligned} Q\mathbf{e}_1 &= W[\varphi_0(z_1), \varphi_0(z_2), \ldots, \varphi_0(z_m)]^T \\ &= \alpha \frac{\mathbf{w}}{\|\mathbf{w}\|^2}. \end{aligned}$$

The upper Hessenberg matrix $H = [h_{j,k}]$ can be determined by solving the following inverse eigenvalue problem:

Definition 12.1 (Inverse eigenvalue problem). *Given the points z_i and the weights w_i, $i = 1, 2, \ldots, m$. Compute an upper Hessenberg matrix H such that*

- *its eigenvalues are z_i,*

- *it is unitarily similar to the diagonal matrix $Z = \mathrm{diag}([z_1, \ldots, z_n])$:*

$$Q^H Z Q = H,$$

- *the first column of Q equals $\alpha \frac{\mathbf{w}}{\|\mathbf{w}\|^2}$ with $|\alpha| = 1$ and $\mathbf{w}$ as defined in (12.7).*

This condition is not sufficient to characterize Q completely. We can fix it uniquely by making the φ_k to have positive leading coefficients. This will be obtained when all the subdiagonal elements of the Hessenberg matrix H are positive, i.e., when $h_{k+1,k}$, $k = 0, 1, \ldots, m - 1$ are positive, and when α is taken equal to one. Since we assumed that none of the weights w_i is zero, the subdiagonal elements $h_{k+1,k}$ are nonzero and therefore this normalization can always be realized. Note that this normalization is not necessary to obtain the recurrence coefficients of a sequence of orthogonal polynomials φ_k, $k = 0, 1, \ldots$ having strict degree k.

In the direct eigenvalue problem, one computes the eigenvalues $\{z_k\}_1^m$ and the eigenvectors Q from the Hessenberg matrix, e.g., with the QR-algorithm. For the inverse eigenvalue problem, the Hessenberg matrix is reconstructed from the spectral data by an algorithm, which could be called an inverse QR-algorithm. This is the Rutishauser-Gragg-Harrod algorithm in case all nodes z_i are on the real line [97, 132] and the unitary inverse QR-algorithm described in [8] for the case of the unit circle. In the Section 12.5, we will give an algorithm that solves the inverse eigenvalue problem and at the same time computes the coefficients of the polynomials solving the least squares approximation problems of strict degrees $0, 1, \ldots, m$. In the next section, we will state this least squares problem in detail.

12.4   Polynomial least squares approximation

In this section, we will study the polynomial least squares problem in detail. We show that it can be solved in terms of the orthogonal polynomials with respect to the discrete inner product. In the next section, we will design an efficient algorithm updating the recurrence coefficients and the coefficients of the least squares solutions when a new data point z_{m+1} and a corresponding weight w_{m+1} are added.

Discrete least squares approximation by polynomials is a classical problem in numerical analysis where orthogonal polynomials play a central role. Given an inner product $\langle \cdot, \cdot \rangle$ defined on $\mathbb{P}_{m-1} \times \mathbb{P}_{m-1}$, the polynomial $p \in \mathbb{P}_n$ of degree at most $n \leq m$, which minimizes the error

$$\|f - p\|, \qquad p \in \mathbb{P}_n$$

is given by

$$p = \sum_{k=0}^{n} \varphi_k a_k, \qquad a_k = \langle f, \varphi_k \rangle$$

when the $\{\varphi_k\}_0^n$ form an orthonormal set of polynomials:

$$\varphi_k \in \mathbb{P}_k - \mathbb{P}_{k-1}, \qquad \mathbb{P}_{-1} = \varnothing, \qquad \langle \varphi_k, \varphi_l \rangle = \delta_{kl},$$

with δ_{kl} the Kronecker delta, i.e., $\delta_{kl} = 0$ when $k \neq l$ and $\delta_{kl} = 1$ when $k = l$. The inner product we shall consider here is of the discrete form (12.4) where the z_i are distinct complex numbers. Note that when $m = n$, the least squares solution is the interpolating polynomial, so that interpolation can be seen as a special case.

To illustrate where the orthogonal polynomials show up in this context, we start with an arbitrary polynomial basis $\{\psi_k\}$, $\psi_k \in \mathbb{P}_k - \mathbb{P}_{k-1}$. Setting

$$p = \sum_{k=0}^{n} \psi_k a_k^{\Psi}, \qquad a_k^{\Psi} \in \mathbb{C},$$

the least squares problem can be formulated as finding the weighted least squares solution, i.e., the coefficients a_k^{Ψ}, of the system of linear equations

$$\sum_{k=0}^{n} w_i \psi_k(z_i) a_k^{\Psi} \approx w_i f(z_i), \quad i = 1, \dots, m.$$

The term "weighted" indicates that we apply a different weight w_i to each of the m equations. Hence, the *weighted* least squares solution turns out to be the least squares solution of

$$W \Psi \mathbf{a}^{\Psi} \approx W \mathbf{f}$$

where $W = \mathrm{diag}([w_0, \dots, w_m])$ and

$$\Psi = \begin{bmatrix} \psi_0(z_1) & \dots & \psi_n(z_1) \\ \vdots & & \vdots \\ \psi_0(z_m) & \dots & \psi_n(z_m) \end{bmatrix}, \quad \mathbf{a}^{\Psi} = \begin{bmatrix} a_0^{\Psi} \\ \vdots \\ a_n^{\Psi} \end{bmatrix}, \quad \mathbf{f} = \begin{bmatrix} f(z_1) \\ \vdots \\ f(z_m) \end{bmatrix}.$$

Note that when $\psi_k(z) = z^k$, the power basis, then Ψ is a rectangular $m \times (n+1)$ Vandermonde matrix. The normal equations for this system are

$$(\Psi^H W^H W \Psi)\mathbf{a}^\Psi = \Psi^H W^H W \mathbf{f}.$$

When the ψ_k are chosen to be the orthonormal polynomials φ_k, then the square coefficient matrix of the previous system becomes very simple, i.e., $\Psi^H W^H W \Psi = I_{n+1}$, and the previous system gives the solution $\mathbf{a}^\Psi = \Psi^H W^H W \mathbf{f}$ immediately.

When the least squares problem is solved by QR-factorization, i.e., when Q is an $m \times m$ unitary matrix such that $Q^H W \Psi = [R^T \ 0^T]^T$, and R is upper triangular, we have to solve the triangular system given by the first $n+1$ rows of

$$\begin{bmatrix} R \\ 0 \end{bmatrix} \mathbf{a}^\Psi = Q^H W \mathbf{f} + \begin{bmatrix} 0 \\ \mathbf{x} \end{bmatrix},$$

where $\mathbf{x}$ is related to the residual vector $\mathbf{r}$ by

$$\begin{bmatrix} 0 \\ \mathbf{x} \end{bmatrix} = Q^H \mathbf{r}, \quad \mathbf{r} = W \Psi \mathbf{a}^\Psi - W \mathbf{f}.$$

Note that the least squares error is $\|\mathbf{x}\| = \|\mathbf{r}\|$. Again, when the ψ_k are replaced by the orthonormal polynomials φ_k, we get the trivial system $(m > n)$

$$\begin{bmatrix} I_{n+1} \\ 0 \end{bmatrix} \mathbf{a}^\Phi = Q^H W \mathbf{f} + \begin{bmatrix} 0 \\ \mathbf{x} \end{bmatrix}.$$

Note that a unitary matrix Q is always related to the orthonormal polynomials φ_k by

$$Q = W \Phi,$$

where

$$\Phi = \Phi_m = \begin{bmatrix} \varphi_0(z_1) & \cdots & \varphi_{m-1}(z_1) \\ \vdots & & \vdots \\ \varphi_0(z_m) & \cdots & \varphi_{m-1}(z_m) \end{bmatrix}$$

since

$$Q^H Q = \Phi^H W^H W \Phi = I_m.$$

12.5 Updating algorithm

In this section, we develop an updating algorithm to compute the extended Hessenberg matrix, i.e., a matrix whose first column contains the coefficients of the least squares polynomial approximants, whose second column contains the transformed weight vector $Q^H \mathbf{w} = \alpha \mathbf{e}_1$, and whose other columns consist of the upper Hessenberg matrix $H = [h_{ij}]$. The updating algorithm will add a new node z_{m+1} and a corresponding weight w_{m+1} and function value $f(z_{m+1})$ and compute the updated extended Hessenberg matrix.

For the least squares problem, we add the function values $f(z_k)$ and when these are properly transformed by the similarity transformations of the inverse QR-algorithm, this will result in the generalized Fourier coefficients of the approximant and some information about the corresponding residual. Indeed, the solution of the approximation problem is given by

$$p = [\varphi_0, \dots, \varphi_n] \mathbf{a}^\Phi, \quad \mathbf{a}^\Phi = \Phi^H W^H W \mathbf{f} = Q^H W \mathbf{f}.$$

Note that the normal equations are never explicitly formed.

The whole scheme can be collected in one table giving the relations

$$Q^H \left[\mathbf{w}_0 \mid \mathbf{w}_1 \parallel Z \right] \begin{bmatrix} I_2 & \\ & Q \end{bmatrix}$$

$$= \begin{bmatrix} & \alpha & h_{00} & \dots & h_{0,m-2} & h_{0,m-1} \\ \mathbf{a}^\Phi & 0 & h_{10} & \dots & h_{1,m-2} & h_{1,m-1} \\ -- & \vdots & & \ddots & \vdots & \vdots \\ \mathbf{x} & 0 & & & h_{m-1,m-2} & h_{m-1,m-1} \end{bmatrix},$$

with $\mathbf{w}_0 = W\mathbf{f}$ and $\mathbf{w}_1 = [w_1, \dots, w_m]^T$ as before. The approximation error is $\|\mathbf{x}\|$. For further reference we shall refer to the matrix of the right-hand side as the extended Hessenberg matrix.

The updating algorithm works as follows. Suppose that $\mathbf{a}^\Phi$ was computed by the last scheme for some data set $\{z_i, f_i, w_i\}_1^m$. We then end up with a scheme of the following form ($n = 3, m = 6$):

$$\begin{array}{cc|cccccc}
\times & \times & \times & \times & \times & \times & \times & \times \\
\times & & \times & \times & \times & \times & \times & \times \\
\times & & & \times & \times & \times & \times & \times \\
\times & & & & \times & \times & \times & \times \\
\hline
\times & & & & & \times & \times & \times \\
\times & & & & & & \times & \times \\
\end{array}$$

A new data triple $(z_{m+1}, f_{m+1}, w_{m+1})$ can be added, for example, in the top line. Note that with respect to the scheme above, a new first row and a new third column are inserted. The three crosses in the top line of the left scheme below represent $w_{m+1} f_{m+1}$, w_{m+1} and z_{m+1}, respectively. The other crosses correspond to the ones we had in the previous scheme.

$$\begin{array}{ccc|cccccc}
\times & \times & \times & & & & & & \\
\times & \times & & \times & \times & \times & \times & \times & \times \\
\times & & & \times & \times & \times & \times & \times & \times \\
\times & & & & \times & \times & \times & \times & \times \\
\times & & & & & \times & \times & \times & \times \\
\times & & & & & & \times & \times & \times \\
\times & & & & & & & \times & \times \\
\end{array} \Rightarrow \begin{array}{cc|ccccccc}
\times & \times & \times & \times & \times & \times & \times & \times & \times \\
\times & & \times & \times & \times & \times & \times & \times & \times \\
\times & & & \times & \times & \times & \times & \times & \times \\
\times & & & & \times & \times & \times & \times & \times \\
\times & & & & & \times & \times & \times & \times \\
\times & & & & & & \times & \times & \times \\
\end{array}$$

This left scheme has to be transformed by unitary similarity transformations into the right scheme which has the same form as the original one but with one extra row and one extra column. This result is obtained by eliminating the (2,2) element by an elementary rotation/reflection in the plane of the first two rows. The corresponding transformation on the columns will influence columns 3 and 4 and will introduce a nonzero element at position (3,3), which should not be there. This is eliminated by a rotation/reflection in the plane of rows 2 and 3, etc. We call this procedure chasing the elements down the diagonal. In the first column of the result, we find above the horizontal line the updated coefficients $\mathbf{a}^\Phi$. When we do not change n, it is sufficient to perform only the operations that influence these coefficients. Thus we could have stopped after we obtained the form

$$
\left[
\begin{array}{cc|ccccccc}
\times & \times & \times & \times & \times & \times & \times & \times & \times \\
\times & & \times & \times & \times & \times & \times & \times & \times \\
\times & & & \times & \times & \times & \times & \times & \times \\
\times & & & & \times & \times & \times & \times & \times \\
\hline
\times & & & & & \times & \times & \times & \times \\
\times & & & & & \times & \times & \times & \times \\
\times & & & & & & & \times & \times \\
\end{array}
\right]
$$

This can be done with $\mathcal{O}(n^2)$ operations per new data point. In the special case of data on the real line or on the unit circle, this reduces to $\mathcal{O}(n)$ operations as we shall see in the next section.

12.6 Special cases

The recurrence relations as well as the algorithm described above simplify considerably when the orthogonal polynomials satisfy a particular recurrence relation.

A classical situation occurs when the $z_i \in \mathbb{R}$, $i = 1, 2, \ldots, m$. Since we can also choose the weights w_i real, the Q and H matrix will be real, which means that we can drop the complex conjugation from our notation. Thus we observe that for $z_i \in \mathbb{R}$, the Hessenberg matrix H satisfies

$$H^T = (Q^T Z Q)^T = Q^T Z Q = H.$$

This means that H is symmetric and therefore tridiagonal. The matrix H reduces to the classical Jacobi matrix, i.e., a tridiagonal matrix whose subdiagonal elements are nonzero:

$$
H = \begin{bmatrix}
a_0 & b_1 & & \\
b_1 & a_1 & \ddots & \\
& \ddots & \ddots & b_m \\
& & b_m & a_m
\end{bmatrix},
$$

containing the coefficients of the three term recurrence relation

$$\varphi_{-1} = 0, \quad z\varphi_k(z) = b_k\varphi_{k-1}(z) + a_k\varphi_k(z) + b_{k+1}\varphi_{k+1}(z), \quad k = 0, 1, \ldots, m-2.$$

A similar situation occurs when the z_i are purely imaginary, in which case the matrix H is skew Hermitian. We shall not discuss this case separately.

The algorithm we described before now needs to perform rotations (or reflections) on vectors of length 3 or 4, which reduces the complexity of the algorithm by an order. This is the basis of the Rutishauser-Gragg-Harrod algorithm [135, 97]. See also [25, 132, 71].

In this context, it was observed only lately [96, 100, 8, 134, 9] that also the situation where the $z_i \in \mathbb{T}$ (the unit circle) leads to a simplification. It follows from

$$H^H H = Q^H Z^H Z Q = Q^H Q = I_m$$

that H is a unitary Hessenberg matrix. The related orthogonal polynomials are orthogonal with respect to a discrete measure supported on the unit circle. The three-term recurrence relation is replaced by a recurrence of Szegő-type

$$z\varphi_{k-1}(z) = \varphi_k(z)\sigma_k + \varphi_{k-1}^H(z)\gamma_k$$

with

$$\varphi_k^H(z) = z^k\overline{\varphi_k(1/\bar{z})} \in \mathbb{P}_k \quad \text{and} \quad \sigma_k^2 = 1 - |\gamma_k|^2, \quad \sigma_k > 0,$$

where the γ_k are the so-called reflection coefficients or Schur parameters. Just like in the case of a tridiagonal matrix, the Hessenberg matrix is built up from the recurrence coefficients γ_k, σ_k. However, the connection is much more complicated. For example, for $m = 4$, H has the form

$$H = \begin{bmatrix} -\gamma_1 & -\sigma_1\gamma_2 & -\sigma_1\sigma_2\gamma_3 & -\sigma_1\sigma_2\sigma_3\gamma_4 \\ \sigma_1 & -\bar{\gamma}_1\gamma_2 & -\bar{\gamma}_1\sigma_2\gamma_3 & -\bar{\gamma}_1\sigma_2\sigma_3\gamma_4 \\ & \sigma_2 & -\bar{\gamma}_2\gamma_3 & -\bar{\gamma}_2\sigma_3\gamma_4 \\ & & -\sigma_3 & \bar{\gamma}_3\gamma_4 \end{bmatrix}.$$

The Schur parameters can be recovered from the Hessenberg matrix by

$$\sigma_j = h_{j,j-1}, \quad j = 1, \ldots, m, \quad \alpha = 1/\varphi_0 = \sigma_0,$$

$$\gamma_j = -h_{0,j-1}/(\sigma_1\sigma_2 \ldots \sigma_{j-1}), \quad j = 1, \ldots, m.$$

The complexity reduction in the algorithm is obtained from the important observation that any unitary Hessenberg matrix H can be written as a product of elementary unitary factors

$$H = G_1 G_2 \ldots G_{m-1} G_m'$$

with

$$G_k = I_{k-1} \oplus \begin{bmatrix} -\gamma_k & \sigma_k \\ \sigma_k & \bar{\gamma}_k \end{bmatrix} \oplus I_{m-k-1}, \quad k = 1, \ldots, m$$

and

$$G_m' = \text{diag}([1, \ldots, 1, -\gamma_m]).$$

This result can be found e.g., in [96, 8].

Now an elementary similarity transformation on rows/columns k and $k + 1$ of H, represented in this factored form, will only affect the factors G_k and part of the factors G_{k-1} and G_{k+1}. Again, these operations require computations on short vectors of length 3, making the algorithm very efficient again. For the details consult [96, 8, 134]. For example, the interpolation problem ($n = m$) is solved in $\mathcal{O}(m^2)$ operations instead of $\mathcal{O}(m^3)$.

Notes and references

This chapter and the next chapter are mainly based on the following article.

☞ A. Bultheel and M. Van Barel. Vector orthogonal polynomials and least squares approximation. *SIAM Journal on Matrix Analysis and Applications*, 16(3):863–885, 1995.

The field of orthonormal functions and, more specifically, orthonormal polynomials is very huge with a lot of connections to other domains of mathematics and its applications. Therefore, we do not intend to give a complete overview of the literature on this topic. For orthogonal polynomials and their computation as such we refer the interested reader to the following two books.

☞ W. Gautschi. *Orthogonal Polynomials: Computation and Approximation*. Oxford University Press, New York, USA, 2004.

☞ B. Simon. *Orthogonal Polynomials on the Unit Circle: Part 1: Classical Theory; Part 2: Spectral Theory*, volume 54 of *Colloquium Publications*. American Mathematical Society, Providence, Rhode Island, USA, 2004.

☞ G. Szegő. *Orthogonal Polynomials*. American Mathematical Society, Providence, Rhode Island, USA, fourth edition, 1975.

For more information on orthogonal polynomials connected to structured matrices, the reader can have a look at the following references.

☞ A. Bultheel, A. M. Cuyt, W. Van Assche, M. Van Barel, and B. Verdonk. Generalizations of orthogonal polynomials. *Journal of Computational and Applied Mathematics*, 179(1–2):57–95, July 2005.

☞ A. Bultheel and M. Van Barel. *Linear Algebra, Rational Approximation and Orthogonal Polynomials*, volume 6 of *Studies in computational mathematics*. North-Holland, Elsevier Science B.V., Amsterdam, Netherlands, 1997.

For a survey of inverse spectral problems, we refer to the following article of Boley and Golub and the book of Chu and Golub.

☞ D. L. Boley and G. H. Golub. A survey of matrix inverse eigenvalue problems. *Inverse Problems*, 3:595–622, 1987.

☞ M. T. Chu and G. H. Golub. *Inverse Eigenvalue Problems: Theory, Algorithms and Applications*. Numerical Mathematics & Scientific Computations. Oxford University Press, New York, USA, 2005.

One of the methods mentioned is the Rutishauser-Gragg-Harrod algorithm. This algorithm can be traced back to Rutishauser and was adapted by Gragg an Harrod with a technique of Kahan, Pal and Walker for chasing a nonzero element in the matrix. The article by Reichel discusses a discrete least squares interpretation of these algorithms. The numerical stability of the inverse eigenvalue algorithm on the real line, is discussed by Reichel and Gragg and Harrod.

☞ W. B. Gragg and W. J. Harrod. The numerically stable reconstruction of Jacobi matrices from spectral data. *Numerische Mathematik*, 44:317–335, 1984.

☞ L. Reichel. Fast QR-decomposition of Vandermonde-like matrices and polynomial least squares approximation. *SIAM Journal on Matrix Analysis and Applications*, 12:552–564, 1991.

☞ H. Rutishauser. On Jacobi rotation patterns. In N. C. Metropolis, A. H. Taub, J. Todd, and C. B. Tompkins, editors, *Experimental Arithmetics, High Speed Computing and Mathematics, Proceedings of Symposia in Applied Mathematics*, volume 15, pages 219–239. American Mathematical Society, Providence, Rhode Island, 1963.

When the z_i are not on the real line but on the unit circle, similar ideas lead to algorithms discussed by Ammar and He and Ammar,Gragg and Reichel for the inverse eigenvalue problem and to Reichel, Ammar and Gragg for a least squares interpretation.

☞ G. S. Ammar, W. B. Gragg, and L. Reichel. Constructing a unitary Hessenberg matrix from spectral data. In G. H. Golub and P. Van Dooren, editors, *Numerical Linear Algebra, Digital Signal Processing and Parallel Algorithms*, volume 70 of *Computer and Systems Sciences*, pages 385–395. Springer-Verlag, Berlin, Germany, 1991.

☞ G. S. Ammar and C. He. On an inverse eigenvalue problem for unitary Hessenberg matrices. *Linear Algebra and its Applications*, 218:263–271, 1995.

☞ L. Reichel, G. S. Ammar, and W. B. Gragg. Discrete least squares approximation by trigonometric polynomials. *Mathematics of Computation*, 57:273–289, 1991.

It is to be preferred over the so-called Stieltjes procedure.

☞ W. Gautschi. On generating orthogonal polynomials. *SIAM Journal on Scientific and Statistical Computation*, 3:289–317, 1982.

☞ W. Gautschi. Computational problems and applications of orthogonal polynomials. In C. Brezinski, L. Gori, and A. Ronveaux, editors, *Orthogonal Polynomials and Their Applications*, volume 9 of *IMACS Annals on Computing and Applied Mathematics*, pages 61–71. Baltzer Science Publishers, Basel, Switzerland, 1991.

Moreover the algorithms for the real line as well as for the unit circle are well suited for implementation in a pipelined fashion on a parallel architecture.

☞ M. Van Barel and A. Bultheel. A parallel algorithm for discrete least squares rational approximation. *Numerische Mathematik*, 63:99–121, 1992.

☞ M. Van Barel and A. Bultheel. Discrete linearized least squares approximation on the unit circle. *Journal of Computational and Applied Mathematics*, 50:545–563, 1994.

In Section 12.5, we only considered the problem of updating, i.e., how to adapt the approximant when one knot is added to the set of data. There also exists a possibility to consider downdating, i.e., when one knot is removed from the set of interpolation points. For the polynomial approximation problem, this was discussed for real data in the following article.

☞ S. Elhay, G. H. Golub, and J. Kautsky. Updating and downdating of orthogonal polynomials with data fitting applications. *SIAM Journal on Matrix Analysis and Applications*, 12(2):327–353, 1991.

For data on the unit circle the downdating procedure is described in the following article.

☞ G. S. Ammar, W. B. Gragg, and L. Reichel. Downdating of Szegő polynomials and data-fitting applications. *Linear Algebra and its Applications*, 172:315–336, 1992.

The procedure can be based on a direct QR-algorithm, which will "diagonalize" the Hessenberg matrix in one row and column (e.g., the last one). This means that the only nonzero element in the last row and the last column of the transformed Hessenberg matrix is z_m on the diagonal. The unitary similarity transformations on the rest of the extended Hessenberg matrix bring out the corresponding weight in its first columns and the leading $m - 1 \times m - 1$ part gives the solution for the downdated problem.

The discrete least squares problem of Section 12.4 is closely related to many other problems in numerical analysis. For example, consider the quadrature formula

$$\int_a^b w(x)f(x)dx \approx \sum_{k=1}^m w_k^2 f(z_k),$$

where $w(x)$ is a positive weight for the real interval $[a, b]$. We get a Gaussian quadrature formula, exact for all polynomials of degree $2m - 1$ by a special choice of the nodes and weights. The nodes z_k are the zeros of the m-th orthogonal polynomial with respect to the inner product $\langle f, g \rangle = \int_a^b f(x)g(x)w(x)dx$. These are also the eigenvalues of the truncated Jacobi matrix which is associated with this orthogonal system. The weights w_i^2 are proportional to q_{i1}^2 where q_{i1} is the first component of the corresponding eigenvector.

☞ G. H. Golub and J. H. Welsch. Calculation of gauss quadrature rules. *Mathematics of Computation*, 23:221–230, 1969.

The inverse eigenvalue problem as studied in Section 12.3 is precisely the inverse of the previous quadrature problem: find the Jacobi matrix, i.e., the unreduced symmetric tridiagonal matrix, when its eigenvalues and the first entries of the normalized eigenvectors are given. The computation of the quadrature formula or the eigenvalue decomposition of the Jacobi matrix are called direct problems, while the inverse spectral problem, and the least squares problem are called inverse problems.

The following book by Chu and Golub discusses extensively inverse problems.

☞ M. T. Chu and G. H. Golub. *Inverse Eigenvalue Problems: Theory, Algorithms and Applications*. Numerical Mathematics & Scientific Computations. Oxford University Press, New York, USA, 2005.

12.7 Conclusions

In this chapter, we have seen that orthogonal polynomials satisfy a recurrence relation corresponding to an upper Hessenberg matrix. This matrix is symmetric tridiagonal for an inner product on the real line and unitary upper Hessenberg for an inner product on the unit circle. To determine these matrices for a discrete inner product we have designed an updating procedure to solve the corresponding inverse eigenvalue problem.

Chapter 13

Orthonormal polynomial vectors

In the previous chapter we have studied orthonormal polynomials and least squares polynomial approximants with respect to a discrete inner product. In this chapter, we investigate the generalization into orthonormal polynomial vectors. In this case the parameters of the recurrence relation are contained in a matrix with a structure that generalizes the upper Hessenberg structure. The details of the structure of the (extended) matrix determine the degree structure of the sequence of orthonormal polynomial vectors.

The chapter is organized as follows. In Section 13.1, we introduce the concept of a polynomial vector approximant with respect to a discrete norm. It will turn out that this approximant is a polynomial vector orthogonal to all the polynomial vectors of 'smaller degree'. In Section 13.2, we first study the case when the degrees of the polynomial elements of the vector approximant are equal. Section 13.3 handles the general case. Section 13.4 investigates when the recurrence relation for the sequence of orthonormal polynomial vectors breaks down.

✎ *Mastering the content of this chapter is not important to understand the following chapter of the book. The essential ideas of this chapter are the following. Orthonormal polynomial vectors satisfy the orthonormality relations (13.5). These polynomial vectors satisfy a recurrence relation. The coefficients of the recurrence relation of orthonormal polynomial vectors can be summarized into a generalization of a Hessenberg matrix, see Subsection 13.3.2. In the beginning of this subsection, it is described how this general Hessenberg matrix can be recovered by solving an inverse eigenvalue problem. This problem can be solved an order of magnitude faster when the points of the inner product are lying all on the real line or all on the unit circle.*

13.1 Vector approximants

In this section, we will generalize the polynomial least squares approximation problem into a least squares problem involving multiple polynomials. In Section 12.4 of the previous chapter, we have studied the polynomial least squares approximation problem: find the polynomial $p \in \mathbb{P}_n$ of degree at most $n \leq m$ that minimizes the error

$$\sum_{i=0}^{m} |w_i(f(z_i) - p(z_i))|^2. \tag{13.1}$$

This can be generalized into the following approximation problem: find two polynomials $p_0 \in \mathbb{P}_{d_0}$ and $p_1 \in \mathbb{P}_{d_1}$ minimizing the error:

$$\sum_{i=0}^{m} |w_{0,i} f_{0,i} p_0(z_i) - w_{1,i} f_{1,i} p_1(z_i)|^2. \tag{13.2}$$

The solution having the error as small as possible is the zero solution $[p_0, p_1] = [0, 0]$. To avoid this trivial solution, we normalize the polynomial vector $[p_0, p_1]$, e.g., by putting the extra condition, that p_0 has to be monic and of strict degree d_0. Note that when we take $w_{0,i} f_{0,i} = w_i f(z_i)$, $w_{1,i} f_{1,i} = w_i$, and p_0 monic of strict degree 0, we get the same error as in (13.1). Instead of considering two polynomials, the previous situation of polynomial approximation can be generalized as follows.

Given $\{z_i; f_{0i}, \ldots, f_{\alpha i}; w_{0i}, \ldots, w_{\alpha i}\}_{i=0}^{m}$, find polynomials $p_k \in \mathbb{P}_{d_k}$, $k = 0, \ldots, \alpha$, such that

$$\sum_{i=0}^{m} |w_{0i} f_{0i} p_0(z_i) + \cdots + w_{\alpha i} f_{\alpha i} p_\alpha(z_i)|^2$$

is minimized. Now it doesn't really matter whether the w_{ji} are positive or not, since the products $w_{ji} f_{ji}$ will now play the role of the weights and the f_{ji} are arbitrary complex numbers. Thus, to simplify the notation, we could as well write w_{ji} instead of $w_{ji} f_{ji}$ since these numbers will always appear as products. Thus the problem is to minimize

$$\sum_{i=0}^{m} |w_{0i} p_0(z_i) + \cdots + w_{\alpha i} p_\alpha(z_i)|^2.$$

Setting $\mathbf{d} = [d_0, \ldots, d_\alpha]^T$, $\mathbb{P}_{\mathbf{d}} = [\mathbb{P}_{d_0}, \ldots, \mathbb{P}_{d_\alpha}]^T$,

$$\mathbf{w}_i = [w_{0i}, \ldots, w_{\alpha i}]^T, \quad \mathbf{p}(z) = [p_0(z), \ldots, p_\alpha(z)]^T \in \mathbb{P}_{\mathbf{d}},$$

we can write this as

$$\min \sum_{i=0}^{m} |\mathbf{w}_i^H \mathbf{p}(z_i)|^2, \quad \mathbf{p} \in \mathbb{P}_{\mathbf{d}}.$$

Of course, this problem has the trivial solution $\mathbf{p} = 0$, unless we require at least one of the $p_i(z)$ to be of strict degree d_i, e.g., by making it monic. This, or any other normalization condition, could be imposed for that matter.

We require in this chapter that p_α is monic of degree d_α, and rephrase this as $p_\alpha \in \mathbb{P}_{d_\alpha}^M$.

To explain the general idea, we restrict ourselves to $\alpha = 1$, the case of a general α being a straightforward generalization which would only increase the notational burden. Thus we consider the problem

$$\min \sum_{i=0}^{m} |w_{0i}p_0(z_i) + w_{1i}p_1(z_i)|^2, \quad p_0 \in \mathbb{P}_{d_0}, p_1 \in \mathbb{P}_{d_1}^M. \tag{13.3}$$

Note that when $w_{0i} = w_i > 0$, $w_{1i} = -w_i f_i$, and $p_1 = 1 \in \mathbb{P}_0^M$, (i.e., $d_1 = 0$), then we get the polynomial approximation problem discussed before.

When we set $w_{0i} = w_i f_{0i}$ and $w_{1i} = -w_i f_{1i}$ with $w_i > 0$, the problem becomes

$$\min \sum_{i=0}^{m} w_i^2 |f_{0i}p_0(z_i) - f_{1i}p_1(z_i)|^2$$

which is a linearized version of the rational least squares problem of determining the rational approximant p_0/p_1 for the data f_{1i}/f_{0i}, or equivalently the rational approximant p_1/p_0 for the data f_{0i}/f_{1i}. Note that in the linearized form, it is as easy to prescribe pole information ($f_{0i} = 0$) as it is to fix a finite function value ($f_{0i} \neq 0$).

The solution of the general case is partly parallel to the polynomial case $d_1 = 0$ discussed before, and partly parallel to another simple case, namely $d_0 = d_1 = n$, which we shall discuss first in Section 13.2. In the subsequent Section 13.3, we shall consider the general case where $d_0 \neq d_1$.

13.2   Equal degrees

In this section we consider the least squares approximation problem having only two polynomials with equal degrees. In Subsection 13.2.1, we solve this problem using orthogonal polynomial 2×2 matrices. In Subsection 13.2.2, it is shown that the recurrence relation for these orthonormal polynomial matrices leads to an inverse eigenvalue problem. Also, the algorithm to solve this inverse eigenvalue problem is developed in this subsection. Subsection 13.2.3 summarizes the results for equal degrees. In the next section, we will handle the case of unequal degrees.

In this section, we consider the case $\alpha = 1$, $d_0 = d_1 = n$. This means that $\mathbb{P}_\mathbf{d}$ is here equal to $\mathbb{P}_n^{2 \times 1}$.

13.2.1   The optimization problem

We have to find

$$\min \sum_{i=0}^{m} |\mathbf{w}_i^H \mathbf{p}(z_i)|^2, \quad p_0 \in \mathbb{P}_n, \quad p_1 \in \mathbb{P}_n^M$$

where $\mathbf{w}_i = [w_{0i} \quad w_{1i}]^T$ and $\mathbf{p}(z) = [p_0(z) \quad p_1(z)]^T \in \mathbb{P}_n^{2 \times 1}$. This problem was considered in [147, 148]. We propose a solution of the form

$$\mathbf{p}(z) = \sum_{k=0}^{n} \varphi_k(z)\mathbf{a}_k,$$

where
$$\varphi_k(z) \in \mathbb{P}_k^{2\times2} - \mathbb{P}_{k-1}^{2\times2}, \quad \mathbf{a}_k \in \mathbb{C}^{2\times1}, \quad k = 0, 1, \ldots, n.$$

Proposing $\mathbf{p}(z)$ to be of this form assumes that the leading coefficients of the block polynomials φ_k[1] are nonsingular. Otherwise this would not represent all possible couples of polynomials $[p_0, p_1]^T \in \mathbb{P}_n^{2\times1}$. We shall call this the regular case and assume for the moment that we are in this comfortable situation. In the singular case, a breakdown may occur during the algorithm, and we shall deal with that separately. Note that the singular case did not show up in the previous scalar polynomial case, unless at the very end when $n = m + 1$, since the weights were assumed to be positive. We shall see below that in this block polynomial situation, the weights are not positive and could even be singular.

When we denote
$$W = \mathrm{diag}([\mathbf{w}_0^T, \ldots, \mathbf{w}_m^T]) \in \mathbb{C}^{(m+1)\times(2m+2)}$$
$$\mathbf{a}_n^\Phi = [\mathbf{a}_0^T, \ldots, \mathbf{a}_n^T]^T \in \mathbb{C}^{(2n+2)\times1}$$
$$\Phi_n = \begin{bmatrix} \varphi_0(z_0) & \cdots & \varphi_n(z_0) \\ \vdots & & \vdots \\ \varphi_0(z_m) & \cdots & \varphi_n(z_m) \end{bmatrix} \in \mathbb{C}^{(2m+2)\times(2n+2)}$$

the optimization problem is to find the least squares solution of the homogeneous linear system
$$W\Phi_n\mathbf{a}_n^\Phi = 0$$

with the constraint that p_1 should be monic of degree n.

For simplicity reasons, suppose that $m + 1 = 2(m' + 1)$ is even. If it were not, we would have to make a modification in our formulations for the index m'. The algorithm, however, does not depend on m being odd or even as we shall see later.

By making the block polynomials φ_k orthogonal so that

$$\sum_{i=0}^m \varphi_k(z_i)^H \mathbf{w}_i \mathbf{w}_i^H \varphi_l(z_i) = \delta_{kl} I_2, \quad k, l = 0, 1, \ldots, m', \tag{13.4}$$

we can construct a unitary matrix $Q \in \mathbb{C}^{(m+1)\times(m+1)}$ by setting

$$Q = W\Phi$$

where $\Phi = \Phi_{m'}$ is a $(2m + 2) \times (m + 1)$ matrix, so that Q is a square matrix of size $m + 1$. We also assume that the number of data points $m + 1$ is at least equal to the number of unknowns $2n + 1$ (recall that one coefficient is fixed by the monic normalization). The unitarity of the matrix Q means that

$$Q^H Q = \Phi^H W^H W \Phi = I_{m+1},$$

---

[1]Although in this section φ_k indicates a matrix, we didn't use a captial letter here. This is done because we want to use the same notation for the scalar orthonormal polynomials as well as the block orthonormal polynomials appearing in Subsection 13.3.3.

and the optimization problem reduces to

$$\min \sum_{i=0}^{m} \mathbf{p}(z_i)^H \mathbf{w}_i \mathbf{w}_i^H \mathbf{p}(z_i) = \min (\mathbf{a}_{m'}^{\Phi})^H \Phi^H W^H W \Phi (\mathbf{a}_{m'}^{\Phi})$$

$$= \min (\mathbf{a}_{m'}^{\Phi})^H (\mathbf{a}_{m'}^{\Phi})$$

$$= \min \sum_{k=0}^{m'} \mathbf{a}_k^H \mathbf{a}_k$$

$$= \min \sum_{k=0}^{m'} (|a_{1k}|^2 + |a_{2k}|^2), \quad \mathbf{a}_k = [a_{1k} \ a_{2k}]^T,$$

with the constraint that $\mathbf{p}(z) \in \mathbb{P}_n^{2 \times 1}$; thus $\mathbf{a}_{n+1} = \cdots = \mathbf{a}_{m'} = 0$, and $p_1 \in \mathbb{P}_n^M$. Since the leading term of p_1 is only influenced by $\varphi_n \mathbf{a}_n$, we are free to choose $\mathbf{a}_0, \ldots, \mathbf{a}_{n-1}$, so that we can set them equal to zero, to minimize the error. Thus it remains to find

$$\min(|a_{1n}|^2 + |a_{2n}|^2)$$

such that

$$\varphi_n(z) \begin{bmatrix} a_{1n} \\ a_{2n} \end{bmatrix} = \begin{bmatrix} p_0(z) \\ p_1(z) \end{bmatrix} \in \begin{bmatrix} \mathbb{P}_n \\ \mathbb{P}_n^M \end{bmatrix}.$$

To monitor the degree of p_1, we shall require that the polynomials φ_k have an upper triangular leading coefficient:

$$\varphi_k(z) = \begin{bmatrix} \alpha_k & \gamma_k \\ 0 & \beta_k \end{bmatrix} z^k + \cdots,$$

with $\alpha_k, \beta_k > 0$. Note that this is always possible in the regular case. The condition $p_1 \in \mathbb{P}_n^M$ then sets $a_{2n} = 1/\beta_n$ and a_{1n} is arbitrary, hence to be set equal to zero if we want to minimize the error.

As a conclusion, we have solved the approximation problem by computing the n-th block polynomial φ_n, orthonormal in the sense of (13.4) and with leading coefficient upper triangular. The solution is

$$\mathbf{p}(z) = \begin{bmatrix} p_0(z) \\ p_1(z) \end{bmatrix} = \varphi_n(z) \begin{bmatrix} 0 \\ a_{2n} \end{bmatrix}, \quad a_{2n} = 1/\beta_n.$$

13.2.2 The algorithm

In this subsection we develop the recurrence relation for the orthonormal polynomial 2×2 matrices φ_k. The coefficients for this recurrence relation can be summarized in a block upper Hessenberg matrix. To compute these recurrence coefficients, we also design an efficient algorithm solving the corresponding inverse eigenvalue problem.

As in the scalar polynomial case, see (12.6), expressing $z\varphi_k(z)$ in terms of $\varphi_0, \ldots, \varphi_{k+1}$ for $z \in \{z_0, \ldots, z_m\}$ leads to the matrix relation

$$\hat{Z} \Phi = \Phi H,$$

where as before $\Phi = \Phi_{m'}$, $Z = \mathrm{diag}([z_0, \ldots, z_m])$, H is a block upper Hessenberg matrix with 2×2 blocks and $\hat{Z} = Z \otimes I_2 = \mathrm{diag}([z_0 I_2, \ldots, z_m I_2])$. Note that in the scalar case, the matrix $\hat{Z}$ is diagonal and the matrix H is upper Hessenberg. If the leading coefficient of φ_k is upper triangular, then the subdiagonal blocks of H are upper triangular. The computational scheme to compute the recurrence coefficients can be summarized in the following formula

$$
Q^H[\mathbf{w}|Z]\begin{bmatrix} I_2 \\ & Q \end{bmatrix} = \begin{bmatrix} \eta_{00} & \eta_{01} & \cdots & \eta_{0m'} & \eta_{0,m'+1} \\ & \eta_{11} & & \eta_{1m'} & \eta_{1,m'+1} \\ & & \ddots & \vdots & \vdots \\ & & & \eta_{m'm'} & \eta_{m',m'+1} \end{bmatrix},
$$

where $\mathbf{w} = [w_0^T, \ldots, w_m^T]^T$ and where all $\eta_{ij}{}^2$ are 2×2 blocks and the η_{ii} are upper triangular with positive diagonal elements. Thus

$$
\varphi_0 = \eta_{00}^{-1}; \quad z\varphi_{k-1}(z) = \varphi_0(z)\eta_{0k} + \cdots + \varphi_k(z)\eta_{kk}, \quad k = 1, \ldots, m'.
$$

The updating after adding the data (z_{m+1}, w_{m+1}), where $w_{m+1} = (w_{0,m+1}, w_{1,m+1})$, makes the transformation with unitary similarity transformations from the left to the right scheme below. The three crosses in the top row of the left scheme represent the new data. So, to the original scheme a first row is added and a third column is inserted containing the new data. This scheme is then updated to the structure as indicated in the right scheme below.

Left scheme:

×	×	×						
×	×		×	×	×	×	×	×
		×	×	×	×	×	×	×
			×	×	×	×	×	×
				×	×	×	×	
					×	×	×	×
						×	×	×

$\Rightarrow$

Right scheme:

×	×	×	×	×	×	×	×	×
	×	×	×	×	×	×	×	×
		×	×	×	×	×	×	×
			×	×	×	×	×	×
				×	×	×	×	×
					×	×	×	×
						×	×	×

The successive elementary transformations eliminate the crosses on the subdiagonal, chasing them down the matrix. This example also illustrates what happens at the end when m is an even number: the polynomial $\varphi_{m'} \in \mathbb{P}_{m'}^{2 \times 1}$ instead of $\varphi_{m'} \in \mathbb{P}_{m'}^{2 \times 2}$. Again, when finishing this updating after φ_n has been computed, it will require only $\mathcal{O}(n^2)$ operations per data point introduced. In the special case of data on the real line or the unit circle, this reduces to $\mathcal{O}(n)$ operations. For the details, we refer to [147, 148].

By the same arguments as in the scalar case, it is still true that H is Hermitian, hence block tridiagonal, when all the z_i are real. Taking into account that the subdiagonal blocks are upper triangular, we obtain in this case that H is pentadiagonal

---

[2]Here each η_{ij} denotes a matrix. However, after updating the block upper Hessenberg matrix H, they can also indicate scalar or vector elements. This is the reason why we do not use a capital letter.

and the extended Hessenberg matrix has the form

$$
\begin{bmatrix}
B_0 & A_0 & B_1^H & & \\
 & B_1 & A_1 & \ddots & \\
 & & \ddots & \ddots & B_{m'}^H \\
 & & & B_{m'} & A_{m'}
\end{bmatrix}
$$

with the 2×2 blocks B_k upper triangular and the A_k Hermitian. This leads to the following block three-term recurrence

$$
\varphi_0 = B_0^{-1}, \quad z\varphi_k(z) = \varphi_{k-1}B_k^H + \varphi_k(z)A_k + \varphi_{k+1}(z)B_{k+1}, \quad 0 \le k < m'.
$$

This case was considered in [147].

Similarly, the case where all z_i lie on the unit circle $\mathbb{T}$, leads to a 2×2 block generalization of the corresponding polynomial case. For example, the extended unitary block Hessenberg matrix takes the form ($m' = 3$)

$$
[H_0|H] =
\begin{bmatrix}
\hat{\Sigma}_0 & -\hat{\Gamma}_1 & -\Sigma_1\hat{\Gamma}_2 & -\Sigma_1\Sigma_2\hat{\Gamma}_3 & \Sigma_1\Sigma_2\Sigma_3 \\
 & \hat{\Sigma}_1 & -\Gamma_1\hat{\Gamma}_2 & -\Gamma_1\Sigma_2\hat{\Gamma}_3 & \Gamma_1\Sigma_2\Sigma_3 \\
 & & \hat{\Sigma}_2 & -\Gamma_2\hat{\Gamma}_3 & \Gamma_2\Sigma_3 \\
 & & & \hat{\Sigma}_3 & \Gamma_3
\end{bmatrix}, \quad \hat{\Gamma}_i, \hat{\Sigma}_i, \Sigma_i, \Gamma_i \in \mathbb{C}^{2\times 2}.
$$

The matrices

$$
U_k = \begin{bmatrix} -\hat{\Gamma}_k & \Sigma_k \\ \hat{\Sigma}_k & \Gamma_k \end{bmatrix}
$$

are unitary: $U_k^H U_k = I_4$. Note that by allowing some asymmetry in the U_k we do not need a $-\hat{\Gamma}_4$ in the last column as we had in the scalar case. We have for $k = 1, \ldots, m'$, the block Szegő recurrence relations

$$
\varphi_k(z)\hat{\Sigma}_k = z\varphi_{k-1}(z) + \varphi'_{k-1}(z)\hat{\Gamma}_k
$$
$$
\varphi'_k(z)\Sigma_k^H = z\varphi_{k-1}(z)\hat{\Gamma}_k^H + \varphi'_{k-1}(z),
$$

which start with $\varphi_0 = \varphi'_0 = \hat{\Sigma}_0^{-1}$.

The block Hessenberg matrix can again be factored as

$$
H = G_1 G_2 \ldots G_{m'},
$$

with

$$
G_k = I_{2(k-1)} \oplus U_k \oplus I_{m-2k-1}, \quad k = 1, \ldots, m'.
$$

The proof of this can be found in [148]. This makes it possible to perform the elementary unitary similarity transformations of the updating procedure only on vectors of maximal length 5, very much like in the case of real points z_i. Thus also here, the complexity of the algorithm reduces to $\mathcal{O}(m^2)$ for interpolation. More details can be found in [148]. For the case of the real line, the algorithm was also discussed in [4], solving an open problem in [25, p. 615]. The previous procedure now solves the problem also for the case of the unit circle.

13.2.3 Summary

The case $\alpha = 1$, $d_0 = d_1 = n$ and also the case $\alpha \geq 1$, $d_0 = d_1 = \cdots = d_\alpha = n$ for that matter, generalizes the polynomial approximation problem by constructing orthonormal polynomials matrices φ_k which are $(\alpha+1) \times (\alpha+1)$ polynomial matrices and these are generated by a block three-term recurrence relation when all $z_i \in \mathbb{R}$ and by a block Szegő recurrence relation when all $z_i \in \mathbb{T}$.

The computational algorithm is basically the same, since it reduces the extended matrix

$$[\mathbf{w}|Z] \in \mathbb{C}^{(m+1)\times(\alpha+m+2)}$$

by a sequence of elementary unitary similarity transformations to an upper trapezoidal matrix

$$Q^H[\mathbf{w}|Z]\begin{bmatrix} I_{\alpha+1} & \\ & Q \end{bmatrix} = [H_0|H]$$

with H block upper Hessenberg with $(\alpha + 1) \times (\alpha + 1)$ blocks and

$$H_0 = Q^H \mathbf{w} = [\eta_{00}^T, 0, \ldots, 0]^T,$$

where $\eta_{00} \in \mathbb{C}^{(\alpha+1)\times(\alpha+1)}$ is upper triangular with positive diagonal elements, as well as all the subdiagonal blocks of H. For $n = m'$, where $(\alpha+1)(m'+1)-1 = m+1$, (which implies that $\Sigma_{m'}$ is of size $\alpha \times (\alpha + 1)$), we solve an interpolation problem. It requires $\mathcal{O}(m^2)$ operations when $z_i \in \mathbb{R}$ or $\in \mathbb{T}$, instead of $\mathcal{O}(m^3)$ when the z_i are arbitrary in $\mathbb{C}$.

13.3 Arbitrary degrees

In this section, we consider the case of two polynomials having unequal degree in general, i.e., $\alpha = 1$ with $d_0 \neq d_1$.

In Subsection 13.3.1, we repeat the polynomial vector approximation problem. Before looking at the recurrence relation for the orthonormal polynomial vectors, we develop the algorithm to solve the corresponding inverse eigenvalue problem in Subsection 13.3.2. The recurrence relation for the 2×1 orthonormal polynomial vectors is developed in Subsection 13.3.3. In Subsection 13.3.4, the polynomial vector approximation problem (13.3) is solved. For more details we refer to [149].

13.3.1 The problem

We suppose without loss of generality that $d_0 = \delta$ and $d_1 = n + \delta$, $n, \delta \geq 0$. We have to find once more

$$\min \sum_{i=0}^m |\mathbf{w}_i^T \mathbf{p}(z_i)|^2, \quad p_0 \in \mathbb{P}_\delta, \quad p_1 \in \mathbb{P}_{n+\delta}^M$$

with $\mathbf{w}_i = [w_{0i} \ w_{1i}]^T$ and $\mathbf{p}(z) = [p_0(z) \ p_1(z)]^T \in \mathbb{P}_\mathbf{d}$, $\mathbf{d} = (d_0, d_1)$. The polynomial approximation problem is recovered by setting $\delta = 0$. The case $d_0 = d_1 = \delta$ is recovered by setting $n = 0$. The simplest approach to the general problem is by starting with the algorithm. In the subsequent subsections, we propose a computational scheme involving unitary similarity transformations. Next we give an interpretation in terms of orthogonal polynomials and finally we solve the approximation problem.

13.3.2 The algorithm

Comparing the cases $\delta = 0$ and $n = 0$, we see that the algorithm applies a sequence of elementary unitary similarity transformations on an extended matrix

$$[\mathbf{w}|Z], \quad \mathbf{w} = [w_0, \ldots, w_m]^T, \quad Z = \mathrm{diag}([z_0, \ldots, z_m])$$

to bring it in the form of an extended (block) upper Hessenberg

$$Q^H[\mathbf{w}|Z] \begin{bmatrix} I_2 & \\ & Q \end{bmatrix} = [H_0|H].$$

When $n = 0$, the transformations were aimed at chasing down the elements of $[\mathbf{w}|Z]$ below the main diagonal, making $[H_0|H]$ upper triangular. Therefore H turned out to be block upper Hessenberg.

When $\delta = 0$, the transformations had much the same objective, but now, there was no attempt to eliminate elements from the first column of $\mathbf{w}$, only elements from the second column were pushed to the lower right part of the matrix. The matrix then turned out to be upper Hessenberg in a scalar sense.

The general case can be treated by an algorithm that combines both of these objectives. We start like in the polynomial case ($n = 0$), chasing only elements from the second column of $\mathbf{w}$. However, once we reached row $n + 1$, we start eliminating elements in the first column too.

Applying this procedure shows that the extended Hessenberg $[H_0|H]$ has the form

This means that the upper left part of H, of size $(n + 1) \times (n + 1)$, will be scalar upper Hessenberg as in the case $n = 0$, while the lower right part of size $(m - n + 2) \times (m - n + 2)$ has the block upper Hessenberg form of the case $\delta = 0$.

The updating procedure works as follows. Starting with (the new data are found in the first row)

$$
\begin{array}{|ccc|ccccccccccc|}
\hline
\times & \times & \times & & & & & & & & & & \\
\hline
\times & \otimes & & \times & \times & \times & \times & \times & \times & \times & \times & \times & \times \\
\times & & & \times & \times & \times & \times & \times & \times & \times & \times & \times & \times \\
\times & & & & \times & \times & \times & \times & \times & \times & \times & \times & \times \\
\times & & & & & \times & \times & \times & \times & \times & \times & \times & \times \\
\times & & & & & & \times & \times & \times & \times & \times & \times & \times \\
\hline
 & & & & & & & \times & \times & \times & \times & \times & \times \\
 & & & & & & & & \times & \times & \times & \times & \times \\
 & & & & & & & & & \times & \times & \times & \times \\
 & & & & & & & & & & \times & \times & \times \\
 & & & & & & & & & & & \times & \times \\
\hline
\end{array}
$$

the element $\otimes$ is chased down the diagonal by elementary unitary similarity transformations operating on two successive rows/columns until we reach the following scheme (where $\odot = 0$ and $\oslash$ and $\ominus$ are the last elements introduced which are in general nonzero):

$$
\begin{array}{|cc|ccccccccccc|}
\hline
\times & \times & \times & \times & \times & \times & \times & \times & \times & \times & \times & \times \\
\times & & \times & \times & \times & \times & \times & \times & \times & \times & \times & \times \\
\times & & & \times & \times & \times & \times & \times & \times & \times & \times & \times \\
\times & & & & \times & \times & \times & \times & \times & \times & \times & \times \\
\times & & & & & \times & \times & \times & \times & \times & \times & \times \\
\otimes & & & & & \ominus & \times & \times & \times & \times & \times & \times \\
\hline
 & & & & & \oslash & \times & \times & \times & \times & \times & \times \\
 & & & & & & \odot & \times & \times & \times & \times & \times \\
 & & & & & & & \times & \times & \times & \times & \times \\
 & & & & & & & & \times & \times & \times & \times \\
 & & & & & & & & & \times & \times & \times \\
\hline
\end{array}
$$

Now the element $\otimes$ in row $n + 1$ is eliminated by a rotation/reflection in the plane of this row and the previous one. The corresponding transformation on the columns will introduce a nonzero element at position $\odot$. Then $\oslash$ and $\odot$ are chased down the diagonal in the usual way until we reach the final situation

$$
\begin{array}{|cc|ccccc|cccccc|}
\hline
\times & \times & \times & \times & \times & \times & \times & \times & \times & \times & \times & \times & \times \\
\times & & \times & \times & \times & \times & \times & \times & \times & \times & \times & \times & \times \\
\times & & & \times & \times & \times & \times & \times & \times & \times & \times & \times & \times \\
\times & & & & \times & \times & \times & \times & \times & \times & \times & \times & \times \\
\times & & & & & \times & \times & \times & \times & \times & \times & \times & \times \\
\hline
 & & & & & \times & \times & \times & \times & \times & \times & \times & \times \\
 & & & & & & \times & \times & \times & \times & \times & \times & \times \\
 & & & & & & & \times & \times & \times & \times & \times & \times \\
 & & & & & & & & \times & \times & \times & \times & \times \\
 & & & & & & & & & \times & \times & \times & \times \\
 & & & & & & & & & & \times & \times & \times \\
\hline
\end{array}
$$

13.3.3 Orthogonal vector polynomials

The unitary matrix Q involved in the previous transformation was for the case $\delta = 0$ of the form $Q = W\Phi_m$ where W was a scalar diagonal matrix of the weights and Φ_m was the matrix with ij-element given by $\varphi_j(z_i)$, with φ_j the j-th orthonormal polynomial.

When $n = 0$, the situation we studied in the previous chapter, then $Q = W\Phi_{m'}$, where W is the block diagonal with blocks being the 2×1 "weights" $\mathbf{w}_i$ and $\Phi_{m'}$ is the block matrix with 2×2 blocks, where the ij-block is given by $\varphi_j(z_i)$, with φ_j the j-th block orthonormal polynomial.

For the general case, we shall have a mixture of both. For the upper left part of the H matrix, we have the scalar situation and for the lower right part we have the block situation.

To unify both situations, we turn to vector polynomials $\boldsymbol{\pi}_k$ of size 2×1. For the block part, we see a block polynomial φ_j as a collection of two columns and set

$$\varphi_j(z) = [\boldsymbol{\pi}_{2j-1}(z)|\boldsymbol{\pi}_{2j}(z)].$$

For the scalar part, we embed the scalar polynomial φ_j in a vector polynomial $\boldsymbol{\pi}_j$ by setting

$$\boldsymbol{\pi}_j(z) = \begin{bmatrix} 0 \\ \varphi_j(z) \end{bmatrix}.$$

In both cases, the orthogonality of the φ_j translates into the orthogonality relation

$$\sum_{i=0}^{m} \boldsymbol{\pi}_k(z_i)^H \mathbf{w}_i \mathbf{w}_i^H \boldsymbol{\pi}_l(z_i) = \delta_{kl} \tag{13.5}$$

for the vector polynomials $\boldsymbol{\pi}_k$. Let us apply this to the situation of the previous algorithm. For simplicity, we suppose that all $z_i \in \mathbb{R}$. For $z_i \in \mathbb{T}$, the situation is similar.

For column number $j = 0, 1, \ldots, n - 1$, we are in the situation of scalar orthogonal polynomials: $Q_{ij} = w_{1i}\varphi_j(z_i) = \mathbf{w}_i^H \boldsymbol{\pi}_j(z_i)$. Setting

$$[H_0|H] = \begin{bmatrix} \times & b_0 & a_0 & \bar{b}_1 & & \\ \vdots & & b_1 & a_1 & \ddots & \\ \times & & & \ddots & \ddots & \ddots \\ & & & & \ddots & \ddots \end{bmatrix}$$

we have for $j = 0, \ldots, n - 2$ the three-term recurrence relation

$$z\varphi_j(z) = \varphi_{j-1}(z)\bar{b}_j + \varphi_j(z)a_j + \varphi_{j+1}(z)b_{j+1}, \quad \varphi_{-1} = 0, \quad \varphi_0 = b_0^{-1}.$$

By embedding, this becomes

$$z\boldsymbol{\pi}_j(z) = \boldsymbol{\pi}_{j-1}(z)\bar{b}_j + \boldsymbol{\pi}_j(z)a_j + \boldsymbol{\pi}_{j+1}(z)b_{j+1}, \quad \boldsymbol{\pi}_0 = [0 \ \varphi_0]^T.$$

Thus, setting
$$\hat{\boldsymbol{\pi}}_j = [\boldsymbol{\pi}_j(z_0)^T, \ldots, \boldsymbol{\pi}_j(z_m)^T]^T$$
we have for the columns $\mathbf{q}_j$ of Q the equality
$$\mathbf{q}_j = W\hat{\boldsymbol{\pi}}_j, \quad j = 0, 1, \ldots, n-1.$$

For the trailing part of Q, i.e., for columns $(n + 2j - 1, n + 2j)$, $j = 0, 1, \ldots$, we are in the block polynomial case. The block polynomials $\varphi_j(z)$ group two vector polynomials
$$\varphi_j(z) = [\boldsymbol{\pi}_{n+2j-1}(z)|\boldsymbol{\pi}_{n+2j}(z)],$$
which correspond to two columns of Q, namely
$$\hat{Q}_j = [\mathbf{q}_{n+2j-1}|\mathbf{q}_{n+2j}].$$

Observe that we have the following relation between $\hat{Q}_j$ and the block orthogonal polynomials
$$\hat{Q}_{ij} = w_i^H \varphi_j(z_i) = \begin{bmatrix} q_{2i,n+2j-1} & q_{2i,n+2j} \\ q_{2i+1,n+2j-1} & q_{2i+1,n+2j} \end{bmatrix},$$
where this time $\mathbf{w}_i^T = [w_{0i} \quad w_{1i}]^T$. As above, denote the vector of function values for $\boldsymbol{\pi}_j$ by $\hat{\boldsymbol{\pi}}_j$. The block column of function values for φ_j is denoted by $\hat{\Phi}_j$. Then clearly
$$\hat{Q}_j = W\hat{\Phi}_j, \quad \hat{\Phi}_j = [\hat{\boldsymbol{\pi}}_{n+2j-1}|\hat{\boldsymbol{\pi}}_{n+2j}].$$

Denoting in the extended Hessenberg matrix
$$[H_0|H] = \begin{bmatrix} \times & b_0 & \ddots & & \ddots & \\ \times & & \ddots & & \ddots & B_0^T \\ 0 & & & B_0 & A_0 & B_1^T \\ 0 & & & & B_1 & A_1 & \ddots \\ \vdots & & & & & \ddots & \ddots \end{bmatrix}, \quad B_0 = \begin{bmatrix} 0 & b_{n-1} \\ 0 & 0 \end{bmatrix},$$
we have the block recurrence
$$z\varphi_j(z) = \varphi_{j-1}(z)B_j^T + \varphi_j(z)A_j + \varphi_{j+1}(z)B_{j+1}, \quad j = 0, 1, \ldots$$

The missing link between the scalar and the block part is the initial condition for this block recurrence. This is related to columns $n-2, n-1$ and n of Q. Because columns $n-2$ and $n-1$ are generated by the scalar recurrence, we know that these columns are $\mathbf{q}_j = W\hat{\boldsymbol{\pi}}_j$, $j = n-2, n-1$, where the $\hat{\boldsymbol{\pi}}_j$ are related to the embedded scalar polynomials. A problem appears in column $\mathbf{q}_n$ where the three-term recurrence of the leading (scalar) part migrates to the block three-term recurrence of the trailing (block) part, i.e., from a three-term to a five-term scalar recurrence. We look at this column in greater detail. Because

$$
\begin{bmatrix} d_0 \\ \vdots \\ d_n \\ 0 \\ \vdots \\ 0 \end{bmatrix} = Q^H \begin{bmatrix} w_{00} \\ \vdots \\ \vdots \\ \vdots \\ \vdots \\ w_{0m} \end{bmatrix} = Q^H W \mathbf{e}, \quad \mathbf{e} = \begin{bmatrix} 1 \\ 0 \\ 1 \\ 0 \\ \vdots \\ 1 \\ 0 \end{bmatrix},
$$

we have

$$
\mathbf{q}_0 d_0 + \cdots + \mathbf{q}_n d_n = W\mathbf{e},
$$

thus

$$
\begin{aligned}
\mathbf{q}_n &= \frac{1}{d_n}(W\mathbf{e} - [\mathbf{q}_0|\ldots|\mathbf{q}_{n-1}]\mathbf{a}_{n-1}^{\Phi}), \quad \mathbf{a}_{n-1}^{\Phi} = [d_0,\ldots,d_{n-1}]^T \\
&= \frac{1}{d_n}(W\mathbf{e} - W[\hat{\boldsymbol{\pi}}_0|\ldots|\hat{\boldsymbol{\pi}}_{n-1}]\mathbf{a}_{n-1}^{\Phi}) \\
&= W\frac{1}{d_n}(\mathbf{e} - [\hat{\boldsymbol{\pi}}_0|\ldots|\hat{\boldsymbol{\pi}}_{n-1}]\mathbf{a}_{n-1}^{\Phi}) \\
&= W\frac{1}{d_n}(\mathbf{e} - \mathbf{p}_{n-1}), \quad \mathbf{p}_{n-1} = [\hat{\boldsymbol{\pi}}_0|\ldots|\hat{\boldsymbol{\pi}}_{n-1}]\mathbf{a}_{n-1}^{\Phi}.
\end{aligned}
$$

Setting $\mathbf{q}_n = W\hat{\boldsymbol{\pi}}_n$, $\hat{\boldsymbol{\pi}}_n = [\boldsymbol{\pi}_n(z_0)^T, \ldots, \boldsymbol{\pi}_n(z_m)^T]^T$, we find that

$$
\boldsymbol{\pi}_n(z) = \frac{1}{d_n}\begin{bmatrix} 1 \\ p_{n-1}(z) \end{bmatrix} \tag{13.6}
$$

where

$$
p_{n-1}(z) = \varphi_0(z)d_0 + \cdots + \varphi_{n-1}(z)d_{n-1}
$$

is the polynomial least squares approximant of degree $n-1$ for the data (z_i, w_i), $i = 0, \ldots, m$.

13.3.4   Solution of the general approximation problem

Now we are ready to solve the general polynomial vector approximation problem (13.3). We start with the degree structure of the polynomial vectors $\boldsymbol{\pi}_j(z)$. Suppose the j-th column of Q is $\mathbf{q}_j$, which we write as

$$
\mathbf{q}_j = W\hat{\boldsymbol{\pi}}_j, \quad \hat{\boldsymbol{\pi}}_j = [\boldsymbol{\pi}_j(z_0)^T, \ldots, \boldsymbol{\pi}_j(z_m)^T]^T
$$

with $W = \text{diag}([\mathbf{w}_0^H, \ldots, \mathbf{w}_m^H])$ and $\boldsymbol{\pi}_j(z) = [\psi_j(z) \quad \phi_j(z)]^T$. Then it follows from the previous analysis that the ϕ_j are the scalar orthogonal polynomials φ_j, and hence the degree of $\phi_j(z)$ is j, for $j = 0, 1, \ldots, n-1$. Moreover, the ψ_j are zero for the same indices (their degree is $-\infty$). For $j = n$, we just found that ψ_n is $1/d_n$, thus of degree 0 and ϕ_n is of degree at most $n-1$, since the latter is proportional to the polynomial least squares approximant of that degree. With the block recurrence relation, we now easily find that the degree structure of the block polynomials

$$\varphi_j = [\boldsymbol{\pi}_{n+2j-1} | \boldsymbol{\pi}_{n+2j}] = \begin{bmatrix} \psi_{n+2j-1} & \psi_{n+2j} \\ \phi_{n+2j-1} & \phi_{n+2j} \end{bmatrix} \quad \text{is} \quad \begin{bmatrix} j-1 & j \\ n+j-1 & n+j-1 \end{bmatrix}$$

for $j = 1, 2, \ldots$, while φ_0 has degree structure

$$\begin{bmatrix} -\infty & 0 \\ n-1 & n-1 \end{bmatrix}.$$

It can be checked that in the regular case, that is when all the subdiagonal elements $b_0, \ldots, b_{n-1}$ as well as d_n are nonzero and when also all the subdiagonal blocks $B_1, \ldots, B'_m$ are regular (upper triangular), then the degrees of $\phi_k = \varphi_k$ are precisely k for $k = 0, 1, \ldots, n-1$ and in the block polynomials φ_j, the entries ψ_{n+2j} and ϕ_{n+2j-1} have the precise degrees that are indicated, i.e., j and $n+j-1$, respectively. Thus, if we propose a solution to our approximation problem of the form (suppose $m \geq n + 2\delta$)

$$\mathbf{p}(z) = \sum_{j=0}^{n+2\delta+1} \boldsymbol{\pi}_j(z) a_j, \quad a_j \in \mathbb{C},$$

then $\mathbf{p}(z) = [p_0(z) \ p_1(z)]^T$ will automatically satisfy the degree restrictions $d_0 \leq \delta$ and $d_1 \leq n + \delta$. We have to find

$$\min(\mathbf{a}_n^{\Pi})^H \Pi_{n'}^H W^H W \Pi_{n'}(\mathbf{a}_n^{\Pi}), \quad n' = n + 2\delta + 1,$$

where

$$\mathbf{a}_n^{\Pi} = [a_0, \ldots, a_{n'}]^T \quad \text{and} \quad \Pi_{n'} = [\hat{\boldsymbol{\pi}}_0 | \ldots | \hat{\boldsymbol{\pi}}_{n'}].$$

Since $W \Pi_{n'}$ form the first $n' + 1$ columns of the unitary matrix Q, this reduces to

$$\min(\mathbf{a}_{n'}^{\Pi})^H (\mathbf{a}_{n'}^{\Pi}) = \min \sum_{j=0}^{n'} |a_j|^2.$$

If we require as before that $p_1(z)$ is monic of degree $n+\delta$, then $a_{n'} = 1/\beta_{n'}$ where β_j is the leading coefficient in ϕ_j. The remaining a_j are arbitrary. Hence, to minimize the error, we should make them all zero. Thus our solution is given by

$$\mathbf{p}(z) = \boldsymbol{\pi}_{n'}(z) a_{n'}, \quad n' = 2n + \delta + 1, \quad a_{n'} = 1/\beta_{n'}.$$

13.4   The singular case

In this section, we investigate the situation when the recurrence relation for the block orthogonal polynomials breaks down.

Let us start by considering the singular case for $d_0 = d_1 = n$. We shall then generate a singular subdiagonal block η_{kk} of the Hessenberg matrix. The algorithm performing the unitary similarity transformations will not be harmed by this situation. However, the sequence of block orthogonal polynomials will break down. From the relation

$$z\varphi_{k-1}(z) = \varphi_0(z)\eta_{0k} + \cdots + \varphi_k(z)\eta_{kk}$$

it follows that if η_{kk} is singular, then this cannot be solved for $\varphi_k(z)$. In the regular case, all the η_{jj} are regular and then the leading coefficient of φ_k is $\eta_{00}^{-1} \cdots \eta_{kk}^{-1}$. Thus, if all the η_{jj} are regular upper triangular, then the leading coefficient of φ_k will also be regular upper triangular. As we have said in the introduction, the singular situation will always occur, even in the scalar polynomial case with positive weights, but there only at the very end where $k = m + 1$. That is exactly the stage where we reached the situation where the least squares solution becomes the solution of an interpolation problem. We show below that this is precisely what will also happen when some premature breakdown occurs.

Suppose that the *scalar* entries of the extended block Hessenberg matrix are $[H_0|H] = [h_{ij}]_{i,j=0,1,\ldots}$. (We use h_{ij} to distinguish them from the block entries η_{ij}.) Suppose that the element h_{kk} is the first element on its diagonal that becomes zero and thus produces some singular subdiagonal block in H. Then it is no problem to construct the successive scalar columns of the matrix $\Phi = \Phi_{m'}$ until the recurrence relation hits the zero entry h_{kk}. If we denote for $j = 0, 1, \ldots, k-1$, the j-th column of Φ as $\hat{\boldsymbol{\pi}}_j$, then we know from what we have seen, that $\hat{\boldsymbol{\pi}}_j$ represents the vector of function values at the nodes $z_0, \ldots, z_m$ of some vector polynomial $\boldsymbol{\pi}_j(z) \in \mathbb{P}^{2 \times 1}$. The problem in the singular case is that $\boldsymbol{\pi}_k(z)$ cannot be solved from

$$z\boldsymbol{\pi}_{k-2} = \boldsymbol{\pi}_k h_{kk} + \boldsymbol{\pi}_{k-1}h_{k-1,k} + \cdots + \boldsymbol{\pi}_0 h_{0k}$$

because $h_{kk} = 0$. However, from

$$Q \left[\, H_0 \mid H \,\right] = \left[\, \mathbf{w}_0 \;\; \mathbf{w}_1 \mid Z \,\right] \left[\begin{array}{c|c} I_2 & 0 \\ \hline 0 & Q \end{array}\right]$$

it follows that

$$\mathbf{w}_0 = \mathbf{q}_0 h_{00}; \qquad \mathbf{w}_1 = \mathbf{q}_0 h_{01} + \mathbf{q}_1 h_{11}$$

and for $k \geq 2$

$$Z\mathbf{q}_{k-2} = \mathbf{q}_0 h_{0k} + \cdots + \mathbf{q}_k h_{kk}$$

where $\mathbf{q}_j$, $j = 0, 1, \ldots$ denotes the j-th column of Q. We shall discuss the case $h_{kk} = 0$ separately for $k = 0$, $k = 1$ and $k \geq 2$ separately. If $h_{00} = 0$, then $\mathbf{w}_0 = 0$. This is a very unlikely situation because then there is only a trivial solution $[p_0, p_1] = [1, 0]$ which fits exactly. Next consider $h_{11} = 0$; then define $\boldsymbol{\pi}_1'$ as

$$\boldsymbol{\pi}_1' = \begin{bmatrix} 0 \\ 1 \end{bmatrix} - \boldsymbol{\pi}_0 h_{01}.$$

Then

$$\begin{aligned} W\hat{\boldsymbol{\pi}}_1' &= (\mathbf{w}_1 - W\hat{\boldsymbol{\pi}}_0 h_{01}) \\ &= (\mathbf{w}_1 - \mathbf{q}_0 h_{01}) \\ &= \mathbf{q}_1 h_{11} = 0. \end{aligned}$$

This means that we get an exact approximation since $w_i \boldsymbol{\pi}'_1(z_i) = 0$, $i = 0, \ldots, m$. For the general case $h_{kk} = 0$, $k \geq 2$, we have that

$$Z\mathbf{q}_{k-2} - \mathbf{q}_0 h_{0k} - \cdots - \mathbf{q}_{k-1} h_{k-1,k} = \mathbf{q}_k h_{kk} = 0.$$

Since $\mathbf{q}_j = W\hat{\boldsymbol{\pi}}_j$ for $j = 0, \ldots, k-1$, we also have

$$\begin{aligned}
0 &= ZW\hat{\boldsymbol{\pi}}_{k-2} - W\hat{\boldsymbol{\pi}}_0 h_{0k} - \cdots - W\hat{\boldsymbol{\pi}}_{k-1} h_{k-1,k} \\
&= W\left(\hat{Z}\hat{\boldsymbol{\pi}}_{k-2} - \hat{\boldsymbol{\pi}}_0 h_{0k} - \cdots - \hat{\boldsymbol{\pi}}_{k-1} h_{k-1,k}\right)
\end{aligned} \tag{13.7}$$

where $\hat{Z} = Z \otimes I_2$. Define the polynomial

$$\boldsymbol{\pi}'_k(z) = z\boldsymbol{\pi}_{k-2}(z) - \boldsymbol{\pi}_0(z)h_{0k} - \cdots - \boldsymbol{\pi}_{k-1}(z)h_{k-1,k}$$

then, $W\hat{\boldsymbol{\pi}}'_k = W[\boldsymbol{\pi}'_k(z_0)^T, \ldots, \boldsymbol{\pi}'_k(z_m)^T]^T$ will be zero since it is equal to the Expression (13.7), which is zero. This means that

$$w_i \boldsymbol{\pi}'_k(z_i) = 0, \quad i = 0, \ldots, m.$$

The latter relations just tell us that this $\boldsymbol{\pi}'_k$ is an exact solution of the approximation problem, i.e., it interpolates.

In the general situation where $d_0 \neq d_1$, we have to distinguish between the scalar and the block part. For the scalar part we can now also have a breakdown in the sequence of orthogonal polynomials since the weights are not positive anymore but arbitrary complex numbers.

Using the notation

$$[H_0|H] = \begin{bmatrix} \times & h_{00} & h_{01} & \cdots & h_{0,n+1} & \\ \vdots & & \ddots & \ddots & \vdots & \\ \times & & & h_{nn} & h_{n,n+1} & \ddots \\ 0 & & & & \ddots & \ddots \end{bmatrix}$$

for the upper left part of the extended Hessenberg matrix, the situation is there as sketched above: whenever some h_{kk} is zero, we will have an interpolating polynomial solution. It then holds that

$$\boldsymbol{\pi}'_k(z) = z\boldsymbol{\pi}_{k-1}(z) - \boldsymbol{\pi}_0(z)h_{0k} - \cdots - \boldsymbol{\pi}_{k-1}(z)h_{k-1,k}$$

and because $W\hat{\boldsymbol{\pi}}'_k = W[\boldsymbol{\pi}'^T_k(z_0), \ldots, \boldsymbol{\pi}'^T_k(z_m)]^T$ is zero, we get

$$w_i \boldsymbol{\pi}'_k(z_i) = 0, \quad i = 0, \ldots, m,$$

identifying $\boldsymbol{\pi}'_k(z)$ as a (polynomial) interpolant.

For the lower right part, i.e., for the block polynomial part, a zero on the second subdiagonal (i.e., when we get a singular subdiagonal block in the Hessenberg matrix), will imply interpolation as we explained above for the block case.

The remaining problem is the case where the bottom element in the first column of the transformed extended Hessenberg matrix becomes zero. That is the element that has previously been denoted by d_n. Indeed, if this is zero, then our derivation, which gave (13.6):

$$\boldsymbol{\pi}_n(z) = \frac{1}{d_n} \left[\begin{array}{c} 1 \\ p_{n-1}(z) \end{array} \right]$$

does not hold anymore. But again, here we will have interpolation, i.e., a least squares error equal to zero. It follows from the derivation in the previous section that when $d_n = 0$,

$$W(\mathbf{e} - \mathbf{p}_{n-1}) = d_n \mathbf{q}_n = 0.$$

Thus

$$\mathbf{w}_i^T \left(\left[\begin{array}{c} 1 \\ 0 \end{array} \right] - \sum_{k=0}^{n-1} \boldsymbol{\pi}_k(z_i) d_k \right) = \mathbf{w}_i^T \left[\begin{array}{c} 1 \\ p_{n-1}(z_i) \end{array} \right] = 0$$

where $p_{n-1}(z) = \sum_{k=0}^{n-1} \varphi_k(z) a_k^{\Phi}$. This is the same as

$$w_{0i} - w_{1i} p_{n-1}(z_i) = 0, \quad i = 0, \ldots, m$$

which means that $(1, p_{n-1}(z))/d'$ with $d' \neq 0$ to normalize $p_{n-1}(z)$ as a monic polynomial, will fit the data exactly.

Notes and references

This chapter and the previous one are based on [36]. Of course, just as the updating procedure for orthonormal polynomials can be generalized to the vector case, also the downdating procedure can be adapted to our general situation. A combination of downdating and updating provides a tool for least squares approximation with a sliding window, i.e., where a window slides over the data, letting new data enter and simultaneously forgetting about the oldest data.

☞ M. Van Barel and A. Bultheel. Updating and downdating of orthonormal polynomial vectors and some applications. In V. Olshevsky, editor, *Structured Matrices in Mathematics, Computer Science, and Engineering II*, volume 281 of *Contemporary Mathematics*, pages 145–162. American Mathematical Society, Providence, Rhode Island, USA, 2001.

13.5 Conclusions

In this chapter we have generalized the concept of orthogonal polynomials into orthogonal vector polynomials. Also the corresponding updating algorithms as developed in Chapter 12 for the corresponding inverse eigenvalue problems are designed for this more general case.

Chapter 14

Orthogonal rational functions

In this chapter we adapt the technique laid down in the previous chapters for polynomial (vector) sequences to a specific set of proper rational functions. The goal is the computation of an orthonormal basis of the linear space $\mathcal{R}_n$ of proper rational functions $\phi(z) = n(z)/d(z)$ with regard to the discrete inner product

$$\langle \phi, \psi \rangle = \sum_{i=0}^{n} |w_i|^2 \overline{\phi(z_i)} \psi(z_i)$$

with given points z_i and weights w_i. The rational function $\phi(z)$ is proper, i.e., the degree of the numerator $n(z)$ is less than or equal to the degree of the denominator $d(z)$. Both degrees are not greater than n and the denominator polynomial $d(z)$ has a prescribed set $\{y_1, \ldots, y_n\}$, $y_i \in \mathbb{C}$, of possible zeros, called the poles of the rational function $\phi(z)$. The orthonormal basis of rational functions can then be used when solving least squares approximation problems with rational functions with prescribed poles. Moreover, it is also closely related with the computation of an orthogonal factorization of Cauchy-like matrices whose nodes are the points z_i and y_i [74, 76].

We prove that an orthonormal basis of $(\mathcal{R}_n, \langle \, \cdot \, , \cdot \, \rangle)$ can be generated by means of a suitable recurrence relation. When the points z_i as well as the points y_i are all real, fast $\mathcal{O}(n^2)$ Stieltjes-like procedures for computing the coefficients of such relation were first devised in [74, 76]. However, like the polynomial (Vandermonde) case [132], these fast algorithms turn out to be quite sensitive to roundoff errors so that the computed functions are far from orthogonal. Therefore, in this chapter we propose a different approach based on the reduction of the considered problem to the following inverse eigenvalue problem (HL-IEP): Find a generator representable Hessenberg-like matrix Z of order $n + 1$, i.e., a matrix whose lower triangular part is the lower triangular part of a rank one matrix, and a unitary matrix Q of order $n + 1$ such that $Q^H \mathbf{w} = \|\mathbf{w}\| \mathbf{e}_1$ and $Q^H D_z Q = Z + D_y$. Here and below $\mathbf{w} = [w_0, \ldots, w_n]^T$, $D_z = \text{diag}([z_0, \ldots, z_n])$ and $D_y = \text{diag}([y_0, \ldots, y_n])$, where y_0 can be chosen arbitrarily. Moreover, we denote by $\mathcal{H}_k$ the class of $k \times k$ Hessenberg-like

447

matrices and by $\mathcal{H}_k^{(g)}$ the class of $k \times k$ generator representable Hessenberg-like matrices. If both Z and Z^H belong to $\mathcal{H}_k$, then Z is a semiseparable matrix.

In Chapter 12, a quite similar reduction to an inverse eigenvalue problem for a tridiagonal symmetric matrix (T-IEP) or for a unitary Hessenberg matrix (H-IEP) was also exploited in the theory on the construction of orthonormal polynomials with regard to a discrete inner product. We also generalized this theory to orthonormal vector polynomials. Since invertible (generator representable) semiseparable matrices are the inverses of (irreducible) tridiagonal ones as we saw in Chapter 1. We find that HL-IEP gives a generalization of T-IEP and, in particular, it reduces to T-IEP in the case where $y_i, z_i \in \mathbb{R}$ and all prescribed poles y_i are equal.

We devise a method for solving HL-IEP which fully exploits its recursive properties. This method proceeds by applying a sequence of carefully chosen Givens transformations to update the solution at the k-th step by adding a new data $(w_{k+1}, z_{k+1}, y_{k+1})$. The unitary matrix Q can thus be determined in its factored form as a product of $\mathcal{O}(n^2)$ Givens transformations at the cost of $\mathcal{O}(n^2)$ arithmetic operations (ops). The complexity of forming the matrix Z depends on the structural properties of its upper triangular part and, in general, it requires $\mathcal{O}(n^3)$ ops. In the case where all the points z_i lie on the real axis, we show that Z is a generator representable semiseparable matrix so that the computation of Z can be carried out using $\mathcal{O}(n^2)$ ops only. In addition to that, the class $\mathcal{H}_{n+1}^{(g)}$ results to be close under linear fractional (Möbius) transformations of the form $z \to (\alpha z + \beta)/(\gamma z + \delta)$. Hence, by combining these two facts together, we are also able to prove that the process of forming Z can be performed at the cost of $\mathcal{O}(n^2)$ ops whenever all points z_i belong to a generalized circle (ordinary circles and straight lines) in the complex plane.

This chapter is organized in the following way. In Section 14.1 we reduce the computation of a sequence of orthonormal rational basis functions to the solution of an inverse eigenvalue problem for matrices of the form $\mathrm{diag}([y_0, \ldots, y_n]) + Z$, with $Z \in \mathcal{H}_{n+1}^{(g)}$. By exploiting this reduction, we also determine relations for the recursive construction of such functions. Section 14.2 provides our method for solving HL-IEP in the general case whereas the more specific situations corresponding to points lying on the real axis, on the unit circle, or on a generic circle in the complex plane are considered in Section 14.3.

✎ *Mastering the content of this chapter is not essential to understanding the other chapters of the book. The essential idea of this chapter is the following. For a discrete inner product as defined in Definition 14.1 the parameters of the relation (14.1) for the corresponding sequence of orthonormal rational functions can be computed by solving an inverse eigenvalue problem for a Hessenberg-like matrix (see Problem 14.3). When the nodes z_i of the inner product are all on the real line or all on the unit circle, the computational complexity can be reduced from $\mathcal{O}(n^3)$ to $\mathcal{O}(n^2)$ with n indicating the number of nodes z_i.*

14.1 The computation of orthonormal rational functions

In this section we will study the properties of a sequence of proper rational functions with prescribed poles that are orthonormal with respect to a certain discrete inner product. We will also design an algorithm to compute such a sequence via a suitable recurrence relation. The derivation of this algorithm follows from reducing the functional problem into a matrix setting to the solution of an inverse eigenvalue problem involving structured rank matrices.

14.1.1 The functional problem

Given the complex numbers $y_1, y_2, \ldots, y_n$ all different from each other. Let us consider the vector space $\mathcal{R}_n$ of all proper rational functions having possible poles in $y_1, y_2, \ldots, y_n$:

$$\mathcal{R}_n = \text{span}\{1, \frac{1}{z - y_1}, \frac{1}{z - y_2}, \ldots, \frac{1}{z - y_n}\}.$$

Note that proper means that the degree of the numerator polynomial cannot be greater than the degree of the denominator polynomial. The vector space $\mathcal{R}_n$ can be equipped with the inner product $\langle \cdot, \cdot \rangle$ defined below:

Definition 14.1 (Bilinear form). *Given the complex numbers $z_0, z_1, \ldots, z_n$ which together with the numbers y_i are all different from each other, and the (in general complex) "weights" $0 \neq w_i$, $i = 0, 1, \ldots, n$, we define a bilinear form $\langle \cdot, \cdot \rangle : \mathcal{R}_n \times \mathcal{R}_n \to \mathbb{C}$ by*

$$\langle \phi, \psi \rangle = \sum_{i=0}^{n} |w_i|^2 \overline{\phi(z_i)} \psi(z_i).$$

Since there is no proper rational function $\phi(z) = n(z)/d(z)$ with $\deg(n(z)) \leq \deg(d(z)) \leq n$ different from the zero function such that $\phi(z_i) = 0$ for $i = 0, \ldots, n$, this bilinear form defines a positive definite inner product in the space $\mathcal{R}_n$.

The aim of this chapter is to develop an efficient algorithm for the solution of the following functional problem:

Problem 14.2 (Computing a sequence of orthonormal rational basis functions). *Construct an orthonormal basis*

$$\boldsymbol{\alpha}_n(z) = [\alpha_0(z), \alpha_1(z), \ldots, \alpha_n(z)]^T$$

of $(\mathcal{R}_n, \langle \cdot, \cdot \rangle)$ satisfying the properties

$$\alpha_j(z) \in \mathcal{R}_j \setminus \mathcal{R}_{j-1} \qquad (\mathcal{R}_{-1} = \emptyset)$$
$$\langle \alpha_i, \alpha_j \rangle = \delta_{i,j} \qquad \text{(Kronecker delta)}$$

for $i, j = 0, 1, 2, \ldots, n$.

We will show later that the computation of such an orthonormal basis $\boldsymbol{\alpha}_n(z)$ is equivalent to the solution of an inverse eigenvalue problem for generator representable Hessenberg-like plus diagonal matrices, i.e., matrices of the form $\mathrm{diag}([y_0, \ldots, y_n]) + Z$, where $Z \in \mathcal{H}_{n+1}^{(g)}$.

14.1.2   The inverse eigenvalue problem

Let $D_y = \mathrm{diag}([y_0, \ldots, y_n])$ be the diagonal matrix whose diagonal elements are $y_0, y_1, \ldots, y_n$, where y_0 can be chosen arbitrarily; analogously, set $D_z = \mathrm{diag}([z_0, \ldots, z_n])$. Furthermore, denote by $\|\mathbf{w}\|$ the Euclidean norm of the vector $\mathbf{w} = [w_0, w_1, \ldots, w_n]^T$.

Our approach to solving Problem 14.2 mainly relies upon the equivalence between that problem and the following inverse eigenvalue problem (HL-IEP):

Problem 14.3 (Solving an inverse eigenvalue problem). *Given the numbers* w_i, z_i, y_i, *find a matrix* $Z \in \mathcal{H}_{n+1}^{(g)}$ *and a unitary matrix* Q *such that*

$$Q^H \mathbf{w} = \|\mathbf{w}\| \mathbf{e}_1, \tag{14.1}$$
$$Q^H D_z Q = Z + D_y. \tag{14.2}$$

Note 14.4. *Observe that, if* (Q, Z) *is a solution of Problem 14.3, then* Z *cannot have zero rows and columns. By contradiction, if we suppose that* $Z\mathbf{e}_j = 0$, *where* $\mathbf{e}_j$ *is the* j-*th column of the identity matrix* I_{n+1} *of order* $n + 1$, *then* $D_z Q\mathbf{e}_j = QD_y\mathbf{e}_j = y_{j-1}Q\mathbf{e}_j$, *from which it would follow* $y_{j-1} = z_i$ *for a certain* i.

Results concerning the existence and the uniqueness of the solution of Problem 14.3 were first proven in the articles [74, 76, 75] for the specific case where $y_i, z_i \in \mathbb{R}$ and Z is a generator representable semiseparable matrix. In particular, under such auxiliary assumptions, it was shown that the matrix Q is simply the orthogonal factor of a QR-decomposition of a Cauchy-like matrix built from the nodes y_i and z_i, i.e., a matrix whose (i, j)-th element has the form $u_{i-1}v_{j-1}/(z_{i-1} - y_{j-1})$ where u_{i-1} and v_{j-1} are components of two vectors $\mathbf{u} = [u_0, \ldots, u_n]^T$ and $\mathbf{v} = [v_0, \ldots, v_n]^T$, respectively. Next we give a generalization of the results of [74, 76, 75] to deal with the more general situation considered here.

Theorem 14.5. *Problem 14.3 has at least one solution. If* (Q_1, Z_1) *and* (Q_2, Z_2) *are two solutions of Problem 14.3, then there exists a unitary diagonal matrix* $F = \mathrm{diag}([1, \mathrm{e}^{\mathrm{i}\theta_1}, \ldots, \mathrm{e}^{\mathrm{i}\theta_n}])$ *such that*

$$Q_2 = Q_1 F, \quad Z_2 = F^H Z_1 F.$$

Proof. It is certainly possible to find two vectors $\mathbf{u} = [u_0, \ldots, u_n]^T$ and $\mathbf{v} = [v_0, \ldots, v_n]^T$ with $v_i, u_i \neq 0$ and $u_i v_0/(z_i - y_0) = w_i$, for $0 \leq i \leq n$. Indeed, it is sufficient to set, for example, $v_i = 1$ and $u_i = w_i(z_i - y_0)$. Hence, let us consider the nonsingular Cauchy-like matrix $C = (u_{i-1}v_{j-1}/(z_{i-1} - y_{j-1}))$ and let $C = QR$ be a QR-factorization of C. From $D_z C - C D_y = \mathbf{u}\mathbf{v}^T$ one easily finds that

$$Q^H D_z Q = R D_y R^{-1} + Q \mathbf{u}\mathbf{v}^T R^{-1} = D_y + Z,$$

where
$$Z = R D_y R^{-1} - D_y + Q \mathbf{u}\mathbf{v}^T R^{-1} \in \mathcal{H}_{n+1}^{(g)}.$$

Moreover, $Q\mathbf{e}_1 = C R^{-1}\mathbf{e}_1 = \mathbf{w}/\|\mathbf{w}\|$ by construction. Hence, the matrices Q and $Z = Q^H D_z Q - D_y$ solve Problem 14.3.

Concerning uniqueness, assume that (Q, Z) is a solution of Problem 14.3 with $Z = (z_{i,j})$ and $z_{i,j} = \tilde{u}_{i-1}\tilde{v}_{j-1}$ for $1 \le j \le i \le n + 1$. As $Z\mathbf{e}_1 \ne 0$, it follows that $\tilde{v}_0 \ne 0$ and, therefore, we may assume $\tilde{v}_0 = 1$. Moreover, from (14.2) it is easily found that

$$D_z Q \mathbf{e}_1 = Q\tilde{\mathbf{u}} + y_0 Q \mathbf{e}_1,$$

where $\tilde{\mathbf{u}} = [\tilde{u}_0, \dots, \tilde{u}_n]^T$. From (14.1) we have

$$\tilde{\mathbf{u}} = Q^H(D_z - y_0 I_{n+1})\frac{\mathbf{w}}{\|\mathbf{w}\|}. \tag{14.3}$$

Relation (14.2) can be rewritten as

$$Q^H D_z Q = \tilde{\mathbf{u}}\tilde{\mathbf{v}}^T + U = \tilde{\mathbf{u}}\tilde{\mathbf{v}}^T + R D_y R^{-1},$$

where U is an upper triangular matrix with diagonal entries y_i and $U = R D_y R^{-1}$ gives its Jordan decomposition, defined up to a suitable scaling of the columns of the upper triangular eigenvector matrix R. Hence, we find that

$$D_z QR - QR D_y = Q\tilde{\mathbf{u}}\tilde{\mathbf{v}}^T R = \mathbf{u}\mathbf{v}^T$$

and, therefore, $QR = C = (u_{i-1}v_{j-1}/(z_{i-1} - y_{j-1}))$ is a Cauchy-like matrix with $\mathbf{u} = Q\tilde{\mathbf{u}}$ uniquely determined by (14.3). This means that all the eligible Cauchy-like matrices C are obtained one from the other by a multiplication from the right by a suitable diagonal matrix. In this way, from the essential uniqueness of the orthogonal factorization of a given matrix, we may conclude that Q is uniquely determined up to multiplication from the right by a unitary diagonal matrix F having fixed its first diagonal entry equal to 1. Finally, the result for Z immediately follows from using again relation (14.2). □

The above theorem says that the solution of Problem 14.3 is essentially unique up to a diagonal scaling. Furthermore, once the weight vector $\mathbf{w}$ and the points z_i are fixed, the determinant of Z is a rational function in the variables $y_0, \dots, y_n$ whose numerator is not identically zero. Hence, we can show that, for almost any choice of $y_0, \dots, y_n$, the resulting matrix Z is nonsingular. The article [74] dealt with this regular case, in the framework of the orthogonal factorization of real Cauchy matrices. In particular, it is shown there that the matrix Z is nonsingular when all the nodes y_i, z_i are real and there exists an interval, either finite or infinite, containing all nodes y_i and none of the nodes z_i.

In what follows we assume that $Z^{-1} = H$ exists. It is known that the inverse of a matrix whose lower triangular part is the lower triangular part of a rank 1 matrix is an irreducible Hessenberg matrix [84]. Hence, we will use the following notation: The matrix $H = Z^{-1}$ is upper Hessenberg with subdiagonal elements $b_0, b_1, \ldots, b_{n-1}$; for $j = 0, \ldots, n-1$, the j-th column $\mathbf{h}_j$ of H has the form

$$\mathbf{h}_j^T = [\tilde{\mathbf{h}}_j^T, b_j, 0], \quad b_j \neq 0.$$

The outline of the remainder of this section is as follows. First we assume that we know a unitary matrix Q and the corresponding matrix Z solving Problem 14.3. Then we provide a recurrence relation between the columns $\mathbf{q}_j$ of Q and, in addition to that, we give a connection between the columns $\mathbf{q}_j$ and the values at the points z_i attained by certain rational functions satisfying a similar recurrence relation. Finally, we show that these rational functions form a basis we are looking for.

14.1.3   Recurrence relation for the columns of Q

Let the columns of Q denoted as follows:

$$Q = [\mathbf{q}_0, \mathbf{q}_1, \ldots, \mathbf{q}_n].$$

Theorem 14.6 (Recurrence relation).   *For $j = 0, 1, \ldots, n$, the columns $\mathbf{q}_j$ satisfy the recurrence relation*

$$b_j(D_z - y_{j+1}I_{n+1})\mathbf{q}_{j+1} = \mathbf{q}_j + ([\mathbf{q}_0, \mathbf{q}_1, \ldots, \mathbf{q}_j] D_{y,j} - D_z [\mathbf{q}_0, \mathbf{q}_1, \ldots, \mathbf{q}_j]) \tilde{\mathbf{h}}_j,$$

with $\mathbf{q}_0 = \mathbf{w}/\|\mathbf{w}\|$, $\mathbf{q}_{n+1} = 0$ and $D_{y,j} = \text{diag}([y_0, \ldots, y_j])$.

Proof.   Since $Q^H\mathbf{w} = \mathbf{e}_1\|\mathbf{w}\|$, it follows that $\mathbf{q}_0 = \mathbf{w}/\|\mathbf{w}\|$. Multiplying relation (14.2) to the left by Q, we have

$$D_zQ = Q(Z + D_y).$$

Multiplying this to the right by $H = Z^{-1}$, gives us

$$D_zQH = Q(I_{n+1} + D_yH). \tag{14.4}$$

Considering the j-th column of the left and right-hand side of the equation above we have the claim.   □

14.1.4   Recurrence relation for the orthonormal functions

In this section we define an orthonormal basis $\boldsymbol{\alpha}_n(z) = [\alpha_0(z), \alpha_1(z), \ldots, \alpha_n(z)]^T$ for $\mathcal{R}_n$ using a recurrence relation built by means of the information contained in the matrix H.

Definition 14.7 (Recurrence for the orthonormal rational functions). *Let us define $\alpha_0(z) = 1/\|\mathbf{w}\|$ and*

$$\alpha_{j+1}(z) = \frac{\alpha_j(z) + ([\alpha_0(z),\ldots,\alpha_j(z)]\, D_{y,j} - z\,[\alpha_0(z),\ldots,\alpha_j(z)])\, \tilde{\mathbf{h}}_j}{b_j(z - y_{j+1})},$$

for $0 \le j \le n-1$.

In the next theorem, we prove that the rational functions $\alpha_j(z)$ evaluated at the points z_i are connected to the elements of the unitary matrix Q. This will allow us to prove in Theorem 14.9 that the rational functions $\alpha_j(z)$ are indeed the orthonormal rational functions we are looking for. In what follows, we use the notation $D_w = \mathrm{diag}([w_0,\ldots,w_n])$.

Theorem 14.8 (Connection between $\alpha_j(z_i)$ and the elements of Q). *Let*

$$\boldsymbol{\alpha}_j = [\alpha_j(z_0),\ldots,\alpha_j(z_n)]^T \in \mathbb{C}^{n+1}, \quad 0 \le j \le n.$$

For $j = 0, 1, \ldots, n$, we have $\mathbf{q}_j = D_w \boldsymbol{\alpha}_j$.

Proof. Replacing z by z_i in the recurrence relation for $\alpha_{j+1}(z)$, we get

$$b_j(D_z - y_{j+1}I_{n+1})\boldsymbol{\alpha}_{j+1} = \boldsymbol{\alpha}_j + ([\boldsymbol{\alpha}_0,\ldots,\boldsymbol{\alpha}_j]\, D_{y,j} - D_z\,[\boldsymbol{\alpha}_0,\ldots,\boldsymbol{\alpha}_j])\, \tilde{\mathbf{h}}_j.$$

Since $\mathbf{q}_0 = \mathbf{w}/\|\mathbf{w}\| = D_w \boldsymbol{\alpha}_0$, the theorem is proved by finite induction on j, comparing the preceding recurrence with the one in Theorem 14.6. □

Now it is easy to prove the orthonormality of the rational functions $\alpha_j(z)$.

Theorem 14.9 (Orthonormality of $\boldsymbol{\alpha}_n(z)$). *The functions $\alpha_0(z),\ldots\alpha_n(z)$ form an orthonormal basis for $\mathcal{R}_n$ with respect to the inner product $\langle\,\cdot\,,\,\cdot\,\rangle$. Moreover, we have $\alpha_j(z) \in \mathcal{R}_j \setminus \mathcal{R}_{j-1}$.*

Proof. Firstly, we prove that $\langle \alpha_i, \alpha_j \rangle = \delta_{i,j}$. This follows immediately from the fact that $Q = D_w[\boldsymbol{\alpha}_0,\ldots,\boldsymbol{\alpha}_n]$ and Q is unitary. Now we have to prove that $\alpha_j(z) \in \mathcal{R}_j \setminus \mathcal{R}_{j-1}$. This is clearly true for $j = 0$ (recall that $\mathcal{R}_{-1} = \emptyset$). Suppose it is true for $j = 0, 1, 2, \ldots, k < n$. From the recurrence relation, we derive that $\alpha_{k+1}(z)$ has the form

$$\alpha_{k+1}(z) = \frac{\text{rational function with possible poles in } y_0, y_1, \ldots, y_k}{(z - y_{j+1})}.$$

Also $\lim_{z\to\infty} \alpha_{k+1}(z) \in \mathbb{C}$ and, therefore, $\alpha_{k+1}(z) \in \mathcal{R}_{k+1}$. Note that simplification by $(z - y_{k+1})$ does not occur in the previous formula for $\alpha_{k+1}(z)$ because $\mathbf{q}_{k+1} = D_w \boldsymbol{\alpha}_{k+1}$ is linearly independent of the previous columns of Q. Hence, $\alpha_{k+1}(z) \in \mathcal{R}_{k+1} \setminus \mathcal{R}_k$. □

In the next theorem, we give an alternative relation among the rational functions $\alpha_j(z)$.

Theorem 14.10 (Alternative relation). *We have*

$$z\boldsymbol{\alpha}_n(z)^T = \boldsymbol{\alpha}_n(z)^T(Z + D_y) + \alpha_{n+1}(z)\mathbf{s}_n^T, \tag{14.5}$$

where $\mathbf{s}_n^T$ is the last row of the matrix Z and the function $\alpha_{n+1}(z)$ is given by

$$\alpha_{n+1}(z) = c\prod_{j=0}^{n}(z - z_j)/\prod_{j=1}^{n}(z - y_j)$$

for some constant c.

Proof. Let $\mathbf{h}_n$ be the last column of $H = Z^{-1}$, and define

$$\alpha_{n+1}(z) = \boldsymbol{\alpha}_n(z)^T(zI_{n+1} - D_y)\mathbf{h}_n - \alpha_n(z). \tag{14.6}$$

Thus, the recurrence relation given in Definition 14.7 can also be written as

$$\boldsymbol{\alpha}_n(z)^T(zI_{n+1} - D_y)H = \boldsymbol{\alpha}_n(z)^T + \alpha_{n+1}(z)\mathbf{e}_{n+1}^T.$$

Multiplying to the right by $Z = H^{-1}$, we obtain the Formula (14.5). To determine the form of $\alpha_{n+1}(z)$ we look at the Definition 14.6. It follows that $\alpha_{n+1}(z)$ is a rational function having degree of numerator at most one more than the degree of the denominator and having possible poles in $y_1, y_2, \ldots, y_n$. Recalling from Theorem 14.8 the notation $\boldsymbol{\alpha}_j = [\alpha_j(z_0), \ldots, \alpha_j(z_n)]^T$ and the equation $Q = D_w[\boldsymbol{\alpha}_0, \ldots, \boldsymbol{\alpha}_n]$, we can evaluate the previous equation at the points z_i and obtain:

$$D_z[\boldsymbol{\alpha}_0, \ldots, \boldsymbol{\alpha}_n]H - [\boldsymbol{\alpha}_0, \ldots, \boldsymbol{\alpha}_n]D_yH = [\boldsymbol{\alpha}_0, \ldots, \boldsymbol{\alpha}_n] + \alpha_{n+1}\mathbf{e}_n^T.$$

Since $D_wD_z = D_zD_w$, multiplying to the left by D_w we obtain

$$D_zQH - QD_yH = Q + D_w\alpha_{n+1}\mathbf{e}_{n+1}^T.$$

From Equation (14.4) we obtain that $D_w\alpha_{n+1}\mathbf{e}_{n+1}^T$ is a zero matrix; hence, it follows that $\alpha_{n+1}(z_i) = 0$, for $i = 0, 1, \ldots, n$, and this proves the theorem. □

Note that $\alpha_{n+1}(z)$ is orthogonal to all $\alpha_i(z)$, $i = 0, 1, \ldots, n$, since $\alpha_{n+1}(z) \notin \mathcal{R}_n$ and its norm is

$$\|\alpha_{n+1}\|^2 = \sum_{i=0}^{n}|w_i\alpha_{n+1}(z_i)|^2 = 0.$$

14.2　Solving the inverse eigenvalue problem

In this section we devise an efficient recursive procedure for the construction of the matrices Q and Z solving Problem 14.3 (HL-IEP). The case $n = 0$ is trivial: It is sufficient to set $Q = w_0/|w_0|$ and $Z = z_0 - y_0$. Let us assume we have already constructed a unitary matrix Q_k and a matrix Z_k for the first $k+1$ points $z_0, z_1, \ldots, z_k$ with the corresponding weights $w_0, w_1, \ldots, w_k$. That is, (Q_k, Z_k) satisfies

$$Q_k^H \mathbf{w}_k = \|\mathbf{w}_k\| \mathbf{e}_1$$
$$Q_k^H D_{z,k} Q_k = Z_k + D_{y,k},$$

where $\mathbf{w}_k = [w_0, \ldots, w_k]^T$, $Z_k \in \mathcal{H}_{k+1}^{(g)}$, $D_{z,k} = \operatorname{diag}([z_0, \ldots, z_k])$ and, similarly, $D_{y,k} = \operatorname{diag}([y_0, \ldots, y_k])$. The idea is now to add a new point z_{k+1} with corresponding weight w_{k+1} and construct the corresponding matrices Q_{k+1} and Z_{k+1}. Hence, we start with the following relations:

$$\begin{bmatrix} 1 & 0 \\ 0 & Q_k^H \end{bmatrix} \begin{bmatrix} w_{k+1} \\ \mathbf{w}_k \end{bmatrix} = \begin{bmatrix} w_{k+1} \\ \|\mathbf{w}_k\| \mathbf{e}_1 \end{bmatrix}$$

$$\begin{bmatrix} 1 & 0 \\ 0 & Q_k^H \end{bmatrix} \begin{bmatrix} z_{k+1} & 0 \\ 0 & D_{z,k} \end{bmatrix} \begin{bmatrix} 1 & 0 \\ 0 & Q_k \end{bmatrix} = \begin{bmatrix} z_{k+1} & 0 \\ 0 & Z_k + D_{y,k} \end{bmatrix}.$$

Then, we find complex Givens transformations $G_i = I_{i-1} \oplus G_{i,i+1} \oplus I_{k-i+1}$,

$$G_{i,i+1} = \begin{bmatrix} c & s \\ -\overline{s} & c \end{bmatrix}, \quad G_{i,i+1}^H G_{i,i+1} = I_2, \tag{14.7}$$

such that

$$G_k^H \cdots G_1^H \begin{bmatrix} 1 & 0 \\ 0 & Q_k^H \end{bmatrix} \begin{bmatrix} w_{k+1} \\ \mathbf{w}_k \end{bmatrix} = \begin{bmatrix} \|\mathbf{w}_{k+1}\| \\ 0 \end{bmatrix},$$

and, moreover,

$$G_k^H \cdots G_1^H \begin{bmatrix} 1 & 0 \\ 0 & Q_k^H \end{bmatrix} \begin{bmatrix} z_{k+1} & 0 \\ 0 & D_{z,k} \end{bmatrix} \begin{bmatrix} 1 & 0 \\ 0 & Q_k \end{bmatrix} G_1 \cdots G_k \in \mathcal{H}_{k+2}^{(g)}.$$

Finally, we set

$$Q_{k+1} = \begin{bmatrix} 1 & 0 \\ 0 & Q_k \end{bmatrix} G_1 \cdots G_k,$$

and

$$Z_{k+1} = G_k^H \cdots G_1^H \begin{bmatrix} z_{k+1} & 0 \\ 0 & Z_k + D_{y,k} \end{bmatrix} G_1 \cdots G_k. \tag{14.8}$$

With the notation

$$\operatorname{trilS} \left(\begin{matrix} [u_0, u_1, \ldots, u_k] \\ [v_0, v_1, \ldots, v_k] \end{matrix} \right)$$

we denote the lower triangular matrix whose nonzero part equals the lower triangular part of the rank 1 matrix $[u_{i-1} v_{j-1}]_{i=0,\ldots,k}^{j=0,\ldots,k}$, i.e.,

$$\operatorname{trilS} \left(\begin{matrix} [u_0, u_1, \ldots, u_k] \\ [v_0, v_1, \ldots, v_k] \end{matrix} \right) = \begin{bmatrix} u_0 v_0 & & & \\ u_1 v_0 & u_1 v_1 & & \\ u_2 v_0 & u_2 v_1 & u_2 v_2 & \\ \vdots & \vdots & \vdots & \ddots \end{bmatrix}.$$

Moreover, with the notation

$$\text{triuR} \left(\begin{array}{c} [\eta_0, \eta_1, \ldots, \eta_{k-1}] \\ [\mathbf{r}_0, \mathbf{r}_1, \ldots, \mathbf{r}_{k-2}] \end{array} \right)$$

we denote the strictly upper triangular matrix whose $(i+1)$-st row, $0 \le i \le k-2$, is equal to $[0, 0, \ldots, 0, \eta_i, \mathbf{r}_i^T]$, i.e.,

$$\text{triuR} \left(\begin{array}{c} [\eta_0, \eta_1, \ldots, \eta_{k-1}] \\ [\mathbf{r}_0, \mathbf{r}_1, \ldots, \mathbf{r}_{k-2}] \end{array} \right) = \left[\begin{array}{cccc} 0 & \eta_0 & \mathbf{r}_0^T & \\ 0 & 0 & \eta_1 & \mathbf{r}_1^T \\ \vdots & \vdots & & \ddots \end{array} \right].$$

Let us describe now in what way Givens transformations are selected in order to perform the updating of Q_k and Z_k. In the first step we construct a Givens transformation working on the new weight. Let $G_{1,2}$ be a Givens transformation as in (14.7), such that

$$G_{1,2}^H \left[\begin{array}{c} w_{k+1} \\ \|\mathbf{w}_k\| \end{array} \right] = \left[\begin{array}{c} \|\mathbf{w}_{k+1}\| \\ 0 \end{array} \right]. \tag{14.9}$$

The matrix Z_k is updated as follows: We know that

$$Z_k = \text{trilS} \left(\begin{array}{c} [u_0, u_1, \ldots, u_k] \\ [v_0, v_1, \ldots, v_k] \end{array} \right) + \text{triuR} \left(\begin{array}{c} [\eta_0, \eta_1, \ldots, \eta_{k-1}] \\ [\mathbf{r}_0, \mathbf{r}_1, \ldots, \mathbf{r}_{k-2}] \end{array} \right).$$

Let

$$Z_{k+1,1} + D_{y,k+1,1} = \left[\begin{array}{cc} G_{1,2}^H & 0 \\ 0 & I_k \end{array} \right] \left[\begin{array}{cc} z_{k+1} & 0 \\ 0 & Z_k + D_{y,k} \end{array} \right] \left[\begin{array}{cc} G_{1,2} & 0 \\ 0 & I_k \end{array} \right],$$

where $Z_{k+1,1}$ and $D_{y,k+1,1}$ are defined as follows:

$$Z_{k+1,1} = \text{trilS} \left(\begin{array}{c} [\hat{u}_0, \tilde{u}_1, u_1, u_2, \ldots, u_k] \\ [\hat{v}_0, \tilde{v}_1, v_1, v_2, \ldots, v_k] \end{array} \right) + \text{triuR} \left(\begin{array}{c} [\hat{\eta}_0, \tilde{\eta}_1, \eta_1, \ldots, \eta_{k-1}] \\ [\hat{\mathbf{r}}_0, \tilde{\mathbf{r}}_1, \mathbf{r}_1 \ldots, \mathbf{r}_{k-2}] \end{array} \right)$$

and

$$D_{y,k+1,1} = \text{diag}([y_0, \tilde{y}_1, y_1, y_2, \ldots, y_k]),$$

with

$$\left[\begin{array}{cc} \alpha & \delta \\ \gamma & \beta \end{array} \right] = G_{1,2}^H \left[\begin{array}{cc} z_{k+1} & 0 \\ 0 & y_0 + u_0 v_0 \end{array} \right] G_{1,2}$$

and

$$\begin{array}{lll} \hat{v}_0 = -\overline{s} v_0 & \hat{u}_0 = (\alpha - y_0)/\hat{v}_0 & \hat{\eta}_0 = \delta \\ \tilde{v}_1 = c v_0 & \tilde{y}_1 = \beta - \tilde{u}_1 \tilde{v}_1 & \tilde{u}_1 = \gamma/\hat{v}_0 \\ \tilde{\eta}_1 = c \eta_0 & \hat{\mathbf{r}}_0 = \left[-s\eta_0, -s\mathbf{r}_0^T \right]^T & \tilde{\mathbf{r}}_1 = c\mathbf{r}_0. \end{array}$$

Observe that $v_0 \neq 0$ from Remark 14.4 and, moreover, $s \neq 0$ since $\|\mathbf{w}_k\| \neq 0$ in the Equation (14.9). Whence, $\hat{v}_0 \neq 0$ and, therefore, all these quantities are well defined.

In the next steps, we are transforming $D_{y,k+1,1}$ into $D_{y,k+1}$. The first of these steps is as follows. If $v_1\tilde{u}_1 - \tilde{\eta}_1 \neq 0$, we choose t such that

$$\bar{t} = \frac{y_1 - \tilde{y}_1}{v_1\tilde{u}_1 - \tilde{\eta}_1},$$

and define the Givens transformation working on the second and third row and column as

$$G_{2,3} = \left[\begin{array}{cc} 1 & t \\ -\bar{t} & 1 \end{array}\right] / \sqrt{1 + |t|^2}.$$

Otherwise, if $v_1\tilde{u}_1 - \tilde{\eta}_1 = 0$, we set

$$G_{2,3} = \left[\begin{array}{cc} 0 & 1 \\ -1 & 0 \end{array}\right].$$

The matrices $Z_{k+1,1}$ and $D_{y,k+1,1}$ undergo the similarity transformation associated to $G_{2,3}$. The transformed matrices $Z_{k+1,2}$ and $D_{y,k+1,2}$ are given by

$$Z_{k+1,2} = \mathrm{trilS}\left(\begin{array}{c} [\hat{u}_0, \hat{u}_1, \tilde{u}_2, u_2, \ldots, u_k] \\ {}[\hat{v}_0, \hat{v}_1, \tilde{v}_2, v_2, \ldots, v_k] \end{array}\right) + \mathrm{triuR}\left(\begin{array}{c} [\hat{\eta}_0, \hat{\eta}_1, \tilde{\eta}_2, \eta_2, \ldots, \eta_{k-1}] \\ {}[\hat{\mathbf{r}}_0, \hat{\mathbf{r}}_1, \tilde{\mathbf{r}}_2, \mathbf{r}_2 \ldots, \mathbf{r}_{k-2}] \end{array}\right),$$

$$D_{y,k+1,2} = \mathrm{diag}([y_0, y_1, \tilde{y}_2, y_2, y_3, \ldots, y_k]),$$

with

$$G_{2,3}^H \left[\begin{array}{c} \tilde{u}_1 \\ u_1 \end{array}\right] = \left[\begin{array}{c} \hat{u}_1 \\ \tilde{u}_2 \end{array}\right], \quad [\tilde{v}_1, v_1]\, G_{2,3} = [\hat{v}_1, \tilde{v}_2].$$

Moreover, $\tilde{y}_2 = \tilde{y}_1$, $\hat{\eta}_1$ is the $(1,2)$-entry of

$$G_{2,3}^H \left[\begin{array}{cc} \tilde{u}_1\tilde{v}_1 + \tilde{y}_1 & \tilde{\eta}_1 \\ u_1\tilde{v}_1 & u_1v_1 + y_1 \end{array}\right] G_{2,3}$$

and

$$G_{2,3}^H \left[\begin{array}{c} \tilde{\mathbf{r}}_1 \\ {}[\eta_1, \mathbf{r}_1] \end{array}\right] = \left[\begin{array}{c} \hat{\mathbf{r}}_1 \\ {}[\tilde{\eta}_2, \tilde{\mathbf{r}}_2] \end{array}\right]. \tag{14.10}$$

At the very end, after k steps, we obtain

$$Z_{k+1,k} = \mathrm{trilS}\left(\begin{array}{c} [\hat{u}_0, \hat{u}_1, \ldots, \hat{u}_k, \tilde{u}_{k+1}] \\ {}[\hat{v}_0, \hat{v}_1, \ldots, \hat{v}_k, \tilde{v}_{k+1}] \end{array}\right) + \mathrm{triuR}\left(\begin{array}{c} [\hat{\eta}_0, \hat{\eta}_1, \hat{\eta}_2, \ldots, \hat{\eta}_k] \\ {}[\hat{\mathbf{r}}_0, \hat{\mathbf{r}}_1, \hat{\mathbf{r}}_2, \ldots, \hat{\mathbf{r}}_{k-1}] \end{array}\right)$$

and

$$D_{y,k+1,k} = \mathrm{diag}([y_0, y_1, \ldots, y_k, \tilde{y}_{k+1}]).$$

Since $Z_{k+1,k} = Z_{k+1} + \mathrm{diag}([0, \ldots, 0, \tilde{y}_{k+1} - y_{k+1}])$, then from (14.8) it follows that $\tilde{u}_{k+1} \neq 0$. Thus, the last step will transform $\tilde{y}_{k+1}$ into y_{k+1} by applying the transformation

$$\hat{u}_{k+1} = \tilde{u}_{k+1}$$
$$\hat{v}_{k+1} = (\tilde{y}_{k+1} - y_{k+1} + \tilde{u}_{k+1}\tilde{v}_{k+1})/\tilde{u}_{k+1}.$$

The computational complexity of the algorithm is dominated by the cost of performing the multiplications (14.10). In general, adding new data $(w_{k+1}, z_{k+1}, y_{k+1})$ requires $\mathcal{O}(k^2)$ ops and hence, computing $Z_n = Z$ requires $\mathcal{O}(n^3)$ ops. In the next section we will show that these estimates reduce by an order of magnitude in the case where some special distributions of the points z_i are considered which lead to a matrix Z with a structured upper triangular part. We stress the fact that, in the light of Theorem 14.5, the above procedure to solve HL-IEP can also be seen as a method to compute the orthogonal factor in a QR-factorization of a suitable Cauchy-like matrix.

14.3   Special configurations of points z_i

In this section we specialize our algorithm for the solution of HL-IEP to cover with the important case where the points z_i are assumed to lie on the real axis or on the unit circle in the complex plane. Under this assumption on the distribution of the points z_i, it will be shown that the resulting matrix Z also possesses a generator representable semiseparable structure. The exploitation of this property allows us to overcome the multiplication (14.10) and to construct the matrix $Z_n = Z$ by means of a simpler parametrization, using $\mathcal{O}(n)$ ops per point, so that the overall cost of forming S reduces to $\mathcal{O}(n^2)$ ops.

14.3.1   Special case: all points z_i are real

When all the points z_i are real, we have that

$$Z + D_y = Q^H D_z Q = (Q^H D_z Q)^H = (Z + D_y)^H.$$

Hence, the matrix Z is semiseparable and can be denoted by S. So, the matrix $Z + D_y$ can be written as

$$Z + D_y = S + D_y = \text{tril}(\mathbf{u}\mathbf{v}^T, 0) + D_y + \text{triu}(\bar{\mathbf{v}}\mathbf{u}^H, 1) = S(\mathbf{u}, \mathbf{v}) + D_y, \quad (14.11)$$

with $\bar{\mathbf{v}}$ the complex conjugate of the vector $\mathbf{v}$. Here we adopt the MATLAB notation $\text{triu}(B, p)$ for the upper triangular portion of a square matrix B, where all entries below the p-th diagonal are set to zero ($p = 0$ is the main diagonal, $p > 0$ is above the main diagonal, and $p < 0$ is below the main diagonal). Analogously, the matrix $\text{tril}(B, p)$ is formed by the lower triangular portion of B by setting to zero all its entries above the p-th diagonal. In particular, the matrix S is a Hermitian semiseparable matrix, and its computation requires only $\mathcal{O}(n)$ ops per point, since its upper triangular part needs not to be computed via (14.10). Moreover, its inverse matrix $T = S^{-1}$ is tridiagonal, hence the vectors $\tilde{\mathbf{h}}_j$ occurring in Definition 14.7 have only one nonzero entry.

When also all the poles y_i (and the weights w_i) are real, all computations can be performed using real arithmetic instead of doing operations on complex numbers. When all the poles are real or come in complex conjugate pairs, also all computations can be done using only real arithmetic. However, the algorithm works then with a block diagonal D_y instead of a diagonal matrix. The details of this algorithm are rather elaborate. So, we will not go into the details here.

14.3.2 Special case: all points z_i lie on the unit circle

The case of points z_i located on the unit circle $\mathbb{T} = \{z \in \mathbb{C} : |z| = 1\}$ in the complex plane can be reduced to the real case treated in the preceding subsection by using the concept of linear fractional (Möbius) transformation [106]. To be specific, a function $\mathcal{M} : \mathbb{C} \cup \{\infty\} \to \mathbb{C} \cup \{\infty\}$ is a Möbius transformation if

$$\mathcal{M}(z) = \frac{\alpha z + \beta}{\gamma z + \delta}, \quad \alpha\delta - \beta\gamma \neq 0, \quad \alpha, \beta, \gamma, \delta \in \mathbb{C}.$$

Interesting properties concerning Möbius transformations are collected in [106]. In particular, a Möbius transformation defines a one-to-one mapping of the extended complex plane into itself and, moreover, the inverse of a Möbius transformation is still a Möbius transformation given by

$$\mathcal{M}^{-1}(z) = \frac{\delta z - \beta}{-\gamma z + \alpha}. \tag{14.12}$$

The Möbius transformation $\mathcal{M}(Z)$ of a matrix Z is defined as

$$\mathcal{M}(Z) = (\alpha Z + \beta I)(\gamma Z + \delta I)^{-1}$$

if the matrix $\gamma Z + \delta I$ is nonsingular. The basic fact relating semiseparable matrices with Möbius transformations is that in a certain sense the semiseparable structure is maintained under a Möbius transformation of the matrix. More precisely, we have that:

Theorem 14.11. Let $Z \in \mathcal{H}_{n+1}^{(g)}$ with $Z = (z_{i,j})$, $z_{i,j} = u_{i-1}v_{j-1}$ for $1 \leq j \leq i \leq n+1$, and $v_0 \neq 0$. Moreover, let $D_y = \mathrm{diag}([y_0, \ldots, y_n])$ and assume that $\mathcal{M}$ maps the eigenvalues of both $Z + D_y$ and D_y into points of the ordinary complex plane, i.e., $-\delta/\gamma$ is different from all the points y_i, z_i. Then, we find that

$$\mathcal{M}(Z + D_y) - \mathcal{M}(D_y) \in \mathcal{H}_{n+1}^{(g)}.$$

Proof. Observe that $Z \in \mathcal{H}_{n+1}^{(g)}$ implies that $RZU \in \mathcal{H}_{n+1}^{(g)}$ for R and U upper triangular matrices. Hence, if we define $R = I - \mathbf{e}_1[0, v_1/v_0, \ldots, v_n/v_0]$, the theorem is proven by showing that

$$R^{-1}(\mathcal{M}(Z + D_y) - \mathcal{M}(D_y))R \in \mathcal{H}_{n+1}^{(g)},$$

which is equivalent to

$$R^{-1}\mathcal{M}(Z + D_y)R - \mathcal{M}(D_y) \in \mathcal{H}_{n+1}^{(g)}.$$

One immediately finds that

$$R^{-1}\mathcal{M}(Z + D_y)R = ((\gamma(Z + D_y) + \delta I)R)^{-1}(\alpha(Z + D_y) + \beta I)R,$$

from which it follows

$$R^{-1}\mathcal{M}(Z + D_y)R = (\gamma v_0 \mathbf{u} \mathbf{e}_1^T + R_1)^{-1}(\alpha v_0 \mathbf{u} \mathbf{e}_1^T + R_2),$$

where R_1 and R_2 are upper triangular matrices with diagonal entries $\gamma y_i + \delta$ and $\alpha y_i + \beta$, respectively. In particular, R_1 is invertible and, by applying the Sherman-Morrison formula we obtain

$$R^{-1}\mathcal{M}(Z + D_y)R = (I - \sigma R_1^{-1}\mathbf{u}\mathbf{e}_1^T)(\alpha v_0 R_1^{-1}\mathbf{u}\mathbf{e}_1^T + R_1^{-1}R_2),$$

for a suitable σ. The thesis is now established by observing that the diagonal entries of $R_1^{-1}R_2$ coincides with the ones of $\mathcal{M}(D_y)$ and, moreover, from the previous relation one gets

$$R^{-1}\mathcal{M}(Z + D_y)R - R_1^{-1}R_2 \in \mathcal{H}_{n+1}^{(g)},$$

and the proof is complete.   □

This theorem has several interesting consequences since it is well known that we can determine Möbius transformations mapping the unit circle $\mathbb{T}$ except for one point onto the real axis in the complex plane. To see this, let us first consider Möbius transformations of the form

$$\mathcal{M}_1(z) = \frac{z + \bar{\alpha}}{z + \alpha}, \quad \alpha \in \mathbb{C} \setminus \mathbb{R}.$$

It is immediately found that $\mathcal{M}_1(z)$ is invertible and, moreover, $\mathcal{M}_1(z) \in \mathbb{T}$ whenever $z \in \mathbb{R}$. For the sake of generality, we also introduce Möbius transformations of the form

$$\mathcal{M}_2(z) = \frac{z - \beta}{1 - \bar{\beta}z}, \quad |\beta| \neq 1,$$

which are invertible and map the unit circle $\mathbb{T}$ into itself. Then, by composition of $\mathcal{M}_2(z)$ with $\mathcal{M}_1(z)$ we find a fairly general transformation $\mathcal{M}(z)$ mapping the real axis into the unit circle:

$$\mathcal{M}(z) = \mathcal{M}_2(\mathcal{M}_1(z)) = \frac{(1 - \beta)z + (\bar{\alpha} - \beta\alpha)}{(1 - \bar{\beta})z + (\alpha - \bar{\alpha}\bar{\beta})}. \tag{14.13}$$

Hence, the inverse transformation $\mathcal{M}^{-1}(z) = \mathcal{M}_1^{-1}(\mathcal{M}_2^{-1}(z))$, where

$$\mathcal{M}_1^{-1}(z) = \frac{\alpha z - \bar{\alpha}}{-z + 1}, \quad \mathcal{M}_2^{-1}(z) = \frac{z + \beta}{\bar{\beta}z + 1},$$

is the desired invertible transformation which maps the unit circle (except for one point) into the real axis.

By combining these properties with Theorem 14.11, we obtain efficient procedures for the solution of Problem 14.3 in the case where all the points z_i belong to the unit circle $\mathbb{T}$. Let $D_y = \text{diag}([y_0, \ldots, y_n])$ and $D_z = \text{diag}([z_0, \ldots, z_n])$ with $|z_i| = 1$. Moreover, let $\mathcal{M}(z)$ be as in (14.13), such that $\mathcal{M}^{-1}(z_i)$ and $\mathcal{M}^{-1}(y_i)$ are finite, i.e., $z_i, y_i \neq (1 - \beta)/(1 - \bar{\beta}) = \mathcal{M}_2(1)$, $0 \leq i \leq n$. The solution (Q, Z) of Problem 14.3 with input data $\mathbf{w}$, $\{\mathcal{M}^{-1}(z_i)\}$ and $\{\mathcal{M}^{-1}(y_i)\}$ is such that

$$Q^H \operatorname{diag}([\mathcal{M}^{-1}(z_0), \ldots, \mathcal{M}^{-1}(z_n)])Q = Z + \operatorname{diag}([\mathcal{M}^{-1}(y_0), \ldots, \mathcal{M}^{-1}(y_n)]),$$

from which it follows that

$$\mathcal{M}(Q^H \operatorname{diag}([\mathcal{M}^{-1}(z_0), \ldots, \mathcal{M}^{-1}(z_n)])Q)$$
$$= \mathcal{M}(Z + \operatorname{diag}([\mathcal{M}^{-1}(y_0), \ldots, \mathcal{M}^{-1}(y_n)])).$$

By invoking Theorem 14.11, this relation gives

$$\mathcal{M}(Q^H \operatorname{diag}([\mathcal{M}^{-1}(z_0), \ldots, \mathcal{M}^{-1}(z_n)])Q) = Q^H D_z Q = \widehat{Z} + D_y, \quad \widehat{Z} \in \mathcal{H}_{n+1}^{(g)},$$

and, therefore, a solution of the original inverse eigenvalue problem with points $z_i \in \mathbb{T}$ is $(\widehat{Q}, \widehat{Z})$ where $\widehat{Q} = Q$ and $\widehat{Z}$ is such that

$$\widehat{Z} + D_y = \mathcal{M}(Z + \operatorname{diag}([\mathcal{M}^{-1}(y_0), \ldots, \mathcal{M}^{-1}(y_n)])). \tag{14.14}$$

Having shown in (14.11) that the matrix Z satisfies

$$Z = S = \operatorname{tril}(\mathbf{u}\mathbf{v}^T, 0) + \operatorname{triu}(\bar{\mathbf{v}}\mathbf{u}^H, 1),$$

for suitable vectors $\mathbf{u}$ and $\mathbf{v}$, we can use (14.14) to further investigate the structure of $\widehat{Z}$. From (14.14) we deduce that

$$\widehat{Z}^H + D_y^H = \tilde{\mathcal{M}}(S^H + \operatorname{diag}([\mathcal{M}^{-1}(y_0), \ldots, \mathcal{M}^{-1}(y_n)])^H).$$

The Möbius transformation $\tilde{\mathcal{M}}$ of a matrix S is defined as

$$\tilde{\mathcal{M}} = (\bar{\gamma}S + \bar{\delta}I)^{-1}(\bar{\alpha}S + \bar{\beta}I)$$

when $\mathcal{M} = (\alpha z + \beta)/(\gamma z + \delta)$. By applying again Theorem 14.11, assuming that all y_i are different from zero, this implies that

$$\widehat{Z}^H + D \in \mathcal{H}_{n+1}^{(g)},$$

for a certain diagonal matrix D. Summing up, we obtain that

$$\widehat{Z} = \operatorname{tril}(\mathbf{u}\mathbf{v}^T, 0) + \operatorname{triu}(\mathbf{p}\mathbf{q}^T, 1), \tag{14.15}$$

for suitable vectors $\mathbf{u}$, $\mathbf{v}$, $\mathbf{p}$ and $\mathbf{q}$. If one or more of the y_i are equal to zero, it can be shown that $\widehat{Z}$ is block lower triangular where each of the diagonal blocks has the desired structure. The proof is rather technical. Therefore, we omit it here.

From a computational viewpoint, these results can be used to devise several different procedures for solving Problem 14.3 in the case of points z_i lying on the unit circle at the cost of $\mathcal{O}(n^2)$ ops. By taking into account the semiseparable structure of $\widehat{S}$ (14.15) we can simply modify the algorithm stated in the previous section in such a way as to compute its upper triangular part without performing multiplications (14.10). A different approach is outlined in the next subsection.

14.3.3   Special case: all points z_i lie on a generic circle

Another approach to deal with the preceding special case that generalizes immediately to the case where the nodes z_i belong to a given circle in the complex plane, $\{z \in \mathbb{C} : |z - p| = r\}$, exploits an invariance property of Cauchy-like matrices under a Möbius transformation of the nodes. Such property is presented in the next lemma for the case of classical Cauchy matrices; the Cauchy-like case can be dealt with by introducing suitable diagonal scalings. With minor changes, all forthcoming arguments also apply to the case where all abscissas lie on a generic line in the complex plane, since the image of $\mathbb{R}$ under a Möbius transformation is either a circle or a line.

Lemma 14.12. *Let z_i, y_j, for $1 \leq i, j \leq n$, be pairwise distinct complex numbers, let*

$$\mathcal{M}(z) = \frac{\alpha z + \beta}{\gamma z + \delta}, \quad \alpha\delta - \beta\gamma \neq 0,$$

be a Möbius transformation and let $C_{\mathcal{M}} = (1/(\mathcal{M}(z_i) - \mathcal{M}(y_j)))$. Then $C_{\mathcal{M}}$ is a Cauchy-like matrix with nodes z_i, y_j.

Proof. Using the notations above, we have

$$\frac{1}{\mathcal{M}(z_i) - \mathcal{M}(y_j)} = \frac{1}{\alpha\delta - \beta\gamma} \frac{(\gamma z_i + \delta)(\gamma y_j + \delta)}{z_i - y_j}.$$

Hence $C_{\mathcal{M}}$ has the form $C_{\mathcal{M}} = (u_i v_j/(z_i - y_j))$.   □

In the next theorem, we show how to construct a Möbius transformation mapping $\mathbb{R}$ onto a prescribed circle without one point, thus generalizing Formula (14.13). Together with the preceding lemma, it will allow us to translate Problem 14.3 with nodes on a circle into a corresponding problem with real nodes. The latter can be solved with the technique laid down in Subsection 14.3.1.

Theorem 14.13. *Let the center of the circle $p \in \mathbb{C}$ and its radius $r > 0$ be given. Consider the following algorithm:*

1. *Choose arbitrary nonzero complex numbers $\gamma = |\gamma|\mathrm{e}^{\mathrm{i}\theta_\gamma}$ and $\delta = |\delta|\mathrm{e}^{\mathrm{i}\theta_\delta}$ such that $\mathrm{e}^{2\mathrm{i}(\theta_\gamma - \theta_\delta)} \neq 1$; moreover, choose $\tilde{\theta} \in [0, 2\pi]$.*

2. *Set $\alpha = p\gamma + r|\gamma|\mathrm{e}^{\mathrm{i}\tilde{\theta}}$.*

3. *Set $\hat{\theta} = \tilde{\theta} + \theta_\gamma - \theta_\delta$.*

4. *Set $\beta = p\delta + r|\delta|\mathrm{e}^{\mathrm{i}\hat{\theta}}$.*

Then the function $\mathcal{M}(z) = (\alpha z + \beta)/(\gamma z + \delta)$ is a Möbius transformation mapping the real line onto the circle $\{z \in \mathbb{C} : |z - p| = r\}$ without the point $\hat{z} = \alpha/\gamma$.

Proof. After simple manipulations, the equation

$$\left|\frac{\alpha z + \beta}{\gamma z + \delta} - p\right|^2 = r^2$$

leads to the equation

$$z^2|\alpha - p\gamma|^2 + 2z\Re((\alpha - p\gamma)\overline{(\beta - p\delta)}) + |\beta - p\delta|^2 = \qquad (14.16)$$

$$= z^2 r^2 |\gamma|^2 + 2z r^2 \Re(\gamma\bar{\delta}) + r^2 |\delta|^2.$$

Here and in the following, $\Re(z)$ denotes the real part of $z \in \mathbb{C}$. By construction, we have $|\alpha - p\gamma| = r|\gamma|$ and $|\beta - p\delta| = r|\delta|$. Moreover,

$$\Re((\alpha - p\gamma)\overline{(\beta - p\delta)}) = r^2|\gamma\delta|\Re(e^{i(\tilde{\theta}-\hat{\theta})})$$
$$= r^2|\gamma\delta|\Re(e^{i(\theta_\delta - \theta_\gamma)})$$
$$= r^2\Re(\gamma\bar{\delta}).$$

Hence Equation 14.16 is fulfilled for any real z. The missing point is given by

$$\hat{z} = \lim_{z \to \infty} \frac{\alpha z + \beta}{\gamma z + \delta} = \frac{\alpha}{\gamma}.$$

It remains to prove that $\alpha\delta - \beta\gamma \neq 0$. Indeed, we have

$$\alpha\delta - \beta\gamma = (p\gamma + r|\gamma|e^{i\tilde{\theta}})\delta - (p\delta + r|\delta|e^{i\hat{\theta}})\gamma$$
$$= r|\gamma|\delta e^{i\tilde{\theta}} - r\gamma|\delta|e^{i\hat{\theta}}$$
$$= r|\gamma\delta|(e^{i(\tilde{\theta}+\theta_\delta)} - e^{i(\hat{\theta}+\theta_\gamma)})$$
$$= r|\gamma\delta|e^{i(\tilde{\theta}+\theta_\delta)}(1 - e^{2i(\theta_\gamma - \theta_\delta)}).$$

Since $e^{2i(\theta_\gamma + \theta_\delta)} \neq 1$ we obtain $\alpha\delta - \beta\gamma \neq 0$. $\qquad\square$

Suppose we want to solve Problem 14.3 with data w_i, z_i, y_i, where $|z_i - p| = r$. As seen from the proof of Theorem 14.5, if we let $C = (w_{i-1}(z_{i-1}-y_0)/(z_{i-1}-y_{j-1}))$ and $C = QR$, then a solution is (Q, Z), with $Z = Q^H D_z Q - D_y$. Let $\mathcal{M}(z) = (\alpha z + \beta)/(\gamma z + \delta)$ be a Möbius transformation built from Theorem 14.13. Recalling the inversion Formula 14.12, let $\tilde{z}_i = \mathcal{M}^{-1}(z_i)$, $\tilde{y}_i = \mathcal{M}^{-1}(y_i)$, $v_i = \gamma\tilde{y}_i + \delta$, and

$$\tilde{w}_i = w_i \frac{z_i - y_0}{\tilde{z}_i - \tilde{y}_0} \frac{\gamma\tilde{z}_i + \delta}{\alpha\delta - \beta\gamma}, \qquad 0 \leq i \leq n.$$

Note that $\tilde{z}_i \in \mathbb{R}$, by construction. From Lemma 14.12, we also have

$$C = \left(\frac{\tilde{w}_{i-1}(\tilde{z}_{i-1} - \tilde{y}_0)v_{j-1}}{\tilde{z}_{i-1} - \tilde{y}_{j-1}}\right).$$

Again from Theorem 14.5, we see that the solution of Problem 14.3 with data $\tilde{w}_i, \tilde{z}_i, \tilde{y}_i$ is $(Q, \tilde{Z})$ where

$$\tilde{Z} = Q^H \mathcal{M}^{-1}(D_z)Q - \mathcal{M}^{-1}(D_y).$$

Let $\widehat{Z} = \widetilde{Z} + \mathcal{M}^{-1}(D_y)$. Observe that $\widehat{Z}$ is a semiseparable plus diagonal matrix [39, 66, 75]. After simple passages, we have

$$Z = \mathcal{M}(\widehat{Z}) - D_y = [\alpha\widehat{Z} + \beta I][\gamma\widehat{Z} + \delta I]^{-1} - D_y.$$

Hence Z can be recovered from $\widetilde{Z}$ by determining the entries in its first and last rows and columns. This latter task can be carried out at a linear cost by means of several different algorithms for the solution of semiseparable plus diagonal linear systems (see, e.g., [39, 66, 111, 156]).

Notes and references

This chapter is based on the following articles.

☞ M. Van Barel, D. Fasino, L. Gemignani, and N. Mastronardi. Orthogonal rational functions and diagonal plus semiseparable matrices. In F. T. Luk, editor, *Advanced Signal Processing Algorithms, Architectures, and Implementations XII*, volume 4791 of *Proceedings of SPIE, Bellingham, Washington, USA*, pages 167–170, 2002.

☞ M. Van Barel, D. Fasino, L. Gemignani, and N. Mastronardi. Orthogonal rational functions and structured matrices. *SIAM Journal on Matrix Analysis and Applications*, 26(3):810–829, 2005.

Proper rational functions are an essential tool in many areas of engineering, as system theory and digital filtering, where polynomial models are inappropriate, due to their unboundedness at infinity. In fact, for physical reasons the transfer functions describing linear time-invariant systems often have to be bounded on the real line. Furthermore, approximation problems with rational functions are in the core of, e.g., the partial realization problem, model reduction problems, robust system identification.

☞ W. B. Gragg and A. Lindquist. On the partial realization problem. *Linear Algebra and its Applications*, 50:277–319, 1983.

☞ A. Bultheel and B. L. R. De Moor. Rational approximation in linear systems and control. *Journal of Computational and Applied Mathematics*, 121:355–378, 2000.

☞ A. Bultheel and M. Van Barel. Padé techniques for model reduction in linear system theory: a survey. *Journal of Computational and Applied Mathematics*, 14:401–438, 1986.

☞ Ph. Delsarte, Y. V. Genin, and Y. Kamp. On the role of the Nevanlinna-Pick problem in circuit and system theory. *International Journal of Circuit Theory and Applications*, 9:177–187, 1981.

☞ B. Ninness and F. Gustafsson. A unifying construction of orthonormal bases for system identification. *IEEE Transactions on Automatic Control*, 42:515–521, 1997.

Recently a strong interest has been brought to a variety of rational interpolation problems where a given function is to be approximated by means of a rational function with prescribed poles.

☞ A. Bultheel, P. González-Vera, E. Hendriksen, and O. Njåstad. Orthogonal rational functions with poles on the unit circle. *Journal of Mathematical Analysis and Applications*, 182(1):221–243, 1994.

☞ M. Van Barel and A. Bultheel. Discrete linearized least squares approximation on the unit circle. *Journal of Computational and Applied Mathematics*, 50:545–563, 1994.

By linearization, such problems naturally lead to linear algebra computations involving structured matrices. Exploiting the close connections between the functional problem and its matrix counterparts generally allows us to take advantage of the special structure of these matrices to speed up the approximation scheme.

☞ R. Bevilacqua, B. Codenotti, and F. Romani. Parallel solution of block tridiagonal linear systems. *Linear Algebra and its Applications*, 104:39–57, 1988.

For example, in this article efficient algorithms are designed for rational function evaluation and interpolation from their connection with displacement structured matrices.

We also mention the recent appearance in the numerical analysis literature of quadrature formulas that are exact for sets of rational functions having prescribed poles.

☞ A. Bultheel. Quadrature and orthogonal rational functions. *Journal of Computational and Applied Mathematics*, 127:67–91, 2001.

☞ W. Gautschi. The use of rational functions in numerical quadrature. *Journal of Computational and Applied Mathematics*, 133:111–126, 2001.

Such formulas provide a greater accuracy than standard quadrature formulas, when the poles are chosen in such a way to mimic the poles present in the integrand. The construction of Gauss-type quadrature formulas is known to be a task closely related to that of orthogonalizing a set of prescribed basis functions. In the polynomial case this fact was explored in the following articles.

☞ L. Reichel. Fast QR-decomposition of Vandermonde-like matrices and polynomial least squares approximation. *SIAM Journal on Matrix Analysis and Applications*, 12:552–564, 1991.

☞ L. Reichel. Construction of polynomials that are orthogonal with respect to a discrete bilinear form. *Advances in Computational Mathematics*, 1:241–258, 1993.

☞ W. B. Gragg and W. J. Harrod. The numerically stable reconstruction of Jacobi matrices from spectral data. *Numerische Mathematik*, 44:317–335, 1984.

Indeed, in these articles the construction of polynomial sequences that are orthogonal with respect to a discrete inner product by means of their three-term recurrence relation is tied to the solution of an inverse eigenvalue problem for symmetric tridiagonal matrices that is equivalent to the construction of Gauss quadrature formulas.

Orthogonal rational functions are also useful in solving multipoint generalizations of classical moment problems and associated interpolation problems. This and several other connections can be found in the book of Bultheel et al.

☞ A. Bultheel, P. González-Vera, E. Hendriksen, and O. Njåstad. *Orthogonal Rational Functions*, volume 5 of *Cambridge Monographs on Applied and Computational Mathematics*. Cambridge University Press, Cambridge, United Kingdom, 1999.

In the previous chapter we generalized the sequence of orthogonal polynomials into a sequence of orthonormal polynomial vectors. In the same way, we can extend the results of this chapter towards orthonormal rational function vectors.

☞ S. Delvaux and M. Van Barel. Orthonormal rational function vectors. *Numerische Mathematik*, 100(3):409–440, May 2005.

14.4   Conclusions

In this chapter we have considered computing the recurrence relation for orthonormal rational functions with prescribed poles. It turned out that solving the corresponding inverse eigenvalue problem means that we have to determine a diagonal plus lower semiseparable matrix. On the real line, this becomes a symmetric semiseparable matrix and on the unit circle, the diagonal plus lower semiseparable matrix is unitary.

Chapter 15

Concluding remarks & software

15.1 Software

In several chapters of this book implementations of various algorithms were discussed. Several of these methods are implemented and freely available for download at the following site:

http://www.cs.kuleuven.be/~mase/books/

The package containing several routines related to semiseparable matrices is called SSPack, the Semiseparable Software Package. The package is still under development, hence it is good to check the site on a regular base.

Important to remark is that currently the MaSe-team is implementing several of the routines in C++, using thereby GLAS (Generic Linear Algebra Software), which exploits BLAS and so forth. The above site will also discuss progress in this project.

Let us provide an example of a routine in MATLAB, to illustrate the global package. For example the routine:

CSS: Construction of a semiseparable matrix

Providing MATLAB the following command:

>> help CSS

will provide the following output:

```
% CSS    Construction of a semiseparable matrix
%
%        [G,d]=CSS(S)
%        produces the representation of a semiseparable
%        matrix from the symmetric matrix S.
%        The fastest Householder implementation is used.
%        G represents a sequence of givens
%        d represents a diagonal
%
%        [G,d]=CSS(S,'h')
%        performs the computation with a more stable, but
```

```
%       slower Householder implementation.
%
%       [G,d]=CSS(S,'g')
%       performs the computation with Givens transformations.
%       This is the most stable but also the slowest.
%
%       Software of the MaSe - Group
%       mase@cs.kuleuven.be
%       Revision Date: 16/12/2003
```

This routine transforms a symmetric matrix into a similar semiseparable one. The package contains many other implementations of proposed algorithms, such as for example the multishift QR-step, the singular value decomposition based on upper triangular semiseparable matrices, the divide-and-conquer algorithm and many others.

For example the routine which computes the eigenvalues and eigenvectors of semiseparable matrices is of the following form.

```
% EIGSS     Eigenvalues and eigenvectors of semiseparable matrices
%
%       E = EIGSS(G,d,cutoff) is a vector containing the eigenvalues of
%       the symmetric semiseparable matrix S constructed with G,d.
%       (Check BSS and CSS for info on the representation G,d.)
%       Remark that the matrix has to be in unreduced form.
%       (Check REDSS for info about unreducedness.)
%
%
%       [V,D] = EIGSS(G,d,cutoff) produces a diagonal matrix D of
%       eigenvalues and a full matrix V whose columns are the
%       corresponding eigenvectors so that S*V = V*D,
%       with S the semiseparable matrix constructed with G,d.
%
%       The variable cutoff stands for the cutting off criterion.
%       If the subdiagonal element is smaller than cutoff the
%       corresponding eigenvalue is separated.
%
%
%       Software of the MaSe - Group
%       mase@cs.kuleuven.be
%       Revision Date: 30/05/2003
```

15.2   Conclusions

The first volume of this book focused onto solving systems of linear equations, involving structured rank matrices. This volume was dedicated to direct and inverse eigenvalue problems related to and based on, structured rank matrices.

The first part of the book discussed different algorithms for transforming matrices to structured rank form. Transitions to semiseparable, semiseparable plus diagonal, Hessenberg-like, upper triangular semiseparable form were explored. Moreover also the convergence properties of these methods were studied extensively, showing how to tune this convergence.

Being able to transform matrices to structured rank form is not enough to compute eigenvalues and/or singular values. The second part of the book discusses therefore all necessary tools and theoretical results for developing QR-algorithms, computing the eigenvalues of the previously mentioned structured rank matrices. An extra chapter also presented a new kind of algorithm for computing the eigendecomposition via a QH-factorization.

The third part of the book briefly discussed some miscellaneous topics such as the reduction to semiseparable form in an iterative (Lanczos-like) way and how to exploit the rank-revealing properties of the reduction algorithms. Also the divide-and-conquer algorithm for computing the eigendecomposition was discussed.

The fourth part of the book discussed some inverse eigenvalue problems, showing relations between structured rank matrices and orthogonal polynomials.

We hope that the two volumes give the reader a thorough introduction in the field of structured rank matrices and can lead to a standardization of the notation in this strongly evolving field of research and applications.

Bibliography

[1] A. Abdallah and Y. Hu, *Parallel VLSI computing array implementation for signal subspace updating algorithm*, IEEE Transactions on Acoustics, Speech and Signal Processing **37** (1989), 742–748. {**64, 471**}

[2] G. S. Ammar, D. Calvetti, W. B. Gragg, and L. Reichel, *Polynomial zerofinders based on Szegő polynomials*, Journal of Computational and Applied Mathematics **127** (2001), 1–16. {**353, 471**}

[3] G. S. Ammar, D. Calvetti, and L. Reichel, *Continuation methods for the computation of zeros of Szegő polynomials*, Linear Algebra and its Applications **249** (1996), 125–155. {**354, 471**}

[4] G. S. Ammar and W. B. Gragg, $\mathcal{O}(n^2)$ *reduction algorithms for the construction of a band matrix from spectral data*, SIAM Journal on Matrix Analysis and Applications **12** (1991), no. 3, 426–432. {**435, 471**}

[5] _____, *Schur flows for orthogonal Hessenberg matrices*, Hamiltonian and Gradient Flows, Algorithms and Control (A. M. Bloch, ed.), vol. 3, American Mathematical Society, Providence, Rhode Island, 1994, pp. 27–34. {**353, 471**}

[6] G. S. Ammar, W. B. Gragg, and C. He, *An efficient QR algorithm for a Hessenberg submatrix of a unitary matrix*, New Directions and Applications in Control Theory (W. Dayawansa, A. Lindquist, and Y. Zhou, eds.), Lecture Notes in Control and Information Sciences, vol. 321, Springer-Verlag, Berlin, Germany, 2005, pp. 1–14. {**354, 471**}

[7] G. S. Ammar, W. B. Gragg, and L. Reichel, *On the eigenproblem for orthogonal matrices*, Proceedings of the 25th IEEE Conference on Decision & Control, IEEE, New York, USA, 1986, pp. 1963–1966. {**353, 471**}

[8] _____, *Constructing a unitary Hessenberg matrix from spectral data*, Numerical Linear Algebra, Digital Signal Processing and Parallel Algorithms (G. H. Golub and P. Van Dooren, eds.), Computer and Systems Sciences, vol. 70, Springer-Verlag, Berlin, Germany, 1991, pp. 385–395. {**419, 424, 426, 471**}

[9] _____, *Downdating of Szegő polynomials and data-fitting applications*, Linear Algebra and its Applications **172** (1992), 315–336. {**424, 427, 471**}

[10] G. S. Ammar and C. He, *On an inverse eigenvalue problem for unitary Hessenberg matrices*, Linear Algebra and its Applications **218** (1995), 263–271. {**426, 472**}

[11] G. S. Ammar, L. Reichel, and D. C. Sorensen, *An implementation of a divide and conquer algorithm for the unitary eigenproblem*, ACM Transactions on Mathematical Software **18** (1992), no. 3, 292–307. {**353, 472**}

[12] S. O. Asplund, *Finite boundary value problems solved by Green's matrix*, Mathematica Scandinavica **7** (1959), 49–56. {**13, 472**}

[13] S. Barnett, *Polynomials and Linear Control Systems*, Monographs and Textbooks in Pure and Applied Mathematics, Marcel Dekker, Inc., New York, USA, 1983. {**327, 352, 472**}

[14] W. W. Barrett, *A theorem on inverses of tridiagonal matrices*, Linear Algebra and its Applications **27** (1979), 211–217. {**6, 9, 472**}

[15] W. W. Barrett and P. J. Feinsilver, *Inverses of banded matrices*, Linear Algebra and its Applications **41** (1981), 111–130. {**6, 9, 472**}

[16] R. Bevilacqua, E. Bozzo, and G. M. Del Corso, *Transformations to rank structures by unitary similarity*, Linear Algebra and its Applications **402** (2005), 126–134. {**360, 472**}

[17] R. Bevilacqua, B. Codenotti, and F. Romani, *Parallel solution of block tridiagonal linear systems*, Linear Algebra and its Applications **104** (1988), 39–57. {**465, 472**}

[18] R. Bevilacqua and G. M. Del Corso, *Structural properties of matrix unitary reduction to semiseparable form*, Calcolo **41** (2004), no. 4, 177–202. {**40, 182, 188, 193, 194, 217, 472**}

[19] D. Bindel, S. Chandrasekaran, J. W. Demmel, D. Garmire, and M. Gu, *A fast and stable nonsymmetric eigensolver for certain structured matrices*, Tech. report, Department of Computer Science, University of California, Berkeley, California, USA, May 2005. {**349, 472**}

[20] D. Bindel, J. W. Demmel, W. Kahan, and O. A. Marques, *On computing Givens rotations reliably and efficiently*, ACM Transactions on Mathematical Software **28** (2002), no. 2, 206–238. {**22, 24, 279, 280, 472**}

[21] D. A. Bini, F. Daddi, and L. Gemignani, *On the shifted QR iteration applied to companion matrices*, Electronic Transactions on Numerical Analysis **18** (2004), 137–152. {**289, 349, 472**}

[22] D. A. Bini, Y. Eidelman, L. Gemignani, and I. C. Gohberg, *Fast QR eigenvalue algorithms for Hessenberg matrices which are rank-one perturbations of unitary matrices*, SIAM Journal on Matrix Analysis and Applications **29** (2007), no. 2, 566–585. {**289, 350, 472**}

[23] D. A. Bini, L. Gemignani, and V. Y. Pan, *Fast and stable QR eigenvalue algorithms for generalized companion matrices and secular equations*, Numerische Mathematik **100** (2005), no. 3, 373–408. {**176, 182, 201, 289, 352, 473**}

[24] D. A. Bini and V. Y. Pan, *Practical improvement of the divide-and-conquer eigenvalue algorithms*, Computing **48** (1992), no. 1, 109–123. {**378, 473**}

[25] D. L. Boley and G. H. Golub, *A survey of matrix inverse eigenvalue problems*, Inverse Problems **3** (1987), 595–622. {**370, 424, 425, 435, 473**}

[26] C. F. Borges, R. Frezza, and W. B. Gragg, *Some inverse eigenproblems for Jacobi and arrow matrices*, Numerical Linear Algebra with Applications **2** (1995), no. 3, 195–203. {**378, 387, 388, 473**}

[27] C. F. Borges and W. B. Gragg, *Divide and conquer for generalized real symmetric definite tridiagonal eigenproblems*, Proceedings of '92 Shanghai International Numerical Algebra and its Applications Conference (E.-X. Jiang, ed.), China Science and Technology Press, Beijing, China, October 1992, pp. 70–76. {**130, 370, 372, 375, 473**}

[28] _____, *A parallel divide and conquer algorithm for the generalized real symmetric definite tridiagonal eigenproblem*, Numerical Linear Algebra and Scientific Computing (L. Reichel, A. Ruttan, and R. S. Varga, eds.), de Gruyter, Berlin, Germany, 1993, pp. 11–29. {**367, 370, 372, 375, 378, 387, 473**}

[29] M. Braun, S. A. Sofianos, D. G. Papageorgiou, and I. E. Lagaris, *An efficient Chebyshev Lanczos method for obtaining eigensolutions of the Schrödinger equation on a grid*, Journal of Computational Physics **126** (1996), no. 2, 315–327. {**85, 473**}

[30] A. Bultheel, *Quadrature and orthogonal rational functions*, Journal of Computational and Applied Mathematics **127** (2001), 67–91. {**465, 473**}

[31] A. Bultheel, A. M. Cuyt, W. Van Assche, M. Van Barel, and B. Verdonk, *Generalizations of orthogonal polynomials*, Journal of Computational and Applied Mathematics **179** (2005), no. 1–2, 57–95. {**425, 473**}

[32] A. Bultheel and B. L. R. De Moor, *Rational approximation in linear systems and control*, Journal of Computational and Applied Mathematics **121** (2000), 355–378. {**464, 473**}

[33] A. Bultheel, P. González-Vera, E. Hendriksen, and O. Njåstad, *Orthogonal rational functions with poles on the unit circle*, Journal of Mathematical Analysis and Applications **182** (1994), no. 1, 221–243. {**465, 473**}

[34] _____, *Orthogonal Rational Functions*, Cambridge Monographs on Applied and Computational Mathematics, vol. 5, Cambridge University Press, Cambridge, United Kingdom, 1999. {**465, 473**}

[35] A. Bultheel and M. Van Barel, *Padé techniques for model reduction in linear system theory: a survey*, Journal of Computational and Applied Mathematics **14** (1986), 401–438. {**464, 474**}

[36] _____, *Vector orthogonal polynomials and least squares approximation*, SIAM Journal on Matrix Analysis and Applications **16** (1995), no. 3, 863–885. {**425, 445, 474**}

[37] _____, *Linear Algebra, Rational Approximation and Orthogonal Polynomials*, Studies in computational mathematics, vol. 6, North-Holland, Elsevier Science B.V., Amsterdam, Netherlands, 1997. {**425, 474**}

[38] S. Chandrasekaran and M. Gu, *Fast and stable eigendecomposition of symmetric banded plus semi-separable matrices*, Linear Algebra and its Applications **313** (2000), 107–114. {**3, 32, 63, 65, 122, 474**}

[39] _____, *A fast and stable solver for recursively semi-separable systems of equations*, Structured Matrices in Mathematics, Computer Science and Engineering, II (V. Olshevsky, ed.), Contemporary Mathematics, vol. 281, American Mathematical Society, Providence, Rhode Island, USA, 2001, pp. 39–53. {**464, 474**}

[40] _____, *A divide-and-conquer algorithm for the eigendecomposition of symmetric block-diagonal plus semiseparable matrices*, Numerische Mathematik **96** (2004), no. 4, 723–731. {**32, 131, 367, 379, 387, 388, 474**}

[41] S. Chandrasekaran, M. Gu, J. Xia, and J. Zhu, *A fast QR algorithm for companion matrices*, Operator Theory: Advances and Applications **179** (2007), 111–143. {**351, 474**}

[42] M. T. Chu and G. H. Golub, *Inverse Eigenvalue Problems: Theory, Algorithms and Applications*, Numerical Mathematics & Scientific Computations, Oxford University Press, New York, USA, 2005. {**425, 427, 474**}

[43] J. K. Cullum and R. A. Willoughby, *Lanczos algorithms for large symmetric eigenvalue computations*, Birkhäuser, Boston, Massachusetts, USA, 1985. {**68, 80, 474**}

[44] J. J. M. Cuppen, *A divide and conquer method for the symmetric tridiagonal eigenproblem*, Numerische Mathematik **36** (1981), 177–195. {**130, 370, 374, 378, 474**}

[45] R. J. A. David and D. S. Watkins, *Efficient implementation of the multishift QR algorithm for the unitary eigenvalue problem*, SIAM Journal on Matrix Analysis and Applications **28** (2006), no. 3, 623–633. {**353, 474**}

[46] Ph. Delsarte, Y. V. Genin, and Y. Kamp, *On the role of the Nevanlinna-Pick problem in circuit and system theory*, International Journal of Circuit Theory and Applications **9** (1981), 177–187. {**464, 474**}

[47] S. Delvaux, *Rank Structured Matrices*, PhD thesis, Department of Computer Science, Katholieke Universiteit Leuven, Celestijnenlaan 200A, 3000 Leuven (Heverlee), Belgium, June 2007. {**361, 475**}

[48] S. Delvaux and M. Van Barel, *Orthonormal rational function vectors*, Numerische Mathematik **100** (2005), no. 3, 409–440. {**466, 475**}

[49] _____, *The explicit QR-algorithm for rank structured matrices*, Tech. Report TW459, Department of Computer Science, Katholieke Universiteit Leuven, Celestijnenlaan 200A, 3000 Leuven (Heverlee), Belgium, May 2006. {**198, 259, 289, 361, 475**}

[50] _____, *Rank structures preserved by the QR-algorithm: the singular case*, Journal of Computational and Applied Mathematics **189** (2006), 157–178. {**182, 195, 197, 361, 475**}

[51] _____, *Structures preserved by matrix inversion*, SIAM Journal on Matrix Analysis and Applications **28** (2006), no. 1, 213–228. {**7, 10, 475**}

[52] _____, *Structures preserved by the QR-algorithm*, Journal of Computational and Applied Mathematics **187** (2006), no. 1, 29–40. {**182, 195, 197, 361, 475**}

[53] _____, *A Givens-weight representation for rank structured matrices*, SIAM Journal on Matrix Analysis and Applications **29** (2007), no. 4, 1147–1170. {**12, 13, 111, 122, 259, 361, 475**}

[54] _____, *A Hessenberg reduction algorithm for rank structured matrices*, SIAM Journal on Matrix Analysis and Applications **29** (2007), no. 3, 895–926. {**65, 122, 361, 475**}

[55] _____, *Eigenvalue computation for unitary rank structured matrices*, Journal of Computational and Applied Mathematics **213** (2008), no. 1, 268–287. {**222, 259, 289, 353, 475**}

[56] _____, *A QR-based solver for rank structured matrices*, SIAM Journal on Matrix Analysis and Applications **30** (2008), no. 2, 464–490. {**158, 175, 181, 361, 475**}

[57] _____, *Unitary rank structured matrices*, Journal of Computational and Applied Mathematics **215** (2008), no. 1, 268–287. {**122, 175, 353, 475**}

[58] J. W. Demmel, *Applied Numerical Linear Algebra*, SIAM, Philadelphia, Pennsylvania, USA, 1997. {**24, 68, 70, 80, 157, 160, 475**}

[59] P. Dewilde and A.-J. van der Veen, *Time-Varying Systems and Computations*, Kluwer Academic Publishers, Boston, Massachusetts, USA, 1998. {**181, 298, 475**}

[60] _____, *Inner-outer factorization and the inversion of locally finite systems of equations*, Linear Algebra and its Applications **313** (2000), 53–100. {**298, 476**}

[61] I. S. Dhillon, *Current inverse iteration software can fail*, BIT **38** (1998), 685–704. {**270, 476**}

[62] I. S. Dhillon and A. N. Malyshev, *Inner deflation for symmetric tridiagonal matrices*, Linear Algebra and its Applications **358** (2003), 139–144. {**267, 268, 476**}

[63] F. Diele, N. Mastronardi, M. Van Barel, and E. Van Camp, *On computing the spectral decomposition of symmetric arrowhead matrices*, Lecture Notes in Computer Science **3044** (2004), 932–941. {**370, 374, 476**}

[64] J. J. Dongarra and D. C. Sorensen, *A fully parallel algorithm for the symmetric eigenvalue problem*, SIAM Journal on Scientific and Statistical Computation **3** (1987), no. 2, 139–154. {**369, 370, 374, 378, 476**}

[65] Y. Eidelman, L. Gemignani, and I. C. Gohberg, *On the fast reduction of a quasiseparable matrix to Hessenberg and tridiagonal forms*, Linear Algebra and its Applications **420** (2007), no. 1, 86–101. {**65, 122, 476**}

[66] Y. Eidelman and I. C. Gohberg, *A look-ahead block Schur algorithm for diagonal plus semiseparable matrices*, Computers & Mathematics with Applications **35** (1997), no. 10, 25–34. {**464, 476**}

[67] _____, *On a new class of structured matrices*, Integral Equations and Operator Theory **34** (1999), 293–324. {**14, 298, 476**}

[68] _____, *A modification of the Dewilde-van der Veen method for inversion of finite structured matrices*, Linear Algebra and its Applications **343–344** (2002), 419–450. {**298, 476**}

[69] Y. Eidelman, I. C. Gohberg, and V. Olshevsky, *Eigenstructure of order-one-quasiseparable matrices. Three-term and two-term recurrence relations*, Linear Algebra and its Applications **405** (2005), 1–40. {**360, 476**}

[70] _____, *The QR iteration method for Hermitian quasiseparable matrices of an arbitrary order*, Linear Algebra and its Applications **404** (2005), 305–324. {**65, 198, 289, 292, 299, 476**}

[71] S. Elhay, G. H. Golub, and J. Kautsky, *Updating and downdating of orthogonal polynomials with data fitting applications*, SIAM Journal on Matrix Analysis and Applications **12** (1991), no. 2, 327–353. {**424, 426, 476**}

[72] D. Fasino, *Rational Krylov matrices and QR-steps on hermitian diagonal-plus-semiseparable matrices*, Numerical Linear Algebra with Applications **12** (2005), no. 8, 743–754. {**222, 354–356, 476**}

[73] D. Fasino and L. Gemignani, *Structural and computational properties of possibly singular semiseparable matrices*, Linear Algebra and its Applications **340** (2001), 183–198. {**3, 477**}

[74] ———, *A Lanczos type algorithm for the QR-factorization of regular Cauchy matrices*, Numerical Linear Algebra with Applications **9** (2002), 305–319. {**447, 450, 451, 477**}

[75] ———, *Direct and inverse eigenvalue problems, for diagonal-plus-semiseparable matrices*, Numerical Algorithms **34** (2003), 313–324. {**387, 450, 464, 477**}

[76] ———, *A Lanczos-type algorithm for the QR factorization of Cauchy-like matrices*, Fast Algorithms for Structured Matrices: Theory and Applications (V. Olshevsky, ed.), Contemporary Mathematics, vol. 323, American Mathematical Society, Providence, Rhode Island, USA, 2003, pp. 91–104. {**447, 450, 477**}

[77] D. Fasino, N. Mastronardi, and M. Van Barel, *Fast and stable algorithms for reducing diagonal plus semiseparable matrices to tridiagonal and bidiagonal form*, Fast Algorithms for Structured Matrices: Theory and Applications, Contemporary Mathematics, vol. 323, American Mathematical Society, Providence, Rhode Island, USA, 2003, pp. 105–118. {**32, 64, 122, 477**}

[78] M. Fiedler, *Basic matrices*, Linear Algebra and its Applications **373** (2003), 143–151. {**13, 477**}

[79] M. Fiedler and T. L. Markham, *Completing a matrix when certain entries of its inverse are specified*, Linear Algebra and its Applications **74** (1986), 225–237. {**5, 6, 9, 477**}

[80] M. Fiedler and Z. Vavřín, *Generalized Hessenberg matrices*, Linear Algebra and its Applications **380** (2004), 95–105. {**13, 477**}

[81] R. D. Fierro and P. Chr. Hansen, *Truncated VSV solutions to symmetric rank-deficient problems*, BIT **42** (2002), no. 3, 531–540. {**404, 407, 409, 477**}

[82] ———, *UTV expansion pack: special-purpose rank-revealing algorithms*, Numerical Algorithms **40** (2005), 47–66. {**407, 409, 477**}

[83] R. D. Fierro, P. Chr. Hansen, and P. S. K. Hansen, *UTV tools, Matlab templates for rank-revealing UTV decompositions*, Numerical Algorithms **20** (1999), 165–194. {**407, 409, 477**}

[84] F. R. Gantmacher and M. G. Kreĭn, *Sur les matrices oscillatoires et complètement non négatives*, Compositio Mathematica **4** (1937), 445–476, (In French). {**452, 477**}

[85] _____, *Oscillation Matrices and Kernels and Small Vibrations of Mechanical Systems*, revised ed., AMS Chelsea Publishing, Providence, Rhode Island, USA, 2002. {**5, 478**}

[86] W. Gautschi, *On generating orthogonal polynomials*, SIAM Journal on Scientific and Statistical Computation **3** (1982), 289–317. {**426, 478**}

[87] _____, *Computational problems and applications of orthogonal polynomials*, Orthogonal Polynomials and Their Applications (C. Brezinski, L. Gori, and A. Ronveaux, eds.), IMACS Annals on Computing and Applied Mathematics, vol. 9, Baltzer Science Publishers, Basel, Switzerland, 1991, pp. 61–71. {**426, 478**}

[88] _____, *The use of rational functions in numerical quadrature*, Journal of Computational and Applied Mathematics **133** (2001), 111–126. {**465, 478**}

[89] _____, *Orthogonal Polynomials: Computation and Approximation*, Oxford University Press, New York, USA, 2004. {**425, 478**}

[90] L. Gemignani, *A unitary Hessenberg QR-based algorithm via semiseparable matrices*, Journal of Computational and Applied Mathematics **184** (2005), 505–517. {**352, 478**}

[91] I. C. Gohberg and V. Olshevsky, *Complexity of multiplication with vectors for structured matrices*, Linear Algebra and its Applications **202** (1994), 163–192. {**388, 478**}

[92] _____, *Fast algorithms with preprocessing for matrix-vector multiplication problems*, Journal of Complexity **10** (1994), no. 4, 411–427. {**388, 478**}

[93] G. H. Golub and W. Kahan, *Calculating the singular values and pseudo-inverse of a matrix.*, SIAM Journal on Numerical Analysis **2** (1965), 205–224. {**44, 55, 170, 478**}

[94] G. H. Golub and C. F. Van Loan, *Matrix Computations*, third ed., Johns Hopkins University Press, Baltimore, Maryland, USA, 1996. {**22–24, 40, 52, 68, 70, 81, 103, 157, 160, 163, 167, 168, 170, 183, 184, 186, 205, 223, 263, 300, 312, 320, 357, 360, 368, 395, 403, 478**}

[95] G. H. Golub and J. H. Welsch, *Calculation of gauss quadrature rules*, Mathematics of Computation **23** (1969), 221–230. {**427, 478**}

[96] W. B. Gragg, *The QR algorithm for unitary Hessenberg matrices*, Journal of Computational and Applied Mathematics **16** (1986), 1–8. {**353, 424, 478**}

[97] W. B. Gragg and W. J. Harrod, *The numerically stable reconstruction of Jacobi matrices from spectral data*, Numerische Mathematik **44** (1984), 317–335. {**419, 424, 425, 465, 478**}

[98] W. B. Gragg and A. Lindquist, *On the partial realization problem*, Linear Algebra and its Applications **50** (1983), 277–319. {**464, 478**}

[99] W. B. Gragg and L. Reichel, *A divide and conquer algorithm for the unitary eigenproblem*, Hypercube Multiprocessors 1987 (M. T. Heath, ed.), SIAM, Philadelphia, Pennsylvania, USA, 1987, pp. 639–647. {**353, 479**}

[100] _____, *A divide and conquer method for unitary and orthogonal eigenproblems*, Numerische Mathematik **57** (1990), 695–718. {**424, 479**}

[101] M. Gu and S. C. Eisenstat, *A stable and efficient algorithm, for the rank-one modification of the symmetric eigenproblem*, SIAM Journal on Matrix Analysis and Applications **15** (1994), no. 4, 1266–1276. {**372, 375, 479**}

[102] _____, *A divide-and-conquer algorithm for the symmetric tridiagonal eigenproblem*, SIAM Journal on Matrix Analysis and Applications **16** (1995), no. 1, 172–191. {**375, 378, 479**}

[103] W. H. Gustafson, *A note on matrix inversion*, Linear Algebra and its Applications **57** (1984), 71–73. {**5, 9, 479**}

[104] P. Chr. Hansen and P. Y. Yalamov, *Computing symmetric rank-revealing decompositions via triangular factorization*, SIAM Journal on Matrix Analysis and Applications **23** (2001), no. 2, 443–458. {**404, 405, 407, 479**}

[105] S. Helsen, A. B. J. Kuijlaars, and M. Van Barel, *Convergence of the isometric Arnoldi process*, SIAM Journal on Matrix Analysis and Applications **26** (2005), no. 3, 782–809. {**80, 479**}

[106] P. Henrici, *Applied and Computational Complex Analysis*, Wiley, West Sussex, United Kingdom, 1974. {**459, 479**}

[107] I. C. F. Ipsen, *Computing an eigenvector with inverse iteration*, SIAM Review **39** (1997), no. 2, 254–291. {**270, 479**}

[108] I. T. Joliffe, *Principal Component Analysis*, Springer-Verlag, Berlin, Germany, 1986. {**85, 479**}

[109] O. D. Kellogg, *The oscillation of functions of an orthogonal set*, American Journal of Mathematics **38** (1916), 1–5. {**4, 479**}

[110] _____, *Orthogonal functions sets arising from integral equations*, American Journal of Mathematics **40** (1918), 145–154. {**4, 479**}

[111] I. Koltracht, *Linear complexity algorithm for semiseparable matrices*, Integral Equations and Operator Theory **29** (1997), no. 3, 313–319. {**464, 479**}

[112] A. B. J. Kuijlaars, *Which eigenvalues are found by the Lanczos method?*, SIAM Journal on Matrix Analysis and Applications **22** (2000), no. 1, 306–321. {**80, 133, 479**}

[113] R.-C. Li, *Solving secular equations stably and efficiently*, LAPACK Working Note 89, April 1993. {**370, 374, 479**}

[114] F. T. Luk and S. Qiao, *A symmetric rank-revealing Toeplitz matrix decomposition*, Journal of VLSI Signal Processing **14** (1996), no. 1, 19–28. {**405, 480**}

[115] N. Mastronardi, S. Chandrasekaran, and S. Van Huffel, *Fast and stable reduction of diagonal plus semi-separable matrices to tridiagonal and bidiagonal form*, BIT **41** (2003), no. 1, 149–157. {**3, 57, 59, 64, 65, 122, 480**}

[116] N. Mastronardi, M. Schuermans, M. Van Barel, R. Vandebril, and S. Van Huffel, *A Lanczos-like reduction of symmetric structured matrices into semiseparable ones*, Calcolo **42** (2005), no. 3–4, 227–241. {**404, 480**}

[117] N. Mastronardi, M. Van Barel, and E. Van Camp, *Divide-and-conquer algorithms for computing the eigendecomposition of symmetric diagonal-plus-semiseparable matrices*, Numerical Algorithms **39** (2005), no. 4, 379–398. {**32, 131, 367, 387, 480**}

[118] N. Mastronardi, M. Van Barel, E. Van Camp, and R. Vandebril, *On computing the eigenvectors of a class of structured matrices*, Journal of Computational and Applied Mathematics **189** (2006), 580–591. {**270, 480**}

[119] N. Mastronardi, M. Van Barel, and R. Vandebril, *Computing the rank revealing factorization by the semiseparable reduction*, Tech. Report TW418, Department of Computer Science, Katholieke Universiteit Leuven, Celestijnenlaan 200A, 3000 Leuven (Heverlee), Belgium, May 2005. {**409, 480**}

[120] A. Melman, *Numerical solution of a secular equation*, Numerische Mathematik **69** (1995), 483–493. {**374, 480**}

[121] _____, *A numerical comparison of methods for solving secular equations*, Journal of Computational and Applied Mathematics **86** (1997), 237–249. {**374, 480**}

[122] _____, *A unifying convergence analysis of second-order methods for secular equations*, Mathematics of Computation **66** (1997), no. 217, 333–344. {**374, 480**}

[123] _____, *Analysis of third-order methods for secular equations*, Mathematics of Computation **67** (1998), no. 221, 271–286. {**374, 480**}

[124] G. Meurant, *A review of the inverse of symmetric tridiagonal and block tridiagonal matrices*, SIAM Journal on Matrix Analysis and Applications **13** (1992), 707–728. {**10, 480**}

[125] B. Ninness and F. Gustafsson, *A unifying construction of orthonormal bases for system identification*, IEEE Transactions on Automatic Control **42** (1997), 515–521. {**464, 480**}

[126] D. P. O'Leary and G. W. Stewart, *Computing the eigenvalues and eigenvectors of symmetric arrowhead matrices*, Journal of Computational Physics **90** (1990), no. 2, 497–505. {**374, 480**}

[127] C. C. Paige, *The Computation of Eigenvalues and Eigenvectors of Very Large Sparse Matrices*, PhD thesis, University of London, London, United Kingdom, 1971. {**395, 481**}

[128] V. Y. Pan, *A reduction of the matrix eigenproblem to polynomial rootfinding via similarity transforms into arrow-head matrices*, Tech. Report TR-2004009, Department of Computer Science, City University of New York, New York, USA, July 2004. {**65, 360, 481**}

[129] B. N. Parlett, *The Symmetric Eigenvalue Problem*, Classics in Applied Mathematics, vol. 20, SIAM, Philadelphia, Pennsylvania, USA, 1998. {**24, 40, 68, 160, 183, 267, 271, 320, 357, 368, 481**}

[130] B. N. Parlett and D. S. Scott, *The Lanczos algorithm with selective reorthogonalization*, Mathematics of Computation **33** (1979), no. 145, 217–238. {**395, 481**}

[131] B. Plestenjak, M. Van Barel, and E. Van Camp, *A cholesky LR algorithm for the positive definite symmetric diagonal-plus-semiseparable eigenproblem*, Linear Algebra and its Applications **428** (2008), 586–599. {**360, 481**}

[132] L. Reichel, *Fast QR-decomposition of Vandermonde-like matrices and polynomial least squares approximation*, SIAM Journal on Matrix Analysis and Applications **12** (1991), 552–564. {**419, 424, 426, 447, 465, 481**}

[133] _____, *Construction of polynomials that are orthogonal with respect to a discrete bilinear form*, Advances in Computational Mathematics **1** (1993), 241–258. {**465, 481**}

[134] L. Reichel, G. S. Ammar, and W. B. Gragg, *Discrete least squares approximation by trigonometric polynomials*, Mathematics of Computation **57** (1991), 273–289. {**424, 426, 481**}

[135] H. Rutishauser, *On Jacobi rotation patterns*, Experimental Arithmetics, High Speed Computing and Mathematics, Proceedings of Symposia in Applied Mathematics (N. C. Metropolis, A. H. Taub, J. Todd, and C. B. Tompkins, eds.), vol. 15, American Mathematical Society, Providence, Rhode Island, 1963, pp. 219–239. {**424, 426, 481**}

[136] Y. Saad, *Numerical Methods for Large Eigenvalue Problems*, Manchester University Press, Manchester, United Kingdom, 1992. {**68, 80, 481**}

[137] B. Simon, *Analysis of the symmetric lanczos algorithm with reorthogonalization methods*, Linear Algebra and its Applications **61** (1984), 101–131. {**395, 481**}

[138] _____, *Orthogonal Polynomials on the Unit Circle: Part 1: Classical Theory; Part 2: Spectral Theory*, Colloquium Publications, vol. 54, American Mathematical Society, Providence, Rhode Island, USA, 2004. {**425, 481**}

[139] H. D. Simon, *The lanczos algorithm, with partial reorthogonalization*, Mathematics of Computation **42** (1984), no. 165, 115–142. {**395, 482**}

[140] G. W. Stewart, *Matrix Algorithms, Volume I: Basic Decompositions*, SIAM, Philadelphia, Pennsylvania, USA, 1998. {**157, 160, 482**}

[141] _____, *The QLP approximation to the singular value decomposition*, SIAM Journal on Scientific and Statistical Computation **20** (1999), no. 4, 1336–1348. {**138, 482**}

[142] _____, *Matrix Algorithms, Volume II: Eigensystems*, SIAM, Philadelphia, Pennsylvania, USA, 2001. {**40, 160, 265, 482**}

[143] M. Stewart, *Stability properties of several variants of the unitary Hessenberg QR-algorithm in structured matrices in mathematics*, Structured Matrices in Mathematics, Computer Science and Engineering, II (V. Olshevsky, ed.), Contemporary Mathematics, vol. 281, American Mathematical Society, Providence, Rhode Island, USA, 2001, pp. 57–72. {**353, 482**}

[144] G. Szegő, *Orthogonal Polynomials*, fourth ed., American Mathematical Society, Providence, Rhode Island, USA, 1975. {**332, 425, 482**}

[145] L. N. Trefethen and D. Bau, *Numerical Linear Algebra*, SIAM, Philadelphia, Pennsylvania, USA, 1997. {**24, 40, 68, 157, 160, 170, 482**}

[146] E. E. Tyrtyshnikov, *Mosaic ranks for weakly semiseparable matrices*, Large-Scale Scientific Computations of Engineering and Environmental Problems II (M. Griebel, S. Margenov, and P. Y. Yalamov, eds.), Notes on Numerical Fluid Mechanics, vol. 73, Vieweg, Braunschweig, Germany, 2000, pp. 36–41. {**14, 298, 482**}

[147] M. Van Barel and A. Bultheel, *A parallel algorithm for discrete least squares rational approximation*, Numerische Mathematik **63** (1992), 99–121. {**426, 431, 434, 435, 482**}

[148] _____, *Discrete linearized least squares approximation on the unit circle*, Journal of Computational and Applied Mathematics **50** (1994), 545–563. {**426, 431, 434, 435, 465, 482**}

[149] _____, *Orthonormal polynomial vectors and least squares approximation for a discrete inner product*, Electronic Transactions on Numerical Analysis **3** (1995), 1–23. {**436, 482**}

[150] _____, *Updating and downdating of orthonormal polynomial vectors and some applications*, Structured Matrices in Mathematics, Computer Science, and Engineering II (V. Olshevsky, ed.), Contemporary Mathematics, vol. 281, American Mathematical Society, Providence, Rhode Island, USA, 2001, pp. 145–162. {**445, 482**}

[151] M. Van Barel, D. Fasino, L. Gemignani, and N. Mastronardi, *Orthogonal rational functions and diagonal plus semiseparable matrices*, Advanced Signal Processing Algorithms, Architectures, and Implementations XII (F. T. Luk, ed.), Proceedings of SPIE, Bellingham, Washington, USA, vol. 4791, 2002, pp. 167–170. {**464, 483**}

[152] _____, *Orthogonal rational functions and structured matrices*, SIAM Journal on Matrix Analysis and Applications **26** (2005), no. 3, 810–829. {**464, 483**}

[153] M. Van Barel, E. Van Camp, and N. Mastronardi, *Orthogonal similarity transformation into block-semiseparable matrices of semiseparability rank k*, Numerical Linear Algebra with Applications **12** (2005), 981–1000. {**40, 407, 483**}

[154] M. Van Barel, R. Vandebril, and N. Mastronardi, *An orthogonal similarity reduction of a matrix into semiseparable form*, SIAM Journal on Matrix Analysis and Applications **27** (2005), no. 1, 176–197. {**27, 40, 78, 88, 98, 132, 483**}

[155] E. Van Camp, *Diagonal-Plus-Semiseparable Matrices and Their Use in Numerical Linear Algebra*, PhD thesis, Department of Computer Science, Katholieke Universiteit Leuven, Celestijnenlaan 200A, 3000 Leuven (Heverlee), Belgium, May 2005. {**40, 88, 95, 98, 387, 483**}

[156] E. Van Camp, N. Mastronardi, and M. Van Barel, *Two fast algorithms for solving diagonal-plus-semiseparable linear systems*, Journal of Computational and Applied Mathematics **164–165** (2004), 731–747. {**3, 171, 174, 175, 181, 280, 298, 464, 483**}

[157] E. Van Camp, M. Van Barel, R. Vandebril, and N. Mastronardi, *An implicit QR-algorithm for symmetric diagonal-plus-semiseparable matrices*, Tech. Report TW419, Department of Computer Science, Katholieke Universiteit Leuven, Celestijnenlaan 200A, 3000 Leuven (Heverlee), Belgium, March 2005. {**130, 197, 219, 222, 250, 259, 483**}

[158] S. Van Huffel and H. Park, *Efficient reduction algorithms for bordered band matrices*, Journal of Numerical Linear Algebra and Applications **2** (1995), no. 2, 95–113. {**64, 483**}

[159] R. Vandebril, *Semiseparable Matrices and the Symmetric Eigenvalue Problem*, PhD thesis, Department of Computer Science, Katholieke Universiteit Leuven, Celestijnenlaan 200A, 3000 Leuven (Heverlee), Belgium, May 2004. {**44, 132, 229, 483**}

[160] R. Vandebril and M. Van Barel, *Necessary and sufficient conditions for orthogonal similarity transformations to obtain the Arnoldi(Lanczos)-Ritz values*, Linear Algebra and its Applications **414** (2006), 435–444. {**80, 483**}

[161] _____, *A note on the nullity theorem*, Journal of Computational and Applied Mathematics **189** (2006), 179–190. {**181, 483**}

[162] R. Vandebril, M. Van Barel, and N. Mastronardi, *A QR-method for computing the singular values via semiseparable matrices*, Numerische Mathematik **99** (2004), 163–195. {**55, 98, 237, 484**}

[163] ———, *An implicit Q theorem for Hessenberg-like matrices*, Mediterranean Journal of Mathematics **2** (2005), 59–275. {**194, 484**}

[164] ———, *An implicit QR-algorithm for symmetric semiseparable matrices*, Numerical Linear Algebra with Applications **12** (2005), no. 7, 625–658. {**130, 194, 217, 287, 320, 484**}

[165] ———, *A note on the representation and definition of semiseparable matrices*, Numerical Linear Algebra with Applications **12** (2005), no. 8, 839–858. {**5, 10, 13, 144, 361, 484**}

[166] ———, *A multiple shift QR-step for structured rank matrices*, Tech. Report TW489, Department of Computer Science, Katholieke Universiteit Leuven, Celestijnenlaan 200A, 3000 Leuven (Heverlee), Belgium, April 2007. {**312, 325, 484**}

[167] ———, *A new iteration for computing the eigenvalues of semiseparable (plus diagonal) matrices*, Tech. Report TW507, Department of Computer Science, Katholieke Universiteit Leuven, Celestijnenlaan 200A, 3000 Leuven (Heverlee), Belgium, October 2007. {**316, 325, 484**}

[168] ———, *A rational QR-iteration*, Tech. Report TW405, Department of Computer Science, Katholieke Universiteit Leuven, Celestijnenlaan 200A, 3000 Leuven (Heverlee), Belgium, September 2007. {**316, 318, 319, 484**}

[169] ———, *Matrix Computations and Semiseparable Matrices, Volume I: Linear Systems*, Johns Hopkins University Press, Baltimore, Maryland, USA, 2008. {**1, 5, 7, 8, 10, 171, 174, 175, 181, 280, 283, 289, 298, 484**}

[170] ———, *A parallel QR-factorization/solver of structured rank matrices*, Electronic Transactions on Numerical Analysis **30** (2008), 144–167. {**290, 484**}

[171] R. Vandebril, E. Van Camp, M. Van Barel, and N. Mastronardi, *On the convergence properties of the orthogonal similarity transformations to tridiagonal and semiseparable (plus diagonal) form*, Numerische Mathematik **104** (2006), 205–239. {**98, 107, 132, 484**}

[172] ———, *Orthogonal similarity transformation of a symmetric matrix into a diagonal-plus-semiseparable one with free choice of the diagonal*, Numerische Mathematik **102** (2006), 709–726. {**33, 40, 95, 125, 484**}

[173] M. E. Wall, P. A. Dyck, and T. S. Brettin, *SVDMAN – singular value decomposition analysis of microarray data*, Bioinformatics **17** (2001), no. 6, 566–568. {**85, 484**}

[174] M. E. Wall, A. Rechtsteiner, and L. M. Rocha, *Singular value decomposition and principal component analysis*, A Practical Approach to Microarray Data Analysis (D. P. Berrar, W. Dubitzky, and M. Granzow, eds.), Kluwer Academic Publishers, Boston, Massachusetts, USA, 2003, pp. 1–20. {**85, 485**}

[175] T. L. Wang, Z. J. and W. B. Gragg, *Convergence of the shifted QR algorithm, for unitary Hessenberg matrices*, Mathematics of Computation **71** (2002), no. 240, 1473–1496. {**353, 485**}

[176] _____, *Convergence of the unitary QR algorithm with unimodular Wilkinson shift*, Mathematics of Computation **72** (2003), no. 241, 375–385. {**353, 485**}

[177] D. S. Watkins, *Understanding the QR algorithm*, SIAM Review **24** (1982), no. 4, 427–440. {**81, 98, 159, 316, 485**}

[178] _____, *Some perspectives on the eigenvalue problem*, SIAM Review **35** (1993), no. 3, 430–471. {**81, 98, 159, 485**}

[179] _____, *QR-like algorithms—an overview of convergence theory and practice*, The Mathematics of Numerical Analysis (J. Renegar, M. Shub, and S. Smale, eds.), Lectures in Applied Mathematics, vol. 32, American Mathematical Society, Providence, Rhode Island, USA, 1996, pp. 879–893. {**81, 88, 96, 98, 103, 159, 316, 360, 485**}

[180] _____, *The Matrix Eigenvalue Problem: GR and Krylov Subspace Methods*, SIAM, Philadelphia, Pennsylvania, USA, 2007. {**160, 485**}

[181] D. S. Watkins and L. Elsner, *Chasing algorithms for the eigenvalue problem*, SIAM Journal on Matrix Analysis and Applications **12** (1991), no. 2, 374–384. {**58, 206, 300, 485**}

[182] _____, *Convergence of algorithms of decomposition type for the eigenvalue problem*, Linear Algebra and its Applications **143** (1991), 19–47. {**81, 84, 88, 93, 96–98, 103, 104, 106, 159, 316, 318, 485**}

[183] _____, *Theory of decomposition and bulge-chasing algorithms for the generalized eigenvalue problem*, SIAM Journal on Matrix Analysis and Applications **15** (1994), no. 3, 943–967. {**58, 206, 485**}

[184] J. H. Wilkinson, *Global convergence of tridiagonal QR-algorithm with origin shifts*, Linear Algebra and its Applications **1** (1968), 409–420. {**205, 485**}

[185] _____, *The Algebraic Eigenvalue Problem*, Numerical Mathematics and Scientific Computation, Oxford University Press, New York, USA, 1999. {**160, 268, 359, 485**}

[186] H. J. Woerdeman, *Matrix and Operator Extensions*, CWI Tract, vol. 68, Centre for Mathematics and Computer Science, Amsterdam, Netherlands, 1989. {**10, 485**}

[187] _____, *A matrix and its inverse: revisiting minimal rank completions*, Operator Theory: Advances and Applications **179** (2008), 329–338. {**10, 486**}

[188] K. S. Yeung, J. Tegner, and J. J. Collins, *Reverse engineering gene networks using singular value decomposition and robust regression*, Proceedings of the National Academy of Sciences of the United States of America **99** (2002), 6163–6168. {**85, 486**}

[189] M.-C. K. S. Yeung and W. L. Ruzzo, *Principal component analysis for clustering gene expression data*, Bioinformatics **17** (2001), no. 9, 763–774. {**85, 486**}

[190] H. Zha, *A two-way chasing scheme for reducing a symmetric arrowhead matrix to tridiagonal form*, Numerical Linear Algebra with Applications **1** (1992), no. 1, 49–57. {**64, 486**}

[191] T. Zhang, G. H. Golub, and K. H. Law, *Subspace iterative methods for eigenvalue problems*, Linear Algebra and its Applications **294** (1999), no. 1–3, 239–258. {**81, 98, 159, 486**}

Author/Editor Index

Subject Index